Textile Science and Clothing Technology

Series Editor

Subramanian Senthilkannan Muthu, Chief Sustainability Officer, Green Story, Canada, Tung Chung, Lantau, Hong Kong

This series aims to broadly cover all the aspects related to textiles science and technology and clothing science and technology. Below are the areas fall under the aims and scope of this series, but not limited to: Production and properties of various natural and synthetic fibres; Production and properties of different yarns, fabrics and apparels; Manufacturing aspects of textiles and clothing; Modelling and Simulation aspects related to textiles and clothing; Production and properties of Nonwovens; Evaluation/testing of various properties of textiles and clothing products; Supply chain management of textiles and clothing; Aspects related to Clothing Science such as comfort; Functional aspects and evaluation of textiles; Textile biomaterials and bioengineering; Nano, micro, smart, sport and intelligent textiles; Various aspects of industrial and technical applications of textiles and clothing; Apparel manufacturing and engineering; New developments and applications pertaining to textiles and clothing materials and their manufacturing methods; Textile design aspects; Sustainable fashion and textiles; Green Textiles and Eco-Fashion; Sustainability aspects of textiles and clothing; Environmental assessments of textiles and clothing supply chain; Green Composites; Sustainable Luxury and Sustainable Consumption; Waste Management in Textiles; Sustainability Standards and Green labels; Social and Economic Sustainability of Textiles and Clothing.

Mohammad Shahid · Mohd Yusuf · Dhiraj Kumar
Editors

Science and Technology of Silk

Volume II: Processing and Applications

Springer

Editors
Mohammad Shahid
Department of Chemistry
Koneru Lakshmaiah Education Foundation
Guntur, Andhra Pradesh, India

Mohd Yusuf
School of Life and Natural Sciences
The Glocal University
Saharanpur, Uttar Pradesh, India

Dhiraj Kumar
Department of Zoology, Centre for Skill
Development and Seri-technology
University of Delhi
Malka Ganj, Delhi, India

ISSN 2197-9863 ISSN 2197-9871 (electronic)
Textile Science and Clothing Technology
ISBN 978-3-032-26879-2 ISBN 978-3-032-26880-8 (eBook)
https://doi.org/10.1007/978-3-032-26880-8

This Springer imprint is published by the registered company Springer Nature Switzerland AG
The registered company address is: Gewerbestrasse 11, 6330 Cham, Switzerland

Preface

Silk has been part of human life for thousands of years, valued for its craftsmanship, elegance and cultural significance. Beyond its aesthetic appeal, silk is a remarkable natural protein material with unique structural, mechanical, and biological properties. In recent decades, scientific advances have significantly expanded the scope of silk research, transforming it from a traditional textile fibre into a versatile material platform for sustainable processing, advanced functional textiles, biomedical engineering, flexible electronics and smart systems. *Science and Technology of Silk, Volume II: Processing and Applications* is conceived in this context—to present a comprehensive and forward-looking account of contemporary developments in silk science and technology.

This volume builds upon fundamental knowledge while emphasizing processing innovations and application-driven research. It brings together contributions from leading researchers and academicians across the globe, reflecting multidisciplinary perspectives that span textile chemistry, materials science, biotechnology, biomedical engineering and wearable technology. The chapters are carefully curated to highlight both established practices and emerging trends that are shaping the future of silk-based materials.

The opening chapters focus on silk processing and coloration, addressing preparatory treatments, dyeing and printing. Advances in eco-friendly degumming, sustainable coloration, functional dyeing and innovative dyeing technologies are discussed with a strong emphasis on environmental responsibility, energy efficiency and preservation of silk's intrinsic qualities. These chapters underline the growing need for greener alternatives to conventional wet processing while maintaining performance and aesthetics.

Subsequent chapters explore silk as a high-value biomaterial, with particular emphasis on fibroin and sericin. The biomedical potential of silk is examined in depth, covering tissue engineering, drug delivery, wound healing, regenerative medicine and implantable devices. The discussions extend to extraction methods, structure–property relationships and challenges associated with scalability, stability and regulatory compliance. Together, these chapters demonstrate how silk proteins are redefining the boundaries between textiles and life sciences.

The latter part of the book is devoted to advanced and emerging applications, including silk-enhanced polymer composites, technical textiles, flexible electronics, energy harvesting, sensing technologies and smart textiles for thermal management. These chapters showcase silk's transformation into a multifunctional material capable of supporting next-generation wearable electronics, sustainable energy solutions and climate-responsive textiles. The integration of silk with nanotechnology, conductive materials and advanced manufacturing techniques highlights its potential in addressing global challenges related to sustainability, healthcare and smart living.

We hope that this volume will serve as a valuable reference for researchers, postgraduate students, industry professionals and policymakers engaged in silk research and allied fields. By presenting both depth and breadth, this book aims to stimulate interdisciplinary collaboration, inspire innovative research and support the translation of laboratory-scale advances into practical and societal applications.

We express our sincere gratitude to all the contributing authors for their scholarly efforts and timely cooperation. We also acknowledge the support of our respective institutions and the publisher for making this volume possible. It is our belief that *Science and Technology of Silk, Volume II: Processing and Applications* will contribute meaningfully to the evolving narrative of silk as a sustainable, high-performance and multifunctional material for the future.

Guntur, India — Mohammad Shahid
Saharanpur, India — Mohd Yusuf
Delhi, India — Dhiraj Kumar

Contents

Editors and Contributors

About the Editors

Dr. Mohammad Shahid Ph.D., is an Assistant Professor in the Department of Chemistry, Koneru Lakshmaiah Education Foundation, India. Before joining KLEF, he worked at various esteemed institutions, including Jamia Millia Islamia, New Delhi; Soochow University, China; Henan University, China; University of Glasgow, UK; Institute of Chemical Technology, Mumbai; Dr. K. N. Modi University, Tonk; and Mewar University, Chittorgarh. He earned his Master's degree in Chemistry from Shibli National College, Azamgarh, in 2006 and his Ph.D. in Chemistry from Jamia Millia Islamia, New Delhi, in 2014. Dr. Shahid has received several prestigious fellowships to support his research pursuits, including grants from the UGC, India, the China Postdoctoral Fellowship, the EU Marie Curie Fellowship, and the Dr. D. S. Kothari Fellowship. He has contributed to over 50 scientific publications with ~5000 citations and an h-index of 29. In addition, has also edited and authored several books with renowned international publishing houses.

Dr. Mohd Yusuf is currently working as an Associate Professor of Chemistry in the School of Science at Glocal University, Saharanpur, India. He received his Ph.D. degree from Jamia Millia Islamia, New Delhi, India, in 2013. He has contributed to academia with about ten plus years of experience, and has published the research findings with esteemed Publishers, such as Elsevier, Springer, Taylor & Francis, Wiley, etc., in the form of research articles, reviews, books and chapters in edited volumes. His research interests include green and sustainable materials, functional biomaterials, composite and applied materials, nanomaterials, wastewater treatment and energy storage technologies. In addition, Dr. Yusuf has been featured in Stanford University's prestigious list of the world's top 2% scientists for the past four consecutive years, 2025, 2024, 2023 and 2022.

Dr. Dhiraj Kumar currently serves as an Assistant Professor in the Department of Zoology at Hansraj College, University of Delhi, India. He is also the Coordinator of the Centre for Skill Development and Seri-technology at Hansraj College. Dr. Kumar is a post-doctorate from Soochow University, China, and later worked as Dr. D. S. Kothari, a Postdoctoral Fellow at the School of Studies in Zoology at Jiwaji University, India. He also served as an Assistant Professor in the Department of Zoology, G. G. Vishwavidyalaya (A Central University), Bilaspur, India. He has published extensively in renowned journals and is the author of over 35 peer-reviewed articles, four books, seven book chapters, three international proceedings and granted a patent of his credit. He is a reviewer of several SCI journals and a member of many scientific bodies. He also received several prestigious awards for his contributions to science and education, including the Research Excellence Award. His research interests include molecular biology, metagenomics, gut microbiome relationships, biomedicine and natural products.

Contributors

Tawsif Sikder Adib Department of Apparel Engineering, Bangladesh University of Textiles, Dhaka, Bangladesh

Mahmuda Akter Department of Apparel Engineering, Bangladesh University of Textiles, Dhaka, Bangladesh

Mila Akter Department of Apparel Engineering, Bangladesh University of Textiles, Dhaka, Bangladesh

Shumya Akther Department of Apparel Engineering, Bangladesh University of Textiles, Dhaka, Bangladesh

M. Amarnath Plasma Research Lab, The Bombay Textile Research Association, Mumbai, India

Deity S. Anand School of Basic and Applied Sciences, Dayananda Sagar University, Bangalore, India

Muhammed İbrahim Bahtiyari Department of Textile Engineering, Faculty of Engineering, Erciyes University, Kayseri, Türkiye

Afreen Begum Plasma Research Lab, The Bombay Textile Research Association, Mumbai, India

Poulami Bhadra Department of Sericulture, Raiganj University, Raiganj, West Bengal, India

Abay Mulusew Birhanu Textile Faculty, Ethiopian Institute of Textile and Fashion Technology, Bahir Dar University, Bahir Dar, Ethiopia;

Textile Science and Engineering Research Cener, Ethiopian Institute of Textile and Fashion Technology, Bahir Dar University, Bahir Dar, Ethiopia

Shristirupa Borah Department of Chemistry, Gauhati University, Guwahati, Assam, India;
Department of Chemistry, Nagaon University, Nagaon, Assam, India

Sohini Chakraborty Department of Chemistry, Stella Maris College (Autonomous), University of Madras, Chennai, India

Patel Disha Chiragkumar School of Basic and Applied Sciences, Dayananda Sagar University, Bangalore, India

Md. Kamrul Hassan Chowdhury Department of Apparel Engineering, Bangladesh University of Textiles, Dhaka, Bangladesh

Tekalgn Mamay Daget Textile Faculty, Ethiopian Institute of Textile and Fashion Technology, Bahir Dar University, Bahir Dar, Ethiopia;
Textile Science and Engineering Research Cener, Ethiopian Institute of Textile and Fashion Technology, Bahir Dar University, Bahir Dar, Ethiopia

Anik Das School of Textile Science and Engineering, Tiangong University, Tianjin, People's Republic of China

Debnirmalya Gangopadhyay Department of Sericulture, Raiganj University, Raiganj, West Bengal, India

Parikshit Gogoi Department of Chemistry, Nagaon University, Nagaon, Assam, India

Vijay Kumar Guna Department of Mechanical Engineering, RNS Institute of Technology, Bengaluru, Karnataka, India

Shah Md. Maruf Hasan Department of Apparel Engineering, Bangladesh University of Textiles, Dhaka, Bangladesh

Md Rashedul Islam Department of Wet Process Engineering, Bangladesh University of Textiles, Dhaka, Bangladesh

Rinku Moni Kalita Department of Botany, Bhattadev University, Bajali, Assam, India

Bantamlak Birlie Kassie Textile Faculty, Ethiopian Institute of Textile and Fashion Technology, Bahir Dar University, Bahir Dar, Ethiopia;
Textile Science and Engineering Research Cener, Ethiopian Institute of Textile and Fashion Technology, Bahir Dar University, Bahir Dar, Ethiopia

Borhan Uddin Khan Department of Textile Engineering, International Standard University, Dhaka, Bangladesh

Ayşegul E. Korlu Department of Textile Engineering, Faculty of Engineering, Ege University, Izmir, Türkiye

Xu Lei School of Textile Science and Engineering, Tiangong University, Tianjin, People's Republic of China

Wanli Liu National Engineering Laboratory for Modern Silk, Jiangsu Engineering Research Center of Textile Dyeing and Printing for Energy Conservation, Discharge Reduction and Cleaner Production, National Textile and Apparel Council Key Laboratory of Natural Dyes, Jiangsu Key Laboratory of Silk Engineering, College of Textile and Clothing Engineering, Soochow University, Suzhou, China

Pankaj Mandal Department of Sericulture, Raiganj University, Raiganj, West Bengal, India

Mohammad Khajeh Mehrizi Department of Textile Engineering, Yazd University, Yazd, Iran

Rony Mia Department of Textile Engineering, Chemistry and Science, Wilson College of Textiles, North Carolina State University, Raleigh, NC, USA

Sunil S. More School of Basic and Applied Sciences, Dayananda Sagar University, Bangalore, India

M. S. A. Muthukumar Nadar Division of Biotechnology, Karunya Institute of Technology and Sciences, Coimbatore, India

O. S. K. Paavana Division of Biotechnology, Karunya Institute of Technology and Sciences, Coimbatore, India

Shital Palaskar Plasma Research Lab, The Bombay Textile Research Association, Mumbai, India

Barsha Roy Department of Sericulture, Raiganj University, Raiganj, West Bengal, India

K. R. Sangeetha Gowda School of Basic and Applied Sciences, Dayananda Sagar University, Bangalore, India

Siam Sarker Department of Apparel Engineering, Bangladesh University of Textiles, Dhaka, Bangladesh

Mehdi Shabgard Department of Textile Engineering, Yazd University, Yazd, Iran

Shubhajit Shaw Department of Sericulture, Raiganj University, Raiganj, West Bengal, India

Md. Mubin Atik Soscho Department of Apparel Engineering, Bangladesh University of Textiles, Dhaka, Bangladesh

S. P. Sruthi Division of Biotechnology, Karunya Institute of Technology and Sciences, Coimbatore, India

Dehenenet Flatie Tassaw Textile Faculty, Ethiopian Institute of Textile and Fashion Technology, Bahir Dar University, Bahir Dar, Ethiopia;

Biorefinery Research Cener, Ethiopian Institute of Textile and Fashion Technology, Bahir Dar University, Bahir Dar, Ethiopia;
Institute of Architecture of Textiles, Faculty of Textiles and Design, Lodz University of Technology, 116 Zeromskiego Str., Poland

Yuyang Zhou National Engineering Laboratory for Modern Silk, Jiangsu Engineering Research Center of Textile Dyeing and Printing for Energy Conservation, Discharge Reduction and Cleaner Production, National Textile and Apparel Council Key Laboratory of Natural Dyes, Jiangsu Key Laboratory of Silk Engineering, College of Textile and Clothing Engineering, Soochow University, Suzhou, China

Chapter 1
Advances in Preparatory Processes for Silk Dyeing, Printing

Mohammad Khajeh Mehrizi, Muhammed İbrahim Bahtiyari, Ayşegul E. Korlu, and Mehdi Shabgard

1.1 Introduction

Silk is an animal fiber whose history dates back to the Old Ages. It is different from hair, like wool or angora, because silk is a secreted fiber in filament form. Cultivated and wild silkworms produce silk. However, about 90% of commercial silk is produced from the cultivated silkworm Bombyx mori. Besides cultivated silk, there are a few other types like Tussah, Muga, Eri, Mussel, and Spider silk, but their production is limited [1]. Silk should be reeled in order to manufacture silk-based textiles. After the reeling process, silk fibers are filaments. However, silk is an expensive textile fiber. Therefore, waste of filament silk is used in staple form for the textile industry. Pretreatment processes of textiles based on both filament and staple silk are discussed in this chapter. Pretreatment is the beginning of silk finishing. Its processes aim to remove foreign substances like sericin and coloring matter from silk and beautify its appearance. Figure 1.1 shows the pretreatment processes of silk textiles.

All the processes and the order of the processes in Fig. 1.1 are not necessary and standard for every type of silk textile. It depends on the application fields of the textile material that will be pretreated and the form of the textile material. Silk-based textiles can be pretreated in different forms, such as fiber, yarn, fabric, and garments. However, all the forms of the textile material only need some of the pretreatment processes (Fig. 1.2). For example, yarn and knitted fabrics are not desized. The

M. K. Mehrizi (✉) · M. Shabgard
Department of Textile Engineering, Yazd University, Yazd, Iran
e-mail: mkhajeh@yazd.ac.ir

M. İ. Bahtiyari
Department of Textile Engineering, Faculty of Engineering, Erciyes University, Kayseri, Türkiye
e-mail: bahtiyari@erciyes.edu.tr

A. E. Korlu
Department of Textile Engineering, Faculty of Engineering, Ege University, Izmir, Türkiye
e-mail: aysegul.ekmekci@ege.edu.tr

M. Shahid et al. (eds.), *Science and Technology of Silk*, Textile Science and Clothing Technology, https://doi.org/10.1007/978-3-032-26880-8_1

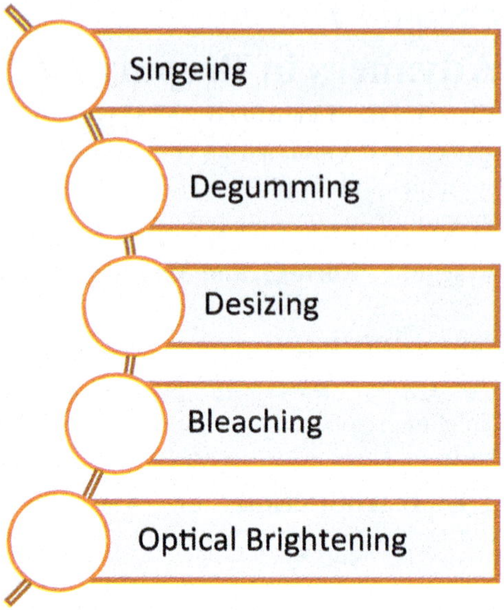

Fig. 1.1 Pretreatment processes of textile materials made of silk [2, 3]

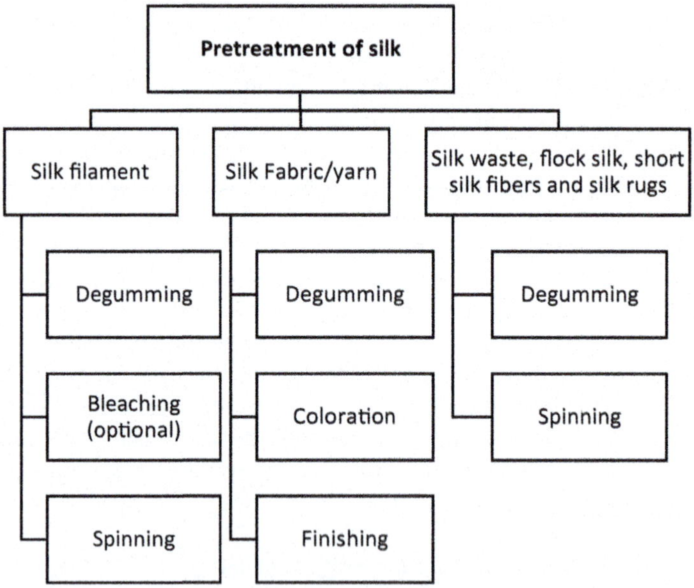

Fig. 1.2 Some manufacturing ways of silk textiles in conventional textile production [4–7]

optical brightening agent is not applied to textile materials that will be dyed. Fabrics that are made of filament silk do not need singeing.

Up to now, silk has been only a raw material for luxury textile products. Silk fabrics like chiffon, georgette, silk satin, and organza have been popular in the production of garments [8]. Silk carpets have been luxury textile products. However, today, silk is used in biomedical, pharmaceutical, therapeutic, cosmetic, smart textiles, and sensor production, unlike conventional textiles, because of the unique silk proteins sericin and fibroin [9–11]. Degumming is also useful for this new application field, in addition to conventional textile production, and as a stage of the process.

1.2 Singeing

Singeing is a mechanical pretreatment process. It aims to eliminate the fibrils on the surface of yarn or fabric for a smooth surface through the burning process. The textile materials consist of filament silk and do not require singeing because there are no fibrils on the surface of the textile material; however, textiles are also produced from staple silk. Silk can be blended with staple fibers like cotton, wool, flax, and ramie in order to reduce fiber cost [12]. In such blends, the singeing process is applied to provide the effects listed below [13]:

- Having better end-use properties and appearance.
- Creating a clean and smooth textile surface.
- Reduce fogginess and make dyed fabrics look brighter.
- Reducing pilling problems in fabrics.
- The process is critical for fabric, which will be printed in intricate and detailed patterns.

Tautkute-Stankuviene et al. (2019) studied the effect of singeing on the pilling resistance of silk/linen (30/70) woven fabric. They singed gray and dyed fabrics. In conclusion, they found that the singeing process changed the pilling resistance of linen/silk fabrics, but without changing the final pilling resistance marks [14]. In their study, Kumpikait˙ et al. (2022) used silk/linen (30/70) fabrics as a standard before bleaching and coloration. Then, they discussed the effect of the coloration processes on fabric color fastness and shrinkage in the direction of warp and weft [15].

1.3 Desizing

Desizing is a pretreatment process for only woven fabrics because warp yarns need sizing agents to increase their weaving performance. The choice of sizing agent depends on the type of fiber and the characteristics of the warp yarn. Indeed, silk warp is not generally sized [13]. However, in the literature, sizing agents are recommended for sizing silk if necessary. Patil and Athalye (2022) stated that bone glue and gelatin are often used to size natural silk and viscose rayon yarns because the amino groups in this type of sizing agent aid in adhesion with protein fibers like silk. However, the sizing agent needs auxiliary agents like grease, surfactants, and preservatives because of its susceptibility to microorganisms, lack of film flexibility, and performance problems [16, 17]. PVA and starch-based sizing agents have moderate sizing performance [14]. Var and Palamutçu (2024) also reported that CMC could be a sizing agent for silk [18]. After the weaving process, removing the sizing agent from the fabric is required since it causes hydrophobicity and harshness in the fabric. The choice of desizing method is determined by the type of sizing agent used. Although there is no study in the literature about the desizing of silk, it is possible to conduct preliminary experiments by using the basic information that starch-based sizing agents can be desized with enzymes, oxidants, alkalis, and acids. CMC is removed from fabric easily during washing because it is a soluble cellulose derivative [18].

1.4 Degumming

Degumming is an essential process for silk among the pretreatment processes. It is not only a critical process of silk and silk waste for the textile industry. Nowadays, sericin and fibroin are popular raw materials for biomedicine, electronics, energy harvesting, cosmetics and skin care products, food packaging, nutraceuticals, etc. Therefore, degumming is vital for suitable and effective raw materials for application fields [9, 11, 19–23]. Degumming is a process that removes sericin between two fibroin filaments in a cocoon and raw silk. During the process, peptide bonds between sericin and fibroin are broken thermochemically [24]. The degumming process has a significant effect on fiber quality and raw material properties for further processes [9, 11, 19–23]. In this chapter, degumming for both textile and other application fields of sericin and fibroin is discussed.

Silk primarily consists of two proteins: fibroin and sericin. Approximately 25% of the raw fiber is sericin, which is present alongside impurities such as waxes, fats, and pigments. Sericin is a brittle and non-elastic amorphous protein. It is a protective layer that hides the brightness of fibroin and glues together twin fibroin filaments. Outer layers of the cocoon contain a higher amount of sericin than inner layers [11]. Sericin is made of three different components called A, B, and C. The inner layer of sericin is C and is next to the fibroin filament. The sericin A is the outer layer. Layer A can be removed from the filament easily because it is the most soluble layer of the sericin. However, sericin is not soluble. Therefore, a hot dilute acid or alkali is needed to remove it from the fiber [25, 26]. If the treated raw silk is treated with water at 120 °C, sericin C is removed from the raw silk. Fibroin is the main protein of silk and 75% of raw filaments. It is not soluble in water and has a crystalline structure [21].

Sericin covers the fibroin. Therefore, it should be removed to dye silk and enhance its bright appearance and soft touch. On the other hand, sericin removal is an important process for extracting fibroin and recovering sericin from wastewater [2, 27, 28]. Raw silk filament is called "hard silk." The name of degummed silk is also "soft silk." During the manufacturing of conventional silk textiles, sericin can be removed from raw silk in different ratios, and different types of silk are obtained [29]. Weight loss of cuite silk is approximately 18–30%, and the sericin has been completely removed. Ecrue silk is silk from which the weight loss is approximately 1–4% due to the removal of too little sericin. Weight loss of souple silk is approximately 6–12%. It has a duller appearance and stiffer handle than completely degummed silk [7].

It is possible to degum both cocoon and fiber because sericin and fibroin can be used outside of the textile industry. However, in textile factories, silk can be degummed in fiber, yarn, and fabric forms. Degumming conditions depend on the technical properties of fibers and textile material. For example, degumming of wild silk is more difficult than mulberry silk [13].

1.4.1 Degumming Methods

The traditional degumming methods can cause some problems such as environmental pollution, cost, energy consumption, and long process time. On the other hand, they can be inadequate for some advanced material production, in which their feedstocks are sericin and fibroin, because the feedstock of advanced materials should have some critical properties. Therefore, researchers focus on new methods to solve the problems of the traditional methods. They study ecological methods that can break the non-covalent bonds between fibroin and sericin with minimal degradation of these proteins [30]. Many methods are offered in the literature and industry with regard to the usage of degumming products and environmental and cost limits. Some of the methods are below:

- Acid degumming
- Alkali degumming
- Biodegumming
- Degumming with amines
- Degumming with ionic liquids
- Degumming with surfactants/detergents/soaps
- Degumming with urea.

1.4.1.1 Acid Degumming

In acidic hydrolysis, organic and inorganic acids are used [24]. During the degumming, the acids hydrolyze protein. Fibroin and sericin are sensitive to acids. The acids aim at specific amino acids of sericin more efficiently than those of fibroin. Diluted acids attack, especially aspartic and glutamic acid [24, 31]. While ranges of aspartic and glutamic acid in fibroin are 1.0–1.9 mol% and 0.8–1.4 mol%, respectively, these values of sericin are 14.6–16.7 mol% and 4.42–7.9 mol% in the same order of fibroin [31]. When the organic acids are applied to silk in the degumming process, acid degumming is safer and more enhancing for silk than alkali degumming [32]. However, if the target product of degumming is sericin, acid degumming is a relatively more risky method than alkali degumming because sericin is intensively hydrolyzed in acidic conditions [20]. In acid degumming, pH and temperature have a significant effect on the success of degumming and material damage [13]. For example, citric acid can be considered an ideal degumming agent because of its biodegradability and low fiber damage [33]. Tartaric acid also supplies better tenacity, strain, and modulus values than soap degumming. On the other hand, wettability, water retention, and dyeability with acid dyes of materials are improved by acid degumming [34]. If the extracted sericin is used in further processes, it needs purification [35].

1.4.1.2 Alkali Degumming

Alkali degumming is a standard sericin-removing method because it is effective, simple, and easy. pH value and temperature have significant effects on product quality [36]. Sericin is removed from silk quickly at pH values higher than 9. However, the process temperature should not be higher than 95 °C to prevent fiber damage. The hydrolyzation degree of fibroin depends on the pH value of the degumming bath [13]. Sodium carbonate is the most popular degumming agent. Sericin adsorbs Na_2CO_3, and it causes the transformation of –COOH to –COONa + in sericin molecules. Na+ combines with sericin. It also leads to sericin being swollen completely, and the solubility of sericin increases because of the high hydrability of Na+. In this way, sericin is removed from silk fibroin (SF) [36, 37]. Extracted sericin needs purification before further processes [35]. There is much research about alkali degumming. Here, the last research will be discussed.

Wang et al. (2019) investigated the impact of Na_2CO_3 and urea degumming methods on silk fibroin properties. They prepared silk fibroin membranes (SFMs) and blank microspheres from silk fibroins obtained via each degumming method, analyzing their molecular weights and viscosities. Alkali degumming resulted in lower fibroin viscosity and significantly affected the thermal stability and mechanical properties of SFMs. The degumming method also influenced microsphere particle size and distribution; alkali-degummed microspheres exhibited smaller and more uniform sizes compared to urea-degummed microspheres [36]. Separately, Hou et al. (2021) produced a flexible broadband microwave absorber by degumming silk cocoons with Na2CO3 solution and subsequently carbonizing the silk fiber mat [38].

Bucciarelli et al. (2021) studied the optimization of the sodium carbonate degumming process and compared it with the reference process. Then, they examined their effects on fibroin. During the reference process, 5 g of cocoons were boiled in 2000 mL Na_2CO_3 solution (2,14 g/L) for 30 min. The results showed that all amounts of sericin could be removed from fibroin with the process, which required less salt, water, and energy, compared with the reference method. They spotlighted that the process was specially optimized for *B. mori* cocoons, and re-optimization was required for different species of cocoons [39].

Yang et al. (2022) studied how different degumming methods affect the structure and characteristics of regenerated silk fibroin films from Antheraea pernyi. They used

Na_2CO_3 ve $NaHCO_3$ during the alkali degumming. $NaHCO_3$ significantly caused an increase in the molecular weights of regenerated silk fibroin (ASF). Neither of the chemicals altered the secondary structure of ASF, but both had a significant impact on the mechanical performance of the ASF films [40]. Saremi et al. (2022) removed serizin with 0.5% (w/v) Na_2CO_3 from cocoons. After the alkali degumming, they produced nanofibers with electrospinning technology. According to the results, degumming conditions had an important effect on the electrospinnability, morphology, and properties of the silk fibers [41]. Wang et al. (2022) were provided raw silk from two provinces in China and degummed it with Na_2CO_3. Silk from Zhejiang, which was degummed with 0.05% of Na_2CO_3, had a fuller structure, higher viscosity, and better protein composition. The silk fibroin is helpful for cell adhesion and proliferation due to sericin was entirely removed after treatment with 0.5% of Na_2CO_3. They stated that the results provided key theoretical and experimental insights for selecting biomaterials in the fabrication of artificial nerve grafts [42]. Uyen et al. (2023) also investigated the effect of two degumming agents, Na_2CO_3 and Marseille soap (MS), on self-dyed silk. The results showed that increasing agent concentration and process time led to an increase in weight loss and color fading. The two degumming methods gave similar results. However, degumming with Na_2CO_3 protected the silk's color better than that of MS. The researchers suggested that fabric-based self-color silk should be produced without pretreatment processes such as degumming and bleaching [37]. Li et al. (2022) degummed raw silk with distilled water and Na_2CO_3 in two ways to prepare silk fibroin (SF)/silk sericin(SS) composite film for bone tissue. The molecular weight of the silk protein obtained by boiling to remove sericin is higher than that obtained with soda, and the SF is less damaged [43]. Bhowmik et al. (2023) studied the degumming of silk fibers as a reinforcement agent to prepare a wheat gluten-based biocomposite. The thermal and physicochemical analysis showed that the composite made with NaOH-degummed waste filature silk (WFS) for 6 h had the best strength, flexibility, and thermal performance [44].

Eliaz et al. (2024) also made an interesting study about the use of NaOH as a degumming agent at low temperatures instead of Na_2CO_3. In this study, to prevent thermal damage of silk, cocoons were degummed with NaOH at room temperature and 40 °C for 10, 15, and 30 min. The reference was the degummed with Na_2CO_3. In conclusion, the secondary structure and crystallinity of protein and fiber morphology were kept and mechanical properties and stability of fibers were improved by degumming with NaOH in contrast to Na_2CO_3 [45]. The other interesting study was about degumming a cocoon with hydrated lime water. Zhao et al. (2018) removed sericin with $Ca(OH)_2$(0.025%) from silk at boiling temperature and obtained a 28% degumming ratio. They suggested that sericin and calcium hydroxide could be recycled after the degumming process, making it a profitable and ecological process. [46].

1.4.1.3 Biodegumming

There is much literature about enzymatic degumming, and it is a popular research area. The main aim of enzymatic degumming is the extraction of sericin and fibroin according to their application fields. After sericin removal, fibroin, which will be used for textile and other application fields, should be a raw material that meets expectations and has the highest productivity in production. It is also valid for sericin. The removed sericin, which is recovered, is a raw material for different application fields. Therefore, the sericin should also have the appropriate qualities for the application area targeted for use.

Enzymes, which are made of amino acids, are commonly used biocatalysts in textile finishing processes due to mild process temperature, pH, and high specificity. Enzymatic processes reduce water and energy consumption. The largest enzyme group among protein-degrading enzymes is proteases. The proteases are of plant, animal, and microbial origins. They can be under basic, acidic, and neutral conditions [33]. Proteases are also an enzyme group used in silk degumming. Proteases are classified according to their domain of action, structure of the active site, and specific reaction mechanisms. They are also called proteinases or peptidases [47]. Protease catalysis involves the hydrolysis of peptide bonds formed by specific amino acids, as well as the cleavage of C–N, C–O, and C–C bonds [48]. Therefore, proteolytic enzymes can be employed in removing sericin [35]. The enzymatic degumming mechanism depends on the broken bonds of sericin, which are peptide and amide. During the enzymatic degumming, these bonds are transformed into amino acids.

However, various enzymes have different efficiencies at different pH values. If it is compared with the nonspecific hydrolytic reaction of conventional methods, it is shown that proteolytic enzymes react only at specific sites [24]. Enzymatic degumming is mainly required for relatively low temperatures under slightly basic or acidic conditions. Enzymatic degummed fibers are less damaged, whiter, and softer than those treated with soap or alkali. However, hydrophobic impurities like natural wax and spinning oil cannot be obliterated because of mild process conditions [24]. Some researchers work on enzymes from many sources and investigate their sericin removal performance and effects on fiber properties. Degumming conditions and obtained effects vary depending on the type of protease used during the process. Although enzymatic degumming is an environmentally friendly process, it has some weak points. It is an expensive method compared to others, with insufficient process optimization and sericin degradation [20].

Trypsin (pancreas) and papain, which are proteases, are interesting in finishing silk and wool. They are mainly used in cocoon cooking, degumming, and biowashing [4]. Bromelain is another enzyme used in silk finishing. It is mainly a cysteine protease, also called thiol endopeptidase, and its toxicity is very low [49]. Ninpetch et al. (2015) degummed Mai 1 silk yarn with bromelain and with sodium carbonate. They removed sericin 96.58% from silk yarn with 2 g/L bromelain in a small-scale procedure. In a large-scale procedure, sericin 5 g/L bromelain caused the removal of 91.84% sericin from the material. The degummed yarn was not damaged. The yarn surface treated with bromelain appeared smoother compared to the yarn that was degummed using sodium carbonate. Bourmalein enhanced the mechanical properties of the fabric according to the Kawabata Evaluation System [50]. Cocoonase, a protease secreted by silk moths to facilitate cocoon emergence [51], presents a promising degumming agent. Recombinantly produced cocoonase exhibited significantly higher specific activity than its natural counterpart. Degumming trials using this enzyme (120 μg/mL in 50 mmol/L buffer, 40 °C for 1 h, repeated twice) achieved 98% efficiency within 1–2 h, surpassing conventional Na_2CO_3 methods. Furthermore, cocoonase simultaneously degrades sericin and bleaches the fibers, streamlining processing and making it a viable alternative for silk treatment [28]. Serine proteases are one of the seven known protease classes [47]. Researchers used serine proteases isolated from Bacillus sp. C4 SS-2013 (C4) in degumming and reported that sericin was wholly removed within 2 h without causing any damage to fibroin, fading the color of the silk and making it ready for dyeing and further processing [28]. Papain, a cysteine proteinase, effectively removes sericin from Bombyx mori silk filaments during degumming at 3.0 g/L under near-neutral conditions, leaving high molecular weight silk fibroin. This is attributed to papain's specific cleavage of L-arginine or L-lysine bonds in sericin. Degumming with papain also resulted in higher tensile strength compared to traditional sodium carbonate methods, demonstrating its potential as a superior alternative for silk degumming [33, 52].

Yu et al. (2022) stated that scientists had developed a type of silk called fluorescent silk protein for use in bioimaging and biosensing. However, conventional methods for processing fluorescent silk could cause it to lose its brightness because silk fibroin is sensitive to temperature and solvent concentration during processing. They tried three different degumming methods. In the first method, silk cocoons were treated with various concentrations and temperatures of sodium carbonate solutions. In the second method, they combined sodium carbonate with an alcalase enzyme for degumming. In the last method, a surfactant was added to a degumming bath. The bath also contained sodium carbonate and alcalase. Their results showed that the degumming temperature is the main reason for the decrease in brightness. The study revealed that brightness loss occurs because of the movements of the protein's atoms and changes in nearby amino acids. Overall, using sodium carbonate and alcalase enzyme at 25 °C keeps the fluorescence in fluorescent silk [29]

Urea alone failed to degum silk at 100 °C, but its combination with papain achieved effective degumming at that temperature. Urea likely facilitated enzyme diffusion and breakdown by expanding the sericin structure [34].

Some bacterial cultures were tried as degumming agents to remove sericin from silk while looking for solutions to high water and chemical consumption and environmental pollution. Pandya and Shetty (2021) examined the degumming activities of nine isolates and compared them with those of Marseille soap. In this study, the quality of degummed silk with microorganisms was better than that of silk treated with the conventional method, and the silk was softer and shinier than others. The crude enzymes obtained from B. subtilis and P. fluorescens isolates demonstrated superior degumming efficiency compared to the conventional method, primarily due to their sericin specificity. These enzymes did not degrade fibroin or other silk components. The degumming process achieved an efficiency rate of 87%. Post-microbial degumming, the silk exhibited improved strength and dyeability. In conclusion, the microbiological method was more effective, cost-effective, faster, and eco-friendly than the conventional method [27].

1.4.1.4 Degumming with Amines

Organic amines like methylamine, ethylamine, diethylamine, and triethylamine were tried as deguming agents. The degumming mechanism of amines is similar to alkalis. The removal of sericin is related to the ionization of amines in water, which causes hydroxyl (OH) ions. Studies about degumming with amines showed that the degumming effect of amines is the same as that of Marseilles soap. However, amines are less sensitive to water hardness. The disadvantage of this method is the low sericin removal ratio [24].

1.4.1.5 Degumming with Ionic Liquids

Ionic liquids are composed of an organic cation and an organic or inorganic anion. Indeed, they are organic salts that can be melted under 100 °C. The researchers focus on ionic liquids to eliminate the conventional volatile solvents and/or catalysts in physical and chemical processes because ionic liquids are generally nonflammable, greener, and designer. They are called "designer solvents" because they are created to improve specific reactions by possessing defined properties [53, 54]. Therefore, some research is about their use in silk degumming. For example, after synthesizing both 1-butyl-3-methylimidazolium chloride [BMIM]Cl and 1-butyl-3-methylimidazolium hydrogen sulfate [BMIM]HSO_4, they tried degumming silk in different conditions. The results showed that the weight loss was 6.29% and 8.29% for [BMIM]Cl and [BMIM]HSO_4, respectively. The values were low. Degumming silk treated with [BMIM]Cl was evaluated as suitable for dyeing because it improved the absorbency of silk. In conclusion, they claimed that [BMIM]Cl might be an effective silk degumming agent [53].

1.4.1.6 Degumming with Surfactants/Detergents/Soaps

Degumming with soap/detergents/Surfactants is a standard method. Soaps behave as a surface-active agent and remove sericin from silk. However, their efficiency depends on their alkalinity and if the degumming bath contains alkali. Neutral soap, which is an anionic surface-active agent, has a slight effect on SF peptide structure and tensile characteristics. The degumming process using regular neutral soap proceeds at a relatively slow rate [19]. Therefore, soap degumming in alkali or weak alkali baths is preferred chiefly depending on the application field of the product after degumming. A weak alkali bath containing soap is commonly used for degumming silk. Different sodium-based soaps are degumming agents for soap degumming purposes [24]. These are degumming agents, but they require auxiliary substances such as sodium carbonate, sodium phosphate, sodium silicate, and sodium hydrosulfite to function effectively [36]. Formation of alkali in hydrolysis of soap causes removal of sericin. It bonds sericin chemically and converts sericin into soluble sodium salt. Due to the emulsifying properties of soap, swollen sericin turns into a soluble form and is removed from silk [24, 37]. As a result, a higher hydrolysis capacity of soap means more efficient degumming.

On the other hand, processing conditions significantly influence the extent of sericin removal [24]. While soap degumming offers benefits such as gentle treatment, minimized sericin loss, and improved softness, whiteness, hydrophilicity, and brightness of the fibers, it also presents drawbacks like prolonged degumming duration, potential yellowing of the fibers, high soap concentration in the solution, and sensitivity to hard water [24, 36]. Water certainly should be softened to prevent the deposition of metallic complexes and appearance of stains [24].

Nonionic surfactants extracted from different plants as an alternative to soaps and detergents are also investigated. Many researchers have used plants such as *Sapindus mukurossi, Citrus limon, Dillenia indica*, and *Musa balbisiana* as biosurfactant sources. They proposed extracting biosurfactants from them as degumming agents [28]. Saponin, which is a nonionic surfactant, is used to remove sericin from silk. Saponin consists of a mixture of natural, nonionic surfactant components. Its hydrophilic portion is known as glycogen, while the hydrophobic segment is referred to as aglycone. Dan et al. (2022) extracted saponin from Acacia concinna (Willd.) DC. using both aqueous and chemical extraction methods. Both saponins, which were extracted in two ways, were used in degumming raw silk. The authors said that they proposed an environmentally friendly degumming method, and the obtained silk fibroin could be useful for medical applications due to its low toxicity [55].

Patil et al. (2023) also extracted saponin from Sapindus. They used it in the degumming of Eri silk fabric and compared it with conventional soda degumming. It resulted in similar weight loss and improved absorbency to the conventional process while causing less mechanical damage. Furthermore, bleaching treatments caused an increase in brightness and whiteness. Regardless of the degumming techniques used, the molecular structure did not change. However, the proposed green method for pretreatment and dyeing of Eri silk offered benefits such as causing less damage to the silk fiber, lowering environmental impact, and utilizing fewer toxic or harmful substances. [3].

Biswal et al. (2023) obtained saponin from the fruit pericarp of *S. laurifolia* Vahl with aqueous and chemical extractions. The two types of saponin obtained by both extraction methods were used in silk degumming. Both increasing saponin concentrations improved the removal of sericin. It was found that the efficacy of extracted fibroin depended on the saponin concentration, the length of the process, and the temperature [56].

A silk peptide surfactant (SPS), prepared from silk amino acid and lauroyl chloride, was introduced as an alternative to sodium carbonate for silk degumming and enhancing fiber quality. SPS achieved a 27% degumming ratio, surpassing neutral soap but falling short of sodium carbonate. However, fibers degummed with SPS exhibited superior tensile properties and less peptide degradation compared to sodium carbonate, suggesting SPS performs similarly to neutral soap and is superior to sodium carbonate [28].

Alkyl polyglycoside (APG), an environmentally friendly natural detergent, is widely used in agricultural and industrial applications. In this study, yellow male and purple female B. mori cocoons were degummed at boiling temperature with 0.25% APG detergent containing two different alkyl chain lengths (C8-10, C8-14), using a material-to-liquid ratio of 1:90 for 30 min. The degumming efficiency, varying between 2 and 28%, was affected by factors such as APG concentration, temperature, boiling duration, and bath ratio. The tensile strength of fibers degummed with APG was comparable to that achieved with neutral soap, and superior to that obtained with alkali degumming. Furthermore, APG degumming preserved the fibroin peptide chains, maintaining their molecular weights around 200 kDa [28].

1.4.1.7 Degumming with Urea

Urea buffer, while moderately effective in removing sericin from silk, can cause some fibroin peptide bond breakage and slightly impacts fiber thermal stability and crystallinity [19]. It extracts higher molecular weight sericin (10 to > 225 kDa) compared to alkali, acid, and heat degumming. SDS-PAGE analysis reveals that urea extraction results in more distinct protein bands, indicating less sericin degradation and allowing for the recovery of sericin with higher molecular weights. Furthermore, urea extraction with 2-mercaptoethanol minimizes sericin degradation and achieves approximately 95% sericin removal without sericin damage. However, this method is costly, time-consuming, and yields sericin that is significantly toxic to cells [57].

1.4.2 Degumming Technologies

With the changing world conditions and the application of new technologies in degumming, the aim is to improve fiber properties and reduce environmental pollution while saving energy, water, chemicals, and time. For this reason, various technologies have been tried in degumming. Some of these technologies are suitable only for the laboratory, while others are suitable for industrial production. However, new devices are needed for these new technologies to be used in the industry [35]. The degumming technologies are below:

- Cold-pad-batch technology
- CO_2 supercritical fluid (CSCF) degumming
- HTHP and steam treatment
- Infrared-assisted degumming
- Low-temperature plasma
- Microwave irradiation
- Ultrasonic degumming.

1.4.2.1 Cold-Pad-Batch Technology

It is a typical wet treatment technique, especially when finishing and dyeing cotton fabrics. Xiang et al. (2011) presented cold-pad-batch technology as an alternative for silk fabric degumming and compared it with the traditional soda–soap degumming method. The fabric was first impregnated with a bath containing tea sapogenin and auxiliaries and then kept at room temperature for 12–48 h. It was found that commercially produced tea sapogenin was effective in mulberry silk fabric degumming. The process parameters, which were batch time, NaOH, tea sapogenin, $Na_2S_2O_4$, and Lamepon A concentrations, had a critical effect on fabric quality. It was observed that the cold-pad-batch process showed no significant differences in whiteness, wettability, breaking strength, and hand feel of degummed silk fabrics compared to the conventional method. [58].

1.4.2.2 CO_2 Supercritical Fluid (CSCF) Degumming

The research on waterless finishing processes of textiles increased because of global warming and predictions about problems in accessing water in the future. One of the research areas is the use of CSCF technology in wet textile processing processes. CSCF technology is evaluated ecologically due to the lack of water, the reduction in chemical consumption, and the recyclability of CO_2 after the process. Supercritical carbon dioxide is a form of carbon dioxide. Its vital temperature and pressure are about 304.25 K and around 7.38 MPa, respectively.. In this state, CO_2 has special properties. It behaves like a gas with low viscosity and high diffusivity while also acting like a liquid with good density and solvating ability. Therefore, supercritical CO_2 can be used in different industries as a strong option for use as a solvent in various fields. After the process, CO_2 can be recovered. However, initial investment is needed for special equipment that can heat and pressurize CO_2 above the critical temperature and pressure [10]. Lo and Chao (2017) removed sericin from raw fiber with CO_2 supercritical fluid. Before degumming, they treated raw fiber with organic acids to clean the fibers from impurities such as oil and dust contamination. Subsequently, the fibers and surfactant were loaded into supercritical carbon dioxide chambers. The chambers were heated to temperatures of 95–127 °C, with CO_2 pressures maintained between 150 and 400 atm, for 45–70 min of processing. They informed that their procedure was efficient and sustainable in a new way [59]. In the study by Lo (2020), woven silk fabric was pretreated with organic acids at a pH range of 2–3 for a duration of 6–8 h. Subsequently, carbon dioxide was introduced into a container, which was then maintained under supercritical conditions for 60 min. Throughout this process, a bio-based surfactant was added to the container to assist the treatment.After supercritical fluid (CSCF) degumming, sericin was removed from the fabric by the aid of ultrasonic treatment. Degummed fabrics were dyed with plant indigo dye. The researcher claimed that degumming and dyeing of silk could be greener, according to the results [60].

1.4.2.3 HTHP and Steam Treatment

Sericin removal is an extraction process. Three important parameters facilitate and accelerate extraction: temperature, chemical substance/enzyme concentration, and process time. These parameters need to be optimized for a suitable, economical, and environmentally friendly extraction process [24]. For this reason, steam degumming under high temperatures and high pressure (HTHP) has become one of the degumming methods that has attracted attention. Under this title, degumming processes performed by both HTHP and steam treatment will be examined.

Steam treatment

As an effective processing method for biomass treatment, steam has higher heat transfer efficiency because it has higher heat capacity and does not reduce the moisture content of processed objects such as wood compared to hot air. Wang et al. (2018) also thought that sericin could be removed by steam because a thermochemical process could easily remove it. In their study, the removal of silk gum was carried out by pressurized steam. Traditional neutral soap (NS) and sodium carbonate (Na_2CO_3) degumming methods were employed as reference standards. In steam degumming, hot pressurized steam containing abundant high-energy water molecules effectively penetrates the raw silk, causing disruption of protein–protein hydrogen bonds. This leads to gradual hydrolysis of sericin, increasing its solubility in water and allowing its separation from fibroin. The study found that steam treatment at a lower pressure (0.14 MPa) with an extended residence time (90 min) did not negatively affect the physicochemical properties of the degummed silk fibers. Moreover, steam degumming was performed without the use of chemicals and required less water, making it a more environmentally friendly alternative compared to conventional chemical-assisted degumming methods. In addition, it had lower processing costs than conventional degumming methods. The steam degumming process did not need a complicated water treatment process because there was no need to use chemicals and replace multiple degumming baths. Sericin could be easily recovered. The results indicated that steam degumming was an effective method for degumming, and sericin could be easily recovered [61].

HTHP treatment

Chemical-free degumming methods are classified into two types: water bath boiling and high-temperature pressurized water degumming. Although the water boiling method involves multiple reprocessing cycles of raw silk, its degumming efficiency remains unsatisfactory[36]. Since no chemicals are used in the HTHP method, the desired extraction degree is achieved at high temperatures under high pressure. Since the method does not use chemicals, it is considered an environmentally friendly [20, 24] and cheap method, but energy consumption and fiber damage should be taken into account. However, it should be noted that this method saves both time and energy compared to degumming at boiling temperature [19]. The application of high temperature and pressure destabilizes the hydrogen bonds between hydroxyl groups, facilitating interactions between water molecules and the hydroxyl groups of polar amino acids, which further promotes the separation of silk's main proteins[35]. In this technique, the sericins that are dissolved in boiling water are called "hydrosylates." They are readily soluble in water and have a low molecular weight (below 20 kDa). Mechanical action and high-pressure values are required for the dissolution of higher molecular weight sericin in water [62]. High temperature, pressure, and long processes are likely to damage fibroin. Therefore, water degumming is generally used in silk reeling, softening, and dissolving of the sericin at 50–60 °C [24]. In addition, the sericin obtained from such a commercial process has better purity [35]; it is used in the cosmetics and pharmaceutical sectors for greater safety [62].

Jaramillo-Quiceno et al. (2021) extracted sericin from silk yarn, high-quality and defective cocoons (DCs) with the HtHP method. Then, they explored the molecular and structural characteristics of extracted SS. The results demonstrated that the characteristics of raw materials and the extent of processing had a significant influence on the amino acid composition and molecular weight distribution of the SS extracts. The SS extracted from raw silk yarn exhibited higher crystallinity and thermal stability compared to those obtained from the two types of cocoon. Overall, these findings provide valuable insights for the development of value-added SS products [63]. Saha et al. (2019) extracted SS powder from waste silk cocoons at 120 °C for 30 min without adding any chemicals. The crystalline index and crystallite diameters of SS were 39.66% and 2.179 nm, respectively. According to the results, SS exhibited a strong scavenging activity against 2,2-diphenyl-1-picrylhydrazyl (DPPH) radicals. The authors reported that the degumming method was both cost-effective and environmentally friendly.They suggested application fields for SS, such as medical textiles, cosmetic and pharmaceutical sectors [64].

Sreenivasa (2023) degummed Eri cocoons without any chemicals. The HTTP method was compared to soap–soda and enzymatic degumming methods. The HTHP method gave the best results in removing sericin from Eri cocoon shells. After the cocoon shells were treated at 130 °C for 30 min with soft water, 15.7% of the sericin was removed, and Eri cocoons were uniformly soft. The soap–soda method caused a weight loss of 14.5%. The enzymatic degumming resulted in 13% weight loss with a concentration of 12% protease. Enzymatic degumming reduced the tensile strength of the fibers significantly. Sreenivasa informed that the best degumming conditions of Eri cocoons were 130 °C for 30 min with soft water [65]

1.4.2.4 Infrared-Assisted Degumming

Infrared (IR) heating is a method that uses infrared radiation to increase the reaction rate, and it is cleaner and quicker than conventional methods. Therefore, it supplies benefits like saving energy, using less space, easy temperature control, and being suitable for various applications. During the process, heat from infrared waves transfers directly to the center of the material. However, it does not have any effect on the air temperature around it. IR technology can be used for degumming without using any chemicals. The amount of removed sericin depends on process time and temperature. On the other side, extracted sericin can be recovered entirely without any decomposition, and the technology can be adapted to large-scale silk processing [28, 34].

1.4.2.5 Low-Temperature Plasma

Plasma treatment technology is gaining increasing popularity in the textile industry. It is an economically and environmentally friendly alternative to traditional wet chemical treatment techniques. The technique's advantages are that it does not require any solvent and a long process time. On the other hand, it is considered a clean and relatively simple method. Plasma, often referred to as the fourth state of matter, is a partially ionized gas classified into hot (thermal) and cold (non-thermal) types. For surface modification of thermosensitive polymers and textiles, only low-temperature plasmas (LTPs) are appropriate. LTPs can be generated either by applying an electric field to a gas between electrodes or by inducing a radiofrequency resonant current in a coil, under both low-pressure and atmospheric-pressure conditions [66]. These plasmas contain reactive species capable of initiating physical and chemical reactions on the surface of substrates [34, 66]. Therefore, the plasma's high-energy components are applied to lower-energy covalent and other chemical bonds, and different chemical reactions can be achieved for surface modification [34]. Plasma treatments modify only the outermost nanometers of a substrate, imparting desired surface functionalities to textile fibers without altering bulk properties. These changes are influenced by gas type, pressure, frequency, power, treatment time, and substrate nature, making plasma processes versatile and multifunctional. Historically, surface cleaning and coating were the primary applications.

More recently, plasma treatments have been used to increase hydrophilicity, and adhesion of both natural and synthetic fibers [66]. Degumming with the LTP technique can also be added to these studies. When the effect of low-pressure argon plasma treatment on degumming was investigated, the variables were argon pressure, discharge power and exposure time. These parameters were optimized for effective degumming of raw silk. A significant degumming effect was observed at pressures above 40 Pa, attributed to the increased formation of oxygen plasma species at higher gas pressures. Consequently, cross-linking reactions occurred between different chains of gummy substances on the silk's outer surface. The proposed method is more ecological due to the reduction of process time, the concentration of process bath, and the energy of the traditional wet chemical treatment method [34].

1.4.2.6 Microwave Irradiation

Microwaves (MV) belong to the electromagnetic spectrum within the frequency range of 0.3 GHz to 300 GHz [67]. Microwave irradiation is an effective non-contact heating technique. It produces heat with the aid of material molecules, which vibrate and rotate in a microwave field. The technology is used in reacting, heating, and drying materials. Microwave energy can penetrate materials, heat all the parts of the material evenly, and reduce heat transfer problems. Therefore, the use of MV technology in silk degumming is interesting [68]. MV-assisted degumming is an extraction process [69]. The principle of the microwave extraction (MAE) process is different from that of the conventional method. With the aid of MV, the extraction process accelerates and the extraction yield increases. Because heat and mass transfer work in the same direction in contrast to conventional extraction methods.

Both transfers occur from inside to outside. However, during the conventional extraction, the directions of the mass and heat transfer are reversed. While mass transfer goes from inside to outside, heat transfer goes from outside to inside of the substrate. During the extraction, the solvent diffuses effectively into the substrate. Then, the solute is dissolved until it reaches a concentration. The properties of the solid are limited in concentration. Ultimately, the solution moves from the surface to the bulk of the solution either through natural or forced convection [70]. Nowadays, some researchers are interested in microwave degumming of silk. Bascou et al. (2022) studied the degumming of silk and the recovery of sericin from wastewater. They treated silk cocoons using four different degumming methods. They are the conventional method with sodium carbonate, hot water, HTHP, and microwave methods. They compared traditional sodium carbonate degumming with chemical-free methods.

The results showed that microwave degumming achieves the highest extraction yield, recovering 99.8% of sericin in water without chemicals. All the methods did not damage fibroin fibers, according to the images of the fibroin. All types of extracted sericin contained similar amounts of nitrogen, carbon, hydrogen, and sulfur. Spectroscopic analysis confirmed that the sericin quality was consistent across all methods and comparable to commercial sericin. The degumming process affected protein size. Conventional alkali and MW degumming methods caused more hydrolyzed sericin because sodium carbonate in the conventional method and high temperature and pressure in the MW method changed the properties. The extracted pure and water-soluble sericin could be used as a biocompatible polymer for various applications. On the other hand, they suggested microwave degumming for extracting sericin from Bombyx mori silk cocoons in water without chemicals and with low energy use. However, they spotlighted the suitability of this process for laboratory or pilot production [71].

Pan et al. (2024) focused on sericin extraction using microwave-assisted steam degumming (MSD). They evaluated four degumming methods: alkali degumming, steam degumming, and microwave-assisted steam degumming. The MSD process involved two steps. First, microwave pretreatment was applied, where silk fabric placed in a container filled with deionized water was treated in a microwave oven at 700 W for 10, 20, and 30 min. Following the pretreatment, steam was applied to the pretreated fabric in a steam pressure cooker at 120 °C under 0.1 MPa pressure for 30 min. The researchers compared the Na_2CO_2 degumming method, steam degumming, and microwave-assisted steam degumming. The results demonstrated that microwave-assisted steam degumming was highly effective in completely removing sericin. Moreover, the microwave treatment created micropores in the cross-section of the silk fibers, enhancing moisture absorption and breathability compared to the other methods. This combined method was clean, energy-efficient, water-saving, and chemical-free. It was also an effective choice for mulberry silk fabrics. They said that they improved practical workshop operation and should design equipment to support its industrial use [72].

1.4.2.7 Ultrasonic Degumming

Ultrasound technology is an irradiation process used in textile pretreatment processes. Sound waves with frequencies higher than 20 kHz are referred to as ultrasound. Ultrasound can be categorized into two types. The first is high-frequency, low-intensity ultrasound, typically ranging from 1 to 10 MHz, which has no significant physical or chemical effects and is mainly used in medical imaging and chemical analysis. The second type is low-frequency, high-intensity ultrasound, with frequencies between 20 kHz and 1 MHz. This type can operate at power levels reaching thousands of watts, making it applicable in various fields such as emulsification, processing, extraction, cleaning, and chemical reactions. It is also interesting for silk degumming. Some researchers focused on degumming silk ultrasonically [73]. Ultrasound technology is an effective method of removing sericin from silk cocoons. During the ultrasonic degumming, soap addition to the bath improves the effect of the process. The ultrasonic method supplied the most favorable outcomes with turpentine soap for degumming [34].

Wei et al. (2022) degummed the cocoon shells of silkworms with both high-temperature and Ca(OH)2-ultrasonic methods. They recovered sericin protein cleanly and then obtained sericins of high and low molecular weight [74]. Rawan et al. (2023) also focused on enzyme-assisted ultrasonic irradiation silk degumming. They combined the enzymatic method with ultrasonic degumming and compared it with soap degumming. Mulberry silk was ultrasonically treated with papain enzyme of 1–5 g/L concentrations at temperatures between 40 and 60 °C for 20–40 min. The results showed that ultrasonic degumming was more effective than soap degumming in removing sericin from mulberry silk. Soap degumming at 85 °C supplied slightly better results than the ultrasonic method. However, it caused a reduction in strength and elongation. Therefore, enzyme-assisted ultrasonic degumming is better than the conventional method [75]

1.5 Bleaching and Optical Brightening of Silk Fibers

Silk fibers have unique visual luster, tactile properties, durability, toughness, biocompatibility, biodegradability, thermal stability, etc. [76]. The finishing processes, starting from degumming and other pretreatment processes to coloration (dyeing-printing) and finishing (antimicrobial, soft handle, drapability, stain repellent, antistatic, and so on), were aimed at improving, saving, or using these characteristic features to attract consumers' attention. Pretreatment is aimed at preparing the fibers for subsequent processes and generally targets the impurities in the fibers. As known, silk is a protein-based fiber, and silk cocoons are composed of protein structures higher than 95% by degumming. The sericin protein structure in the fiber has been removed, and the fibroin has been left [77]. Silks are essentially pure protein, with only low levels of sugar and minerals in some structures [78]. The impurities in silk cocoons other than the sericin are the waxes, mineral salts, and ash, which constitute about 4–5% of the fiber [77]. The silk fibers can be in cream or yellowish shades/colors even after the degumming process, which is targeting to remove sericin-related gum hosting the yellow, yellow-green, and brown pigments such as chlorophyll, xanthophyll, and carotin. It occurs due to the residual pigments and sericin adsorbed by fibroin even after degumming [13, 79]. In this respect, it is necessary to mention bleaching as a pretreatment process. So, bleaching of silk fibers is another common pretreatment process unless dyeing in dark shades is planned. It is often preferable to do this with degumming in the same bath [80, 81]. However, it can also be conducted as a separate process.

Textile bleaching is a chemical process that aims to obtain an optimum level of whiteness degree by removing unwanted natural colors from textile materials. For this purpose, different chemicals/methods can be employed according to the fiber type [82–85]. Bleaching agents generally act in an oxidative or reductive manner. They break down chromophoric groups—colored structures into smaller, more soluble units which can be readily removed during bleaching, or they modify chromophoric groups—colored structures and so the absorbance wavelength shifts out of the visible region [86, 87]. Silk fibers can be bleached by the use of either reductive or oxidative agents (Fig. 1.3). However, after reductive bleaching, the original color may be returned to its original color because of re-oxidization. Hence, oxidative bleaching is mostly preferred. Chlorine-based bleaching agents, on the other hand, are also questionable but are not preferred because they tend to chlorinate fibroin. In Fig. 1.3, some commonly used bleaching agents for silk were listed [79, 88]. Among them, the most useful and preferable bleaching agent for silk bleaching is hydrogen peroxide [79, 81, 89]. Hydrogen peroxide-based silk bleaching can be applied at higher temperatures and alkali pHs than wool bleaching [88].

It is a generalization that hydrogen peroxide (H_2O_2) is the most widespread bleaching agent in textile finishing because its decomposition products are not dangerous. So, toxicologically and environmentally, it is more acceptable [90]. The natural color of the silk can be varied; for example, tussah (tassar) silk has a fast fawn color, and it can hold its original color even after a bleaching process [91]. So, depending on the silk type, the recipe can be varied; 10–20 ml/L H_2O_2 (50%) is used with a stabilizing agent at a temperature between 75 and 90 °C, and the pH of the bleaching bath should be arranged to around 9. In this way, the perhydroxyl ion or even atomic oxygen is responsible for the oxidation, so bleaching of the silk occurs [13].

On the other hand, reductive bleaching of silk is also possible, as reported above. In reductive bleaching, the agents destroy the coloring matter by reductive reactions. Different chemicals are available for this purpose, as shown in Fig. 1.3. However, sodium dithionite, also called hydrosulfite, is more convenient and preferable. The

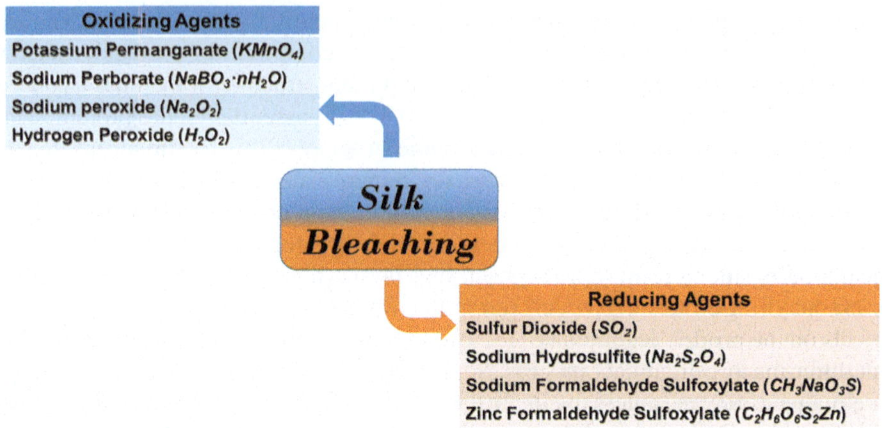

Fig. 1.3 Some common chemicals used in silk bleaching [79, 88]

reductive bleaching of degummed silk can be carried out with 4 g/L $Na_2S_2O_4$ at around 50 °C [79, 91, 92].

On the other hand, reductive bleaching can be applied as a sequential process to oxidative bleaching. For example, in a study by Sharma et al. (1999), the effect of degumming on mulberry and tasar silk fabrics was investigated after sequential oxidative and reductive bleaching processes. Meanwhile, for oxidative bleaching, hydrogen peroxide with sodium silicate (stabilizer) in the presence of a nonionic surfactant, and for reductive bleaching, sodium hydrosulfite with a nonionic surfactant was used [93]. In another study, bleaching and dyeing conditions of silk fabric were investigated in terms of abrasion resistance, tear strength, and whiteness values. In this respect, oxidative (alkaline hydrogen peroxide) and reductive (thiourea dioxide-EDTA) bleaching processes were compared. The results showed that silk bleached with H_2O_2 has better mechanical features and higher whiteness values than reductively bleached silk. Meanwhile, the effects of crosslinkers and their concentration in the dyeing of silk were investigated, too [94]. As a common and preferred bleaching agent, "*hydrogen peroxide*" has also been tested for bleaching wild silk species. For this purpose, the hydrogen peroxide-based oxidative bleaching of *Gonometa postica* silk, which has a rich tan color, was investigated. The results showed that acceptable bleaching results could be achieved in *Gonometa postica* silk fabric with hydrogen peroxide-based oxidative bleaching under appropriate time, pH, and temperature conditions. However, disadvantages in mechanical properties occur [95]. In a recent study by Patil et al. (2023), a sustainable process was investigated for the Eri silk pretreatment and dyeing. Principally, *Sapindus* extract and the conventional products were compared in terms of degumming performance. Additionally, natural dyeings were used in that study. Meanwhile, hydrogen peroxide-based oxidative bleaching was carried out for 90 min at 70 °C in the context of the study, and the bleaching treatments were reported as providing additional benefits such as higher brightness and whiteness [3].

Instead of the common bleaching agents, some different agents and chemicals were also investigated in the bleaching and pretreatment of the silk fibers. Ozone gas was one of the agents used. Due to its high oxidation potential (2.07 V), ozone is a powerful oxidizing agent and has therefore been investigated for the pretreatment of various textile materials [96]. For instance, Bahtiyari and Benli (2020) introduced the ozone-based bleaching method as a cold bleaching method for cotton [97]. In terms of the usage of ozone gas in silk pretreatment, the studies by Sargunamani and Selvakumar are good examples [98–100]. They reported the effects of ozonation on the mulberry silk by comparing ozone application with soap degumming and H_2O_2-based bleaching on raw and degummed silk. Ozone was reported as having significant effects on the properties of raw and degummed silk; however, ozone application did not differ much from hydrogen peroxide treatment except for yellowness, resulting in a higher yellowness index [98]. Similar findings were also reported in the other study about the investigation of ozone on raw and degummed tassar silk fabrics [99]. Additionally, the mulberry and tassar silk fabrics were compared in terms of the effects of ozone. It was declared that ozone application has a noticeable effect on the features of raw silk compared with degummed ones. Mulberry silk was found to be more sensitive and less resistant to ozone treatment compared to tassar silk [100].

In another study, ozone and plasma were applied to silk fabrics separately and one after another. Different features of the silk were measured and in terms of the whiteness indexes which are important for the evaluation of a bleaching process and related to the content of this caption, it was reported that applying plasma treatment on raw silk fabrics generally provided higher whiteness degrees than applying solely ozone or a combination of ozone and plasma. Meanwhile, the effect of ozone on whiteness was reported as positive for the lowest ozonation duration (5 min) [101].

Optical brightening should also be considered in the case of talking about silk fiber bleaching. This brightening–whitening process is based on the usage of an optical brightening agent (OBA). They are also known as fluorescent whitening agents (FWA). These agents have been used in different industries, such as the textile industry, paper industry, plastic industry, and detergent industry [102]. They have been used to make the products whiter and brighter [103]. In terms of textile usage, the optical brightening process can be carried out as a separate process. However, the OBAs can also be used as an additive to the bleach bath [88]. So, the optical brightening of silk fibers can be managed along with bleaching with the addition of optical brightening agents (OBA). In this way, clear white and brilliant light shades can come out [80, 81, 91]. However, this initial bright white can turn into a severely yellowed state over time, which is one of the most significant disadvantages of using OBAs. Durability is another important issue; an adequate fastness against washing is a sought-after feature [88]. Optical brightening agents are colorless to weakly colored organic compounds. They absorb light in the UV range and emit it again in the visible range. So they cause a brightening—an increase in the whiteness of the material [102–104]. They show their features in the presence of UV light. In Fig. 1.4, the photos of silk fabrics with and without OBA under D65 and UV light are given. A light booth (color-matching cabinet) was employed for this purpose. As seen in Fig. 1.4, the presence of OBAs makes no sense under D65. However, under UV light,

OBAs act as described above and cause changes in the appearance of the fabric. So, under sunlight where UV-A and UV-B are present, color changes and whiteness/brightness changes do occur, but certainly not as much or as noticeable as under just UV light alone.

Actually, optical brightening agents are also dyestuffs, but in place of the chromophoric group, they have a fluorescing system [92]. They typically have a conjugated double bond system and should be substantially more planar. Moreover, the chemical system contains both electron-donating groups (such as OH and NH_2) and electron-accepting groups (such as NO_2 and $-N = N-$) [13]. OBAs are commonly based on stilbenes, coumarins, pyrazolins, naphthalimides, and naphthooxazoles [105]. The OBAs can be classified into six groups, as shown in Fig. 1.5 [92, 106].

The chemical structures of OBAs contain aliphatic carbon–carbon double bonds, which contribute to their substantivity as well as their sensitivity to factors such as weathering, sunlight, and oxidation [92, 106]. The majority of optical brighteners produced are stilbene-based [13, 106]. Optical brightening can be applied by batch or continuous processes in textile finishing [102], and fabrics may occasionally contain 0.5% OBA finally [105]. The optical brightening of the degummed silk can

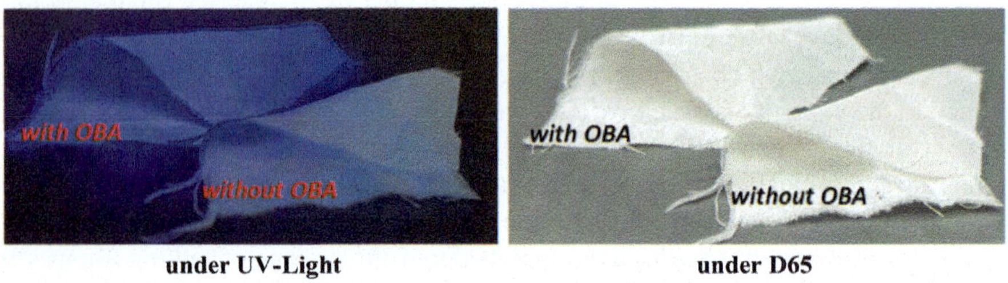

Fig. 1.4 The photos of silk fabrics with and without OBA under D65 and UV light

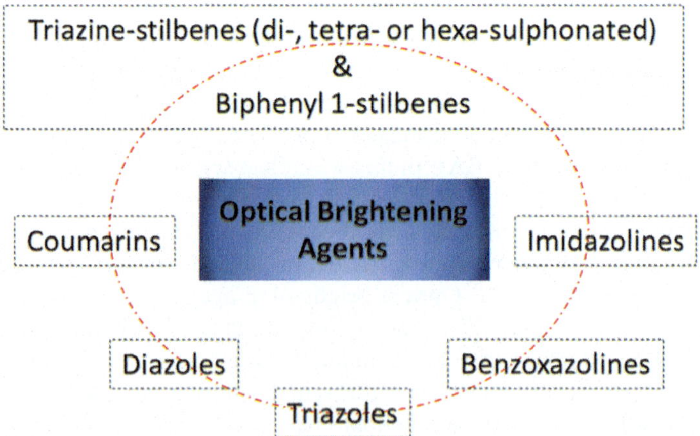

Fig. 1.5 Classification of optical brightening agents [92, 106]

be managed after peroxide bleaching by adding the OBA (0.5–4%) to the subsequent reductive bleaching bath. Generally, the usage of direct brightening agents can be preferred in silk, although acid/basic types of OBAs can be applicable [13]. However, as with many other chemicals, it should not be forgotten that their use should also be considered from an environmental perspective. Because most OBAs do not biodegrade easily and, therefore, can adversely affect water quality, animal, and plant life. There is also the potential to cause allergic skin reactions or eye irritation in sensitive individuals [105].

1.6 Conclusion and New Trends

Silk is a very important luxury fiber with an ancient history. It even gave its name to a trade route [107, 108]. However, changing times, expectations, and problems of this age affect the use of silk fibers and their by-products today. Nowadays, there are critical issues to consider in the production of textile products made of silk fibers. The different topics/issues that are thought to shape the future of silk are listed below:

- Animal welfare issue [107, 108].
- Pretreatment of silk fibers with environmentally friendly methods.
- Pretreatment of silk fibers and by-products, such as sericin and silk waste, in a way that ensures they are transformed into high-value-added products without deterioration.

To secure silk's future role in the textile industry, ensuring peace silk production is of primary importance. However, this issue has not yet been fully addressed by manufacturers. For instance, in a study conducted by Ozkavruk Adanir et al. (2022), it was found that among all the families engaged in sericulture in Hatay, Türkiye, only one family has been producing peace silk for over half a century, and they have done so without being aware that their method is referred to as peace silk or non-violent silk [107]. This is critical for customer groups with ethical sensitivity. Peace silk production is not enough in sericulture. At the same time, the finishing and dyeing/printing processes must be environmentally friendly. Today, with the developments in biotechnology, the use of technologies such as enzymes, ultrasound, and plasma during finishing processes has reduced the consumption of water, energy, and chemicals [35]. Among pretreatment processes, degumming is an important process for silk. It is not only a critical process for silk and silk waste for the textile industry. Nowadays, sericin and fibroin are popular raw materials for biomedical, electronics, energy harvesting, cosmetics and skin care products, food packaging, nutraceuticals, etc. Therefore, degumming is vital for suitable and effective raw materials for application areas [9, 11, 19–23]. As a result, silk will be a preferred raw material in the future with its fibers, wastes, and by-products with conscious pretreatment processes in peaceful silk production.

References

1. Kadam V, Shanmugam N (2023) Wool and silk fibres from sustainable standpoint. In: Nayak R (ed) Sustainable fibres for fashion and textile manufacturing. Woodhead Publishing, Kidlington
2. Radhakrishnan S (2022) Biotechnological approaches for sustainable pretreatment of textile materials. In: Applications of Biotechnology for Sustainable Textile Production. Woodhead Publishing, Kidlington, pp 99–133.
3. Patil H, Deshmukh I, Athalye A (2023) Sustainable process for the pretreatment and dyeing of Eri silk. J Indian Chem Soc 100:100925. https://doi.org/10.1016/j.jics.2023.100925
4. Padaki N, Das B, Thirumalesh R (2015) Enzyme applications in silk processing. In: Advances in Silk Science and Technology. Woodhead Publishing, Cambridge, pp 111–120
5. Yin R, Xiang Y, Zhang Z et al (2020) Cleaner production of mulberry spun silk yarns via a shortened and gassing-free production route. J Clean Prod 278:123690. https://doi.org/10.1016/j.jclepro.2020.123690
6. Czaplicki Z, Gliścińska E, Machnowski W (2021) Natural silk—an unusual fibre: origin, processing and world production. Fibres Text East Eur 29:22–28. https://doi.org/10.5604/01.3001.0014.9291
7. Veit D (2022) Silk. In: Fibers. Springer, Cham, pp 351–380.
8. Nayak R, Jajpura L, Khandual A (2023) Traditional fibres for fashion and textiles: associated problems and future sustainable fibres. In: Nayak R (ed) Sustainable Fibres for fashion and textile manufacturing. Woodhead Publishing, Kidlington, pp. 3–25
9. El-Sayed HEZ, Taleb MA, Mowafi S (2021) Potential applications of textile wastes and by-products in preparation of textile auxiliaries. Egypt J Chem 64:4433–4447. https://doi.org/10.21608/ejchem.2021.79398.3899
10. DeBari MK, King CI, Altgold TA et al (2021) Silk fibroin as a green material. ACS Biomater Sci Eng 7:3530–3544. https://doi.org/10.1021/acsbiomaterials.1c00493
11. Chand S, Chand S, Raula B (2023) Usage of silkworm materials in various ground of science and research. J Nat Fibers. https://doi.org/10.1080/15440478.2022.2139328
12. Saikia S, Konwar P, Saikia M (2022) Blending of silk. In: Singh P, Saini V (eds) Research trends in home science and extension, vol 10. AkiNik Publications, Delhi, pp 41–53
13. Karmakar SR (1999) Chemical technology in the pre-treatment processes of textiles. Elsevier Science B.V, Amsterdam
14. Tautkute-Stankuviene I, Simanavicius L, Kumpikaite E (2019) Influence of singeing of linen/silk fabrics on pilling resistance. Paper presented at AUTEX2019—19th world textile conference on textiles at the crossroads, Ghent, Belgium, 11–15 June 2019
15. Kumpikaitė E, Tautkutė-Stankuvienė I, Milašienė D et al (2022) Analysis of color fastness and shrinkage of dyed and printed Linen/Silk fabrics. Coatings 12:408. https://doi.org/10.3390/coatings12030408
16. Patil H, Athalye A (2022) Developments in sizing chemicals and application techniques. Colourage:27–34
17. Tari S, Athalye A (2023) Valorising desizing textile effluent. IJFTE 3:7–12. https://doi.org/10.54105/ijfte.b2408.113223
18. Var C, Palamutçu S (2024) Sustainable approaches in textile-sizing process. In: Muthu SS (ed) Sustainable manufacturing practices in the textiles and fashion sector. Springer, Cham, pp 55–74
19. Wang H, Zhang Y, Wei Z (2021) Dissolution and processing of silk fibroin for materials science. Crit Rev Biotechnol 41:406–424. https://doi.org/10.1080/07388551.2020.1853030
20. Kalita M (2022) Extraction and characterization of silk sericin and its potential application. Dissertation, Deakin University
21. Xing T, He A, Huang Z et al (2023) Silk-based flexible electronics and smart wearable textiles: progress and beyond. J Chem Eng 474:145534. https://doi.org/10.1016/j.cej.2023.145534
22. Oduor EO, Ciera LW, Kamalha E (2021) Applications of silk in biomedical and healthcare textiles. In: Kumar B (ed) Textiles for functional applications. Intechopen, London

23. Sarma M (2023) Silk protein sericin: as a versatile biomaterial. Chem Axis 22:23–28
24. Wang K, Ma Q, Zhou H et al (2022) Review on Fabrication and application of regenerated bombyx mori silk fibroin materials. Autex Res J 23:164–183. https://doi.org/10.2478/aut-2021-0059ej.2023.145534
25. Shaw JTB, Smith SG (1951) Amino-acids of silk sericin. Nature 168:745. https://doi.org/10.1038/168745a0
26. The Food and Agriculture Organization (1999) Characteristics of the cocoon. https://www.fao.org/4/x2099e/x2099e03.htm. Accessed 2 Apr 2025
27. Pandya B, Shetty S (2021) Bacterial degradation of sericin for degumming of silk fibers—a green approach. J Appl Biol Biotechnol. https://doi.org/10.7324/jabb.2021.9513
28. Reddy N (2020) Silk: materials, processes, and applications. Woodhead Publishing, Kidlington
29. Yu CY, Nettey-Oppong EE, Effah E et al (2022) Effect of degumming conditions on the fluorescence intensity of fluorescent silk cocoons: a combined experimental and molecular dynamics study. Int J Ind Entomol 45:56–69. https://doi.org/10.7852/IJIE.2022.45.2.56
30. Kalita M, Sarma C, Kalita M et al (2024) Efficient extraction of sericin protein from mulberry and non-mulberry silk fibers: utilizing aqueous ionic liquid solutions as green solvents. ChemistrySelect 9:e202304064. https://doi.org/10.1002/slct.202304064
31. Gulrajani ML Jr, Sethi S, Gupta S (1992) Some studies in degumming of silk with organic acids. JSDC 108:79–86. https://doi.org/10.1111/j.1478-4408.1992.tb01420.x
32. Aniş P, Çapar G, Toprak T et al (2016) Sericin removal from silk fibers with eco-friendly alternative methods. Text Appar 26:368–374
33. Zhu L, Lin J, Pei L et al (2022) Recent advances in environmentally friendly and green degumming processes of silk for textile and non-textile applications. Polymers 14:659. https://doi.org/10.3390/polym14040659
34. Biswal B, Dan AK, Sengupta A et al (2022) Extraction of silk fibroin with several sericin removal processes and its importance in tissue engineering: a review. J Polym Environ 30:2222–2253. https://doi.org/10.1007/s10924-022-02381-w
35. Silva AS, Costa EC, Reis S et al (2022) Silk sericin: a promising sustainable biomaterial for biomedical and pharmaceutical applications. Polymers 14:4931. https://doi.org/10.3390/polym14224931
36. Wang Z, Yang H, Li W et al (2018) Effect of silk degumming on the structure and properties of silk fibroin. J Text Inst 110:134–140. https://doi.org/10.1080/00405000.2018.1473074
37. Uyen TNT, Thao HT, Huong BM (2023) Study on the influence of degumming technological parameters on self-dyed silk. JST-UD 21:34–37. https://doi.org/10.31130/ud-jst.2023.082e
38. Hou Y, Quan J, Su X et al (2021) Carbonized silk fiber mat: a flexible and broadband microwave absorber, and the length effect. ACS Sustain Chem Eng 9:12747–12754. https://doi.org/10.1021/acssuschemeng.1c02857
39. Bucciarelli A, Greco G, Corridori I et al (2021) A design of experiment rational optimization of the degumming process and its impact on the silk fibroin properties. ACS Biomater Sci Eng 7:1374–1393. https://doi.org/10.1021/acsbiomaterials.0c01657
40. Yang H, Guo Y, Zhang Q et al (2022) Effect of degumming methods on structure and properties of regenerated antheraea pernyi silk fibroin films. J Nat Fibers 19:11114–11123. https://doi.org/10.1080/15440478.2021.2009404
41. Saremi J, Zarei-Behjani Z, Vojoudi E et al (2022) Evaluation of viability and cell attachment of human endometrial stem cells on electrospun silk scaffolds prepared under different degumming conditions and solvents. Regen Eng Transl Med 8:593–606. https://doi.org/10.1007/s40883-022-00258-x
42. Wang Y, Liang Y, Huang J et al (2022) Proteomic analysis of silk fibroin reveals diverse biological function of different degumming processing from different origin. Front Bioeng Biotechnol 9:777320. https://doi.org/10.3389/fbioe.2021.777320
43. Li M, Tian W, Zhang Y et al (2022) Enhanced silk fibroin/sericin composite film: preparation, mechanical properties and mineralization activity. Polymers 14:2466. https://doi.org/10.3390/polym14122466

44. Bhowmik P, Kant R, Singh H (2023) Effect of degumming duration on the behavior of waste filature silk-reinforced wheat gluten composite for sustainable applications. ACS Omega 8:6268–6278. https://doi.org/10.1021/acsomega.2c05963
45. Eliaz D, Kellersztein I, Miali ME et al (2024) Fine structural analysis of degummed fibroin fibers reveals its superior mechanical capabilities. Chemsuschem 18:e202401148. https://doi.org/10.1002/cssc.202401148
46. Zhao Z, Li W, Wang F et al (2017) Using of hydrated lime water as a novel degumming agent of silk and sericin recycling from wastewater. J Clean Prod 172:2090–2096. https://doi.org/10.1016/j.jclepro.2017.11.213
47. Bansal P, Sharma AK, Kumar et al. (2023) Serine proteases: classification, catalytic mechanism, types, and industrial applications. In Sharma AK, Bansal P (eds) Serine proteases: role in human health and disease. Walter de Gruyter GmbH, Berlin/Boston, pp. 1–16.
48. Dash AK, Sahoo SK (2021) Role of enzymes in textile processing. In Thatoi H, Mohapatra S, Das SK (eds) Bioprospecting of enzymes in industry, healthcare and sustainable environment. Springer, Singapore, pp 395–410.
49. Venetikidou M, Lykartsi E, Adamantidi T et al (2025) Proteolytic enzyme activities of bromelain, ficin, and papain from fruit by-products and potential applications in sustainable and functional cosmetics for skincare. Appl Sci 15:2637. https://doi.org/10.3390/app15052637
50. Ninpetch U, Tsukada M, Promboon A (2015) Mechanical properties of silk fabric degummed with bromelain. J Eng Fibers Fabr 10. https://doi.org/10.1177/155892501501000319
51. Kafatos FC, Tartakoff AM, Law JH (1967) Cocoonase. J Biol Chem 242:1477–1487. https://doi.org/10.1016/s0021-9258(18)96117-x
52. Storer AC, Ménard R (2012) Papain. In: Rawlings ND, Salvesen g (eds) Handbook of proteolytic enzymes, vol 2. Elsevier, Oxford, pp 1858–1861
53. Vyas SK, Shukla SR (2015) Degumming of eri silk using ionic liquids and optimization through response surface methodology. J Text Inst 107:1096–1111. https://doi.org/10.1080/00405000.2015.1086196
54. Nasirpour N, Mohammadpourfard M, Heris SZ (2020) Ionic liquids: promising compounds for sustainable chemical processes and applications. Process Saf Environ Prot 160:264–300. https://doi.org/10.1016/j.cherd.2020.06.006
55. Dan AK, Biswal B, Das M et al (2022) Aqueous and chemical extraction of saponin of Acacia concinna (Willd.) Dc.: an effective Bio-surfactant solution to extract silk fibroin from muga silk cocoons. J Mol Liq 360:119547. https://doi.org/10.1016/j.molliq.2022.119547
56. Biswal B, Dan AK, Aamna B et al (2023) Self-assembled micellar saponin from *sapindus laurifolia* Vahl. investigations on the surfactant activity on the extraction of fibroin from silk cocoons. J Polym Environ 31:3803–3813. https://doi.org/10.1007/s10924-023-02861-7
57. Huang L, Shi J, Zhou W et al (2023) Advances in preparation and properties of regenerated silk fibroin. Int J Mol Sci 24:13153. https://doi.org/10.3390/ijms241713153
58. Xiang W, Quan QY, Ding J et al (2011) Study on mulberry silk degumming process with cold-pad-batch using tea sapogenin. Adv Mater Res 233–235:909–914. https://doi.org/10.4028/www.scientific.net/amr.233-235.909
59. Lo C, Chao Y (2017) Degumming of silk fibers by CO_2 supercritical fluid. J Mater Sci Chem Eng 5:1–8. https://doi.org/10.4236/msce.2017.54001
60. Lo C (2020) Degumming silk by CO_2 supercritical fluid and their dyeing ability with plant indigo. Int J Cloth Sci Technol 33:465–476. https://doi.org/10.1108/ijcst-06-2019-0072
61. Wang R, Zhu Y, Shi Z et al (2018) Degumming of raw silk via steam treatment. J Clean Prod 203:492–497. https://doi.org/10.1016/j.jclepro.2018.08.286
62. Javali U, Padaki N, Das B et al (2015) Developments in the use of silk by-products and silk waste. In: Basu A (ed) Advances in silk science and technology. Woodhead Publishing, Amsterdam, pp 261–270
63. Jaramillo-Quiceno N, Callone E, Dirè S et al (2021) Boosting sericin extraction through alternative silk sources. Polym J 53:1425–1437. https://doi.org/10.1038/s41428-021-00539-2
64. Saha J, Mondal MIH, Sheikh MRK et al (2019) Extraction, structural and functional properties of silk sericin biopolymer from bombyx mori silk cocoon waste. J Textile Sci Eng 9:1000390. https://doi.org/10.4172/2165-8064.1000390

65. Sreenivasa NVP (2023) Development of eco-friendly method for degumming of eri cocoons. Text Leather Rev 6:742–759. https://doi.org/10.31881/tlr.2023.107
66. Jelil RAA (2015) Review of low-temperature plasma treatment of textile materials. J Mater Sci 50:5913–5943. https://doi.org/10.1007/s10853-015-9152-4
67. Wulffson RL (2024) Microwave technology. In: Research starters home. EBSCO. https://www.ebsco.com/research-starters/science/microwave-technology. Accessed 2 Apr 2025
68. Mahmoodi NM, Moghimi F, Arami M et al (2010) Silk degumming using microwave irradiation as an environmentally friendly surface modification method. Fibers Polym 11:234–240. https://doi.org/10.1007/s12221-010-0234-2
69. Wang Y, Gao J, Yang Y et al (2025) A review on the extraction methods, bioactivities, and application in foods of silk sericin. J Food Biochem 2025:2155701. https://doi.org/10.1155/jfbc/2155701
70. Veggi PC, Martinez J, Meireles MAA (2013) Fundamentals of microwave extraction. In: Chemat F, Cravotto G (eds) Microwave-assisted extraction for bioactive compounds: theory and practice. Food engineering series, vol 4. Springer, New York, pp 15–52
71. Bascou R, Hardouin J, Mlouka MAB et al (2022) Detailed investigation on new chemical-free methods for silk sericin extraction. Mater Today Commun 33:104491. https://doi.org/10.1016/j.mtcomm.2022.104491
72. Pan M, Jin Y, Ye Y et al (2024) An efficient and eco-friendly method for removing sericin using microwave-assisted steam degumming. Environ Technol Innov 35:103674. https://doi.org/10.1016/j.eti.2024.103674
73. Körlü A, Bahtiyari Mİ (2021) Ultrasound-based wet processes in textile industry. In: Rather LJ, Haji A, Shabbir M (eds) Innovative and emerging technologies for textile dyeing and finishing. Scrivener Publishing LLC, Massachusetts
74. Wei ZZ, Weng YJ, Zhang YQ (2022) Enhancing the in vitro biological activity of degraded silk sericin and its analog metabolites. Biomolecules 12:161. https://doi.org/10.3390/biom12020161
75. Rawan S, Zhang S, Dhavale A et al (2023) Optimization of enzyme-assisted ultrasonic irradiation silk degumming process. J Inst Eng (India) E 104:237–250. https://doi.org/10.1007/s40034-023-00275-7
76. Asakura T, Kametani S, Suzuki Y (2018) Silk. In: Encyclopedia of polymer science and technology. John Wiley & Sons, Inc., New York. https://doi.org/10.1002/0471440264.pst339.pub2
77. Padaki NV, Das B, Basu A (2015) Advances in understanding the properties of silk. In: Basu A (ed) Advances in silk science and technology. Elsevier Ltd., Woodhead Publishing, Cambridge. https://doi.org/10.1016/B978-1-78242-311-9.00001-X
78. Matsumoto A, Kim HJ, Tsai IY et al (2007) Silk. In: Lewin M (ed) Handbook of fiber chemistry, 3rd edn. CRC Press, Taylor & Francis Group, Boca Raton
79. Babu KM (2013) The dyeing of silk. In: Silk-processing, properties and applications. Woodhead Publishing Limited, Cambridge. https://doi.org/10.1533/9781782421580.117
80. Das M, Basak S, Samanta KK et al (2014) Silk yarn processing: an industrial review. Int J Bioresour Sci 1(1):65–71
81. Lacasse K, Baumann W (2004) Pretreatment. In: Textile chemicals: environmental data and facts. Springer, Heidelberg. https://doi.org/10.1007/978-3-642-18898-5_4
82. Islam S, Alam SMM, Akter S (2019) Identifying the values of whiteness index, strength and weight of cotton spandex woven fabric in peroxide bleaching of different concentration. Fibres Text 4:96–107
83. Mamun MHA, Hossain MDA, Zakaria Ahmed MK et al (2017) Effect of different types scouring against different types of bleaching process on dyeing of cotton fabric with monochlorotriazine (hot brand) reactive dye. Int J Text Sci 6(5):128–134. https://doi.org/10.5923/j.textile.20170605.02
84. Le Marechal AM, Križanec B, Vajnhandl S et al (2012) Textile finishing industry as an important source of organic pollutants, In: Puzyn T, Mostrag-Szlichtyng A (eds) Organic pollutants ten years after the stockholm convention–environmental and analytical update. InTech, Croatia

85. Bahtiyari Mİ, Benli H (2016) Ozone bleaching of cotton fabrics with the aid of ultrasonic humidifier. Cellulose 23:2715–2725. https://doi.org/10.1007/s10570-016-0978-y
86. Farr JP, Smith WL, Steichen DS (2003) Bleaching agents. In: Kirk-Othmer encyclopedia of chemical technology. John Wiley & Sons, Inc., New York, pp 43–81. https://doi.org/10.1002/0471238961.1921182206011818.a01.pub2
87. Hage R, Lienke A (2006) Applications of transition-metal catalysts to textile and wood-pulp bleaching. Angew Chem Int Ed 45:206–222. https://doi.org/10.1002/anie.200500525
88. Cardamone JM, Marmer WN (1995) The whitening of textiles. In: Carr CM (ed) Chemistry of the textiles industry. Springer, Dordrecht. https://doi.org/10.1007/978-94-011-0595-8_2
89. Shukla SR (2001) Environment-friendly processing of protein fibres. Indian J Fibre Text Res 26:116–124
90. Körlü A, Bahtiyari Mİ, Duran K (2014) Effect of fiber type on hydogen peroxide stabilization during bleaching. Ind Textila 65:75–79
91. Das S (1992) The preparation and processing of tussah silk. JSDC 108:481–486. https://doi.org/10.1111/j.1478-4408.1992.tb01393.x
92. Choudhury AKR (2011) Pre-treatment and preparation of textile materials prior to dyeing. In: Clark M (ed) Handbook of textile and industrial dyeing principles, vol 1. Processes and types of dyes. Woodhead Publishing Limited, Cambridge, pp 64–149
93. Sharma IC, Chattopadhyay DP, Mukhopadhyay A et al (1999) Effect of degumming followed by sequential oxidative and reductive bleaching on mulberry and tasar silk fabrics. Indian J Fibre Text Res 24:290–293
94. Leksophee T, Supansomboon S, Sombatsompop N (2004) Effects of crosslinking agents, dyeing temperature, and pH on mechanical performance and whiteness of silk fabric. J Appl Polym Sci 91(2):1000–1007. https://doi.org/10.1002/app.13256
95. Steyn HJH, Olivier Y (2010) The evaluation of hydrogen peroxide bleaching of gonometa postica silk. J Fam Ecol Consum Sci 38:84–91
96. Bahtiyari Mİ, Körlü A (2021) Ozone-Based finishing of textile materials. In: Rather LJ, Haji A, Shabbir M (eds) Innovative and emerging technologies for textile dyeing and finishing. Scrivener Publishing LLC, Massachusetts
97. Bahtiyari Mİ, Benli H (2020) Comparison of ozone-based cold bleaching processes with conventional pretreatment of cotton. Ozone: Sci Eng 42(5):450–460. https://doi.org/10.1080/01919512.2019.1704218
98. Sargunamani D, Selvakumar N (2006) A study on the effects of ozone treatment on the properties of raw and degummed mulberry silk fabrics. Polym Degrad Stab 91:2644–2653. https://doi.org/10.1016/j.polymdegradstab.2006.05.001
99. Sargunamani D, Selvakumar N (2007) Effects of ozone treatment on the properties of raw and degummed tassar silk fabrics. J Appl Polym Sci 104:147–155. https://doi.org/10.1002/app.24761
100. Sargunamani D, Selvakumar N (2011) Comparative analysis of the effect of ozone treatment on the properties of mulberry and tassar silk fabrics. J Text Inst 102(10):870–874. https://doi.org/10.1080/00405000.2010.525814
101. Balcı N, Ömeroğulları Z, Kut D et al (2015) Effects of plasma and ozone treatments on tensile and whiteness properties of 100 % silk. Uludağ Univ J Fac Eng 20:43–56
102. Kaschig, J (2003) Optical Brighteners. In: Hunger K (ed) Industrial dyes: chemistry, properties, applications. Wiley-VCH Verlag GmbH & Co. KGaA, Weinheim
103. Perincek S (2016) Ozonation: a new patterning method for optical bleached fabrics. Text Appar 26:76–83
104. Mustalish RA (2000) Optical brighteners: history and technology. Stud Conserv 45(sup1):133–136. https://doi.org/10.1179/sic.2000.45.Supplement-1.133
105. Chakrabarti R, Vignesh A (2011) The chemistry of optical brightening agent and its environmental impact. Fiber2fashion. https://www.fibre2fashion.com/industry-article/5710/the-chemistry-of-optical-brightening-agent. Accessed 2 Apr 2025
106. Sabale AG, Rane VM (2009) Whiter whites! Brighter brights!! J Text Assoc 70:188–193

107. Ozkavruk Adanir E, İleri B, Can F et al (2022) An ethical approach to sericulture: production of peace silk in Hatay/Turkey. Textile 22:20–30. https://doi.org/10.1080/14759756.2022.2036072
108. Karthik T, Rathinamoorthy R (2017) Sustainable silk production. In: Muthu SS (ed) Sustainable fibers and textiles. Woodhead Publishing, Kidlington, pp 135–170

Chapter 2
Silk Colouration: Innovations in Sustainable and Functional Dyeing

Shristirupa Borah, Rinku Moni Kalita, and Parikshit Gogoi

2.1 Introduction

Silk holds an influential and enduring role in the textile industry, as it influences not only global trade routes but also advancements in textile technology and fashion. The art of sericulture raising of silk worms and the spinning of silk into fabric originated in China, where it evolved into an essential industry. The Chinese government imposed strict control on silk production and exports, imposing severe penalties on those who broke the rules. During the sixth century, Benedictine monks secretly smuggled silkworms and their eggs across China, after which other regions were able to gain access to silk production. The dissemination of silk technology across different regions of the world had a great influence on global trade, connecting the East and West via the Silk Road and turning the silk trade into one of the most influential commercial enterprises of the ancient and medieval period [1]. The emergence of synthetic fibres in the twentieth century led to a decrease in traditional silk production as these fibres provided comparable properties at a lower price. However, silk has never been completely replaced and is valued for its smoothness, lightness and natural breathability. In the present day, silk plays a prominent role in luxury fashion, home décor and medical applications due to its strength, durability and hypoallergenic properties [2]. Silk is widely used in clothing, scarves, curtains, carpets and upholstery fabrics. However, a common concern among consumers about silk is its weak antibacterial properties. Addressing this issue, natural compounds like polyphenols and certain dyes demonstrated effective ability to suppress bacterial proliferation on

S. Borah
Department of Chemistry, Gauhati University, Guwahati, Assam, India

R. M. Kalita
Department of Botany, Bhattadev University, Bajali, Assam, India

S. Borah · P. Gogoi (✉)
Department of Chemistry, Nagaon University, Nagaon, Assam, India
e-mail: parikk100@gmail.com

M. Shahid et al. (eds.), *Science and Technology of Silk*, Textile Science and Clothing Technology, https://doi.org/10.1007/978-3-032-26880-8_2

silk, offering a promising alternative for antimicrobial effect on silk [3, 4]. Moreover, employing inorganic nanoparticles and nanocomposites presents an innovative approach for producing antimicrobial properties in silk fabric [5].

The textile industry predominantly depends on synthetic dyes in response to increasing global demand for textile colouration. This may be due to their low cost, broader spectrum of vibrant colours and enhanced colourfastness when compared to natural dyes. These traditional dyeing methods often depend upon synthetic chemicals and detrimental processes, which can lead to water pollution, high energy consumption and environmental degradation [6]. Synthetic dyes are manufactured from petrochemicals, and some of these may contain amines linked to cancer. The use of these dyes not only presents significant health risks but also disrupts the ecological balance of the environment. Moreover, many countries have imposed strict environmental regulations on the use of these synthetic dyes. As a result, there has been a significant drive towards exploring greener alternatives, like agricultural residues [7]. This shift towards sustainability has intensified interest in natural dyes, which are seen as an effective way to produce textile dyeing processes. Since the investigation of new natural sources commenced, many natural substances have been identified and documented. These sources produce a rich palette of vibrant colours ranging from yellow to black. In the early stages of human civilization, plants were invaluable not only for food, fuel, fibre and shelter but also as essential sources for dyeing fabrics, crafting designs and cave paintings. Dyes obtained from botanical, animal and mineral origins are gaining significant popularity owing to their renewable, biodegradable and non-toxic properties as compared to synthetic dyes [8].

Natural dye sources must be dried, crushed and ground before they can be used (Fig. 2.1). Various techniques have been developed to extract these colourants, particularly from plant sources. The techniques comprise aqueous, solvent and acid/alkaline extraction, as well as advanced techniques such as ultrasonic-assisted, microwave-assisted, supercritical fluid, enzymatic extraction and fermentation. Significant development has occurred to enhance the colour properties, fastness and durability of naturally dyed fabrics. Mordants vitalize dyeing, facilitating a strong coordination bond between dye molecules and the fabric surface, which enables the yield of vivid colours and improves functional properties across the dye fabric. The metal ions in mordants function as electron acceptors, forming coordination interactions with electron-donating functional groups of dyes, resulting in the formation of stable dye-metal-fibre complexes [7]. Mordants generally consist of metallic compounds, including aluminium potassium sulphate (alum), copper (II) sulphate, iron (II) sulphate, tin (II) chloride, zinc chloride, and potassium dichromate. These mordants can be applied to silk fabric via three methods, depending on the time of their usage, which are pre-mordanting, meta-mordanting and post-mordanting before dyeing operations. Following mordanting and dyeing, the fabrics undergo washing, and then air-drying to increase their enhanced functionality [7, 9]. Thus, these shift towards environment-friendly alternatives creates a balance between aesthetics and environmental sustainability.

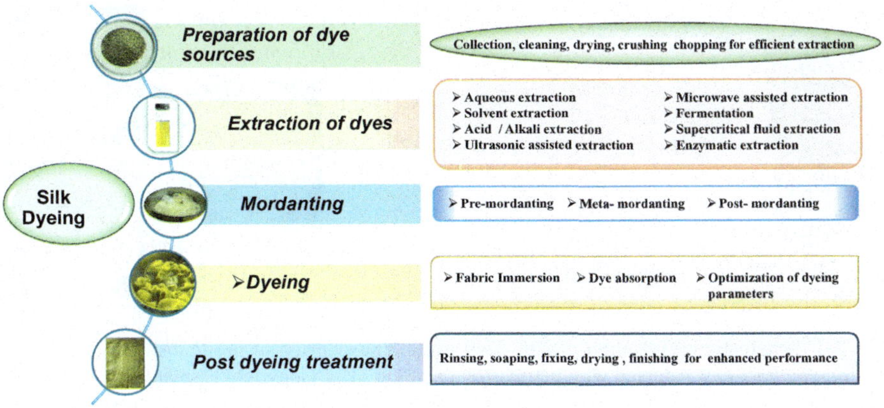

Fig. 2.1 Sequential process of silk dyeing using natural dye

This chapter focuses on the conventional methods of silk colouration using green technologies. It investigates the innovative approaches with an emphasis on improved efficiency and sustainability. Functionalization of silk using enzyme-assisted technologies and nanotechnology has been explored. The economic, environmental and aesthetic consequences of implementing these innovations have also been studied.

2.2 Developments in Natural and Sustainable Dyeing Methods

2.2.1 Classification of Dyes According to Their Sources

Recent advancements in natural and sustainable dyeing methods are transforming the textile industry, spurred by the increasing demand for eco-friendly practices. Traditionally used synthetic dyes, which are associated with harmful impacts on both human health and the environment, are gradually being replaced by natural alternatives. These eco-friendly dyes extracted from microorganisms, animals and plants can provide a more sustainable and biodegradable option, paving the way for a cleaner, greener future in fabric dyeing [8]. Some of these eco-friendly dyes extracted from different sources are discussed below.

2.2.1.1 Dyes from Microorganisms

An array of bacterial species, such as *Agrobacterium aurantiacum*, *Flavobacterium multivorum*, *Gordonia jacobeae*, *Paracoccus carotinifaciens*, *Rhodobacter sphaeroides* and *Sphingomonas sp.*, naturally synthesizes a biologically important

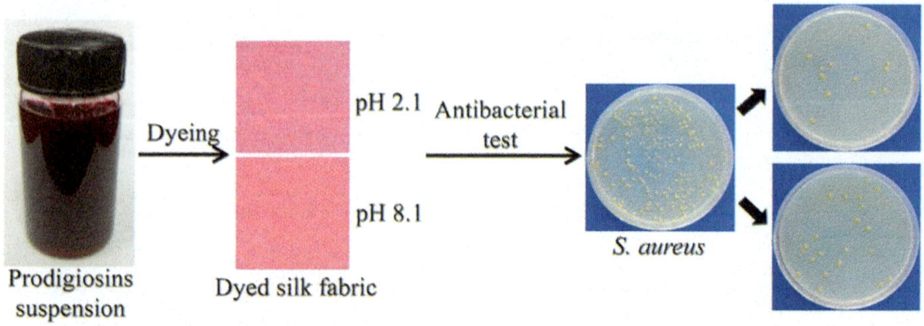

Fig. 2.2 Schematic presentation of silk dyed using a suspension of prodigiosin produced from *Serratia marcescens*. Adapted with permission from Ref [12], Copyright Elsevier, 2018

class of pigments called carotenoids. These bacteria synthesize a class of organic carotenoids that exhibit a wide range of orange, red and yellow colours [10]. Seventy per cent ethanol was employed as the extraction solvent of strain J4 grown on broken mill rice through solid-state fermentation to produce an actinobacterial dye for silk dyeing. The coloured extract exhibited good colour durability and washing fastness (rating between 3 and 5), along with good sensitivity to light (between 3 and 4) [11]. A study investigated the suspension of prodigiosins, produced via microbial fermentation of *Serratia marcescens*, as a novel, environment-friendly dye for silk dyeing (Fig. 2.2). The study examined the effect of dyeing temperature, dyeing time and pH of the dye bath on both the colour characteristics and antibacterial properties of dyed silk fabric. The dyeing results were optimum at a temperature of 80°C, time of 30 min and at a pH of 2.1. The dyed silk exhibited good colour strength, colour fastness and satisfactory antibacterial activity against *Staphylococcus aureus* [12].

2.2.1.2 Dyes from Animal Origin

Certain insects and mollusks contain colourants that can be extracted from their bodies for use in dyeing. Laccaic acid, carminic acid and kermesic acid are such colourants that are obtained from the desiccated remains of insects, including lac, cochineal and kermes, respectively [13]. Carminic acid (C.I. Natural Red 4) is derived from bodies of female cochineals and their eggs, a scale insect known as *Dactylopius coccus*. This colourant has a vast array of applications, including food colouration, pharmaceuticals and textile dyeing. Carminic acid can yield a variety of colours, like pink, crimson, scarlet red and blue, depending upon the type of mordants applied. *Kermes vermilio* (Kermes), a species of scale insect is valued for its natural colorants predominantly accumulated in the bodies of female insects and eggs [9]. Researchers explored the use of an insect-derived lac dye, from *Laccifer lacca Kerr*, in the colouration and functionalization of silk (Fig. 2.3). The optimal conditions of dye extraction were achieved at an ethanol-to-water ratio of 60:40, a solid–liquid ratio of 1:10, an

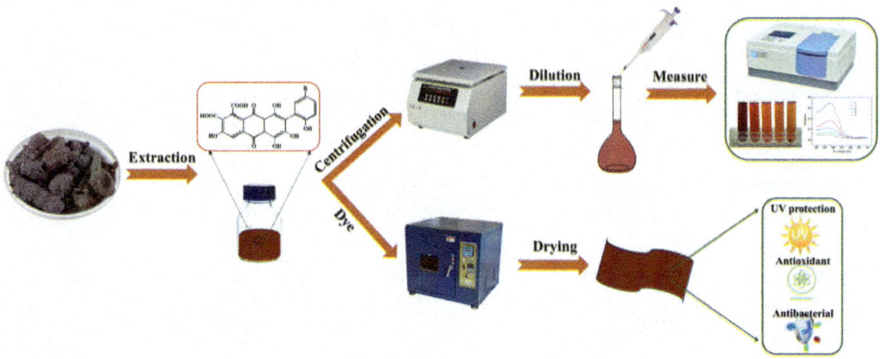

Fig. 2.3 Schematic illustration of lac pigment extraction and dyeing process. Adapted from Huang et al. [14] published under CC BY 4.0 license

extraction time of 1 h, a temperature of 80°C and a neutral pH. The dyed silk exhibited excellent colourfastness to washing, rubbing and light (rating > 4). In addition to promising UV protection, the treated fabrics displayed notable antibacterial and antioxidant activities [14].

2.2.1.3 Dyes from Plant Origin

Plants facilitate primary natural colourants as different parts of plant species, such as roots, stems, twigs, bark, fruits, seeds, leaves, hulls, husks, rind and trunks, are rich in natural pigments [15]. The colour produced by these plants can vary depending upon several factors like age of the plant, environmental conditions, stage of growth and the season in which it is harvested. In India, approximately four hundred and fifty plant species have been identified as potential sources of dyes [16]. Silk dyed using some of these natural dyes extracted from plant sources has been discussed in Table 2.1. Silk dyed using turmeric and pomegranate as natural colourants exhibited excellent colour strength and colour retention properties through assorted mordants [17, 18]. A study focused on the colouration and functional treatment of mulberry using natural colourants derived from teak (*Tectona grandis*) leaves (Fig. 2.4). A combination of methanol, water and HCl emerged as the most effective solvent for extracting bioactive compounds like polyphenols, flavonoids and antioxidants. Maximum dyeing efficiency was observed at pH 2.5, temperature 80°C and a time duration of 45 min. Dyed silk exhibited excellent colourfastness properties to washing, rubbing and light, along with enhanced UV protection properties and significant antibacterial activity against *Staphylococcus epidermis* MTCC43 [19].

Table 2.1 Several studies reported on silk dyeing using natural dyes from plant sources

Source	Botanical name	Treatment	Reference
Mangrove bark	*Rhizophora apiculata Blume*	Effective dyeing conditions were observed at 90°C for 60 min at pH 3	[20]
Hambo hambo	*Cassia singueana*	Optimum dyeing conditions were obtained at 70°C for a time of 60 min and at pH 4	[21]
Alkanet	*Alkanna tinctoria*	Superior colour strength was obtained at 55°C for 55 min and 7g/100ml NaCl used as the exhausting agent	[22]
Green tea	*Camellia sinensis*	Optimum dyeing results were obtained at 40°C	[23]
Vine tea	*Ampelopsis grossedentata*	Maximum extraction efficiency was observed at 100°C, time 0.5 h and a bath ratio of 1:20	[24]
Black tea leaves	*Camelia sinensis*	Dyeing results were optimum at 55°C for 45 min, and at pH 3	[25]
Teak leaves	*Tectona grandis*	Highest colour strength was achieved at 80°C, pH 2.5 for time 45min	[19]
Mango leaves	*Mangifera indica*	Optimum colour depth was attained at pH 10	[7]
Yerbe mate	*Ilex paraguariensis*	Optimum dyeing results were obtained at 20g/L^{-1} dye stuff concentration, at pH 3 and temperature 90°C	[26]

Fig. 2.4 Schematic illustration of the extraction and dyeing process using teak leaves. Adapted with permission from Ref. [19], Copyright Elsevier, 2024

2.2.2 Enzymatic Dyeing

Enzyme application exemplifies industrial biotechnology, offering environmentally friendly alternatives that enhance fibre treatment and improve end-product quality. The application of enzymes is at different phases in textile manufacturing, where they play an essential role in modifying fibre characteristics or functionalizing their surfaces. These biotechnological advancements can foster more sustainable manufacturing practices by substituting harsh chemicals with natural enzymatic processes [27]. Enzymatic treatment is an environmentally friendly, safe and effective method for modifying the surface of fabrics (Fig. 2.5). It effectively improves the final product quality by enhancing the fabric's water retention capacity, while also significantly altering its softness and texture. This becomes beneficial in silk dyeing, which can facilitate better dye uptake and improve overall fabric appearance. Several research efforts have focused on optimizing the methods for preparation, dyeing and finishing silk using natural dyes in combination with different enzymatic treatments (Table 2.2). These enzymatic processes have been found to modify the physiological properties of silk, thereby enhancing the absorption of natural colourants [28].

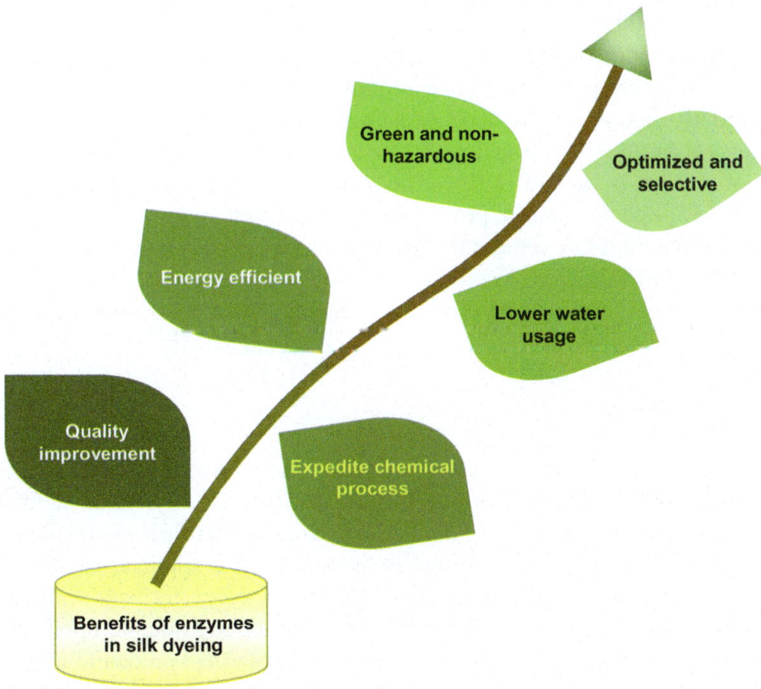

Fig. 2.5 Advantages of enzyme-assisted dyeing

Table 2.2 Previous literature on the application of enzymes used in silk dyeing

Enzyme	Natural dye/colourant used in the study	Findings	Reference
Protease, amylase, xylanase, pectinase, phytase	Rubia cordifolia	Enzymatic treatment gave good colour strength to the fabric	[37]
Protease, lipase, amylase, diastase	Delonix regia	Dyed silk exhibited good wash and light fastness	[38]
Protease	*Rhus coriaria, Terminalia chebula, Quercus alba, Quercus infactoria*	Exhibited good wash and light fastness	[39]
Horseradish peroxidase	Gallic acid	Dye silk exhibited excellent colourfastness	[40]
Tyrosinase	Caffeic acid	Dyed silk exhibited excellent colour and staining fastness	[29]
Laccase	Catechin, Gallic acid	Dyed silk exhibited good colour strength, antioxidant, and UV protection properties	[41]

Polyphenol oxidase, a group of enzymes belonging to the class of oxidases, catalyse the oxidation of phenolic compounds, resulting in coloured polymeric products. These coloured pigments offer a promising alternative to traditional synthetic dyes in silk dyeing, providing benefits such as mild processing conditions, reduced waste generation and lower toxicity [29]. Although enzyme application in the natural dyeing of textiles is limited in scientific literature, it holds a significant promise. Enzymatic dyeing can enhance process efficiency, fabric quality, optimize resource utilization and reduce environmental impact [27]. Recent advancements in the enzymatic applications can suggest a promising future in textile manufacturing. A concise overview of several enzymes employed in the textile industry is provided below.

2.2.2.1 Cellulases

Cellulases are a widely utilized group of hydrolytic enzymes in wet processing of textiles and laundry that function as finishing agents to modify cellulose fibres in a controlled and desired manner. Cellulose plays a significant role in the textile industry as they are the third largest group of enzymes used in these applications. Cellulose is used in household laundry applications as it enhances the performance of detergents and improves the colour vibrancy of textiles. One of the notable applications of cellulase biotechnology is achieving the stone-washed effect on denim garments, in addition to cleaning and rejuvenating fabric surfaces compromised by microfibrils, fuzz and protruding strands [30]. Cellulase can have varying effects on synthetic fibres that include rayon, lyocell (Tencel) and cellulose acetate. Research has shown that cellulase pretreatment can enhance the rate of dye absorption for natural dyes, while maintaining their colourfastness properties [31].

2.2.2.2 Proteases

As biocatalysts, protease enzymes provide a sustainable and effective approach for preventing felting in wool, replacing chemical treatments. Natural proteases can effectively hydrolyse the scale layer of wool fibres, but their tendency to penetrate deep inside the fibre structure can cause undesirable consequences, such as excessive damage and a decrease in colour depth in pigmented textiles. To overcome these challenges, a two-step method involving mild reduction followed by controlled protease incubation was introduced [32]. The protease enzyme derived from microorganisms resulted in improved quality parameters of silk fabrics. When protease enzymatic treatment was extended to 48 h, the dyeing properties of the fabrics improved. However, no significant mechanical alterations were observed in the fabric [33].

2.2.2.3 Lipases

The potential of lipase in enhancing the hydrophilicity of polyester fabrics was studied. It was observed that lipases could improve the wetting and absorbency properties under optimum conditions compared to alkaline hydrolysis. The enzyme hydrolysis proceeds under mild operational parameters such as reduced enzyme concentration, moderate temperature and reduced reaction time compared to aqueous hydrolysis. The impact of enzyme treatment with lipase on both the weight reduction and surface alterations in polyester fabrics was examined. This study analyzed the modifications in functional groups induced by enzyme treatment and assessed its impact on dye uptake. The findings indicate that the enzymatic process causes minimal surface abrasion and weight reduction, boosts moisture retention, dye absorbability and strengthens shear properties [34].

2.2.2.4 Xylanases

Xylanases are recognized as one of the most significant microbial enzymes that can catabolize xylan residues. It is highly effective in several textile wet processing stages, including desizing, bioscouring and bio-finishing of cellulosic fabrics, without the need for additional additives [35]. Cellulose pulps and textile fabrics are created with the application of xylanase enzymes that are free of cellulase. The need for cellulases in eliminating protruding superficial fabrics was reduced by combining pectinase and xylanases. The treated fabric surfaces were smoother than those of untreated samples (control), indicating that the enzyme combination proved more effective than cellulase treatment alone [36].

2.3 Industrial Insights and Development in Silk Colouration

2.3.1 Emerging Technologies in Silk Colouration

Recent developments in silk colouration have been marked by a shift towards sustainability and innovations in dyeing technologies. The integration of biotechnology in silk colouration is playing a transformative role in improving dye uptake and reducing chemical usage using enzyme-based treatments. The silk colouration industry is increasingly leveraging nanotechnology to boost both performance and efficiency. The industry is adopting technologies like ultrasonic, plasma treatment, supercritical carbon dioxide, microwave, electrochemical and nano-dye technology for large-scale production. These methods significantly reduce water consumption and energy usage as compared to conventional methods [42]. A brief overview of the emerging technologies in silk colouration has been discussed below (Fig. 2.6):

2.3.1.1 Ultrasonic Textile Treatment

Ultrasonic-assisted dyeing facilitates enhanced dispersion, removes gases and accelerates the diffusion rate of dyes or finishing agents into the fibre matrix [43]. In this process, dye molecules are dispersed individually, increasing their activation energy and enabling faster diffusion throughout the fibre structure, thereby improving the

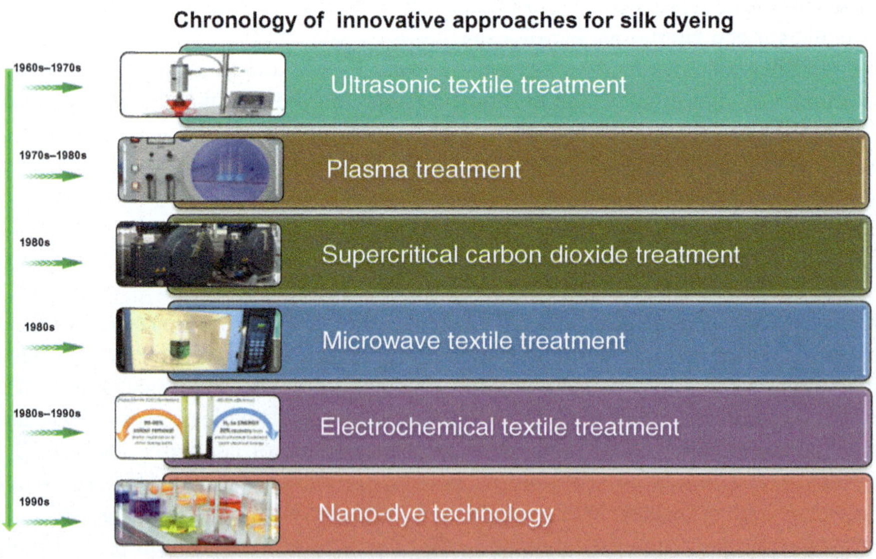

Fig. 2.6 Innovative approaches for dyeing silk fabrics

overall dye uptake [44]. This technique is applicable to various preparatory wet processes, including desizing, bleaching, scouring and washing [45].

2.3.1.2 Plasma Treatment

Plasma treatment enables surface-level chemical and physical modifications of polymeric and textile substrates, leading to enhanced hydrophilicity and improved surface energy. These modifications contribute to increasing the dye affinity of fibres, resulting in better dye-fibre interaction. Plasma technology offers a sustainable, effluent-free alternative to traditional methods [43, 46]. It provides significant environmental benefits, including lower energy consumption, reduction in water and chemical usage and the elimination of drying steps required in chemical wet processing [47].

2.3.1.3 Treatment Using Supercritical Carbon Dioxide

The use of supercritical carbon dioxide treatment represents a sustainable substitute for conventional dyeing processes in the textile sector [48]. This process involves lower energy requirements as compared to traditional techniques. During this process, CO_2 is heated to 120°C and pressurized to 250 bars, enabling it to deeply penetrate the fibre, facilitating enhanced dye diffusion and uptake during the dyeing process [43, 49].

2.3.1.4 Microwave Textile Treatment

Microwave lies within the electromagnetic spectrum between radio waves and infrared light, corresponding to frequencies from 300 MHz to 300 GHz with wavelengths between 1 and 1 mm. This technology provides rapid, uniform and effective heating, which increases the mobility and penetration of dye molecules into polymer matrices [50–52].

2.3.1.5 Electrochemical Textile Treatment

Electrochemical presents an effective method for the redox processing of vat and sulphur dyes, used in the colouration of cellulosic fibres. The indirect electrochemical reduction of indigo dye offers low-operational costs and high cyclic stability. These advantages contribute significantly to improving the sustainability and environmental impact of denim dyeing processes [53].

2.3.1.6 Nano-Dye Technology

Nano-dye technology is an innovative dyeing technology that uses nanotechnology to enhance the bonding and performance of dyes in textiles. This technology offers multiple advantages, including the elimination of salt usage, reductions in water and energy consumption up to 75% and a high dye exhaustion rate up to 97% [43].

2.4 Application of Nanotechnology in Textiles

Nanotechnology is an emerging field with its potential impact across various domains of science and technology, from materials science and mechanics to electronics, medicine, energy, aerospace, plastics and textiles. Even at an early stage, this technology is already showing its value in boosting textile performance and attracting significant global interest. The innovative use of nanotechnology in textiles offers a broad spectrum of properties, paving the way for improved functionality and entirely new applications in various products [54]. Nanotechnology offers significant potential for commercial applications in the textile industry. The reason is that conventional approaches to imparting fabric properties to the fabric commonly lack long-term durability and lose their functionality after multiple washings or prolonged use. The integration of nanotechnology has paved the way for new possibilities in fabric finishing, facilitating the development of advanced finishes and innovative methods for their application [55].

Modern textiles have evolved to exhibit a range of higher-performance features, including hydrophobicity, UV resistance, antimicrobial properties and resistance to wrinkles and stains. Researchers are now focusing on advanced fabrication methods and surface finishing techniques by incorporating nanotechnology to develop textiles with smart, innovative applications. Their primary objective was to develop new applications that offer high efficiency without compromising the fabric's comfort, flexibility and lightweight properties. In recent decades, a variety of functionalities have been explored, including bacterial resistance, hydrophobicity, flame retardancy, UV protection, colourfastness and self-cleaning properties [56].

2.4.1 Antimicrobial Properties in Silk with the Use of Nanoparticles

Silk fibres are prone to microbial attack because of their pronounced hydrophilicity and extensive fibre surface, which foster favourable conditions for the microorganisms to thrive. Since silk is an expensive fashion fibre, it needs to be handled with proper care. To preserve the intrinsic properties of silk and improve its durability, antimicrobial finishing serves as a key strategy. Consequently, many researchers have explored antimicrobial treatments for silk to mitigate fabric degradation and issues related to microbial growth [57]. A wide variety of antimicrobial agents, including antibiotics and biocides of both organic and inorganic sources are used to impart antimicrobial functionality in the textile sector. The use of numerous antimicrobial substances has been limited due to their toxicity and harmful effects. Inorganic nanoparticles and nanocomposites offer a promising solution as an alternative. In recent years, the development of antimicrobial textiles using these nanoparticles has gained considerable interest among both researchers and consumers [5]. Inorganic nanostructured materials exhibit effective antimicrobial properties making them suitable for use in textile applications. These include a diverse range of antimicrobial agents such as TiO_2 nanocomposites, silver, gold, zinc oxide and copper nanoparticles, along with nano clays, carbon nanotubes, gallium-based composites, dendrimer-based metallic and inorganic nanocomposites and cyclodextrins-embedded nanoparticles [58]. A study focussed on the extraction of natural dye from the alkanet plant using microwave-assisted methods reducing energy consumption and processing time. Silk fibres were treated using both synthesized and commercially available titanium oxide (TiO_2) nanoparticles. The impact of TiO_2 nanoparticles on the functionality silk fibres including dye uptake, antibacterial activity against *Escherichia Coli* (Gram-negative) and *Staphylococcus aureus* (Gram-positive) and self-cleaning properties was assessed. The post-treatment with the synthesized TiO_2 nanoparticles resulted in satisfactory antimicrobial activity [59]. The effect of treatment conditions with plasma or in combination with TiO_2: ZnO nanoparticles on the antimicrobial activity of the printed silk using natural dyes was investigated. The printed silk treated with TiO_2: ZnO nanoparticles (50:50 wt%) without plasma exhibited enhanced antimicrobial activity against both Gram-positive and Gram-negative bacterial strains [60]. Nano silver, one of the metal-based nanomaterial, is widely used in areas like medicine, electrochemistry, biochemistry, optics and textiles [61]. Silver nanoparticles synthesized through chemical reduction of silver were applied onto silk fabric. The main antimicrobial mechanism of silver nanoparticles involves causing oxidative stress, which leads to the degradation of lipids, protcins and DNA of microorganisms [5]. CeO_2 nanoparticles were effectively anchored on the surface of silk fabric which demonstrated outstanding antibacterial properties [62].

2.4.2 UV Protection Properties in Silk Using Nanoparticles

Enhanced UV protection in fabrics is achieved by incorporating nanomaterials improves UV resistance by shielding against UV A and UV B rays. Ultraviolet Protection Factor (UPF) assesses how effectively a fabric shields against UV radiation which depends on structure and composition of fabric used. Titanium oxide (TiO_2) and zinc oxide (ZnO) nanoparticles can effectively scatter or absorb UV radiation due to their optical properties. These materials are known for their heat resistance, safety and ability to maintain durability under high temperature conditions. The interaction of UV radiation with these nanoparticles is critically dependent on both particle size and the wavelength of the incident light. These parameters govern the scattering efficiency and optical extinction characteristics of the nanoparticles [56]. ZnO and TiO_2 nanoparticles have been used to enhance functionality in silk fibres. These nanomaterials impart UV protection activities to the silk, protecting it from photodegradation and turning yellow in sunlight [63–65]. It has been reported that nano-sized titanium dioxide and zinc oxide particles were more effective in absorbing and scattering ultraviolet (UV radiation) compared to larger, conventional counterparts making them superior UV blocking agents. The incorporation of titanium dioxide nanoparticles into printing pastes or dyeing notably enhanced the Ultraviolet Protection Factor (UPF) of the silk fabric. This improvement may be attributed to the high-surface area of nanoparticles per unit mass and volume which enhanced their ability to block UV radiation more efficiently as compared to traditional materials [66]. Silk fabrics were functionally enhanced by dyeing with natural indigo followed by coating using ZnO nanoparticles. The colourfastness and various functional attributes of fabrics undergone dyeing and coating treatments were investigated. Indigo-dyed silk fabrics coated with ZnO nanoparticles demonstrated excellent UV protection properties along with enhanced colour fastness, heat resistance and antibacterial properties [67]. Cerium oxide (CeO_2) nanoparticles have gained significant interest due to their diverse applications, including use as catalysts, catalyst supports, superconducting buffer layers, fuel cell electrolytes, UV absorption and oxygen storage [68]. Compared to TiO_2 and ZnO, CeO_2 possesses a lower-band gap energy, which enhances its UV protection ability, particularly when used in textile fibre coatings. CeO_2 exhibits superior UV absorption within the UV A range (315–400 nm). These findings are consistent with the previous literature and demonstrate that CeO_2 nanoparticles are promising materials for functionalizing silk fibres to enhance UV protection ability [62, 69]. Silk sericin served both as a structural scaffold and a reducing medium to facilitate the synthesis and deposition of silver nanoparticles (AgNPs) on silk fabric surfaces. The functionalized fabrics exhibited enhanced UV protection ability in addition to antimicrobial activity [70].

2.5 Conclusion

The work described in this chapter demonstrates the field of smart textiles that has made remarkable progress in recent years, yet there is significant potential for the further development of innovative products. While traditional dyeing techniques hold significant cultural and artistic values, they face numerous limitations in meeting the environmental and industrial demands of the textile industry. Although natural dyes are eco-friendly, they tend to offer limited colourfastness as compared to synthetic alternatives, making them less viable for large-scale manufacturing. Traditional dyeing processes are largely manual, time-consuming and lack consistency in colour quality, which requires challenges to meet the scalability and efficiency required by today's fast-moving fashion market. These challenges face the need for innovations in dyeing technologies. Advancements in technologies like nanotechnology, enzyme-assisted dyeing and bioengineered dyes provide sustainable alternatives, reducing the ecological impact and enhancing the efficiency, colour accuracy and material adaptability. Developing skills to implement innovative dyeing techniques in silk involves a multidisciplinary approach that blends traditional textile industry with modern scientific advancements. Nanotechnology has emerged as a vital innovation, revitalizing material science, driving the development of smart materials, including polymers and textiles. It holds a strong potential to bring transformative changes in the field of technical textile applications. Nanotechnology enables the development of innovative approaches that provide fabrics with diverse functionalities, including antiseptic properties, UV and flame protection, enhanced durability, antimicrobial activity and resistance to water, oil and dirt, along with self-cleaning resistance. Advancements in nanomaterial-based production techniques, along with growing market demand, and broader research opportunities have drawn considerable interest of new researchers to this area.

Declaration AI tools were used only to enhance readability and language. All contributors have reviewed and approved the final content of the chapter and take full responsibility for it.

References

1. Howitt FO (1951) Silk- an historical survey with special reference to the past century. J Text Inst Proc 42:339–360. https://doi.org/10.1080/19447015108663848
2. Cheng TH, Liu ZJ, Yang JY, Huang YZ, Tang RC, Qiao YF (2019) Extraction of functional dyes from tea stem waste in alkaline medium and their application for simultaneous coloration and flame retardant and bioactive functionalization of silk. ACS Sustain Chem Eng 7:18405–18413. https://doi.org/10.1021/acssuschemeng.9b04094
3. Shahid-ul-Islam, Mohammad F (2015) Natural colorants in the presence of anchors so-called mordants as promising coloring and antimicrobial agents for textile materials. ACS Sustain Chem Eng 3:2361–2375. https://doi.org/10.1021/acssuschemeng.5b00537
4. Zhou Y, Tang RC (2017) Natural flavonoid-functionalized silk fiber presenting antibacterial, antioxidant, and UV protection performance. ACS Sustain Chem Eng 5:10518–10526. https://doi.org/10.1021/acssuschemeng.7b02513
5. Nadiger VG, Shukla SR (2016) Antibacterial properties of silk fabric treated with silver nanoparticles. J Text Inst 107:1543–1553. https://doi.org/10.1080/00405000.2015.1129756
6. El-Nagar K, Sanad SH, Mohamed AS, Ramadan A (2005) Mechanical properties and stability to light exposure for dyed Egyptian cotton fabrics with natural and synthetic dyes. Polym Plast Technol Eng 44:1269–1279. https://doi.org/10.1080/03602550500207816
7. Uddin MG (2015) Extraction of eco-friendly natural dyes from mango leaves and their application on silk fabric. Text Cloth Sustain 1:7. https://doi.org/10.1186/s40689-015-0007-9
8. Jose S, Gurumallesh H, Ammayappan L (2017) Eco-friendly dyeing of silk and cotton textiles using combination of three natural colorants. J Nat Fibers 14:40–49. https://doi.org/10.1080/15440478.2015.1137530
9. Uddin MA, Rahman MM, Haque ANMA, Smriti SA, Datta E, Farzana N, Chowdhury S, Haider J, Muhammad Sayem AS (2022) Textile colouration with natural colourants: a review. J Clean Prod 349:131489. https://doi.org/10.1016/j.jclepro.2022.131489

10. Mumtaz R, Bashir S, Numan M, Shinwari ZK, Ali M (2019) Pigments from soil bacteria and their therapeutic properties: a mini review. Curr Microbiol 76:783–790. https://doi.org/10.1007/s00284-018-1557-2
11. Urtgam S, Sumpradit T, Thurnkul N (2023) The antibacterial activity and silk dyeing of the crude pigment extract from J4 actinobacteria strain. J Curr Sci Technol 13:455–464. https://doi.org/10.59796/jcst.V13N2.2023.262
12. Ren Y, Gong J, Fu R, Zhang J, Fang K, Liu X (2018) Antibacterial dyeing of silk with prodigiosins suspension produced by liquid fermentation. J Clean Prod 201:648–656. https://doi.org/10.1016/j.jclepro.2018.08.098
13. Yusuf M, Mohammad F, Shabbir M (2017) Eco-friendly and effective dyeing of wool with anthraquinone colorants extracted from Rubia cordifolia roots: optimization, colorimetric and fastness assay. J King Saud Univ Sci 29:137–144. https://doi.org/10.1016/j.jksus.2016.06.005
14. Huang Q, Wang Z, Zhao L, Li X, Cai H, Yang S, Yin M, Xing J (2024) Environmental dyeing and functionalization of silk fabrics with natural dye extracted from lac. Molecules 29:2358. https://doi.org/10.3390/molecules29102358
15. Saxena S, Raja ASM (2014) Natural dyes: sources, chemistry, application and sustainability issues. In: Roadmap to Sustainable Textiles and Clothing. Textile Science and Clothing Technology. Springer, Singapore, pp 37–80
16. Yusuf M, Shabbir M, Mohammad F (2017) Natural colorants: historical, processing and sustainable prospects. Nat Prod Bioprospect 7:123–145. https://doi.org/10.1007/s13659-017-0119-9
17. Yang R, Li J, Cheng G, Inta A, Yang L (2023) Textiles dyeing with pomegranate (*Punica granatum*) peel extract using natural mordant. J Nat Fibers 20. https://doi.org/10.1080/15440478.2023.2282056
18. Ghoreishian SM, Maleknia L, Mirzapour H, Norouzi M (2013) Antibacterial properties and color fastness of silk fabric dyed with turmeric extract. Fibers Polym 14:201–207. https://doi.org/10.1007/s12221-013-0201-9
19. Borah S, Kalita RM, Bhuyan PM, Sarma B, Hazarika S, Gogoi P (2024) Application of Tectona grandis leaf extract for efficient and stabilized coloration of mulberry silk fabric. Fibers Polym 25:4755–4770. https://doi.org/10.1007/s12221-024-00752-9
20. Punrattanasin N, Nakpathom M, Somboon B, Narumol N, Rungruangkitkrai N, Mongkholrattanasit R (2013) Silk fabric dyeing with natural dye from mangrove bark (*Rhizophora apiculata* Blume) extract. Ind Crops Prod 49:122–129. https://doi.org/10.1016/j.indcrop.2013.04.041
21. Teklemedhin TB, Gopalakrishnan LH (2018) Environmental friendly dyeing of silk fabric with natural dye extracted from *Cassia singueana* plant. J Text Sci Eng 01. https://doi.org/10.4172/2165-8064.S3-001
22. Fazal-ur- AS, Liaqat S, Hussaan M, Mia R, Ahmed B, Wafa H (2022) Environmental friendly bio-dyeing of silk using *Alkanna tinctoria* based Alkannin natural dye. Ind Crops Prod 186:115301. https://doi.org/10.1016/j.indcrop.2022.115301
23. Vankar PS (2017) Low temperature optimized dyeing of cotton, wool and silk with extract of *Camellia sinensis* (tea leaves). J Text Eng Fashion Technol 2:. https://doi.org/10.15406/jteft.2017.02.00045
24. Zhang W, Wang X, Zhang Y, Wu S, Liu R (2022) Flavonoid dyes from vine tea (*Ampelopsis grossedentata*) have excellent bioactive properties for dyeing and finishing of silk fabrics. Sustain Chem Pharm 28:100708. https://doi.org/10.1016/j.scp.2022.100708
25. Hayat T, Adeel S, Fazal-ur- BF, Amin N, Ahmad T, Ozomay M (2022) Waste black tea leaves (*Camelia sinensis*) as a sustainable source of tannin natural colorant for bio-treated silk dyeing. Environ Sci Pollut Res 29:24035–24048. https://doi.org/10.1007/s11356-021-17341-5
26. Giacomini F, Menegazzo MAB, dos Santos JCO, Arroyo PA, de Barros MASD (2017) Ecofriendly dyeing of silk with extract of yerba mate (*Ilex paraguariensis*). Text Res J 87:829–837. https://doi.org/10.1177/0040517516641357
27. Araújo R, Casal M, Cavaco-Paulo A (2008) Application of enzymes for textile fibres processing. Biocatal Biotransformation 26:332–349. https://doi.org/10.1080/10242420802390457

28. Padaki NV, Das B, Thirumalesh RM (2015) Enzyme applications in silk processing. In: Advances in Silk Science and Technology. Woodhead Publishing, Cambridge, pp 111–120.
29. Sun S-S, Xing T, Tang RC (2013) Simultaneous coloration and functionalization of wool, silk, and nylon with the tyrosinase-catalyzed oxidation products of caffeic acid. Ind Eng Chem Res 52:8953–8961. https://doi.org/10.1021/ie303350z
30. Vankar PS, Shanker R, Verma A (2007) Enzymatic natural dyeing of cotton and silk fabrics without metal mordants. J Clean Prod 15:1441–1450. https://doi.org/10.1016/j.jclepro.2006.05.004
31. Zhao Q, Feng H, Wang L (2014) Dyeing properties and color fastness of cellulase-treated flax fabric with extractives from chestnut shell. J Clean Prod 80:197–203. https://doi.org/10.1016/j.jclepro.2014.05.069
32. Zhang J, Li K, Wang J, Xu B, Liu Y, Wang Q, Wang P (2025) Enzymatic anti-felting finishing of the dyed woolen textiles through the collaborative action of mild reduction and protease treatment. Int J Biol Macromol 307:142160. https://doi.org/10.1016/j.ijbiomac.2025.142160
33. Kim SJ, Cha MK, Oh ET, Kang SM, So JS, Kwon YJ (2005) Use of protease produced by *Bacillus* sp. SJ-121 for improvement of dyeing quality in wool and silk. Biotechnol Bioprocess Eng 10:186–191. https://doi.org/10.1007/BF02932011
34. Kumar JA, Senthil Kumar M (2020) A study on improving dyeability of polyester fabric using lipase enzyme. Autex Res J 20:243–249. https://doi.org/10.2478/aut-2019-0030
35. Abd El AA, Saleh SAA, Eid BM, Ibrahim NA, Mostafa FA (2018) Thermodynamics characterization and potential textile applications of *Trichoderma longibrachiatum* KT693225 xylanase. Biocatal Agric Biotechnol 14:129–137. https://doi.org/10.1016/j.bcab.2018.02.011
36. Carmona EC, Pizzirani-Kleiner AA, Monteiro RTR, Jorge JA (1997) Xylanase production by *Aspergillus versicolor*. J Basic Microbiol 37:387–393. https://doi.org/10.1002/jobm.3620370602
37. Vankar PS, Shukla D, Wijayapala S, Samanta AK (2017) Innovative silk dyeing using enzyme and *Rubia cordifolia* extract at room temperature. Pigment Resin Technol 46:296–302. https://doi.org/10.1108/PRT-06-2016-0065
38. Vankar PS, Shanker R (2009) Eco-friendly pretreatment of silk fabric for dyeing with *Delonix regia* extract. Color Technol 125:155–160. https://doi.org/10.1111/j.1478-4408.2009.00189.x
39. Akçakoca EP, Atav R, Bahtiyari MI (2009) Effects of alkali proteases on dyeing properties of various proteinous materials with natural dyes. Text Res J 79:517–525. https://doi.org/10.1177/0040517508090775
40. Baek N, Fan X, Yuan J, Xu J, Wang Q (2021) Polymerization and dyeing properties of gallic acid on silk fabric catalyzed by horseradish peroxidase. Fibers Polym 22:2145–2155. https://doi.org/10.1007/s12221-021-0673-y
41. Borah S, Kalita RM, Hazarika S, Gogoi P (2024) Coloration and functionalization of mulberry silk fabrics with laccase-catalyzed oxidation products of catechin and gallic acid. Ind Eng Chem Res 63:15351–15364. https://doi.org/10.1021/acs.iecr.4c00497
42. Khatri M, Hussain N, El-Ghazali S, Yamamoto T, Kobayashi S, Khatri Z, Ahmed F, Kim IS (2020) Ultrasonic-assisted dyeing of silk fibroin nanofibers: an energy-efficient coloration at room temperature. Appl Nanosci 10:917–930. https://doi.org/10.1007/s13204-019-01191-2
43. Lara L, Cabral I, Cunha J (2022) Ecological approaches to textile dyeing: a review. Sustainability 14:8353. https://doi.org/10.3390/su14148353
44. Jajpura L, Nayak R (2020) Ultrasound applications in textiles and apparels. In: Sustainable Technologies for Fashion and Textiles. Woodhead Publishing, Cambridge, pp 143–161
45. Amesimeku J, Fan L, Jakpa W, Wang C (2021) Dyeing properties of meta-aramid fabric dyed with basic dye using ultrasonic-microwave irradiation. J Clean Prod 285:124844. https://doi.org/10.1016/j.jclepro.2020.124844
46. Kan CW (2015) Plasma treatments for sustainable textile processing. In: Sustainable Apparel: Production, Processing and Recycling. Woodhead Publishing, Cambridge, pp 49–118.
47. Zille A (2020) Plasma technology in fashion and textiles. In: Sustainable Technologies for Fashion and Textiles. Woodhead Publishing, Cambridge, pp 117–142.

48. Shahid-ul-Islam, Mohammad F (2014) Emerging green technologies and environment friendly products for sustainable textiles. In: Roadmap to Sustainable Textiles and Clothing. Textile Science and Clothing Technology. Springer, Singapore, pp 63–82.
49. Arputharaj A, Raja ASM, Saxena S (2016) Developments in sustainable chemical processing of textiles. In: Green Fashion: Environmental Footprints and Eco-design of Products and Processes. Springer, Singapore, pp 217–252.
50. Öner E, Büyükakinci Y, Sökmen N (2013) Microwave-assisted dyeing of poly(butylene terephthalate) fabrics with disperse dyes. Color Technol 129:125–130. https://doi.org/10.1111/cote.12014
51. Al-Mousawi SM, El-Apasery MA, Mahmoud HM (2012) A facile synthesis of arylazonicotinates for dyeing polyester fabrics under microwave irradiation and their biological activity profiles. Molecules 17:11495–11506. https://doi.org/10.3390/molecules171011495
52. Elgarahy AM, Elwakeel KZ, Elshoubaky GA, Mohammad SH (2019) Microwave-accelerated sorption of cationic dyes onto green marine algal biomass. Environ Sci Pollut Res 26:22704–22722. https://doi.org/10.1007/s11356-019-05417-2
53. Li X, Wang K, Wang M, Zhang W, Yao J, Komarneni S (2020) Sustainable electrochemical dyeing of indigo with Fe(II)-based complexes. J Clean Prod 276:123251. https://doi.org/10.1016/j.jclepro.2020.123251
54. Khan FR, Misra SK, García-Alonso J, Smith BD, Strekopytov S, Rainbow PS, Luoma SN, Valsami-Jones E (2012) Bioaccumulation dynamics and modeling in an estuarine invertebrate following aqueous exposure to nanosized and dissolved silver. Environ Sci Technol 46:7621–7628. https://doi.org/10.1021/es301253s
55. Hosne Asif AKMA, Hasan MdZ (2018) Application of nanotechnology in modern textiles: a review. Int J Curr Eng Technol 8(2):227–231. https://doi.org/10.14741/ijcet/v.8.2.5
56. Shah MA, Pirzada BM, Price G, Shibiru AL, Qurashi A (2022) Applications of nanotechnology in smart textile industry: a critical review. J Adv Res 38:55–75. https://doi.org/10.1016/j.jare.2022.01.008
57. Zhang G, Liu Y, Gao X, Chen Y (2014) Synthesis of silver nanoparticles and antibacterial property of silk fabrics treated by silver nanoparticles. Nanoscale Res Lett 9:216. https://doi.org/10.1186/1556-276X-9-216
58. Dastjerdi R, Montazer M (2010) A review on the application of inorganic nano-structured materials in the modification of textiles: focus on anti-microbial properties. Colloids Surf B Biointerfaces 79:5–18. https://doi.org/10.1016/j.colsurfb.2010.03.029
59. El-Khatib EM, Ali NF, Nassar SH, El-Shemy NS (2022) Functionalization of natural fibers properties by using TiO2 nanoparticles to improve its antimicrobial activity. Biointerface Res Appl Chem 12:4177–4191. https://doi.org/10.33263/BRIAC123.41774191
60. Ahmed H, El-Sayad H, Raslan W, Rashed U, El-Halwagy A (2022) Natural dyes printability of modified silk fabric with plasma/nano particles of metal oxides. Egypt J Chem 65(12):385–396. https://doi.org/10.21608/ejchem.2022.113249.5145
61. Nischala K, Rao TN, Hebalkar N (2011) Silica–silver core–shell particles for antibacterial textile application. Colloids Surf B Biointerfaces 82:203–208. https://doi.org/10.1016/j.colsurfb.2010.08.039
62. Lu Z, Mao C, Meng M, Liu S, Tian Y, Yu L, Sun B, Li CM (2014) Fabrication of CeO_2 nanoparticle-modified silk for UV protection and antibacterial applications. J Colloid Interface Sci 435:8–14. https://doi.org/10.1016/j.jcis.2014.08.015
63. Xu J, Su H, Han J, Chen Y, Song W, Gu Y, Moon W-J, Zhang D (2012) In situ deposition of flower-like ZnO on silk fibroin fibers. Appl Phys A 108:235–238. https://doi.org/10.1007/s00339-012-6882-x
64. Li G, Liu H, Zhao H, Gao Y, Wang J, Jiang H, Boughton RI (2011) Chemical assembly of TiO_2 and TiO_2@Ag nanoparticles on silk fiber to produce multifunctional fabrics. J Colloid Interface Sci 358:307–315. https://doi.org/10.1016/j.jcis.2011.02.053
65. Khanjani S, Morsali A, Joo SW (2013) In situ formation deposited ZnO nanoparticles on silk fabrics under ultrasound irradiation. Ultrason Sonochem 20:734–739. https://doi.org/10.1016/j.ultsonch.2012.09.013

66. El-Molla MM, El-Khatib EM, El-Gammal MS, Abdel-Fattah SH (2011) Nanotechnology to improve coloration and antimicrobial properties of silk fabrics. Indian J Fibre Text Res 36:266
67. Tangkawanit S, Keawsri P (2023) Silk fabrics coated ZnO nanoparticles and dyeing with natural indigo to enhance functional textiles. Suranaree J Sci Technol 30:030090. https://doi.org/10.55766/sujst-2023-01-e01905
68. Paun C, Safonova OV, Szlachetko J, Abdala PM, Nachtegaal M, Sa J, Kleymenov E, Cervellino A, Krumeich F, van Bokhoven JA (2012) Polyhedral CeO_2 nanoparticles: size-dependent geometrical and electronic structure. J Phys Chem C 116:7312–7317. https://doi.org/10.1021/jp300342b
69. Duan W, Xie A, Shen Y, Wang X, Wang F, Zhang Y, Li J (2011) Fabrication of superhydrophobic cotton fabrics with UV protection based on CeO_2 particles. Ind Eng Chem Res 50:4441–4445. https://doi.org/10.1021/ie101924v
70. Chitichotpanya P, Chitichotpanya C (2017) In vitro assessment of sericin-silver functionalized silk fabrics for enhanced UV protection and antibacterial properties using experimental design. Coatings 7:145. https://doi.org/10.3390/coatings7090145

Chapter 3
Innovative Approaches to Silk Coloration

Wanli Liu and Yuyang Zhou

3.1 Introduction

In recent years, as living standards have improved, eco-friendly textiles have gained significant popularity. Silk, with its soft and smooth texture, offers a comfortable wearing experience and provides skin health benefits, including UV protection. Silk is often referred to as the "second skin" for its eco-friendly and functional properties. As biological and medical research proceeds, the use of silk has expanded beyond the textile industry. Its excellent biocompatibility and biodegradability have been widely acknowledged, leading to increased applications in areas such as medicine, biomedical materials, health products, and cosmetics [1–4]. Synthetic dyeing is a conventional dyeing method that consumes a lot of energy and water and requires sewage treatment. Some dyes, especially direct dyes and acid dyes used for silk dyeing, are banned due to the presence or decomposition of carcinogenic aromatic amines under certain conditions. Some dyes also contain heavy metal elements, which are harmful to the ecological environment and human health. In contrast, natural dyes derived from nature with low or no toxicity are environmentally friendly. Their inherent medicinal properties can also endow dyeing products with health benefits such as antibacterial, anti-moth, antioxidant, and UV protection [5–7]. In recent years, the dyeing and printing industry has taken "green environmental protection, energy conservation, and emission reduction" as its development theme. New dyeing technologies such as natural dyeing, supercritical dyeing, and structural coloration

W. Liu · Y. Zhou (✉)
National Engineering Laboratory for Modern Silk, Jiangsu Engineering Research Center of Textile Dyeing and Printing for Energy Conservation, Discharge Reduction and Cleaner Production, National Textile and Apparel Council Key Laboratory of Natural Dyes, Jiangsu Key Laboratory of Silk Engineering, College of Textile and Clothing Engineering, Soochow University, Suzhou, China
e-mail: yuyangzhou@suda.edu.cn

W. Liu
e-mail: 19873926709@163.com

M. Shahid et al. (eds.), *Science and Technology of Silk*, Textile Science and Clothing Technology, https://doi.org/10.1007/978-3-032-26880-8_3

of silk fabrics, as well as sericulture feeding methods, have become research hotspots [8–10]. This article reviews the research progress of natural dyeing and innovative dyeing technologies for silk fabric.

3.2 Non-Conventional Dyes for Silk Dyeing

3.2.1 Application of Natural Dyes

Natural dyes are the pigments derived from natural sources like plants, animals, and minerals, used for coloring textiles. These dyes, based on their origin, are categorized into plant, animal, and mineral types. Traditional plants are often the source for plant-based dyes, and efforts are being made to enhance both the extraction process and the dyeing results using modern techniques. Additionally, there is a focus on developing new plant dyes, such as those from blueberries, spirulina, wisteria, and purslane, which have only recently been explored in depth. Another emerging trend, driven by the principles of circular economy and sustainability, involves extracting dyes from plant waste materials, including grape skins, black plums, and tea stems. Some natural dyes are directly dissolved in water for silk dyeing and form hydrogen bonds and van der Waals forces with silk fibers.

3.2.2 Dyeing Methods

3.2.2.1 Direct Dyeing

Akhtar et al. [11] synthesized acid dyes by diazotizing substituted aromatic amines and substituted naphthalene sulfonic acid as coupling components and dyed silk with significant dyeing effects. Guo et al. [12] introduced an energy-saving and fast reaction silk dyeing technology, which directly synthesizes azo dyes on fabrics through a three-step chemical modification process at room temperature, thereby achieving excellent color fastness. The mechanism is shown in Fig. 3.1.

3.2.2.2 Reducing Dyeing

Certain natural dyes, like indigo, are not soluble in water and should be converted into water-soluble forms through a reduction process with a reducing agent under alkaline conditions. Once the silk is dyed with vat dye, it is exposed to air or oxidizing agents, which revert it to its insoluble form and fasten it onto the silk.

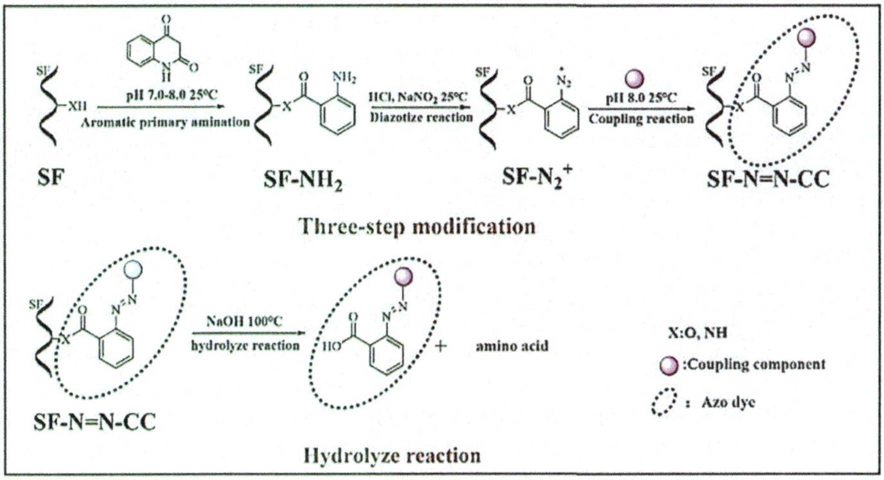

Fig. 3.1 In situ production of azo dye on silk and its alkaline hydrolysis [12]. Reproduced with permission from Ref. [12], Copyright Elsevier, 2024

3.2.2.3 Mordant Dyeing

Mordant dyeing technique is primarily categorized into three methods: pre-dyeing, post-dyeing, and co-bath dyeing. In the pre-dyeing method, silk is initially treated with a mordant, enabling the metal ions to bond with the functional groups on the silk fibers, followed by dyeing with natural dyes. For instance, treating silk with aluminum mordant followed by dyeing with madder results in a vibrant red color due to the stable complexes formed between aluminum ions and the madder pigments. Naheed et al. [13] demonstrated that using microwave-assisted extraction and dyeing with log-derived gibberellins yield superior color intensity and durability on silk, particularly when employing sustainable mordants, thus enhancing the eco-friendliness of this process. The dyeing procedure is illustrated in Fig. 3.2. The post-dyeing method entails silk dyeing with natural dyes first, followed by a mordant treatment. An example is silk dyeing with purple violet dye and subsequently being immersed in an iron mordant solution, which transforms the color from its initial purplish-red to a deeper purple-black, along with significant improvement on color fastness. In the co-bath dyeing method, silk, natural dyes, and mordants are all introduced into the dye bath simultaneously. Using sappanwood as a dye and aluminum salt as a mordant in this method ensures even coloring, producing a distinctive brownish-yellow shade.

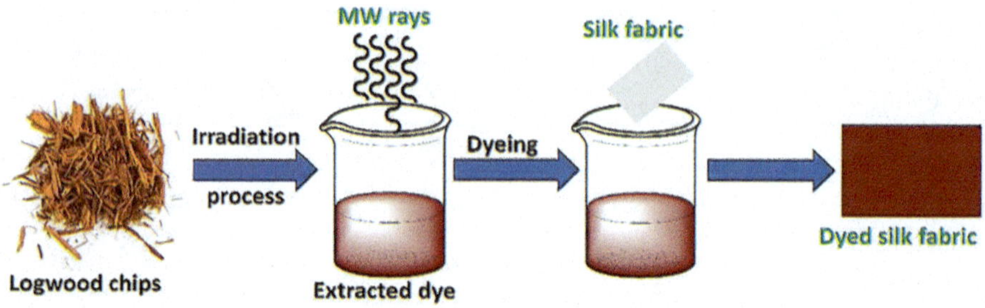

Fig. 3.2 Schematic diagram of the microwave-assisted silk dyeing using logwood extractions [13]. Reproduced with permission from Ref [13], Copyright Elsevier, 2024

3.2.3 Environmentally Friendly Synthetic Dyes

The environmental friendliness of synthetic dyes can be improved mainly in terms of low toxicity, biodegradability, and no heavy metal residue. Adeel S et al. [14] used microwave radiation and acid blue 07 to dye polyamide-based protein fabrics and physically modify the fabric surface to improve absorption capacity. Yuan et al. [15] used CI Disperse Blue-77 and CI Disperse Yellow E-3G dyes, with no azo molecular structure, which avoids the decomposition and production of carcinogenic aromatic amines. The resultant silk fabric shows excellent washing and rubbing fastness.

3.2.4 New Functional Dyes

The growing need for specialized textiles has made multifunctional silk treatments crucial for enhancing product value. While reactive dyes demonstrate superior uptake on silk in acidic environments, their bonding stability diminishes at lower pH levels. Alkaline conditions prove essential for establishing durable chemical linkages between silk proteins and reactive dye molecules. Specifically, vinyl sulfone functional groups in these dyes participate in nucleophilic addition reactions with cellulose hydroxyl groups when exposed to alkaline media. Building upon these principles, Zhou's research group [16] developed a water-soluble UV-absorbing compound featuring vinyl sulfone moieties to modify curcumin, thereby enhancing its solubility and photostability. This modified curcumin derivative demonstrated superior performance when applied to silk, exhibiting enhanced color retention alongside notable antimicrobial effects and UV-blocking capabilities.

Azo dyes present distinct advantages, including straightforward synthesis, economic viability, stable performance, and diverse coloration options. In a separate investigation, Ibrahim's group [17] created four innovative rhodanine-based disperse dyes incorporating cyanuric chloride groups, utilizing N-aminorhodanine and various aromatic amine precursors. These novel dyes delivered uniform, intense coloration on both silk and nylon substrates while achieving excellent fastness properties: light fastness (rating 4), wash fastness (4–5), and rub fastness (5). Beyond coloration, these rhodanine derivatives displayed antimicrobial characteristics, positioning them as promising candidates for producing hygienic textile finishes. Jin and colleagues [18] introduced an alternative approach for simultaneous dyeing and functional enhancement of silk using riboflavin sodium phosphate (RSP) immersion. This treatment imparts a vibrant yellow hue to protein-based silk while conferring multiple performance benefits. Post-treatment analysis revealed that RSP-modified silk exhibits a distinct yellow–orange shade under visible light while emitting fluorescent green when exposed to 365 nm UV radiation, demonstrating clear photochromic behavior. Additionally, the treated fabric displayed enhanced bacterial resistance and achieved B1-grade flame retardancy, meeting stringent requirements for protective decorative textiles.

3.3 Novel Dyeing Techniques

3.3.1 Energy-Saving and Emission-Reducing Technologies

3.3.1.1 Enzyme-Catalyzed Polymerization Dyeing

Enzymatic polymerization has always been considered an environmentally friendly synthesis process, generally carried out under relatively mild pH and temperature conditions. Enzymatic polymerization is an eco-friendly oxidative polymerization mechanism that polymerizes the hydroxyl groups of natural phenolic substances and the amino groups of pyrrole monomers.

This innovative approach broadens the utilization potential of plant-derived phenolic compounds in textile coloration and functional modification, presenting novel opportunities for sustainable enzymatic fabrication of biologically active fabrics. Researchers led by Jia [19] investigated laccase-driven oxidative transformation of DOPA (3,4-dihydroxyphenylalanine) for silk fabric coloration. Their work compared two distinct enzymatic coloration protocols: (1) sequential adsorption of DOPA followed by laccase treatment and (2) concurrent polymerization and coloration processes. Both methodologies yielded deep black coloration with superior UV-blocking characteristics on silk substrates. In subsequent research [20], the team employed laccase-catalyzed oxidative systems in acetate buffer (pH 5.0, 50 °C) for silk modification. Comprehensive surface analysis through SEM, AFM, ATR-FTIR, and XPS techniques revealed that the coloration mechanism involves oxidative crosslinking of tyrosine moieties within silk proteins, generating conjugated molecular structures and durable yellow–red surface layers, predominantly in the amorphous regions of silk fibers. Parallel investigations by Baek's group [21, 22] explored horseradish peroxidase (HRP)-mediated polymerization of gallic acid (GA) with silk textiles, establishing an eco-conscious dyeing protocol under ambient conditions. The HRP-catalyzed GA polymerization system produces stable dark brown hues through molecular interactions between silk's amino/hydroxyl groups and GA's phenolic hydroxyls. Comparative studies demonstrated HRP's superior performance over laccase in achieving enhanced color fastness properties. These enzymatic polymerization techniques represent significant advancements in sustainable textile coloration, offering low-environmental-impact alternatives to conventional dyeing processes for silk fabrics.

3.3.1.2 Ultrasonic and Microwave-Assisted Dyeing

Ultrasonic and microwave dyeing methods refer to the use of ultrasound and microwave as auxiliary tools and physical radiation technology to intensify the movement of dye molecules, increase molecular kinetic energy, accelerate the diffusion of dyes in fabrics, and promote the dye to dye fibers, thereby achieving the goal of improving dye uptake and color fastness. Compared with conventional dyeing methods, ultrasound and microwave-assisted dyeing can improve dye utilization, reduce pollutant emissions, and decrease the use of additives. Yin and colleagues [23] employed an ultrasonic-enhanced ammonium sulfate/ethanol dual-phase extraction technique to isolate anthocyanin pigments from purple sweet potato powder for textile applications. Their research demonstrated that mordant-assisted dyeing of silk with these natural pigments resulted in 40.6% greater color retention than conventional direct dyeing approaches. The treated fabrics exhibited excellent durability, achieving Grade 4 performance in both dry/wet abrasion resistance and laundering stability tests. Adeel et al. [24] studied the extraction of safflower red pigment from safflower and its stimulation after microwave treatment, and then chemically dyed and bio-mordanted silk. The results showed that after the microwave treatment for 3 min, a darker color tone was obtained. The untreated silk fabric was dyed for 60 min using saffron acid extraction solution ($pH = 2$), which endowed it with high dyeing intensity. The color fastness evaluation shows that 7% curcumin and benzoquinone, as biological mediators, not only provide high color depth but also excellent fastness performance compared to using chemical mordants. It was found that microwave treatment not only improved the dyeing performance of safflower, but also enhanced its color characteristics. Akram et al. [14] used microwave radiation as a sustainable technology to dye protein fabrics with acid blue 07 dye. The study found that the microwave-treated dye solution could achieve excellent results in dyeing microwave-treated silk and wool fabrics without changing the chemical properties of the fabric. Instead, the surface of the fabric was physically modified, which not only enhanced its adsorption capacity and improved its color characteristics but also reduced dyeing time and temperature, making the dyeing process sustainable, economical, and energy-saving.

3.3.1.3 Nanoparticle Processing Technology

The application of noble metal nanoparticles, including gold and silver variants, on silk textiles enables chromatic modulation through localized surface plasmon resonance effects. Varying the compositional ratios of these nanoparticles generates diverse color profiles, significantly broadening the chromatic spectrum achievable. This nanoparticle-based coloration method operates under ambient conditions, contrasting sharply with traditional dyeing techniques that demand substantial energy inputs. Silk fibers demonstrate exceptional affinity for silver nanoparticles during the coloration phase. Beyond chromatic effects, nanoparticle deposition enhances textile functionality. Research by Zhang's team [25] involved modifying silk surfaces with tetrabutyl titanate followed by CI Reactive Blue 19 application, yielding substantial UV protection enhancement. While the TiO_2 film reduced K/S values initially, subsequent removal of the coating restored color intensity while improving light fastness properties. Tang's research group [26] developed antimicrobial colored silk by combining silver nanoparticles with chromatic nanoparticles (blue, red, yellow). Remarkably, silver nanoparticle deposition achieves rapid coloration (<10 min) at room temperature without compromising fiber integrity or tactile qualities. Subsequent polydimethylsiloxane (PDMS) coating ensured excellent wash durability of the colored textiles. In a separate investigation, Zhang et al. [27] achieved dual functionality by in situ synthesis of AuNPs and AgNPs on silk substrates, subsequently integrating conventional dyes. This hybrid approach preserved nanoparticle morphology while expanding the color gamut and enhancing K/S values. The resultant fabrics exhibited both vibrant coloration and potent antimicrobial characteristics.

3.3.1.4 Supercritical CO_2 Fluid Dyeing

Supercritical fluid technology does not require water sources in dyeing and finishing. When dyeing and finishing are completed, unused dyes and media can be recycled to achieve recycling. It is a simple, efficient, environmentally friendly, and economical waterless dyeing technology. However, silk is made by mature silkworms secreting silk fibroin, which is soft and lightweight. Its dyeing effect in supercritical CO_2 is extremely poor, which is related to the triangular or semi-elliptical coating structure of silk. Therefore, in order to meet the dyeing standards, pre-treatment or synthesis of a new type of dispersed active dye is needed for silk supercritical carbon dioxide waterless environmentally friendly dyeing. Sawada et al. [28] studied the dyeing of natural fabrics in supercritical carbon dioxide using a non-ionic reverse micelle system. Sodium diethylhexyl benzenesulfonate was added to the CO_2 fluid, and unmodified silk and wool fabrics were successfully dyed with conventional acidic dyes. The dyeing amount of silk was higher than that of wool, because on the one hand, the hydrophilic groups on the surface of silk were more than those on wool, and on the other hand, the scales on the surface of wool fibers hindered further diffusion of dyes. Fan et al. [29] successfully designed and synthesized a novel red pigment dispersion dye containing anthraquinone-based chromophores and multifunctional bridging groups by utilizing copper-mediated Ullmann cross-coupling reaction in a ligand-free system, followed by the introduction of dichlorotriazine-type active groups through condensation reaction. The dye was successfully applied to cotton, silk, and wool fabrics in supercritical carbon dioxide, resulting in a bright red color tone. Yan et al. [30] utilized an intermediate of 1-chloroanthraquinone and N-(2-aminoethyl)-3-methoxyaniline, based on ligand-free Ullmann reaction, with copper metal as a promoter, to develop a precursor of the target dye. Then, through nucleophilic condensation reaction, the dichlorotriazine-type active group was grafted onto a special bridging group of the dye precursor, successfully synthesizing a special dispersed active SCFX-AnB3L dye (synthesis route shown in Fig. 3.3), whose red color involves three parts of anthraquinone precursor, bridging group, and dichlorotriazine-type active group. Apply this dye to the ecological dyeing of silk and compare the dyeing of cotton and wool in SCF-CO_2. The dyeing results showed that compared with the control dye, the synthesized special dye had good adsorption, adsorption, and fixation properties on silk fabrics. Its fading fastness could reach up to 4–5 levels, dyeing fastness could reach up to 5 levels, and dry wet friction fastness could reach up to 5 levels. Especially, under the same dyeing conditions, the color intensity obtained on silk substrates is the highest among all tested substrates.

3.3.2 Structural Coloring

Structural coloring does not rely on dyes and additives, is pollution-free, and does not change color, making it a sustainable, ecological dyeing process technology. The earliest discovery of structural coloration was by Isaac Newton. Nassau also proposed that the essence of physical coloring is to produce color effects through the dispersion, refraction, diffraction, and interference of light. Japanese scientists S Kinoshita and S Yoshioka classified physical coloration, also known as structural coloration, into three types: thin-film interference, multilayer-film interference, and photonic crystals.

Recent scientific investigations have identified five primary mechanisms responsible for structural coloration in biological systems: thin-film interference, multilayer interference, diffraction phenomena, light scattering effects, and both ordered (photonic crystals) and disordered structural arrangements [31]. This environmentally sustainable coloration approach offers significant advantages over conventional dyeing techniques for textile applications. In a notable study, Sha utilized three polyphenolic compounds (tannic acid, gallic acid, and tea polyphenols) sharing identical pyrogallol configurations as base materials. These substances were subjected to Schiff base reactions with polyethyleneimine in mildly alkaline environments, facilitating rapid nanofilm formation at air–liquid interfaces. Subsequent deposition onto silk substrates enabled precise color modulation through temporal control of film thickness, yielding fabrics exhibiting diverse structural coloration effects. This method of creating structural colors is not specific to the substrate and can quickly

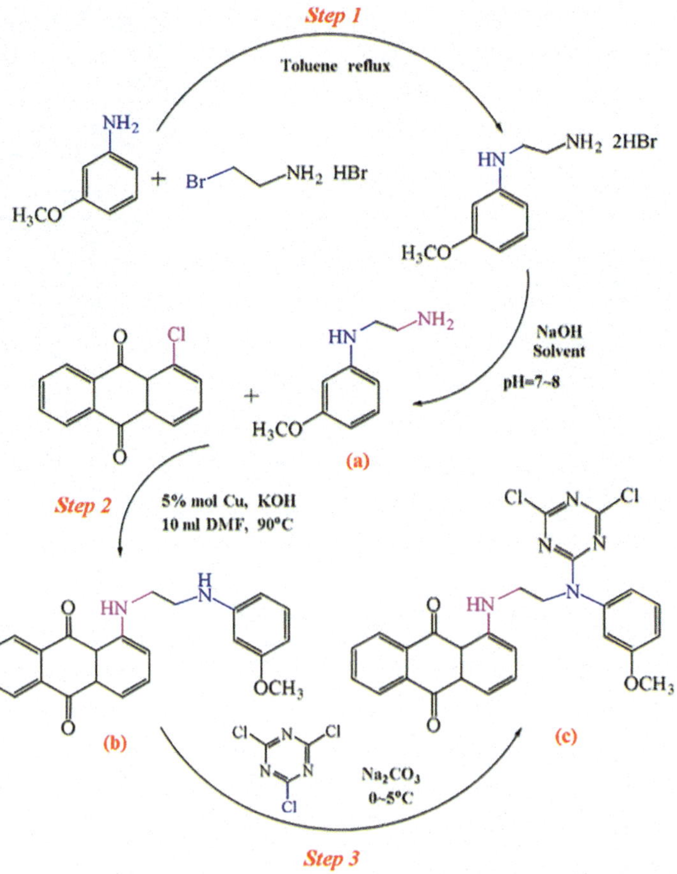

Fig. 3.3 Synthesis route for the disperse reactive red SCFX-AnB3L dye[30]. Reproduced with permission from Ref [30], Copyright Elsevier, 2020

generate structural colors on various textile substrates. In addition, structured silk fabrics based on plant polyphenols have antibacterial properties. Diao et al. [32] designed and constructed a dual color structure of silk protein photonic crystal with inverse opal structure in a single inverse opal photonic structure by biomimetic binary structure color reflection and self-assembly method using a special photon band structure. By adjusting the structural parameters of silk protein inverse opal, the wavelength of the dual structure color reflection peak was controlled. The special property of silk fiber, humidity-induced cyclic shrinkage, was used to control the wavelength of the dual structure color reflection peak. The prepared silk protein inverse opal film can produce structural colors in the UV/Vis, UV/IR, and Vis/IR regions. This anti-opal structured photonic crystal has potential applications in ecological dyeing and multifunctional silk fabrics. Kim et al. [33] used polymethyl methacrylate (PMMA) nanospheres as templates to fabricate silk fibroin inverse opal structured films. The

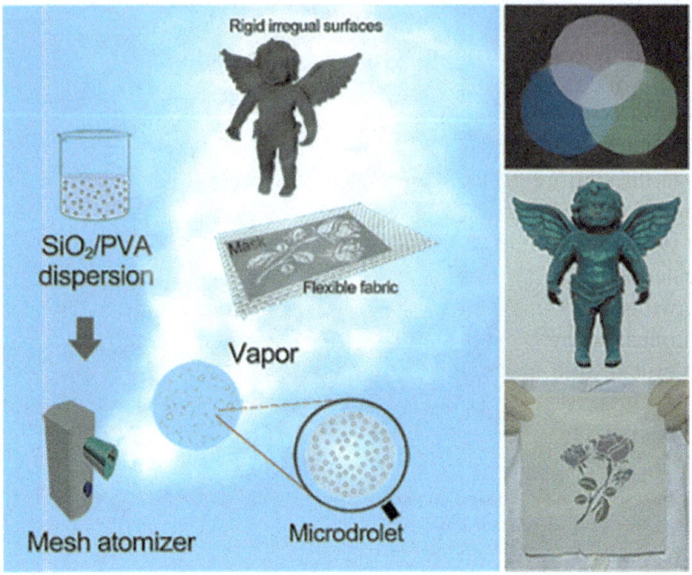

Fig. 3.4 Preparation of photonic crystals with iridescent colors by atomized deposition of amorphous SiO_2 nanoparticles [34]. Reproduced with permission from Ref [34], Copyright American Chemical Society, 2024

aqueous solution of silk fibroin extracted from silkworm cocoons is filled into the gaps of PMMA nanosphere templates and solidified to form an amorphous self-supporting silk fibroin film. Peel off the silk fibroin film from the silicon substrate and immerse it in acetone to remove PMMA microspheres, thereby obtaining a colored silk fibroin film with a reverse opal structure. Zhang et al. [34] reported the formation of a brilliant structural color by co depositing SiO_2 nanoparticles and polyvinyl alcohol coating on silk fabrics through atomization. The precise control of atomization deposition was used to manufacture non-photonic crystal structures layer by layer, making it easy to achieve the addition and mixing of large-area non-iridescent structural colors (as shown in Fig. 3.4), and the coating exhibited stronger friction resistance.

3.3.2.1 Application of Biotechnology in Silk Dyeing

Various techniques exist for creating colored luminescent silk: altering silkworm genetics, adjusting the nutritional intake of silkworms, or post-production treatments. Conventional silk dyeing methods create wastewater containing dyes, requiring significant time and causing environmental harm, making eco-friendly alternatives essential. Research by Kurioka and Tsuchida [35, 36] demonstrated the accumulation of carotenoids and flavonoids from mulberry leaves within cocoons. Nevertheless, these pigments were eliminated during degumming as they primarily accumulated in the sericin layer instead of the fibroin component. Tansil [37] mixed fluorescent dyes rhodamine and fluorescein from the diet of silkworms into silk fibroin to produce colored silkworm cocoons.

Nisal et al. [38] conducted experiments using seven structurally analogous azo dyes with systematically adjusted partition coefficients, administering them to silkworm larvae through treated feed solutions. Their research successfully identified Acid Orange II, Mordant Black 17, and Direct Acid Fast Red as effective colorants, producing vibrant orange, pale violet, and rose-colored cocoons, respectively. The study revealed that dye absorption into silk cocoons primarily depends on molecular mass, with compounds exceeding 400 g/mol failing to permeate the digestive tract's peritoneal barrier. The investigation demonstrated that lipophilic characteristics play a crucial role in dye–protein interactions within the silk gland. Effective dye transfer from the larval gut to hemolymph and ultimately to silk glands requires an optimal equilibrium between hydrophobic and hydrophilic properties. Nevertheless, questions remain regarding the practical viability of this technique, particularly concerning dye incorporation efficiency and its potential for large-scale industrial implementation. The use of genetic engineering to produce colored silk is achieved through gene editing or biotechnology, which allows silkworms to directly produce colored silk. Genetically modified colored silk has economic advantages because it increases the new value of silk, reduces the cost of water waste, and is more conducive to water conservation in environmental protection. Due to the fact that the fluorescence of green fluorescent protein (GFP) does not require exogenous substrates or cofactors, and it is often used as a marker in organisms and cells. Tamura et al. [39] first prepared transgenic silkworms using piggyBac transposase. Silkworms carry enhanced GFP (EGFP, maximum absorption 489nm/maximum emission 509 nm), a GFP variant with improved stability in eukaryotic cells and strong fluorescence cDNA, which guides constitutive gene expression under the control of cytoplasmic actin A3 promoter sequence. Tomita et al. [40] successfully produced transgenic silkworms for spinning fluorescent silk. We constructed the cDNA of the FLC and EGFP fusion protein controlled by the FLC gene promoter sequence and integrated it into the PiggyBac donor vector to establish the transgenic silkworm as described above.

3.4 The Current Situation and Challenges of Silk Dyeing

At present, the industrialization of silk dyeing is still dominated by traditional technologies, and the green transformation and application of emerging technologies are in a transitional period. In the future, it is necessary to accelerate the large-scale implementation of innovative dyeing technology through technology cost reduction, policy guidance, and consumer market education and promote the upgrading of the silk industry toward "zero pollution and high added value." Although natural dyes have ecological and cultural value in dyeing silk, the industrialization process is slow due to issues such as color fastness, cost, process efficiency, single color spectrum, and environmental disputes. Breakthroughs require the assistance of biotechnology, the development of green mordants, and new dyeing processes to balance environmental friendliness and scale requirements. Although the research and development of new silk dyeing technologies have made progress in environmental protection and functionalization, they still face multiple industrialization bottlenecks: at the economic level, the high pressure requirements of supercritical CO_2 dyeing equipment, the significant increase in investment thresholds due to the high cost of nanomaterials and enzyme preparations; in terms of process stability, enzyme-catalyzed polymerization dyeing is susceptible to environmental interference, and the parameter regulation of ultrasound/microwave-assisted technology is complex. The uneven dispersion of structural coloring nanoparticles leads to insufficient dyeing uniformity and color fastness; on the level of technological limitations, the supercritical CO_2 staining chromatography is narrow, the color of genetically modified silk is single and regulated, and biotechnology faces challenges in public acceptance; at the level of standardization and scale, issues such as the volatility of natural dyes, precision of gene editing targets, and durability of dynamic structural coloration constrain process stability, and most technologies still remain in the laboratory or small-scale trial stage. How to balance cost, performance, environmental protection, and scale is still the core proposition that urgently needs to be solved in the innovation of silk dyeing.

3.5 Summary and Prospect

In recent years, silk dyeing technology has made significant progress in the application of natural dyes, new dyeing processes, and energy-saving and emission reduction technologies. Natural dyes have received widespread attention due to their environmental friendliness and functionality, but still face problems such as low color fastness, limited chromatography, and high industrialization costs. Green dyeing technologies such as ultrasound, microwave, enzyme catalysis, and supercritical CO_2 have shown potential in reducing pollution and conserving water resources, but large-scale industrialization is still limited by challenges such as high-equipment costs, insufficient process stability, and uncertain market acceptance. In addition, structural coloring technology, as a dye-free dyeing method, provides a new solution in the field of sustainable dyeing, but its durability, mass production, and application promotion still need further optimization.

In the future, the development of silk dyeing technology needs to be further promoted in the following aspects: (1) optimizing the extraction and application process of natural dyes, improving color fastness, and expanding available chromatography; (2) promote the industrialization of green dyeing technology, reduce costs, and improve process stability; (3) strengthen the integration of biotechnology, nanotechnology, and silk dyeing, and develop intelligent and high-performance new dyeing systems. Through technological innovation and industrial upgrading, the silk dyeing industry will develop toward a more environmentally friendly, efficient, and sustainable direction.

References

1. Dayioglu H, Kut D, Merdan N, et al (2015) The effect of dyeing properties of fixing agent and plasma treatment on silk fabric dyed with natural dye extract obtained from Sambucus ebulus L. plant. Procedia - Soc Behav Sci, 195, 1609–1617
2. Lee AR, Hong JU, Yang YA et al (2010) Dyeing properties and antimicrobial activity of silk fabric with extract of unripe *Citrus Unshiu* fruits. Fibers and Polymers 11:982–988
3. Yasukawa A, Chida A, Kato Y et al (2017) Dyeing silk and cotton fabrics using natural blackcurrants. Text Res J 87(19):2379–2387
4. Khan A, Hussain MT, Jiang H (2018) Dyeing of silk fabric with natural dye from camphor (Cinnamomum camphora) plant leaf extract. Color Technol 134(4):266–270
5. Uddin MG (2014) Effects of different mordants on silk fabric dyed with onion outer skin extracts. J Textes 2014(1):405626
6. Mongkholrattanasit R, Saiwan C, Rungruangkitkrai N et al (2015) Ecological dyeing of silk fabric with lac dye by using padding techniques. J Texte Inst 106(10):1106–1114
7. Jahan N, Datta E (2015) A comparative study on dyeing of cotton and silk fabric using madder as a natural dye. IOSR J Poly Text Eng 2(2):5–9
8. Teklemedhin TB, Gopalakrishnan LH (2018) Environmental friendly dyeing of silk fabric with natural dye extracted from Cassia singueana. Plant. J Textile Sci Eng S, 2018, 3(2).
9. Ghoreishian SM, Maleknia L, Mirzapour H et al (2013) Antibacterial properties and color fastness of silk fabric dyed with turmeric extract. Fibers and Polymers 14:201–207
10. Punrattanasin N, Nakpathom M, Somboon B et al (2013) Silk fabric dyeing with natural dye from mangrove bark (Rhizophora apiculata Blume) extract. Ind Crops Prod 49:122–129
11. Akhtar Z, Ali S I, Farooq M, et al (2018) Application of acid dyes on silk fabric and fastness properties Part II. J Chem Soc Pak, 40(2):283–287
12. Guo Q, Shang Z, Chen W et al (2024) In situ generation of azo dyes on silk fibroin through three-step chemical modification. Dyes Pigm 228:112245
13. Naheed S, Haider S, Adeel S et al (2024) Eco-friendly utilization of microwaves for extraction of dye from logwood and its application onto silk. Results Eng 24:103372
14. Adeel S, Akram H, Usman M et al (2023) Eco-friendly microwave assisted sustainable coloration of silk and wool fabric with Acid Blue 07 dye. Environ Sci Pollut Res 30(30):74939–74951
15. Yan J, Zheng LJ (2011) Study on dyeing of the plasma modification silk fabric in supercritical carbon dioxide. Adv Mater Res 175:661–666
16. Zhou Y, Tang RC (2016) Modification of curcumin with a reactive UV absorber and its dyeing and functional properties for silk. Dyes Pigm 134:203–211
17. Ibrahim SA, Mohamed FA, Ibrahim HM et al (2024) Synthesis, characterization, antibacterial activity, and applications of new reactive disperse rhodanine dyes for silk and Nylon Fabrics. Fibers Polym 25(1):243–255
18. Jin WJ, He WL, Lehner S, et al (2024) Edible squid ink proteoglycans as a multifunctional sustainable additive for silk coatings, Fire protection, UV shielding and as a colorant. Chem Eng J, 495, 153320.5
19. Jia W, Wang Q, Fan X, et al (2017) Laccase-mediated in situ oxidation of dopa for bio-inspired coloration of silk fabric. RSC advances, 7(21), 12977–12983
20. Jia W, Wang Q, Fan X et al (2018) Mechanism and analysis of laccase-mediated coloration of silk fabrics. Fibers and Polymers 19:868–876
21. Baek NW, Fan XR, Yuan JG, Wang Q (2020) Polymerization and dyeing properties of gallic acid on silk fabric catalyzed by horseradish peroxidase. Fibers Polym 22(8):2145–2155

22. Baek NW, Fan XR, Zhang X, Wang D, Yuan JG (2021) Horseradish peroxidase-catalyzed "template" polymerization of gallic acid for the functionalization of silk fabrics. J Nat Fibers 19(14):9486–9499
23. Yin Y, Jia J, Wang T et al (2017) Optimization of natural anthocyanin efficient extracting from purple sweet potato for silk fabric dyeing. J Clean Prod 149:673–679
24. Adeel S, Naseer K, Javed S et al (2020) Microwave-assisted improvement in dyeing behavior of chemical and bio-mordanted silk fabric using safflower (Carthamus tinctorius L) extract. J Nat Fibers 17(1):55–65
25. Zhang H, Zhang XT (2014) Modification and dyeing of silk fabric treated with tetrabutyl titanate by hydrothermal method. J Nat Fibers 11(1):25–38
26. Tang B, Li J, Hou X et al (2013) Colorful and antibacterial silk fiber from anisotropic silver nanoparticles. Ind Eng Chem Res 52(12):4556–4563
27. Zhang Z, Lv X, Chen Q et al (2019) Complex coloration and antibacterial functionalization of silk fabrics based on noble metal nanoparticles. J Eng Fibers Fabr 14:1558925019866943
28. Sawada K, Takagi T, Jun JH et al (2002) Dyeing natural fibres in supercritical carbon dioxide using a nonionic surfactant reverse micellar system. Color Technol 118(5):233–237
29. Fan Y, Zhang YQ, Yan K et al (2019) Synthesis of a novel disperse reactive dye involving a versatile bridge group for the sustainable coloration of natural fibers in supercritical carbon dioxide. Adv Sci 6(1):1801368
30. Yan K, Zhang YQ, Xiao H, et al (2020) Development of a special SCEX-AnB3L dye and its application in ecological dyeing of silk with supercritical carbon dioxide. J CO2 Util, 35, 67–78
31. Kinoshita S, Yoshioka S, Miyazaki J (2008) Physics of structural colors. Rep Prog Phys 71(7):076401
32. Diao YY, Liu XY, Toh GW et al (2013) Multiple structural coloring of silk-fibroin photonic crystals and humidity-responsive color sensing. Adv Func Mater 23(43):5373–5380
33. Kim S, Mitropoulos AN, Spitzberg JD et al (2012) Silk inverse opals. Nat Photonics 6(12):818–823
34. Li Q, Zhang Y, Shi L et al (2018) Additive mixing and conformal coating of noniridescent structural colors with robust mechanical properties fabricated by atomization deposition. ACS Nano 12(4):3095–3102
35. Kurioka A, Yamazaki M (2002) Purification and identification of flavonoids from the yellow green cocoon shell (Sasamayu) of the silkworm, Bombyx mori. Biosci Biotechnol Biochem 66(6):1396–1399
36. Tsuchida K, Sakudoh T (2015) Recent progress in molecular genetic studies on the carotenoid transport system using cocoon-color mutants of the silkworm. Arch Biochem Biophys 572:151–157
37. Tansil NC, Li Y, Koh LD et al (2011) The use of molecular fluorescent markers to monitor absorption and distribution of xenobiotics in a silkworm model. Biomaterials 32(36):9576–9583
38. Nisal A, Trivedy K, Mohammad H et al (2014) Uptake of azo dyes into silk glands for production of colored silk cocoons using a green feeding approach. ACS Sugtainable Chem & Eng 2(2):312–317
39. Takasu Y, Kobayashi I, Beumer K et al (2010) Targeted mutagenesis in the silkworm Bombyx mori using zinc finger nuclease mRNA injection. Insect Biochem Mol Biol 40(10):759–765
40. Tomita M, Munetsuna H, Sato T et al (2003) Transgenic silkworms produce recopgbinant human type III procollagen in cocoons. Nat Biotechnol 21(1):52–56
41. Tamura T, Thibert C, Royer C, et al (2000) Germline transformation of the silkworm Bombyx mori L. using a piggyBac transposon-derived vector. Nat Biotechnol, 18(1), 81–84
42. Chu M, Liu G (2008) Fluorescent silkworm silk prepared via incorporation of green, yellow, red, and near-infrared fluorescent quantum dots. IEEE Trans Nanotechnol 7(3):308–315

Chapter 4
Utilization of Silk in the Field of Medicine and Biotechnology

O. S. K. Paavana, S. P. Sruthi, and M. S. A. Muthukumar Nadar

4.1 Introduction to the Molecular Structure of Silk

The protein fibers that silkworms and spiders spin have exceptional mechanical properties attributed to their diverse molecular structures. Their strength, elasticity, and possible uses can be gleaned from these structures.

Silkworm Silk (*Bombyx mori*): Silk made by silkworms is composed of two main proteins: fibroin and sericin. Fibroin is the major structural element, and sericin is the glue-like protein that combines the fibroin fibers. H-fibroin, the heavy chain of fibroin, is a polypeptide with a molecular weight of approximately 350 kDa. It is covalently attached to the light chain (L-fibroin, with a molecular weight of around 26 kDa) through a disulfide bond. This bond contributes to the stability and mechanical properties of the fiber. The amino acid composition of H-fibroin, particularly its high content of glycine (45.9 mol%) and alanine (30.3 mol%), enables the formation of β-sheet structures. These structures enhance the strength and rigidity of silk fibers [1, 2].

Spider Silk: Spider silk, particularly the dragline silk, is renowned for its impressive strength and elasticity. The two main proteins, major ampullate spidroins (MaSp1 and MaSp2), have molecular weight > 250–350 kDa or higher. Spider silk proteins, similar to silkworm silk, are rich in glycine and alanine, which favor β-sheet formation. These β-sheets are further structured into nanocrystals that are scattered between amorphous regions, imparting the hierarchical architecture that underlies the silk's mechanical characteristics [2].

Comparative Insights: The basic structure of silkworm and spider silks is a hierarchical assembly of nanofibrils with diameters between 20 and 60 nm. Compared to silkworm silk fibers, which are usually 10–20 μm in diameter, spider dragline silk

O. S. K. Paavana · S. P. Sruthi · M. S. A. Muthukumar Nadar (✉)
Division of Biotechnology, Karunya Institute of Technology and Sciences, Coimbatore, India
e-mail: research.muthukumar74@gmail.com

M. Shahid et al. (eds.), *Science and Technology of Silk*, Textile Science and Clothing Technology, https://doi.org/10.1007/978-3-032-26880-8_4

Table 4.1 General differences between silkworm silk and spider silk

Feature	Silkworm silk	Spider silk
Major protein	Fibroin	Spidroin
Secondary structure	Predominantly β-sheets	Combination of β-sheets & α-helices
Elasticity	Relatively low	High
Tensile strength	High	Extremely high
Biological purpose	Cocoon formation	Web construction, prey capture, safety lines
Production method	Sericulture (commercial)	Natural production, difficult to harvest in bulk

fibers have a diameter ranging from 3 to 5 μm. In addition to lacking sericin, spider silk possesses a glycoprotein layer that safeguards it against bacterial and fungal infections [3]. Table 4.1 lists some general differences between silkworm silk and spider silk.

The molecular packing of silk fibers is hierarchical; each level is salient to its mechanical performance, from molecular arrangements to macroscopic structures. This complex structure, involving β-sheet nanocrystals aligning and interacting in semi-amorphous matrices, provides silk with its unique balance of strength, flexibility, and toughness [4].

Thus, the aforementioned molecular and structural features observed in silkworm and spider silks have stimulated research toward the engineering of advanced biomaterials that exploit silk for multiple applications.

4.2 Silk Protein Structure Analysis and Importance

Silk is a natural fibrous protein that exhibits a unique combination of structural complexity, mechanical strength, and biocompatibility. The study of its molecular architecture is essential for understanding its multifunctional properties and for guiding its application in textiles, biotechnology, and biomedical engineering. A range of analytical techniques, including spectroscopic and diffraction-based methods, are used to investigate the composition, secondary structure, and conformational transitions of silk proteins (Fig. 4.1) [1, 5–10].

A few important techniques for analyzing silk protein structure are listed in Table 4.2.

Exceptional mechanical properties of silk originate from its molecular architecture, consisting of well-organized β-sheet crystalline domains interspersed within a flexible, amorphous matrix [18–22]. These hierarchical structures provide an optimal balance of strength, toughness, and elasticity. Comparative analyses across species, such as *Bombyx mori* and *Nephila*, reveal that variations in amino acid composition, repetitive motifs, and post-translational modifications contribute to differences in crystallinity, elasticity, and molecular organization [20–22]. The relationship between silk's molecular structure and its macroscopic properties underscores the

•Designing High-Performance Biomaterials

- Silk fibroin is valued for tensile strength, toughness, and elasticity.
- Structural features allow fabrication of tissue engineering scaffolds, drug delivery systems, and surgical sutures.
- Increasing β-sheet content improves mechanical strength and stability, suitable for load-bearing medical implants

•Enhancing Biocompatibility and Biodegradability

- Silk proteins exhibit biocompatibility, making them ideal for biomedical applications.
- Structural knowledge enables surface property modification and tailored biodegradability.
- Protein structure–degradation relationship allows creation of scaffolds that degrade at controlled rate

•Developing Novel Silk-Based Materials

- Requires understanding of hierarchical protein structure.
- Enables genetic modification or property tuning of silk proteins.
- Key strategy for producing recombinant spider silk proteins (large-scale production not possible with spiders)

•Understanding Natural Silk Production

- Structural insights explain the natural spinning process in silkworms and spiders.
- Replicating natural silk spinning could enable energy-efficient, eco-friendly manufacturing methods.
- Potential to produce high-performance materials under sustainable conditions

•Improving Textile Industry Applications

- Structural analysis supports retention of strength, flexibility, and appearance of silk textiles.
- Controlling protein composition during processing yields enhanced textile properties.
- Applications include clothing, upholstery, and industrial uses

•Studying Evolution and Species Variations

- Comparative studies reveal evolutionary adaptations in silk proteins across species.
- Molecular basis of silk properties (e.g., spider silk toughness) is both fundamental and applied.
- Insights guide design of biomimetic materials with species-inspired functionalities

•Fundamental Scientific Understanding

- Silks provide a model for studying protein folding and self-assembly.
- Demonstrate how hierarchical structures yield strength and resilience.
- Illustrate nature's ability to derive remarkable performance from simple protein building blocks

Fig. 4.1 Applications of silk protein structural analysis [1, 5–10]

importance of ordered β-sheet nanocrystals embedded in flexible amorphous matrices [23–28]. The dynamic arrangement of these domains and their ability to undergo conformational adjustments in response to external stimuli explain silk's exceptional mechanical adaptability. Silk protein folding and fiber formation are governed by complex thermodynamic and kinetic mechanisms involving self-assembly, micelle formation, and sol–gel transitions [29–32]. These processes are influenced by environmental factors such as pH, temperature, and concentration, which determine the extent of molecular alignment and β-sheet formation. Understanding these mechanisms provides insights for designing silk-based materials with tunable structural and functional characteristics, particularly in biomaterials and drug delivery systems [33].

Table 4.2 Common analytical techniques for structural characterization of silk proteins

Technique	Application	Ref
Amino acid composition analysis	Determines the primary structure of silk by hydrolyzing proteins and analyzing amino acid content using HPLC or MS; helps relate amino acid composition to strength, elasticity, and biocompatibility	[11, 12]
Fourier transform infrared spectroscopy (FTIR)	Examines the secondary structure of silk proteins by detecting amide I and II bands, revealing α-helix, β-sheet, and random coil contents, and monitoring conformational changes during processing or environmental variation	[13]
Raman spectroscopy	Complements FTIR by identifying vibrational modes and molecular orientation, particularly useful for assessing β-sheet formation and structural organization of silk proteins	[14]
X-ray diffraction (XRD)	Identifies crystalline β-sheet regions and assesses crystallinity, molecular orientation, and order using WAXS and SAXS to correlate structural features with mechanical and biocompatibility properties.	[15]
Nuclear magnetic resonance (NMR) spectroscopy	Provides detailed molecular-level information on both crystalline and amorphous regions, probing local environments, dynamics, and conformational transitions using ^{13}C and ^{15}N labeling	[15, 16]
Amino acid profiling (advanced technique)	Offers high-resolution quantification and identification of amino acids and degradation products via NMR without chemical modification, useful for comparing natural and engineered silk proteins	[17]

Silk's mechanical behavior rivals that of many synthetic fibers due to its nanoscale organization and β-sheet crystallinity, offering high strength, flexibility, and energy dissipation capacity [34–36]. Environmental factors such as humidity and temperature significantly influence its stiffness, extensibility, and stability, emphasizing the need for careful consideration in material design and application [37, 38]. Beyond its physical properties, silk demonstrates excellent biocompatibility and biodegradability. Its proteinaceous nature allows favorable interactions with biological tissues, supporting cell attachment and growth while minimizing immune responses. Functional modifications can further enhance specific biological activities, reinforcing silk's potential in tissue engineering, regenerative medicine, and controlled drug delivery applications [39].

4.3 General Applications in the Medical Field

Silk fibroin, renowned for its exceptional biocompatibility, holds great promise in the biomedical field as a versatile material for sutures, scaffolds, and drug delivery systems, among other applications (Fig. 4.2). Its proteinaceous nature enables it to be integrated with biological tissues without triggering powerful immune responses.

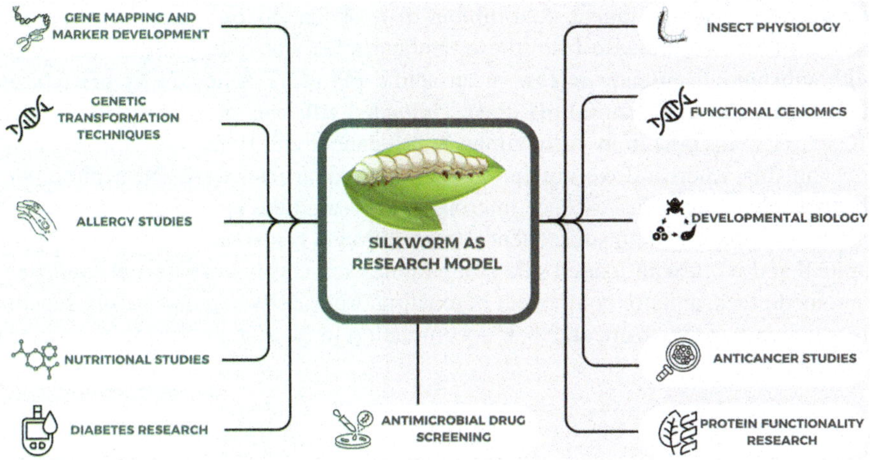

Fig. 4.2 Biomedical applications of silk

Silk degradation in vivo generates non-toxic amino acids, which can be easily metabolized and thus do not cause adverse reactions. Additionally, the distinctive surface chemistry and protein structure of silk contribute to the material's biocompatibility, facilitating cell adhesion and growth, and resulting in beneficial interactions with cells and extracellular matrices. Moreover, silk can be functionalized with bioactive molecules to enhance specific interactions, such as promoting angiogenesis and preventing bacterial colonization. Due to this tunability and biocompatibility, silk, an appealing biomaterial, holds great promise for applications in tissue engineering and regenerative medicine [40].

4.3.1 Silk in Drug Delivery Systems

It highlights the growing interest in silk-based drug delivery systems, which can be used to encapsulate and release therapeutic agents. The amphiphilic silk fibroin is capable of stabilizing both hydrophobic and hydrophilic drugs and can provide controlled release profiles [41]. Targeted and sustained drug delivery systems have been explored, including silk nanoparticles, hydrogels, and films. These systems are highly immunocompatible and biodegradable, making them ideal for long-term therapeutic delivery. Moreover, modifications of silk at the molecular level can increase drug loading efficiency and regulate release kinetics, showing potential applications such as in cancer therapy and wound healing [42]. A study highlighted that the production of uniform silk nanoparticles was achieved through the acetone precipitation of digested silk, yielding particles with an average diameter of ~98 nm and a low polydispersity index (0.109) and a moderately negative surface charge (-33.6 ± 5.8 mV). These silk nanoparticles were adept at encapsulating the chemotherapeutic agent doxorubicin at a 40 ng doxorubicin per microgram of silk ratio. Drug release from these nanoparticles was highly dependent on pH, with substantially greater release at an acidic pH (4.5), followed by pH 6.0, and the lowest release at neutral pH (7.4). Unloaded silk nanoparticles were not toxic (IC50 > 120 μg/mL) in in vitro human breast cancer cell lines with drug resistance mechanisms, whereas doxorubicin-loaded silk nanoparticles were able to circumvent these mechanisms. Fluorescence microscopy revealed the uptake of silk nanoparticles by the cells and their subsequent accumulation in lysosomes. The results demonstrate that doxorubicin-loaded silk nanoparticles serve as lysosomotropic anticancer nanomedicines, a major component of existing drugs delivered using a pH-sensitive drug delivery mechanism, and their accumulation in lysosomes enhances anticancer efficacy [43].

4.3.2 Tissue Engineering Scaffolds

The structural integrity and biocompatibility of silk scaffolds make them a crucial type of scaffold in tissue engineering. Silk fibroin can also be prepared into a porous structure that resembles the extracellular matrix, facilitating cell attachment, proliferation, and differentiation [44]. Silk-based scaffolds have demonstrated success in a range of applications, including bone and cartilage regeneration, as well as skin and vascular tissue engineering. In addition, the mechanical integrity of silk scaffolds can be tailored to suit load-bearing applications, and chemical functionalization enables the immobilization of growth factors, allowing for the modulation of cellular responses. One of the novel characteristics of silk fibroin (SF) is that it maintains in vitro interaction with human cells to supply their adequate requirements. In a biological environment, it exhibits remarkable biocompatibility, as it integrates into the host tissues and does not elicit a foreign body response. Moreover, SF can be dissolved in aqueous solutions, enabling the fabrication of diverse morphologies, including films. Significantly, it also degrades at a controllable rate in both in vitro and in vivo systems, which is often essential for biodegradable scaffolds where a steady influx of tissue is ideal, thus making this a highly versatile and useful biomaterial for tissue engineering [45].

Due to its high surface area and porosity, silk nanofibers have garnered interest for applications in tissue regeneration and wound healing. Electrospinning is a successful method for forming silk nanofibers. The electrospun silk fibers possess good mechanical properties and biocompatibility, as well as an environment conducive to cell growth. The fiber diameter and morphology can be controlled by varying the electrospinning parameters to optimize scaffold performance. Co-spinning silk and synthetic polymers was also shown to improve the electrospun fiber stability and bioactivity, thereby increasing their potential usefulness in biomedical applications [46, 47].

Due to its versatility, silk can be used in 3D printing to manufacture scaffolds with complex architectures, making them ideal for specialized applications in tissue engineering. Using silk fibroin-derived bioinks enables the design of 3D supports with superior mechanical properties and biocompatibility [48]. By imitating the natural extracellular matrix, it promotes cellular infiltration and tissue regeneration. The use of advanced techniques, such as digital light processing and extrusion-based printing, enhances the precision and bioactivity of scaffolds. Bioactive molecules incorporated into 3D silk-printed scaffolds facilitate cellular differentiation, making them beneficial for the engineering of bone, cartilage, and vascular tissues [49].

A study developed hydrocolloid inks by incorporating porogens into silk fibroin protein solutions to create porous hydrogel scaffolds. These inks are highly printable, enabling the fabrication of complex geometries with hierarchical porosity, guided by porogen templates (40–200 μm in size, with more than 50% exceeding 100 μm in size) and macroscale features established by the 3D printing process (200 μm or greater). Based on tests performed with human bone marrow mesenchymal stem cells (hMSCs) and mouse embryonic osteoblast precursor cells (MC3T3-E1), it was demonstrated that cells adhere to, infiltrate, and proliferate on and within the surface of the hydrogels. When the hydrogels were implanted under the skin of rats, they were biocompatible and promoted the migration and growth of the animals' own cells. In addition, these microscale biomimetic features significantly promoted bone restoration in a rat femoral bone defect model compared with control groups, including silk hydrogels with smaller pore sizes (approximately 21.39 μm) as well as other 3D-printed scaffolds containing a thickening agent (around 20.78 μm). This work presents a promising strategy for efficient bone defect repair in future clinical applications, utilizing advanced multiscale silk 3D biomimetic scaffolds [50].

4.3.3 Bone Regeneration

Silk-based composites demonstrate strong potential for bone regeneration, providing suitable mechanical strength and enhancing cell adhesion, differentiation, and mineralization, making them promising for orthopedic and maxillofacial applications. One instance of this is osteoinductive silk–silica composites, where silk fibroin is combined with silica-based materials to form a tissue-compatible scaffold that simulates the native bone extracellular matrix. Osteoinduction is mediated by the influence of silica on the differentiation of bone cells and mineral deposition. Due to its unique mechanical properties, the composite biomaterial can be utilized for load-bearing bone regeneration. Such composites not only support osteointegration but also facilitate the healing process by inducing an osteogenic microenvironment [51].

4.3.4 Cartilage Repair

Silk scaffolds have been proposed as effective strategies for supporting chondrogenic cell proliferation and extracellular matrix deposition in cartilage repair. Silk has tunable mechanical properties, and cartilage-like elastic scaffolds can be fabricated. The addition of growth factors and hydrogel matrices to the silk scaffolds promotes chondrogenesis, tissue integration, and repair. Silk-based constructs have been proven to withstand mechanical stress, such as load-bearing, and have been demonstrated to be suitable for joint cartilage regeneration [52].

To improve cartilage regeneration, the researchers designed a silk fibroin (SF) porous scaffold for sequential, cell-free drug delivery. In this system, the BMSC–affinity peptide PFSSTKT (abbreviated as PFS) is coated on the surface of SF scaffolds, and kartogenin (KGN) and anti-inflammatory drug dexamethasone (DEX) are encapsulated in the interior of the scaffold structure. The release profile is precisely designed: PFS is released quickly in the first 10 days, promoting the recruitment and migration of bone marrow mesenchymal stem cells (BMSCs) in the early stage. In comparison, KGN and DEX are released progressively over a 28-day period to promote chondrogenic differentiation and sustain anti-inflammatory effects. In vitro evaluations revealed that the sequential release of PFS and KGN from the scaffold actively promoted the migration of BMSCs and their differentiation into the chondrogenic lineage. The programmed drug release of the SF scaffold effectively recruited endogenous BMSCs and significantly promoted in situ cartilage regeneration in rat models with knee cartilage defects. In this way, an integrated silk fibroin-based and silk-based, fully environmentally friendly approach combining biocompatibility, biodegradability, and tunable drug release for tissue engineering was introduced. In summary, the study demonstrated a readily clinically translatable functional scaffold that has potential for in situ cartilage repair, with promising advantages for regenerative medicine [53].

4.3.5 Nerve Regeneration

Biomaterials based on silk are crucial in nerve regeneration, as they offer a biocompatible matrix that supports neural cell proliferation and axonal regeneration. Silk fibroin-based conduits have been demonstrated to facilitate axonal regeneration and promote functional recovery in cases of peripheral nerve injury. Silk's natural biocompatibility and tunable degradation rates reduce the inflammatory response, and chemical modifications enhance neurotrophic support. Adding conductive materials to silk also potentially enhances its use for transmitting neural signals, which could aid in the development of bioengineered nerve grafts to repair nerve damage [54, 55].

4.4 Applications of Silk Nanoparticles in Biomedical Applications

Silk nanoparticles (SNPs) functionalization is essential for obtaining the desired physicochemical properties for specific biomedical applications. A commonly used approach is surface modification with polyethylene glycol (PEG), also known as PEGylation, which enhances colloidal stability and prolongs circulation time by reducing opsonization [56]. In another strategy, targeting ligands (such as peptides or antibodies) are conjugated to the surface of the SNP, enabling the targeted delivery of a drug to a specific cell type or tissue. For instance, the targeting of tumor-homing peptides has been found to enhance the targeting of SNPs with respect to cancer cells [57]. Moreover, SNPs functionalized with cationic polymers, such as polyethyleneimine, have a high potential for improving the effect of gene delivery by enhancing endosomal escape and protecting genetic material from degradation [58].

Silk nanoparticles (SNPs) demonstrate significant potential as drug delivery vehicles due to their biocompatibility and stability under a wide range of environmental conditions. Previous research has demonstrated that SNPs are highly stable under aqueous conditions over a wide range of temperatures (4–37 °C) for extended periods [56]. They are highly stable in physiological conditions, such as phosphate-buffered saline (PBS), and their stability can be significantly improved by surface modifications, like PEGylation, to prevent aggregation [59]. Furthermore, SNPs are robust to freeze-drying without/minimal usage of cryoprotectants, and exhibit excellent colloidal stability after rehydration [60].

The biodistribution of SNPs has been elucidated through in vivo studies, which are crucial for their consideration as efficient and safe drug carriers. Silk sericin-based nanoparticles loaded with the antineoplastic agent doxorubicin were also remarkable, as evidenced by the fact that mouse models showed reduced doxorubicin presence in the heart, which was associated with decreased cardiotoxicity [60]. Similarly, FNPs labeled with fluorophores have been used to monitor the distribution profiles, and the data demonstrate preferential accumulation within tumor tissue, followed by the enhanced penetration and retention (EPR) effect. A drug delivery system based on functionalized silk nanoparticles has been developed for the targeted delivery of drugs, aiming to achieve selective accumulation in tumor tissue through the EPR effect [61]. Such disposition, attributed to the distinct nature of tumor vasculature, allows nanoparticles to enter the tumor microenvironment after extravasation and remain there continuously [62]. Silk fibroin nanoparticles have also been reported to confer the specific delivery of therapeutic molecules to tumor sites upon functionalization with the appropriate fibroins, thereby enhancing therapeutic efficacy and reducing systemic toxicity [61]. These advances demonstrate the potential applications of SNPs in targeted drug delivery, aiming to reduce off-target effects and enhance therapeutic efficacy.

4.5 Applications of Silk in Wound Healing: Mechanisms and Efficacy

Silk proteins have been extensively researched as a biomaterial for wound healing due to their biocompatibility, biodegradability, and ability to elicit a lower immune response. Silk fibroin promotes cellular adhesion, proliferation, and differentiation, thereby facilitating tissue regeneration. Oxygen–nutrient exchange is preserved, considering the porous nature, and its mechanical properties provide structural support. Silk fibroin has been demonstrated to enhance wound healing by facilitating wound closure, promoting collagen deposition, and facilitating re-epithelialization. Silk hydrogels, films, and scaffolds have also shown promise in the controlled release of drugs and the treatment of chronic wounds. Silk-derived biomaterials thus provide a potential avenue for promoting wound healing [63].

Wound healing is a multifaceted process that involves complex cellular and molecular events [64]. These characteristics enable its biomedical applications due to its biocompatibility and biodegradability. Sericin can enhance wound healing mechanisms by promoting collagen synthesis, increasing the re-epithelialization process, and exhibiting antioxidant as well as anti-inflammatory activities. It has been demonstrated that the use of sericin enhances both the closure rates and overall healing outcomes of wounds. For example, a clinical investigation using sericin with silver sulfadiazine cream for burn wounds demonstrated improved healing efficacy. Moreover, sericin-based hydrogels, when combined with natural extracts, demonstrated enhanced healing in animal models, achieving 86% healing within 11 days. This highlights the use of sericin as an active agent in wound care applications [65, 66]. Due to the unique properties of spider silk, it has gained attention for various wound healing applications. The mechanism of action of spider silk mediates its physiologic cue effect, promoting cell migration and proliferation to facilitate faster wound healing. It is commonly used in many medical applications due to its biocompatibility and mechanical stability. Research has shown that recombinant spider silk proteins dramatically accelerate wound healing compared to the control group. Moreover, improved healing was also observed due to the presence of nanoparticles in the spider silk dressings. Such observations highlight the potential of spider silk-based materials in facilitating effective wound healing [67].

4.5.1 Role of Cytokines Produced During Wound Healing Influenced by Silk

Cytokines are small proteins that play a vital role in cell communication during the complex process of wound healing. Silk-based biomaterials, including silk fibroin and silk sericin, respond to wound beds and affect the secretion and regulation of a wide range of cytokines, thus regulating all stages of healing: inflammation, proliferation, and remodeling. During the initial inflammatory stage of wound healing, inflammatory cytokines such as TNF-α, IL-1β, and IL-6 are produced to attract immune cells, including neutrophils and macrophages, to the injured site. These cytokines are crucial for clearing debris and pathogens, and preparing the wound for the later stages of healing. Research indicates that silk biomaterials can modulate this inflammatory response by preventing the overexpression of pro-inflammatory cytokines. In particular, silk fibroin is known to reduce the levels of TNF-α and IL-6, thereby mediating a more controlled inflammatory response, which is beneficial for advancing the wound healing process into the proliferative phase [68].

With resolution of inflammation, there is an increase in anti-inflammatory cytokines such as IL-10 and TGF-β. Their main function is to inhibit pro-inflammatory cytokine activity and promote tissue repair. Silk-based materials further facilitate this process by upregulating the expression of IL-10 and TGF-β, which are effective in resolving inflammation and promoting tissue repair. This aspect of silk often leads to its utilization as a biomaterial, especially in wound dressings or tissue engineering applications, where the aim is to achieve faster healing with minimal scarring. Formation and remodeling of tissue are addressed during the proliferative phase. Angiogenesis, fibroblast proliferation, collagen synthesis, and epithelialization are all promoted through the action of growth factors [VEGF (Vascular Endothelial Growth Factor), PDGF (Platelet-Derived Growth Factor), FGF (Fibroblast Growth Factor), and EGF (Epidermal Growth Factor)]. Silk fibroin-derived scaffolds and hydrogels have been shown to promote the expression of growth factors, including VEGF and FGF, thereby facilitating vascularization and collagen deposition at the wound site. This characteristic of silk renders it particularly suitable for managing chronic wounds, where promoting tissue regeneration is of paramount importance [63, 68].

The remodeling phase involves the degradation and reorganization of the extracellular matrix (ECM), primarily mediated by matrix metalloproteinases (MMPs) and tissue inhibitors of metalloproteinases (TIMPs). A proper equilibrium between MMPs and TIMPs is necessary for the formation of functional healed tissue with a supportive structure. Silk fibroin scaffolds upregulate NF-κB at the pro-inflammatory M1 macrophage phase, which facilitates tissue remodeling and promotes high-magnitude immune responses, potentially leading to long-term increases in angiogenesis and reepithelization. FDA-approved silk-based materials help regulate this balance, resulting in optimal collagen remodeling and diminished scar formation [68]. Silk, with its remarkable biocompatible properties, antioxidant capability, and sustained-release profiles for cytokines or growth factors, may help facilitate the effects of these factors. Silk sericin, in particular, exhibits antioxidant properties that help reduce oxidative stress and inflammation. In addition, the biocompatibility and bioactivity of silk fibroin, as well as its versatility for use in scaffolds, films, hydrogels, and nanoparticles, offer unique capabilities in regulating cytokine activity to promote wound healing [69, 70].

4.5.2 Effect of Moisture Retention on Healing Rates

To further clarify the healing rates of silk-based biomaterials, several studies have investigated the role of moisture retention on silk fibroin and sericin proteins [66]. Silk fibroin exhibits an outstanding ability to retain moisture. It maintains a hydrated atmosphere at the wound sites, an essential aspect required for effective healing. Moreover, silk fibroin has inbuilt antibacterial properties that counteract infection risks, provide favorable gas exchange for cell proliferation at the wound site, and improve hemostatic mechanisms. These properties collectively make silk fibroin a promising biomaterial for wound healing [71]. Another silk protein, sericin, also has excellent moisturizing effects. Sericin is also shown to create a semiocclusive film in wounds, which maintains moisture levels and aids in wound healing. It performs better than typical moisture-retaining agents, such as glycerin and hyaluronic acid. Some studies have explored sericin's moisture-retaining ability, and in some of them, its performance has been contrasted with that of common moisturizers, such as glycerin and hyaluronic acid [72].

Studies have shown that sericin exhibits exceptional hygroscopicity, with a 3% solution demonstrating moisture absorption comparable to a 60% glycerol solution, indicating its high efficiency at low concentrations [73]. It also shows superior moisture retention compared to common moisturizing agents, including hyaluronic acid [74]. These findings emphasize sericin's strong potential as an effective natural moisturizer. Silk fibroin-based dressings have been reported to accelerate wound healing due to their anti-inflammatory and proangiogenic effects, outperforming several commercial materials [75]. Moreover, a combination of silk sericin and Aloe vera gel resulted in notable wound healing in diabetic patients, underscoring sericin's therapeutic role in skin repair [76]. Recent advances have led to the development of silk-based wound care products such as **Sylke**, an adhesive wound closure made of 99.9% medical-grade silk fibroin that is biocompatible, breathable, water-resistant, and highly elastic, offering 14 days of comfortable, secure, and flexible post-surgical healing even in high-mobility areas [77].

4.5.3 Patient Demographics Influencing Wound-Healing Outcomes

Wound-healing outcomes can vary significantly depending on individual patient characteristics. Younger patients generally exhibit faster healing due to robust cellular proliferation and collagen production. Silk fibroin (SF) biomaterials have been shown to promote cellular adhesion and fibroblast proliferation, thereby accelerating wound closure in this group [78]. In contrast, older adults often experience slower healing due to reduced skin elasticity and vascularity; however, incorporating SF biomaterials may facilitate neovascularization and help mitigate these age-related deficits [78]. Comorbidities such as diabetes and peripheral vascular disease can further impede healing by inhibiting new blood vessel formation; however, SF biomaterials have demonstrated the ability to enhance angiogenesis, which is particularly beneficial for patients with such conditions [78]. Additionally, immune responses should be carefully considered, as certain silk-based dressings containing silk sericin (SS) may cause hypersensitivity reactions in some individuals [78]. Socioeconomic factors also influence treatment success, as high-quality silk-based dressings may not be accessible to all patients, despite their demonstrated effectiveness in promoting wound healing; expanding accessibility remains a crucial goal [79]. Overall, while silk-derived biomaterials offer promising benefits for wound management, factors such as age, comorbidities, immune responsiveness, and socioeconomic status must be considered, and individualized treatment strategies are essential for maximizing therapeutic outcomes.

4.5.4 Comparative Analyses Between Different Dressing Materials

Comparative analyses between different dressing materials have demonstrated that silk-based biomaterials exhibit significant potential as wound dressings. Studies comparing silk fibroin dressings with conventional materials have shown that silk fibroin-based biomaterials promote cellular adhesion, fibroblast proliferation, neovascularization, and re-epithelialization in cutaneous wounds. Preclinical investigations further indicate that these materials exhibit low cytotoxicity and immunogenicity, outperforming several standard dressings [78]. In another study, silk biomaterials functionalized with epidermal growth factor (EGF) and silver sulfadiazine were found to enhance wound healing in a mouse model, accelerating re-epithelialization, dermal proliferation, and collagen synthesis, while reducing scar formation compared to commercial dressings such as Tegaderm Hydrocolloid [80]. A bi-layered silk fibroin–gelatin dressing has also been developed, exhibiting superior mechanical properties and promoting fibroblast proliferation, wound closure, and collagen deposition more effectively than clinically used dressings [81]. Furthermore, a clinical trial comparing pure knitted silk (Dressilk) with a synthetic skin substitute (Suprathel) for partial-thickness burns revealed comparable outcomes in terms of pain management, exudate control, and infection rates, with no significant difference in scar assessment after 12 months. Notably, Dressilk was about twenty times more cost-effective, suggesting its potential as a practical alternative for clinical applications [82].

4.5.5 Clinical Applications Beyond Wound Healing, Such as Organ Repair

The natural biopolymer silk fibroin has received considerable attention in biomedical science for applications beyond wound healing. Its excellent characteristics (biocompatibility, biodegradability, and mechanical) make it a multipurpose substrate for a variety of clinical applications, e.g., organ and tissue regeneration.

- **Cartilage and Bone Regeneration**: Silk fibroin scaffolds are one of the most useful carriers for cartilage tissue engineering. Their structural properties encourage chondrocyte expansion and differentiation critical in cartilage regeneration. These silk-based materials have demonstrated osteoconductivity and collective potential for mechanical strength in bone regeneration, where they have been combined with hydroxyapatite to facilitate new bone formation [83].
- **Cardiovascular Applications**: Silk fibroin hydrogels have been studied for the reconstruction of small-diameter blood vessels. Studies also show that it has good thrombogenicity and patency periods, thereby developing great potential for use in coronary artery bypass grafts [84].
- **Nerve Regeneration**: Silk fibroin scaffolds have been utilized in nerve tissue engineering, as they promote the adherence and proliferation of nerve cells. Their supportive properties for nerve regeneration make them potential candidates for treating peripheral nerve injuries [85].
- **Liver Tissue Engineering**: Silk fibroin matrices have also been explored for use in liver tissue engineering. Their promising biocompatibility and tunable mechanical properties support hepatocyte function, suggesting potential for use in liver tissue regeneration strategies [84].

Such applications highlight the multifaceted applications of silk fibroin as a biomaterial for informing regenerative medicine approaches to not only wound healing but also organ repair.

4.6 Immunomodulatory Effects of Silk in Healing Processes

Rich in bioactivity, silk-based biomaterials, specifically silk fibroin, have been of great interest in the biomedical research field due to their immunomodulatory effects, which promote the healing cascade [78]. Silk fibroin has been reported to affect a variety of immune cells, with macrophages being a particular target. Silk-based scaffolds can modulate macrophage activity in vitro, resulting in a significant decrease in pro-inflammatory markers (TNF-α, CXCL10, and CD197) and TNF-α secretion. This suggests that silk materials can reduce inflammatory responses by directly affecting macrophage activity [86]. Silk fibroin modulates macrophage polarization, facilitating a transition from the pro-inflammatory M1 phenotype to the anti-inflammatory M2 phenotype. This switch is crucial in resolving inflammation and promoting tissue repair. Silk regulates macrophage activity through the NF-κB signaling pathway, an important signaling pathway involved in various cellular behaviors during wound healing [63]. Recent evidence suggests that silk-based biomaterials can modulate cytokine profiles by inhibiting/downregulating pro-inflammatory cytokines and upregulating anti-inflammatory cytokines. Broadly, this cytokine modulation leads to a balanced inflammatory response that facilitates appropriate healing and tissue regeneration [68]. Silk fibroin possesses distinctive immunomodulatory properties as compared to other biomaterials [87]. In contrast to other well-known biocompatible materials, such as chitosan and alginate, silk fibroin exhibits chemokine and cytokine modulatory properties that enhance angiogenesis while reducing scarring. Furthermore, silk is biocompatible with significantly tunable mechanical properties and degradation rates, which would be beneficial compared to other types of biomaterials in wound healing applications [63]. Silk-based materials possess immunomodulatory properties, making them promising candidates for the treatment of chronic inflammatory disorders. Silk biomaterials may be used for therapeutic interventions aimed at resolving chronic wounds and other chronic inflammatory diseases by promoting balanced immune responses and supporting tissue regeneration. They play a considerable role in controlling macrophage activity, cytokine profiles, and their level of activity is especially important in regulating chronic inflammation [68].

4.7 Silk-Based Hydrogels: Design and Applications

4.7.1 Mechanisms Behind the Swelling Behavior of Silk Hydrogels

The swelling property of silk hydrogels may significantly impact their performance in biomedical applications. This phenomenon is primarily regulated by factors such as pH, temperature, and ionic strength. For example, Dinerman et al. found specific swelling behavior in silk-elastin-like protein polymer hydrogels (genetically engineered biopolymers that combine the properties of silk and elastin) as the environmental conditions changed, with an impact on drug diffusion rates and drug release profiles. Such insights are crucial for the development of application-specific hydrogels with optimal timing in both drug delivery and tissue engineering approaches [88].

4.7.2 Use of Hydrogels in Controlled Drug Release Applications

Hydrogels present unique features as drug carriers, such as the capability to encapsulate therapeutic agents and release them in a sustained manner, making these systems attractive candidates for controlled drug delivery. They have a well-developed porosity, and their high water content allows for local drug diffusion, while their tunable properties enable the tailoring of release kinetics. For example, Li and Mooney (2016) provided an overview of design considerations for drug delivery hydrogels, emphasizing the role of network structure and cross-linking density in regulating the rate of drug release. These parameters can be tuned to design hydrogels capable of sustained and targeted delivery of a wide range of therapeutics, thereby improving efficacy while minimizing side effects [89].

4.7.3 Evaluation Methods for Mechanical Properties of Hydrogels

Due to their tunable mechanical properties and biocompatibility, silk-based hydrogels are being examined for use in biomedical applications. Assessing the properties is crucial for achieving functional tissue engineering and drug delivery devices. Standard practices, such as viscoelastic and compressive modulus (for stiffness and damping), are applied to these materials [90]. Tensile testing provides information on the elasticity and fracture resistance of the hydrogel. Atomic force microscopy (AFM) is a crucial method for characterizing the mechanical properties of silk-based hydrogels at the nanoscale. It enables us to measure surface stiffness and nanoindentation properties, providing information about the material's elasticity and its overall structure. One example is Hansma's work, which demonstrates the use of AFM in studying the mechanics of silk fibers, highlighting its relevance in hydrogel analysis [91]. Another important technique is dynamic mechanical analysis (DMA), which assesses a hydrogel's viscoelastic properties under cyclic loading conditions. Such parameters include storage and loss moduli, as well as the damping factor, which are important for characterizing the material's behavior under mechanical loads. Zhang et al. (2023) employed DMA to study the viscoelastic properties of tunable-stiffness hydrogels, highlighting its feasibility in silk-based hydrogel profiling [92].

4.7.4 Biocompatibility Studies Comparing Different Hydrogel Formulations

Biocompatibility is a crucial parameter for silk-based hydrogels in biomedical applications, including wound healing and drug delivery. The studies most frequently comprise cytotoxicity assays to determine cell vitality. In vivo behavior in animal models also provides information regarding immunogenicity and tissue integration. For instance, Zhang et al. (2023) investigated the use of tunable-stiffness viscoelastic hydrogels to control a wide range of cell behaviors, highlighting the significant impact of mechanical properties on biocompatibility. Moreover, the addition of bioactive molecules, such as growth factors, to silk-based hydrogels has been shown to enhance cell proliferation and mitigate inflammatory responses. The standardization of comparative studies is necessary to achieve the optimal formulations of polymers that retain mechanical integrity while facilitating favorable biological interactions. Zhang et al. (2023) highlighted the importance of dynamic mechanical properties in the design of hydrogels that can closely mimic the native ECM to improve biocompatibility [92].

4.8 Antimicrobial Properties of Silk Dressing

The antimicrobial properties of silk dressings, attributed to the structural and biochemical characteristics of silk fibroin and sericin, have garnered considerable attention. In this regard, SF, in particular, exerts an inhibitory effect on bacterial colonization, which can be enhanced by the crystallization of specific portions of SF, as evidenced by the enhanced hydration of the surrounding environment, resulting in an adverse condition for microbial growth [75]. Additionally, silk sericin consists of bioactive peptides with antimicrobial activity, thereby contributing to the protective function of silk. Studies have shown that silk-based dressings are effective in promoting wound healing and preventing infections. Material modifications or optimization are additional approaches to further enhance the antimicrobial efficacy of silk dressings [93].

4.8.1 Mechanisms by Which Silk Interacts with Microbial Membranes

The antimicrobial properties of silk are largely due to its interaction with the microbial membranes. Silk fibroin has been shown to interfere with bacterial cell membranes through electrostatic interactions, thereby rendering bacterial membranes unstable and compromising cellular integrity. In addition, silk sericin contains active peptides that firmly bind to bacterial surfaces, acting in contact to interfere with critical cellular processes. This mechanism effectively prevents the growth of both Gram-positive and Gram-negative bacteria. Furthermore, silk fibroin's high molecular weight and hydrophobic character further enhance its capacity to form a physical barrier around various interacting microbe components [94].

4.8.2 Comparative Studies on Antimicrobial Efficacy Against Various Pathogens

Silk-based dressings have been tested for their antimicrobial activity against a range of pathogens, such as *Staphylococcus aureus*, *Escherichia coli*, *Pseudomonas aeruginosa*, and *Candida albicans*. The bacteria include Gram-positive *S. aureus* and Gram-negative *E. coli* and *P. aeruginosa*, where the activity toward Gram-positive bacteria is moderate, and the activity toward Gram-negative bacteria is low. As an example, SF/PEI (fibroin/polyethyleneimine) electrospun nanofibers showed good antibacterial activity against *S. aureus* and *P. aeruginosa* within 5 h, while they were not cytotoxic against fibroblast cells [95]. To widen the antibacterial spectrum, hybrid silk dressings were designed by incorporating additional antibacterial agents. Specifically, silk fibroin hydrogels embedded with gallium ions (SF/Ga) demonstrated effective antibacterial activity against *P. aeruginosa* in vitro and in vivo, and promoted accelerated wound healing in infected models [96]. In a related study, hydrogels composed of silk sericin/poly (vinyl alcohol) incorporated with azithromycin demonstrated augmented antimicrobial activity against *S. aureus, P. aeruginosa, E. coli,* and *C. albicans*, suggesting that composites containing silkworm products may be effective against a broad spectrum of pathogens [97].

4.8.3 Development of Hybrid Dressings Combining Silk with Other Materials

Novel silk-based dressings with enhanced antimicrobial activity have been developed by incorporating silk fibroin or sericin with antimicrobials, including silver nanoparticles, chitosan, graphene oxide, and antibiotics. These combinations work together to enhance microbial inhibition and facilitate tissue regeneration. For example, chitosan, silver (Ag) nanocrystals, and graphene oxide were synthesized into a nanocomposite (ChAgG) for antimicrobial actions and potential use in bioactive wound dressings [98]. Likewise, silver nanoparticles loaded on chitosan-based scaffolds demonstrated better antibacterial activity and biocompatibility, which could be applied to wound healing [99]. The incorporation of antimicrobial nanocomposites into bioactive electrospun fibers has been successful in enhancing wound healing, owing to the synergistic advantages of its components, such as improved mechanical strength and a lower risk of infection [100].

The versatility of silk as a biomaterial and its potential, especially when combined with other antimicrobial agents for controlled release, are highlighted in these studies. This combination enables the development of multifunctional wound dressings that inhibit infections and promote tissue repair.

4.9 Silkworm as a Research Model in Biotechnology

The silkworm (*Bombyx mori*) has been utilized as a platform for high-throughput biopharmaceutical research due to its well-annotated genome, ease of growth, and high economic value in the silk industry. It has a large progeny size, a short generation time, and rich genetic resources, which contribute to its wide range of research applications [101]. The development of more sophisticated genetic manipulation techniques, such as transposon-mediated transgenesis and genome-editing tools like CRISPR/Cas9, has facilitated both precise functional studies and recombinant protein production [102]. Silkworms have been utilized for drug screening, vaccine development, and environmental monitoring, underscoring their versatility in life sciences research [101]. Figure 4.3 shows an overview of various applications of silkworm as research model.

4.9.1 Diabetes Research

The silkworm (*Bombyx mori*) has become a powerful model system in diabetes research, in part due to its physiological and genetic resemblance to higher eukaryotes, particularly in terms of insulin signaling and carbohydrate metabolism. Mice and rats are the classic models for in vivo studies, but they are more expensive, raise ethical concerns, and take longer to develop as experimental organisms than silkworms. Their ease of maintenance and short life cycle enable the high-throughput screening of anti-diabetic drugs and genetic studies. The insulin-like peptide (ILP) pathway in silkworms exhibits striking similarities to the mammalian

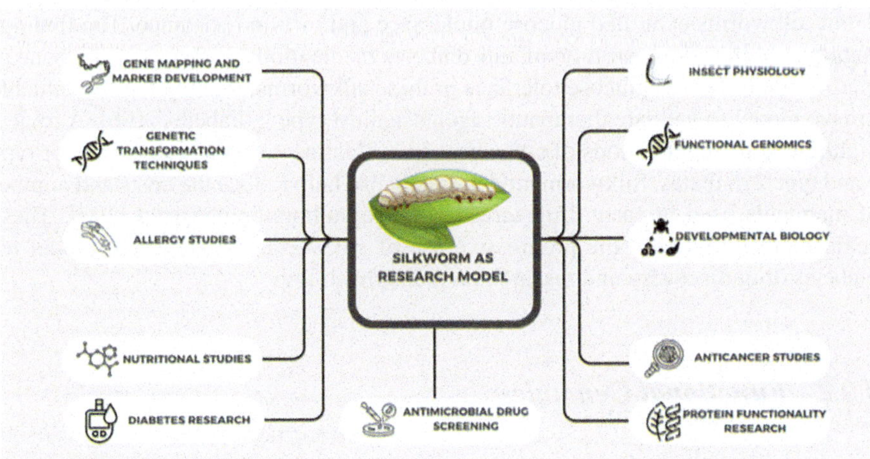

Fig. 4.3 Silkworm as research model

insulin signaling pathway. As this pathway is involved in glucose homeostasis, silkworms provide an accessible model for studying insulin-resistant and diabetes-related metabolic abnormalities in diabetes. Moreover, silkworms also undergo glucose-induced lipogenesis, which may help in understanding diabetes-driven lipid metabolism. Multiple natural and synthetic compounds have been shown to regulate blood glucose levels in silkworms and, therefore, serve as useful tools in preliminary drug screening. The relative simplicity of their immune and metabolic systems also enables researchers to isolate specific molecular mechanisms that may be involved in triggering diabetes. Collectively, the silkworm model presents a powerful alternative for diabetes research, allowing for faster, more cost-effective, and more ethical studies that complement murine models. Further study of this model could significantly enhance our understanding of diabetes and facilitate the discovery of new therapeutic agents.

In diabetic mice induced by streptozotocin, researchers analyzed silkworm powder and documented a marked reduction in blood glucose and vasopressin expression in the hypothalamus. The research suggests the possibility of using nourishments derived from silkworms to alleviate symptoms of diabetes [103]. Development of a hyperglycemic silkworm model for the identification of anti-diabetic drugs. The researchers found that silkworms exposed to a high-glucose diet showed increased sugar levels in their hemolymph, and these levels were effectively decreased through human insulin treatment, thus confirming that their model could be used to screen for hypoglycemic agents [104]. Scientists established transgenic silkworms expressing human insulin receptors using the GAL4/UAS system. Experimental human insulin injection into these transgenic silkworms lowered hemolymph sugar levels and promoted Akt phosphorylation in the fat body, consistent with the biological activity of human insulin on the insulin receptor [105]. Researchers developed a silkworm model of type II diabetes by maintaining silkworms on a high-glucose diet for extended periods, resulting in a hyperlipidemic phenotype. In fact, these hyperlipidemic silkworms exhibited glucose intolerance and insulin resistance. The findings suggest that the administration of anti-diabetes medications, such as pioglitazone or metformin, improves glucose tolerance in these silkworms, making them a valuable in vivo model to validate therapeutic agents against type 2 diabetes [106]. A review highlighted the applications of silkworms in evaluating anti-diabetic drugs for type 1 and type 2 diabetes. Silkworm models could also help reduce the costs and number of mammals used in early drug screenings, the authors emphasized [107]. These achievements highlight the progressive use of silkworms as a relevant model for diabetes drug discovery and research on pathophysiology.

4.9.2 Functional Genomics

Due to their well-annotated genome and ease of genetic manipulation, silkworms are highly useful in functional genomics. Functional genomics in silkworms is an excellent system for studying gene function and gene-gene interactions on a large scale. It was sequenced in 2004 and revealed ~14,000 protein-coding genes. Specific gene knockdown through RNA interference (RNAi) has also been utilized to investigate the roles of specific genes in silk protein production, as well as their associated metabolic pathways. For example, a comparative transcriptomic analysis identified an epidermis-specific gene, BmCPG25, and explored its function by utilizing RNAi [108, 109]. CRISPR–Cas9 technology has also been used to induce gene knockouts, enabling functional experimentation. In one study using CRISPR–Cas9 to knock out BmE2F1, the number of silk gland cells was reduced by 7.68%, and the silk yield was reduced by 22%, indicating a critical role in silk production [110]. Transcriptomic characterization has provided insights into gene expression dynamics across various physiological states, including molting and silk production. One example of this is a study performed on the fibroin gene cluster, which shows key regulatory elements involved in silk protein synthesis. This has not only helped broaden our understanding of silkworm biology but also enabled the improvement of silk production through genetic engineering. Such functional genomics in silkworms also plays a crucial role in agricultural biotechnology, leading to the identification of resistance genes against pathogens to enhance sericulture practices. The potential of silkworms as a model for functional genomics studies is greatly amplified by the advent of high-throughput sequencing technologies and bioinformatics.

4.9.3 Anticancer Studies

Silkworms are also highly sensitive to human pathogenic microorganisms, pathogenic fungi, antibiotics, and pesticides, according to studies. The silkworm (*Bombyx mori*), chosen for this study, has shown great potential as a model organism due to its sensitivity to many bioactive compounds and its similarities with mammalian systems in specific disease modeling. With its open circulatory system, it is also possible to administer compounds directly into the hemolymph, making studies on drug uptake and metabolism more efficient. Hence, using silkworm models to investigate human tumors is an emerging area of study [101]. Table 4.3 gives an overview of various silkworm-derived anticancer compounds.

Carboxypeptidase inhibitor: A study reported that silkworm carboxypeptidase inhibitor presented abundant expression in the silkworm silk gland, and can inhibit the expression of proto-oncogene Myc, significantly inhibiting the proliferation of gastric cancer cells. In light of the anticancer activity of the silkworm carboxypeptidase inhibitor, the effective peptide of silkworm carboxypeptidase inhibitor, as a lead compound for tumor therapy, was designed, and its potent drug potential for gastric cancer was identified in this study [111].

Antimicrobial Peptides: Cecropin-XJ is an antimicrobial peptide from silkworms that reportedly inhibits growth and induces apoptosis in esophageal cancer cells, especially in the presence of chemotherapeutic agents, and has lower toxicity to normal cells than traditional drugs [111]. In a study, the authors engineered transgenic

Table 4.3 Silkworm-derived anticancer compounds [101, 111, 112]

Application/Compound	Cancer type studied	Mechanism
Carboxypeptidase inhibitor	Gastric cancer	Inhibits EGF/EGFR pathway, suppresses c-Myc
Cecropin-XJ (antimicrobial peptide)	Esophageal cancer	Inhibits proliferation, induces apoptosis
Sericin (silk protein)	Colon cancer	Induces apoptosis, cell cycle arrest
Silk fibroin peptides	Lung cancer	Suppresses proliferation, induces apoptosis
Transgenic TA (tumor antigen) protein production	Head and neck cancer	Enables immune response studies for immunotherapy

silkworms to produce a tumor antigen protein (MAGE-A4 and wild-type p53) for individual in vitro evaluation of antitumor immune responses. This method enables safe, large-scale production of high-quality tumor antigens for the advancement of cancer immunotherapy development [112].

Several properties are manipulable in nature and allow for further optimization routes of silk, particularly within the parameters of cancer drug delivery. Researchers have successfully leveraged some of silk's unique features, including its first-order drug release characteristics and pH responsiveness, for intratumoral applications in the cancer tumor microenvironment. This approach enables sustained and pH-responsive drug release from silk nanoparticles [113]. Sericin can decrease the viability of colon cancer cells (SW480) by triggering apoptosis via caspase-3 activation and downregulation of Bcl-2 expression. Colon cancer prevention available as a dietary supplement—Sericin inhibits 53BP1 and RPA for cell cycle arrest at S phase [114].

Silk fibroin is a sustainable and biodegradable biomaterial for various applications in both in vivo and in vitro cancer drug delivery [113]. Early research has shown that silk fibroin peptides can inhibit proliferation, as well as induce apoptosis and cell cycle arrest in human lung cancer cells, providing further evidence of the potential anticancer effects of silkworm proteins in general [115].

4.9.4 Nutritional Studies

Silkworms (*Bombyx mori*) have become important model organisms for nutritional research, as they are sensitive to dietary alterations, which directly affect their growth, silk yield, and metabolism. For example, researchers have recently used silkworms to investigate the metabolic effects of different diets and their health impact in studies with implications for human health. Silkworm feeding silkworms data of research indicates that the absence of an essential amino acid diet can directly harm silk protein synthesis [116, 117]. The authors found that the PI3K-Akt-SGF1-Dimm signaling pathway mediates the nutritional regulation of silk protein genes. Changes in this pathway resulting from amino acid deprivation caused downregulation of silk protein gene expression, signaling metabolic stress [118]. To investigate the effects of plant-derived polyphenols and vitamins on lipid metabolism, silkworms have been utilized. For example, plant polyphenols can regulate lipid metabolism through various mechanisms, including modulation of lipid absorption and metabolism pathways [119]. Silkworms are used as a good model for the evaluation of the hypolipidemic activity of functional foods. Feeding silkworms high-fat diets, followed by the administration of potential lipid-lowering substances, could also serve as a useful model to study how these compounds interact with lipid metabolic pathways. For example, some peptides isolated from silkworm pupae have been shown to inhibit adipogenesis in 3 T3-L1 cells and prevent body weight gain in rats fed a high-fat diet [120].

These findings highlight the utility of silkworms as model organisms for nutritional studies, providing insights into the metabolic effects of diverse dietary components and facilitating the development of functional foods for the prevention of metabolic disorders.

4.9.5 Antimicrobial Drug Screening

Silkworms (*Bombyx mori*) have emerged as a useful in vivo system to screen antimicrobial agents because they possess an innate immune system with major characteristics similar to those of mammalian innate immunity. This enables the evaluation of pathogenicity and drug validation in a physiologic environment, which has several advantages over conventional bacterial cultures. Silkworms have a relatively strong innate immune system, despite lacking adaptive immunity. AMPs from different sources have been tested using silkworms. For example, feeding heat-killed *Pseudomonas aeruginosa* cells orally as an immune challenge significantly enhanced silkworm AMP production, thereby increasing their resistance to infections. This suggests that silkworms are a valid model for studying immune responses against specific bacterial pathogens [121]. Silkworm infection models are also used to test the effectiveness of known antibiotics. In studies of *Staphylococcus aureus* and *Pseudomonas aeruginosa* infections, the administration of antibiotics such as vancomycin and daptomycin resulted in improved survival, confirming findings obtained in mammalian models. Glycopeptide antibiotics (vancomycin, teicoplanin, dalbavancin) were also challenged in the silkworm model against *Staphylococcus epidermidis* infection; these antibiotics showed impressive results, significantly lowering mortality and immune responses [122–124]. Overall, the model provides a rapid and reproducible in vivo system for assessing antimicrobial molecules, serving as an alternative to in vitro and mammalian assays [125].

4.9.6 Protein Functionality Research

Since they are widely used for recombinant protein expression and silk protein analysis, silkworms (*Bombyx mori*) have emerged as an attractive model in protein research [126]. They have different processes of protein synthesis that have complicated their genetic regulation and protein folding. Human therapeutic proteins have been expressed for a long time from the silkworm baculovirus expression system. For instance, Maeda et al. Specifically, in the late 1980s, human interferon-alpha (IFN-α) was successfully expressed in silkworm larvae using a recombinant *Bombyx mori nucleopolyhedrovirus* (BmNPV) vector. In addition, this system has also been developed for the production of various recombinant proteins, including antibodies, fusion proteins, and cytokines, thereby elucidating the potential of the silkworm as a bioreactor [127, 128].

4.9.7 Genetic Transformation Techniques

Genetic transformation methods in silkworms, such as transgenesis with piggyBac transposon, CRISPR–Cas9 for gene editing, and RNA interference (RNAi) have promoted research in various biological fields. Here are some of the leading studies and sources that document these advances:

The piggyBac (PB) transposon system has proven to be an indispensable tool for generating and propagating stable germline transformation in silkworms [129]. For example, one study utilized piggyBac to construct a gene transfer vector containing both the GFP gene (gfp) and the neomycin resistance gene (NeoR), which facilitates the high-efficiency generation of transgenic silkworms [130]. Nonetheless, concerns have been raised about the stability of the PB transposon after integration. In research, transgene silkworms exhibited loss of reporter gene traits, highlighting the need for further investigation into the regularity and stability of PB-mediated transformations [131].

With CRISPR–Cas9 technology, in-depth functional genomics studies, as well as the enhancement of traits, have been facilitated in silkworms. For example, a knockout of the Fib-L gene using the CRISPR/Cas9 system produced homozygous mutants with thin layers of cocoon silk, which were unable to complete the pupation process, thereby confirming the role of Fib-L in silk production [131]. Use of CRISPR/Cas9-mediated editing of the gene sequence of the BmEcKL1 gene altered silk features, suggesting its potential role in enhancing silk features [132].

RNA interference (RNAi) is a powerful tool for gene knockdown in silkworms, which has greatly aided in elucidating gene function and regulation. In a recent study, RNAi was shown to be effective after oral feeding in silkworms, achieving overall efficiencies significantly higher than those observed in studies where injections were the sole delivery method [133]. In addition, RNA interference (RNAi)-mediated silencing successfully induced significant morphological alterations of animals, indicating the use of RNAi in functional analysis as a powerful approach [134].

4.9.8 Insect Physiology Studies

Silkworms (*Bombyx mori*) have been used as a model insect, allowing for the investigation of physiological properties due to the availability of developmental stages and ease of maintenance, as well as physiological similarity with other insect groups [135]. This area of research has provided important new insights into molting, metamorphosis, digestion, and reproduction, among other physiological processes. It has also been demonstrated in silkworms that hormonal pathways are crucial regulators of molting, as well as metamorphosis. One example is that the microRNA let-7 alters the expression of ecdysone-related genes and regulates these processes. Through this approach, we were able to isolate transgenic silkworms expressing a miRNA sponge targeting let-7 and characterize a developmental arrest phenotype observed during larval transitions, which supports the involvement of let-7 within the ecdysone signaling repertoire [136]. Moreover, the enzyme phenol oxidase (PO) has been implicated as essential for molting of the silkworm in a hormone (ecdysone)-regulated process [137]. Extensive studies have been conducted on the digestive system of the silkworm to characterize digestive enzymes, including proteases, amylases, and lipases. These enzymes, produced by the gut microflora of silkworm larvae, allow them to easily digest the food they consume (most commonly mulberry leaves). In particular, protease-producing bacteria predominate, as do microbes that degrade lipids and polysaccharides [138]. Subsequently, lipolytic bacteria clinging to the silkworm gut are characterized; the diet of silkworms modifies the range of lipolytic microorganisms. It has also been shown that silkworms fed on different host plant leaves, such as mulberry leaves and tricuspid cudrania leaves, have distinct lipase-producing bacterial communities, indicating a network between diet, gut microbiota, and digestive enzyme activity [135]. In addition, genome-wide analysis identified 38 lipase genes in the silkworm genome, 12 of which were expressed in the midgut. The temporal and spatial expression of these genes also suggests their involvement in lipid digestion at different stages of development [139].

4.9.9 Gene Mapping and Marker Development

Research on genetic mapping and marker development in *Bombyx mori* has contributed to an understanding of silk-producing traits, disease resistance, and morphological traits. With the availability of a high-quality reference genome, QTLs (Quantitative Trait Loci) and molecular markers can be identified, thereby promoting breeding programs and genetic studies. Genetic background of silk yield, QTL analysis has identified several genes associated with silk protein synthesis, and many of these genes co-localize in the QTL region associated with silk yield [140]. The identity of BmGlcNase1 indicates its relevance, as this gene has been linked with sericin synthesis and cocoon construction, suggesting the functionality of silk protein production. Microsatellite markers, also known as simple sequence repeats (SSRs), have been established to evaluate the genetic diversity of silkworm strains, which can assist breeding and conservation programs. By using 26 microsatellite markers analyzed in 31 silkworm strains, a total of 188 alleles were recorded, indicating a considerably high level of genetic diversity [141]. Marker-assisted selection (MAS) has been used to develop silkworm lines with favorable characteristics, such as increased silk yield and resistance to *densonucleosis virus* (DNV-Z), through Simple Sequence Repeat (SSR) markers associated with the *nsd-Z gene* [142]. Recent advances in MAS have opened up avenues for discovering individual traits associated with superior silk production and disease resistance, facilitating breeding programs for the silkworm [143]. KAIKObase is a comprehensive genome database for *Bombyx mori*, containing gene sets, genetic maps, and genome annotations, and is useful for various genetic research applications [144]. These developments highlight the integration of molecular genetics into sericulture, which can help improve economically important traits in silkworms.

4.9.10 Allergy Study

Since silkworms possess a functional innate immune system similar to that of mammalian immune systems, they are suitable for examining immune responses to allergens. The silkworm has been widely used as a model organism for evaluating the toxicity and immunogenicity of drugs and natural compounds. Their reactions can provide clues to possible allergenic effects. For example, silkworms have been used to study and characterize adverse proteins such as tropomyosin, which is well known to induce human allergic reactions [145]. It can also provide a scoring system to predict the allergenic potential of plant-based compounds prior to transitioning to more advanced mammalian equivalents [101]. The silkworm (*Bombyx mori*) is a valuable model for studying allergic sensitization through ingestion or inhalation, making it an effective screening tool for allergens. Studies were conducted systematically by researchers who evaluated potential allergens for the different life stages of the silkworm, including the larva, pupa, and moth, as well as silkworm metabolites such as silk, slough, and feces. They obtained the results using two-dimensional gel electrophoresis (2DE) and Western blot (WB) techniques and identified 45 possible IgE-binding proteins, demonstrated by LC-MS/MS analysis. These developmentally stage-specific allergens indicate that the expression of allergenic proteins changes during metamorphosis. By development stage, pupae and moths contained significantly more allergens than larvae, likely due to newly emerging proteins as these organisms progressed through development. In addition, several allergenic proteins discovered in silkworms have demonstrated considerable sequence homology with well-characterized allergens found in mites, mosquitoes, fungi, and grains, warranting a high risk of cross-reactivity. This means that silkworm allergens could help explain allergic reactions to other types of silkworms. The detailed characterization of silkworm allergens and their cross-reactivity further highlights their significance as a powerful model organism in allergen discovery and screening, particularly in assessing food and environmental allergens of concern to human health [146].

4.9.11 Developmental Biology

Silkworms (*Bombyx mori*) are an attractive model organism in developmental biology (Table 4.4) due to their well-defined life cycle, large body size, ease of aerial and terrestrial movement, and genetic tractability. The field has primarily focused on genetic and environmental determinants that modulate development and metamorphosis. A landmark study focused on regulating transcription factors during embryogenesis and larval development. Through RNA interference (RNAi) and CRISPR–Cas9 gene editing, the researchers identified master regulatory genes that controlled the differentiation of tissues and the formation of functional organs. For example, knocking down genes that govern wing development caused deformities, a reminder that morphogenesis is not a mere post-mitotic phenomenon [147]. Another focus of research is the development of the silk gland. The fibroin and sericin genes are differentially expressed in the fifth instar larval stage, a process regulated by both hormonal and genetic factors. Such knowledge could be utilized to enhance silk production through genetic methods [148]. In addition, transgenic silkworms expressing fluorescent proteins have enabled real-time visualization of development events, further establishing silkworms as a powerful tool in the field of developmental biology [149].

Table 4.4 Contribution of silkworm in developmental biology

Developmental topic	Contribution of silkworm studies
Embryogenesis	Stage-specific gene expression profiling, germ layer formation
Molting & metamorphosis	Hormonal control by ecdysone & juvenile hormone
Programmed cell death (PCD)	Studied in salivary glands and midgut during metamorphosis
Tissue differentiation	Silk gland as a model for secretory tissue differentiation
Hormonal pathways	Molecular analysis of ecdysteroid signaling and gene regulation
Genetic expression timing	Time-specific transcriptomics using microarrays and RNA-Seq

4.10 Conclusion

Currently, silk is experiencing rapid development in biotechnological and medical applications. Due to its outstanding biocompatibility, biodegradability, and mechanical strength, silk is gaining momentum for state-of-the-art applications, including, but not limited to, tissue engineering, drug delivery systems, wound healing, and biosensing. New genetic engineering and innovative techniques are opening doors to silk's potential, paving the way for tailored properties in targeted medical applications. With ongoing research, silk-based materials are likely to hold significant importance in the advancement of safer and more efficient biomedical devices and therapies. Ongoing interdisciplinary collaboration will be crucial to realizing the full potential of silk and transforming the biomedical arena.

AI usage declaration AI tools were utilized **solely for language refinement and grammatical correction** in the preparation of this chapter. The authors **take full responsibility for the accuracy, interpretation, and originality** of the final content presented herein.

References

1. Ki CS, Jin H-J, Park YH (2009) Silk protein as a fascinating biomedical polymer: structural fundamentals and applications. Macromol Res 17:935–942. https://doi.org/10.1007/bf03218639
2. Mu X, Amouzandeh R, Vogts H et al (2023) A brief review on the mechanisms and approaches of silk spinning-inspired biofabrication. Front Bioeng Biotechnol 11:1252499. https://doi.org/10.3389/fbioe.2023.1252499
3. Belbéoch C, Salaün F, Vroman P, Lejeune J (2021) Silkworm and spider silk electrospinning: a review. Environ Chem Lett 19:1737–1763. https://doi.org/10.1007/s10311-020-01147-x
4. Wang Y, Guo J, Zhou L et al (2018) Design, fabrication, and function of silk-based nanomaterials. Adv Funct Mater 28:1805305. https://doi.org/10.1002/adfm.201805305
5. Uddin MG, Allardyce BJ, Rashida N, Rajkhowa R (2021) Mechanical, structural and biodegradation characteristics of fibrillated silk fibres and papers. Int J Biol Macromol 179:20–32. https://doi.org/10.1016/j.ijbiomac.2021.02.211
6. Huang W, Omenetto FG, Kaplan DL et al (2018) Silkworm silk-based materials and devices generated using bio-nanotechnology. Chem Soc Rev 47:6486–6504. https://doi.org/10.1039/c8cs00187a
7. Numata K, Kaplan DL (2010) Silk-based delivery systems of bioactive molecules. Adv Drug Deliv Rev 62:1497–1508. https://doi.org/10.1016/j.addr.2010.03.009
8. Peng X, Deng H, Lv X et al (2024) Influence of spider silk protein structure on mechanical and biological properties for energetic material detection. Molecules 29:1025. https://doi.org/10.3390/molecules29051025
9. Li G, Liu H, Li T, Wang J (2011) Surface modification and functionalization of silk fibroin fibers/fabric toward high performance applications. Mater Sci Eng C 32:627–636. https://doi.org/10.1016/j.msec.2011.12.013
10. Sun W, Yu H, Shen Y et al (2012) Phylogeny and evolutionary history of the silkworm. Sci China Life Sci 55:483–496. https://doi.org/10.1007/s11427-012-4334-7
11. Din MMU, Ahmad N, Salman M, Amin F, Ud-Din S (2024) Exploring the intricate biochemistry of silk protein: from structure to function. Pak J For 72(2):63–68

12. Zhou J, Pan L, Zheng H et al (2023) Molecular evidence of structural changes in silk using unlimited degradation mass spectrometry. ACS Omega 8:34410–34419. https://doi.org/10.1021/acsomega.3c02254
13. Zhong J, Tang Y, Zhou X et al (2021) Using FTIR imaging to investigate silk fibroin-based materials. Methods Mol Biol (Clifton, NJ) 2347:207–219. https://doi.org/10.1007/978-1-0716-1574-4_18
14. Mcgill M, Holland GP, Kaplan DL (2018) Experimental methods for characterizing the secondary structure and thermal properties of silk proteins. Macromol Rapid Commun 40:1800390. https://doi.org/10.1002/marc.201800390
15. Liu X, Zhang K-Q (2014) Silk fiber—molecular formation mechanism, structure-property relationship and advanced applications. In: Oligomerization of Chemical and Biological Compounds. IntechOpen.
16. Asakura T (2024) Structural analysis of silk using solid-state NMR. Magn Reson Lett 4:200111. https://doi.org/10.1016/j.mrl.2024.200111
17. Blamires SJ, Rawal A, Edwards AD et al (2023) Methods for silk property analyses across structural hierarchies and scales. Molecules (Basel, Switzerland) 28:2120. https://doi.org/10.3390/molecules28052120
18. Liu J, Shao R, Ge X et al (2022) Challenges and opportunities of silk protein hydrogels in biomedical applications. Mater Adv 3:2291–2308. https://doi.org/10.1039/d1ma00960e
19. Mcdougall C, Woodcroft BJ, Degnan BM (2016) The widespread prevalence and functional significance of silk-like structural proteins in metazoan biological materials. PLoS One 11:e0159128. https://doi.org/10.1371/journal.pone.0159128
20. Hakimi O, Knight DP, Vollrath F, Vadgama P (2007) Spider and mulberry silkworm silks as compatible biomaterials. Compos Part B 38:324–337. https://doi.org/10.1016/j.compositesb.2006.06.012
21. Dos Santos-Pinto JRA, Hardy JG, Priewalder H et al (2014) Structure and post-translational modifications of the web silk protein spidroin-1 from Nephila spiders. J Proteome 105:174–185. https://doi.org/10.1016/j.jprot.2014.01.002
22. Sirichaisit J, Young RJ, Vollrath F, Brookes VL (2003) Analyis of structure/property relationships in silkworm (Bombyx mori) and spider dragline (Nephila edulis) silks using Raman spectroscopy. Biomacromolecules 4:387–394. https://doi.org/10.1021/bm0256956
23. Nova A, Keten S, Redaelli A et al (2010) Molecular and Nanostructural mechanisms of deformation, strength and toughness of spider silk fibrils. Nano Lett 10:2626–2634. https://doi.org/10.1021/nl101341w
24. Keten S, Buehler MJ (2010) Nanostructure and molecular mechanics of spider dragline silk protein assemblies. J R Soc Interface 7:1709–1721. https://doi.org/10.1098/rsif.2010.0149
25. Keten S, Buehler MJ (2010) Atomistic model of the spider silk nanostructure. Appl Phys Lett 96:153701. https://doi.org/10.1063/1.3385388
26. Keten S, Buehler MJ, Xu Z, Ihle B (2010) Nanoconfinement controls stiffness, strength and mechanical toughness of β-sheet crystals in silk. Nat Mater 9:359–367. https://doi.org/10.1038/nmat2704
27. Inoue S, Tanaka K, Arisaka F et al (2000) Silk fibroin of Bombyx mori is secreted, assembling a high molecular mass elementary unit consisting of H-chain, L-chain, and P25, with a 6:6:1 molar ratio. J Biol Chem 275(51):40517–40528. https://doi.org/10.1074/jbc.m006897200
28. Jafari MJ, Schmuck B, Arndt T et al (2023) Force-induced structural changes in spider silk Fibers introduced by ATR-FTIR spectroscopy. ACS Appl Polym Mater 5:9433–9444. https://doi.org/10.1021/acsapm.3c01892
29. Lu Q, Zhu H, Zhang C et al (2012) Silk self-assembly mechanisms and control from thermodynamics to kinetics. Biomacromolecules 13:826–832. https://doi.org/10.1021/bm201731e
30. Lin Z, Huang W, Yang D et al (2009) Solution structure of eggcase silk protein and its implications for silk fiber formation. Proc Natl Acad Sci 106:8906–8911. https://doi.org/10.1073/pnas.0813255106

31. Matsumoto A, Kim U-J, Chen J et al (2006) Mechanisms of silk fibroin sol−gel transitions. J Phys Chem B 110:21630–21638. https://doi.org/10.1021/jp056350v
32. Hovanová V, Hovan A, Žoldák G et al (2023) Global analysis of kinetics reveals the role of secondary nucleation in recombinant spider silk self-assembly. Protein Sci: Publ Protein Soc 32. https://doi.org/10.1002/pro.4722
33. Germershaus O, Werner V, Kutscher M, Meinel L (2014) Deciphering the mechanism of protein interaction with silk fibroin for drug delivery systems. Biomaterials 35:3427–3434. https://doi.org/10.1016/j.biomaterials.2013.12.083
34. Altman GH, Diaz F, Jakuba C et al (2003) Silk-based biomaterials. Biomaterials 24:401–416. https://doi.org/10.1016/s0142-9612(02)00353-8
35. Lu W, Yu W, Zhang B et al (2021) Kevlar fibers reinforced straw wastes-polyethylene composites: combining toughness, strength and self-extinguishing capabilities. Compos Part B 223:109117. https://doi.org/10.1016/j.compositesb.2021.109117
36. Oshkovr SA, Eshkoor RA, Taher ST et al (2012) Crashworthiness characteristics investigation of silk/epoxy composite square tubes. Compos Struct 94:2337–2342. https://doi.org/10.1016/j.compstruct.2012.03.031
37. Min S, Teramoto A, Kai T, Abe K (2010) Effects of temperature and humidity on the solubility and degradability of fibroin membrane. J Sericultural Sci Japan 66:387–392. https://doi.org/10.11416/kontyushigen1930.66.387
38. Hu W, Yanagi Y, Hirabayashi K et al (2010) Effect of humidity on the degradation of silk. J Sericultural Sci Japan 56:127–130. https://doi.org/10.11416/kontyushigen1930.56.127
39. Koh L-D, Cheng Y, Teng C-P et al (2015) Structures, mechanical properties and applications of silk fibroin materials. Prog Polym Sci 46:86–110. https://doi.org/10.1016/j.progpolymsci.2015.02.001
40. Temesgen AG, Turşucular ÖF, Turşucular ED (2024) A review on silk Fibers and their medical applications. Recent Prog Mater 06:1–9. https://doi.org/10.21926/rpm.2401008
41. Li AB, Kluge JA, Guziewicz NA et al (2015) Silk-based stabilization of biomacromolecules. J Control Release 219:416–430. https://doi.org/10.1016/j.jconrel.2015.09.037
42. Wani SUD, Masoodi MH, Alshlowi A et al (2022) Silk fibroin as an efficient biomaterial for drug delivery, gene therapy, and wound healing. Int J Mol Sci 23:14421. https://doi.org/10.3390/ijms232214421
43. Seib FP, Rnjak-Kovacina J, Jones GT et al (2013) pH-dependent anticancer drug release from silk nanoparticles. Adv Healthc Mater 2:1606–1611. https://doi.org/10.1002/adhm.201300034
44. Sah MK, Pramanik K (2010) Regenerated silk fibroin from B. Mori silk cocoon for tissue engineering applications. Int J Environ Sci Dev 1:404–408. https://doi.org/10.7763/ijesd.2010.v1.78
45. Sobajo C, Behzad F, Yuan X-F, Bayat A (2008) Silk: a potential medium for tissue engineering. Eplasty 8:438–446
46. Zhang F, Lu Q, Kaplan DL et al (2012) Mechanisms and control of silk-based electrospinning. Biomacromolecules 13:798–804. https://doi.org/10.1021/bm201719s
47. Phillips DM, Long HCD, Drummy LF et al (2005) Regenerated silk fiber wet spinning from an ionic liquid solution. J Mater Chem 15:4206. https://doi.org/10.1039/b510069k
48. Chawla S, Ghosh S, Midha S, Sharma A (2018) Silk-based bioinks for 3D bioprinting. Adv Healthc Mater 7:1701204. https://doi.org/10.1002/adhm.201701204
49. Wei L, Wu S, Kuss M et al (2019) 3D printing of silk fibroin-based hybrid scaffold treated with platelet rich plasma for bone tissue engineering. Bioact Mater 4:256–260. https://doi.org/10.1016/j.bioactmat.2019.09.001
50. Liu Q, Chen L, Liu H et al (2025) Promotion of bone defect repairs using multiscale 3D printed silk porous hydrogel scaffolds. Acta Biomater 198:24–36. https://doi.org/10.1016/j.actbio.2025.04.027
51. Mieszawska AJ, Fourligas N, Georgakoudi I et al (2010) Osteoinductive silk–silica composite biomaterials for bone regeneration. Biomaterials 31:8902–8910. https://doi.org/10.1016/j.biomaterials.2010.07.109

52. Fazal N, Latief N (2018) Bombyx mori derived scaffolds and their use in cartilage regeneration: a systematic review. Osteoarthr Cartil 26:1583–1594. https://doi.org/10.1016/j.joca.2018.07.009
53. Xiao M, Sun L, Wu K et al (2025) A sequential drug delivery system based on silk fibroin scaffold for effective cartilage repair. Bioact Mater 49:255–270. https://doi.org/10.1016/j.bioactmat.2025.03.005
54. Yonesi M, González-Nieto D, Guinea GV et al (2021) Silk fibroin: an ancient material for repairing the injured nervous system. Pharmaceutics 13:429. https://doi.org/10.3390/pharmaceutics13030429
55. Doblado LR, Pradas MM, Martínez-Ramos C (2021) Biomaterials for neural tissue engineering. Front Nanotechnol 3. https://doi.org/10.3389/fnano.2021.643507
56. Wongpinyochit T (2014) Silk nanoparticles for drug delivery applications. Master's thesis, University of Strathclyde, Glasgow, UK.
57. Howard FH, Mansor HB, Muthana M et al (2023) Silk fibroin nanoparticles: a biocompatible multi-functional polymer for drug delivery. Intechopen
58. Niu L, Chen G, Feng Y et al (2021) Polyethylenimine-modified Bombyx mori silk fibroin as a delivery carrier of the ING4-IL-24 Coexpression plasmid. Polymers 13:3592. https://doi.org/10.3390/polym13203592
59. Wongpinyochit T, Totten JD, Johnston BF, Seib FP (2019) Microfluidic-assisted silk nanoparticle tuning Nanoscale advances 1:873–883. https://doi.org/10.1039/c8na00208h
60. Hu D, Li T, Xu Z et al (2018) Self-stabilized silk sericin-based nanoparticles: in vivo biocompatibility and reduced doxorubicin-induced toxicity. Acta Biomater 74:385–396. https://doi.org/10.1016/j.actbio.2018.05.024
61. Florczak A, Marszalek A, Kramer E et al (2020) Functionalized silk spheres selectively and effectively deliver a cytotoxic drug to targeted cancer cells in vivo. J Nanobiotechnol 18. https://doi.org/10.1186/s12951-020-00734-y
62. Wu J (2021) The enhanced permeability and retention (EPR) effect: the significance of the concept and methods to enhance its application. J Pers Med 11:771. https://doi.org/10.3390/jpm11080771
63. Vidya M, Rajagopal S (2021) Silk fibroin: a promising tool for wound healing and skin regeneration. Int J Polym Sci 2021:1–10. https://doi.org/10.1155/2021/9069924
64. Kochhar D, Abbott RD, Debari MK (2021) The Materiobiology of silk: exploring the biophysical influence of silk biomaterials on directing cellular Behaviors. Front Bioeng Biotechnol 9. https://doi.org/10.3389/fbioe.2021.697981
65. Aramwit P, Muangman P, Palapinyo S et al (2013) Silk sericin ameliorates wound healing and its clinical efficacy in burn wounds. Arch Dermatol Res 305:585–594. https://doi.org/10.1007/s00403-013-1371-4
66. Ersel M, Yigitturk G, Ozcete E et al (2016) Effects of silk Sericin on incision wound healing in a dorsal skin flap wound healing rat model. Med Sci Monit: Int Med J Exp Clin Res 22:1064–1078. https://doi.org/10.12659/msm.897981
67. Setooni Z, Simi F, Bargahi A et al (2018) Evaluation of wound dressing made from spider silk protein using in a rabbit model. Int J Low Extrem Wounds 17:71–77. https://doi.org/10.1177/1534734618782827
68. Tian Z, Chen H, Zhao P (2025) Compliant immune response of silk-based biomaterials broadens application in wound treatment. Front Pharmacol 16. https://doi.org/10.3389/fphar.2025.1548837
69. Downer M, Parker JB, Griffin M et al (2023) Current biomaterials for wound healing. Bioengineering 10:1378. https://doi.org/10.3390/bioengineering10121378
70. Reeves ARD, Spiller KL, Freytes DO et al (2015) Controlled release of cytokines using silk-biomaterials for macrophage polarization. Biomaterials 73:272–283. https://doi.org/10.1016/j.biomaterials.2015.09.027
71. Lehmann T, Liechty KW, Seal S et al (2022) Silk fibroin-based therapeutics for impaired wound healing. Pharmaceutics 14:651. https://doi.org/10.3390/pharmaceutics14030651

72. Aad R, Dragojlov I, Vesentini S (2024) Sericin protein: structure, properties, and applications. J Funct Biomater 15:322. https://doi.org/10.3390/jfb15110322
73. Sheng JY, Sun DQ, Zhuang Y et al (2013) Study on the application of Sericin in cosmetics. Adv Mater Res 796:416–423. https://doi.org/10.4028/www.scientific.net/amr.796.416
74. Polaskova J, Pavlackova J, Egner P (2015) Effect of vehicle on the performance of active moisturizing substances. Skin Res Technol 21:403–412. https://doi.org/10.1111/srt.12206
75. Mazurek Ł, Szudzik M, Konop M, Rybka M (2022) Silk fibroin biomaterials and their beneficial role in skin wound healing. Biomolecules 12:1852. https://doi.org/10.3390/biom12121852
76. Tariq M, Nadeem J, Raza C et al (2021) Silk derived formulations for accelerated wound healing in diabetic mice. Peer J 9:e10232. https://doi.org/10.7717/peerj.10232
77. SYLKE® Adhesive Wound Closure. In: SYLKE®. https://www.sylke.com/. Accessed 25 Aug 2025
78. Kamalathevan P, Ooi PS, Loo YL (2018) Silk-based biomaterials in cutaneous wound healing: a systematic review. Adv Skin Wound Care 31:565–573. https://doi.org/10.1097/01.asw.0000546233.35130.a9
79. Zhang W, Zeng M, Ouyang H et al (2017) Silk fibroin biomaterial shows safe and effective wound healing in animal models and a randomized controlled clinical trial. Adv Healthc Mater 6:1700121. https://doi.org/10.1002/adhm.201700121
80. Gil ES, Kaplan DL, Panilaitis B, Bellas E (2012) Functionalized silk biomaterials for wound healing. Adv Healthc Mater 2:206–217. https://doi.org/10.1002/adhm.201200192
81. Kanokpanont S, Ratanavaraporn J, Aramwit P, Damrongsakkul S (2012) An innovative bi-layered wound dressing made of silk and gelatin for accelerated wound healing. Int J Pharm 436:141–153. https://doi.org/10.1016/j.ijpharm.2012.06.046
82. Schiefer JL, Andreae J, Bagheri M et al (2021) A clinical comparison of pure knitted silk and a complex synthetic skin substitute for the treatment of partial thickness burns. Int Wound J 19:178–187. https://doi.org/10.1111/iwj.13613
83. Qadir J, Afreen S, Pavankumar S, Aryan S (2024) Potential of silk fibroin in biomedical science. Int J Adv Biochem Res 8:1179–1190. https://doi.org/10.33545/26174693.2024.v8.i8o.1947
84. Koczoń P, Maj K, Masztakowski J et al (2023) Applications of silk fibroin in human and veterinary medicine. Materials 16:7128. https://doi.org/10.3390/ma16227128
85. Babu PJ, Suamte L (2023) Applications of silk-based biomaterials in biomedicine and biotechnology. Eng Regen 5:56–69. https://doi.org/10.1016/j.engreg.2023.11.002
86. Zakeri Siavashani A, Huber L, Mohammadi J et al (2020) Silk based scaffolds with immunomodulatory capacity: anti-inflammatory effects of nicotinic acid. Biomater Sci 8:148–162. https://doi.org/10.1039/c9bm00814d
87. Apte A, Liechty KW, Zgheib C (2023) Immunomodulatory biomaterials on chemokine signaling in wound healing. Front Pharmacol 14. https://doi.org/10.3389/fphar.2023.1084948
88. Dinerman AA, Cappello J, Ghandehari H, Hoag SW (2002) Swelling behavior of a genetically engineered silk-elastinlike protein polymer hydrogel. Biomaterials 23:4203–4210. https://doi.org/10.1016/s0142-9612(02)00164-3
89. Li J, Mooney DJ (2016) Designing hydrogels for controlled drug delivery. Nat Rev Mater 1. https://doi.org/10.1038/natrevmats.2016.71
90. Yao D, Wang T, Li M et al (2021) Viscoelastic silk fibroin hydrogels with Tunable strength. ACS Biomater Sci Eng 7:636–647. https://doi.org/10.1021/acsbiomaterials.0c01348
91. Hansma HG, Hansma PK, Keller RA (1993) Potential applications of atomic force microscopy of DNA to the human genome project, Proc SPIE - Int. Soc Opt Eng (USA). 1891:66–70
92. Zhang Y, Sun Q, Wang Z et al (2023) Dynamic hydrogels with viscoelasticity and Tunable stiffness for the regulation of cell behavior and fate. Materials 16:5161. https://doi.org/10.3390/ma16145161
93. Aramwit P, Sangcakul A (2007) The effects of Sericin cream on wound healing in rats. Biosci Biotechnol Biochem 71:2473–2477. https://doi.org/10.1271/bbb.70243

94. Schäfer S, Aavani F, Köpf M, et al (2022) Silk proteins in reconstructive surgery: do they possess an inherent antibacterial activity? A systematic review. Wound repair and regeneration: official publication of the Wound Healing Society [and] the European Tissue Repair Society 31:99–110. https://doi.org/https://doi.org/10.1111/wrr.13049
95. Ghalei S, Handa H (2021) A review on antibacterial silk fibroin-based biomaterials: current state and prospects. Mater Today Chem 23:100673. https://doi.org/10.1016/j.mtchem.2021.100673
96. Zheng H, Huang Z, Chen T et al (2022) Gallium ions incorporated silk fibroin hydrogel with antibacterial efficacy for promoting healing of Pseudomonas aeruginosa-infected wound. Front Chem 10. https://doi.org/10.3389/fchem.2022.1017548
97. Bakadia BM, Lamboni L, Qaed Ahmed AA et al (2022) Antibacterial silk sericin/poly (vinyl alcohol) hydrogel with antifungal property for potential infected large burn wound healing: systemic evaluation. Smart Mater Med 4:37–58. https://doi.org/10.1016/j.smaim.2022.07.002
98. Estévez-Martínez Y, Villarreal-Gómez LJ, Vázquez Mora R et al (2023) Antibacterial nanocomposite of chitosan/silver nanocrystals/graphene oxide (ChAgG) development for its potential use in bioactive wound dressings. Sci Rep 13. https://doi.org/10.1038/s41598-023-29015-y
99. Alven S, Aderibigbe BA (2024) Chitosan-based scaffolds incorporated with silver nanoparticles for the treatment of infected wounds. Pharmaceutics 16:327. https://doi.org/10.3390/pharmaceutics16030327
100. Reyes-Guzmán VL, Villarreal-Gómez LJ, Vázquez-Mora R et al (2024) Integrating an antimicrobial nanocomposite to bioactive electrospun fibers for improved wound dressing materials. Sci Rep 14. https://doi.org/10.1038/s41598-024-75814-2
101. Meng X, Zhu F, Chen K (2017) Silkworm: a promising model organism in life science. J Insect Sci 17(5):97. https://doi.org/10.1093/jisesa/iex064
102. Xu H, O'Brochta DA (2015) Advanced technologies for genetically manipulating the silkworm Bombyx mori, a model lepidopteran insect. Proc R Soc B Biol Sci 282:20150487. https://doi.org/10.1098/rspb.2015.0487
103. Kim M-J, Hong S-J, Yang J, Kim H-K (2007) Silkworm (Bombyx mori L.) reduces vasopressin expression in the hypothalamus of streptozotocin-induced diabetic mice. Neurol Res 29:72–77. https://doi.org/10.1179/016164107x172374
104. Matsumoto Y, Sumiya E, Sugita T, Sekimizu K (2011) An invertebrate Hyperglycemic model for the identification of anti-diabetic drugs. PLoS One 6:e18292. https://doi.org/10.1371/journal.pone.0018292
105. Matsumoto Y, Ishii M, Ishii K et al (2014) Transgenic silkworms expressing human insulin receptors for evaluation of therapeutically active insulin receptor agonists. Biochem Biophys Res Commun 455:159–164. https://doi.org/10.1016/j.bbrc.2014.10.143
106. Matsumoto Y, Miyazaki S, Sumiya E et al (2015) Diabetic silkworms for evaluation of therapeutically effective drugs against type II diabetes. Sci Rep 5. https://doi.org/10.1038/srep10722
107. Matsumoto Y, Sekimizu K (2016) Evaluation of anti-diabetic drugs by using silkworm, Bombyx mori. Drug Discov Ther 10:19–23. https://doi.org/10.5582/ddt.2016.0101
108. Wang Y, Du T, Li A et al (2022) Establishment and application of a silkworm CRISPR/Cas9 tool for conditionally manipulating gene disruption in the epidermis. Insect Biochem Mol Biol 151:103861. https://doi.org/10.1016/j.ibmb.2022.103861
109. Li Y, Wang J, Tian J et al (2012) Transcriptome analysis of the silkworm (Bombyx mori) by high-throughput RNA sequencing. PLoS One 7:e43713. https://doi.org/10.1371/journal.pone.0043713
110. Tong X, Xiang Z, Xu J et al (2022) High-resolution silkworm pan-genome provides genetic insights into artificial selection and ecological adaptation. Nat Commun 13. https://doi.org/10.1038/s41467-022-33366-x
111. Ye J, Li J, Zhao P (2023) The silkworm carboxypeptidase inhibitor prevents gastric cancer cells' proliferation through the EGF/EGFR Signaling pathway. Int J Mol Sci 24:1078. https://doi.org/10.3390/ijms24021078

112. Yamada K, Tatematsu K-I, Ida S et al (2021) In vitro assessment of antitumor immune responses using tumor antigen proteins produced by transgenic silkworms. J Mater Sci Mater Med 32. https://doi.org/10.1007/s10856-021-06526-6
113. Blake S, Kim NY, Kong N et al (2021) Silk's cancer applications as a biodegradable material. Mater Today Sustain 13:100069. https://doi.org/10.1016/j.mtsust.2021.100069
114. Kaewkorn W, Sutheerawattananonda M, Tiyaboonchai W et al (2012) Effects of silk sericin on the proliferation and apoptosis of colon cancer cells. Biol Res 45:45–50. https://doi.org/10.4067/s0716-97602012000100006
115. Wang M-S, Huang H-M, Zhao H-P et al (2018) Silk fibroin peptide suppresses proliferation and induces apoptosis and cell cycle arrest in human lung cancer cells. Acta Pharmacol Sin 40:522–529. https://doi.org/10.1038/s41401-018-0048-0
116. Ito T, Arai N (1966) Nutrition of the silkworm, Bombyx mori—XI. Requirements for aspartic and glutamic acids. J Insect Physiol 12:861–869. https://doi.org/10.1016/0022-1910(66)90036-9
117. Murugesh KA, Aruna R, Chozhan K (2021) Influence of amino acids on the economic characters of silkworm, Bombyx mori L. Madras Agric J 108. https://doi.org/10.29321/maj.10.000516
118. Cao J, Tao C, Qin X et al (2024) PI3K-Akt-SGF1-Dimm pathway mediates the nutritional regulation of silk protein synthesis in Bombyx mori. Int J Biol Macromol 278:134650. https://doi.org/10.1016/j.ijbiomac.2024.134650
119. Wei H, Rui J, Yan X et al (2025) Plant polyphenols as natural bioactives for alleviating lipid metabolism disorder: mechanisms and application challenges. Food Res Int (Ottawa, Ont) 203:115682. https://doi.org/10.1016/j.foodres.2025.115682
120. Lee SH, Yang G, Park D et al (2011) Silk and silkworm pupa peptides suppress adipogenesis in preadipocytes and fat accumulation in rats fed a high-fat diet. Eur J Nutr 51:1011–1019. https://doi.org/10.1007/s00394-011-0280-6
121. Nesa J, Mandal AK, Kati A et al (2020) Antimicrobial peptides from Bombyx mori: a splendid immune defense response in silkworms. RSC Adv 10:512–523. https://doi.org/10.1039/c9ra06864c
122. Montali A, Cappellozza S, Marinelli F et al (2022) A Bombyx mori infection model for screening antibiotics against Staphylococcus epidermidis. Insects 13:748. https://doi.org/10.3390/insects13080748
123. Montali A, Mastore M, Brivio MF et al (2020) A silkworm infection model for in vivo study of Glycopeptide antibiotics. Antibiotics 9:300. https://doi.org/10.3390/antibiotics9060300
124. Kaito C, Akimitsu N, Watanabe H, Sekimizu K (2002) Silkworm larvae as an animal model of bacterial infection pathogenic to humans. Microb Pathog 32:183–190. https://doi.org/10.1006/mpat.2002.0494
125. Panthee S, Sekimizu K, Paudel A, Hamamoto H (2017) Advantages of the silkworm As an animal model for developing novel antimicrobial agents. Front Microbiol 8. https://doi.org/10.3389/fmicb.2017.00373
126. Tan H, Ji Y, Lei H et al (2023) Large-scale and cost-effective production of recombinant human serum albumin (rHSA) in transgenic Bombyx mori cocoons. Int J Biol Macromol 245:125527. https://doi.org/10.1016/j.ijbiomac.2023.125527
127. Chen J, Zhang Y-Z, Wu X-F (2006) Expression, purification and characterization of human GM-CSF using silkworm pupae (Bombyx mori) as a bioreactor. J Biotechnol 123:236–247. https://doi.org/10.1016/j.jbiotec.2005.11.015
128. Maeda S, Obinata M, Furusawa M et al (1985) Production of human alpha-interferon in silkworm using a baculovirus vector. Nature 315:592–594. https://doi.org/10.1038/315592a0
129. Ye L, You Z, Qian Q et al (2015) TAL effectors mediate high-efficiency transposition of the piggyBac transposonin silkworm Bombyx mori L. Sci Rep 5. https://doi.org/10.1038/srep17172
130. Zhou W, Xue R, Ye A et al (2012) Germline transformation of the silkworm Bombyx mori L. by sperm-mediated gene Transfer1. Biol Reprod 87. https://doi.org/10.1095/biolreprod.112.102822

131. Ye X, Tang X, Zhao S et al (2021) Mechanism of the growth and development of the posterior silk gland and silk secretion revealed by mutation of the fibroin light chain in silkworm. Int J Biol Macromol 188:375–384. https://doi.org/10.1016/j.ijbiomac.2021.08.009
132. Li S, Wang F, Shen G et al (2024) CRISPR/Cas9-mediated editing of BmEcKL1 gene sequence affected silk gland development of silkworms (Bombyx mori). Int J Mol Sci 25:1907. https://doi.org/10.3390/ijms25031907
133. Liu J, Yang Y, Yang Q et al (2025) Successful oral RNA interference efficiency in the silkworm Bombyx mori through nanoparticle-shielded dsRNA delivery. J Insect Physiol 161:104749. https://doi.org/10.1016/j.jinsphys.2025.104749
134. Yang H, Qin X, Guo Y et al (2025) Bmgsb directly activates Bmubxn-4 to inhibit the DNA endoreplication and affect the cell fate in the silk gland of Bombyx mori. Int J Biol Macromol 308:142335. https://doi.org/10.1016/j.ijbiomac.2025.142335
135. Feng W, Wan Y-J, Zhou W et al (2011) Isolation and characterization of lipase-producing bacteria in the intestine of the silkworm, *Bombyx mori*, reared on different forage. J Insect Sci 11:1–10. https://doi.org/10.1673/031.011.13501
136. Ling L, Ge X, Li Z et al (2014) MicroRNA Let-7 regulates molting and metamorphosis in the silkworm, Bombyx mori. Insect Biochem Mol Biol 53:13–21. https://doi.org/10.1016/j.ibmb.2014.06.011
137. Wang M-X, Cai Z-Z, Niu Y-S et al (2012) Phenol oxidase is a necessary enzyme for the silkworm molting which is regulated by molting hormone. Mol Biol Rep 40:3549–3555. https://doi.org/10.1007/s11033-012-2428-8
138. Kalpana S, Lakshmanaperumalsamy P, Hatha AAM (1994) Gut microflora of the larva of silkworm, Bombyx mori. Int J Trop Insect Sci 15:499–502. https://doi.org/10.1017/s1742758400015873
139. Shen Y, Zhao S, Chen G, Wu X (2022) Genome-wide identification of lipases in silkworm (Bombyx mori) and their spatio-temporal expression in larval midgut. Gene 813:146121. https://doi.org/10.1016/j.gene.2021.146121
140. Gao R, Li C, Zhou A et al (2024) QTL analysis to identify genes involved in the trade-off between silk protein synthesis and larva-pupa transition in silkworms. Genet Sel Evol 56. https://doi.org/10.1186/s12711-024-00937-z
141. Li M, Sun P, Hou C et al (2005) Genetic diversity among silkworm (*Bombyx mori*L., Lep., Bombycidae) germplasms revealed by microsatellites. Genome 48:802–810. https://doi.org/10.1139/g05-053
142. Hou CX, Huang YP, Sun PJ et al (2013) Marker-assisted selection in breeding silkworm strains with high silk production and resistance to the densonucleosis virus. Genet Mol Res 12:4171–4178. https://doi.org/10.4238/2013.october.7.3
143. Kankanawadi N, As T, Aramani K et al (2024) Biotechnological advances in sericulture: improving silkworm breeding and disease management. Int J Adv Biochem Res 8:1093–1097. https://doi.org/10.33545/26174693.2024.v8.i11n.3030
144. Yang C-C, Yamamoto K, Yokoi K, Jouraku A (2021) An update of KAIKObase, the silkworm genome database. Database 2021:baaa099. https://doi.org/10.1093/database/baaa099
145. Jeong KY, Lee J-H, Park KH et al (2017) Role of tropomyosin in silkworm allergy. Mol Med Rep 15:3264–3270. https://doi.org/10.3892/mmr.2017.6373
146. He W, Li S, He K et al (2021) Identification of potential allergens in larva, pupa, moth, silk, slough and feces of domestic silkworm (Bombyx mori). Food Chem 362:130231. https://doi.org/10.1016/j.foodchem.2021.130231
147. Mei X, Huang T, Liu W et al (2024) BmC/EBPZ gene is essential for the larval growth and development of silkworm, Bombyx mori. Front Physiol 15:1298869. https://doi.org/10.3389/fphys.2024.1298869
148. Chen T, Ma Y, Huang L et al (2020) A transcriptome atlas of silkworm silk glands revealed by PacBio single-molecule long-read sequencing. Mol Gen Genomics 295:1227–1237. https://doi.org/10.1007/s00438-020-01691-9
149. Inagaki Y, Matsumoto Y, Ishii M et al (2015) Fluorescence imaging for a noninvasive in vivo toxicity-test using a transgenic silkworm expressing green fluorescent protein. Sci Rep 5:11180. https://doi.org/10.1038/srep11180

Chapter 5
Properties, Functions, and Biomedical Applications of Silk Sericin

Barsha Roy, Poulami Bhadra, Shubhajit Shaw, Pankaj Mandal, and Debnirmalya Gangopadhyay

5.1 Introduction

In sericulture, silk sericin, a natural protein primarily produced by the mulberry silkworm, *Bombyx mori*, constitutes a significant component of the cocoon. Silkworms in their final larval stage spin cocoons composed mainly of two proteins: fibroin and sericin. Fibroin, a hydrophobic fibrous protein, makes up about 70–75% of the cocoon and forms the structural core of the silk thread, whereas sericin, a hydrophilic gum-like protein, surrounds fibroin fibers, acting as a binding agent and accounting for 25–30% of the silk protein [1]. Although non-mulberry or wild silkworms such as Tasar (*Antheraea mylitta, Antheraea pernyi*), Eri (*Samia ricini, Antheraea yamamai*), and Muga (*Antheraea assamensis*) also produce sericin, their production volume is far smaller compared to mulberry silk. During silk processing, sericin is removed in the degumming process to enhance the luster and texture of silk threads, which are subsequently spun into yarns and fabrics. Annually, nearly 4 lakh MT of dried cocoons yield about 0.5 lakh MT of sericin, the majority of which is discarded into effluent streams. This not only represents a significant loss of a valuable biomaterial but also leads to severe water pollution and environmental degradation [2].

Although frequently regarded as waste, sericin exhibits exceptional characteristics that make it highly attractive for biomedical applications. Properties such as biocompatibility, biodegradability, antioxidant and anti-inflammatory activities, immunocompatibility, and resistance to ultraviolet radiation highlight its potential in diverse applications [3]. Its hydrophilic nature, organic composition, solubility, and abundance of polar amino acids enable the formation of hydrogen bonds with water, supporting gel formation for medical and pharmaceutical purposes. Importantly, biocompatibility ensures that sericin does not trigger adverse immune responses, but instead facilitates positive biological interactions [4]. As a biodegradable protein,

B. Roy · P. Bhadra · S. Shaw · P. Mandal · D. Gangopadhyay (✉)
Department of Sericulture, Raiganj University, Raiganj, West Bengal, India
e-mail: dng@raiganjuniversity.ac.in

M. Shahid et al. (eds.), *Science and Technology of Silk*, Textile Science and Clothing Technology, https://doi.org/10.1007/978-3-032-26880-8_5

it naturally breaks down into non-toxic by-products, thereby eliminating the need for removal after therapeutic use and reducing patient discomfort while protecting newly regenerated tissues [5].

Beyond these advantages, sericin demonstrates additional functional properties. Its antioxidant activity neutralizes harmful free radicals, reducing cellular damage; its antibacterial effects inhibit microbial growth and promote wound healing; its UV resistance provides skin protection; and its hydrophilic behavior contributes moisturizing effects. Collectively, these features establish sericin as a promising biomaterial for wound healing, tissue engineering, regenerative medicine, and cosmetic formulations [6]. The use of silk in medicine has a long history, particularly in China and India, where it has been applied in wound care for thousands of years [7]. Silk sutures remain widely used in modern medicine due to their strength, flexibility, and biocompatibility. Both fibroin and sericin proteins have shown therapeutic potential in addressing metabolic, cardiovascular, and neurological disorders. Although the medical relevance of silk is ancient, recent advances in biomedical science have renewed interest in sericin as a resource for next-generation medical devices and regenerative healthcare solutions.

This chapter provides an overview of the fundamental characteristics of sericin, highlighting its unique properties and the reasons it should no longer be regarded as a mere by-product of silk processing. It further explores extraction and purification strategies, functional attributes, and emerging biomedical applications across various fields, including tissue engineering, pharmaceuticals, and cosmetics. By highlighting sericin's reutilization and expanding role in biomedical innovation, this work aims to strengthen the understanding of its potential as a sustainable and versatile biomaterial for future healthcare solutions.

5.2 Global Trends in Sericin-Based Research Studies

Research on sericin, covering its properties, structure, extraction methods, and biomedical applications, has been reported from 1926 to 2025. A meta-analysis based on ScienceDirect data was conducted to track trends in its medicinal use, covering the period from 2007 to 2024 and depicted in Fig. 5.1.

The analysis reveals a steady increase in sericin-focused studies, particularly in the biomedical sciences. Over recent decades, publications have expanded their recognition as a valuable biomaterial in pharmaceutical and biomedical industries, supported by increasingly advanced experimental approaches. The dataset includes review papers, original research, book chapters, and mini-reviews, all of which have enriched scientific knowledge and broadened the scope of sericin studies. These contributions have also strengthened insights into biotechnology and materials science, highlighting sericin's relevance across disciplines. Research activity has concentrated on tissue engineering, wound healing, cosmetics, and drug delivery, with additional interest in food and textiles. Collectively, these findings underscore sericin's growing

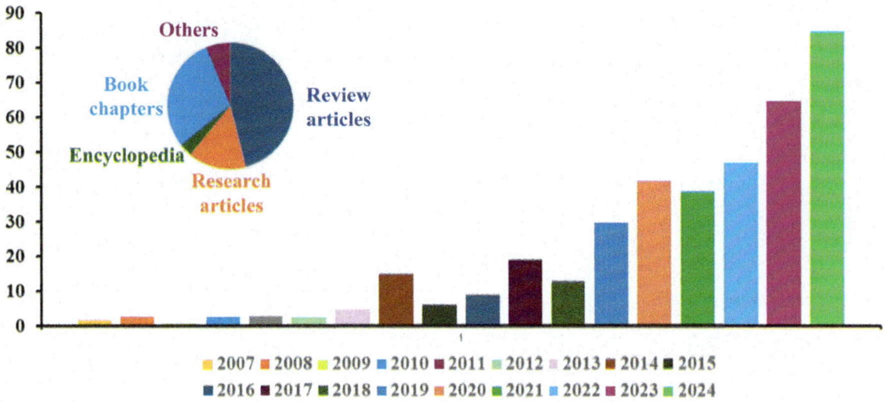

Fig. 5.1 Trend of publications on sericin research (2007–2024); in the inset: types of publications

scientific and sustainable importance, reflecting its multidisciplinary impact and its promise for future applications in medicine and biomedical engineering.

5.3 Chemical Composition and Structural Properties of Silk Sericin

Sericin binds two fibroin filaments together. It is a glue-like, globular protein composed of random coils and β-sheets. The transition between these structures is affected by temperature, humidity, and mechanical stress. At temperatures above 50 °C, sericin dissolves in water; however, at lower temperatures, its solubility decreases, promoting the transformation of random coils into β-sheets and leading to gel formation [8]. The addition of poloxamers accelerates this transition, with gelation occurring more rapidly at higher temperatures and sericin concentrations. Chromatographic analyses further indicate that sericin is soluble in water in its native glandular form but becomes less soluble after spinning and drying, due to the formation of β-sheets during these processes.

Chemically, sericin is hydrophilic, enriched with hydroxyl and carboxyl groups, and contains 18 amino acids, of which serine is most abundant. Other amino acids include alanine, glycine, valine, threonine, and lysine, among others [9]. In *B. mori*, sericin is produced in paired labial glands, anatomically divided into anterior, middle, and posterior silk glands (Fig. 5.2). The middle silk gland secretes three types of sericin, namely Sericin A, forming the outermost water-soluble layer; Sericin B, the intermediate layer; and Sericin C, the innermost layer, insoluble in hot water and lower in nitrogen [10].

Extraction methods, including degumming, enzymatic, chemical, and thermal processes, affect sericin yield, molecular weight, and functional properties. Its molecular weight ranges from 20 to 400 kDa, with lower molecular weight fractions being

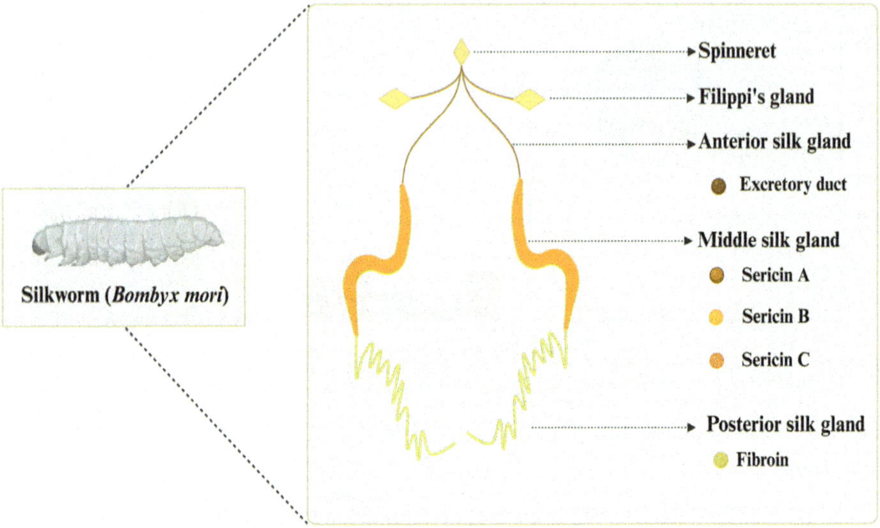

Fig. 5.2 The structure of the silk gland of *Bombyx mori*

suited for cosmetics and pharmaceuticals, while higher molecular weight fractions are useful in biomedical engineering, hydrogels, and membranes [11]. Functional groups, such as hydroxyl, carboxyl, and amide groups, enable crosslinking, copolymerization, or blending with natural and synthetic polymers, thereby enhancing the mechanical properties of sericin-based biomaterials [12].

The physicochemical attributes of sericin, including solubility, thermal stability, biodegradability, and biocompatibility, support its biomedical relevance. It is readily soluble in hot water; degrades into non-toxic by-products; and can be processed into films, fibers, hydrogels, or nanoparticles for use in drug delivery, wound dressings, tissue engineering, and other medical applications. Thermogravimetric studies reveal that sericin stability varies across species, with non-mulberry sericins, particularly from *S. ricini*, showing higher thermal stability than mulberry sericin [13]. Moreover, the presence of proteolytic enzymes facilitates sericin degradation in wound healing, while its non-toxic and non-immunogenic nature ensures compatibility with living tissues, reinforcing its role as a promising biomaterial in biomedical engineering.

5.4 Extraction and Purification of Sericin for Biomedical Applications

5.4.1 Extraction Methods

The extraction method of sericin from mulberry and non-mulberry cocoons strongly influences its physicochemical and biological properties [13]. Different approaches are presented below and summarized in Table 5.1.

5.4.1.1 Traditional Degumming Methods

Hot water extraction is the oldest and most common technique. In this method, the gum-like sericin dissolves in water, releasing fibroin fibers to produce raw silk yarns. However, the purity of sericin is generally lower compared with chemical, thermal, or biological methods [11]. In the silk industry, sericin has been removed using soap (e.g., Marseille), detergents, or sodium carbonate (Na_2CO_3) solutions. Although this process produces clean fibroin for textiles, the recovered sericin is often degraded, resulting in a reduced molecular weight and impaired functional properties [21]. Residual soap also contaminates the extract, limiting its biomedical and pharmaceutical applications [22]. Thermal methods utilize boiling water (80–100 °C) under atmospheric or pressurized conditions, sometimes employing autoclaving [19]. Heat and pressure disrupt hydrogen bonding, facilitating the separation of fibroin and sericin. Extraction parameters, such as duration, temperature, and pressure, strongly affect the molecular weight of sericin [16]. Acids (citric, tartaric, and succinic) and bases (sodium carbonate, phosphate, silicate, and hydrosulfite) are widely used to hydrolyze sericin, breaking peptide bonds and releasing soluble protein [23, 24]. Urea with 2-mercaptoethanol can extract up to 95% of sericin without structural damage, though the method is costly, time consuming, and may induce cytotoxicity. Chemical extraction generally requires additional purification to remove residual chemicals [25]. Proteolytic enzymes such as alcalase, savinase, papain, and trypsin have also been applied to degum cocoons. These enzymes cleave peptide bonds between amino acids, releasing sericin. However, the yield is relatively low and the extracted protein is often degraded [26–28].

5.4.1.2 Modern Extraction Methods

Recent techniques aim to improve sustainability and protein quality. Infrared heating, microwave, ultrasound, steam treatment, and supercritical CO_2 extraction reduce water use and energy consumption while producing sericin of varying molecular weights [19, 29]. Infrared heating achieves complete removal with high protein quality, while microwave degumming requires less time and avoids chemical

Table 5.1 Comparative analysis of different extraction methods

Degumming methods	Approaches	Advantages	Disadvantages	References
Physical	Hot water	Use of natural polymers	Extracted sericin contains low purity. Sericin can damage fibroin fibers	[11, 14, 15]
Conventional	Detergents, soaps (e.g., Marseille), sodium bicarbonate	Simple process. It is effective in extraction of sericin	Sericin is highly degraded. Recovery of sericin is difficult Time-consuming process. It is not environment friendly and causes effluent problems	[1, 16]
Thermal	Boiling of water, HTHP (high temperature high pressure by autoclaving)	Simple and low cost. No purification steps are required. Environmentally friendly; therefore, causes no effluent problems	Sericin is degraded when used in high temperature. Time-consuming process. By removing the outer layer of sericin it can damage the fibroin fibers which produces raw silk yarns	[16, 17]
Chemical	Alkaline solutions, acid solutions, and urea (with or without 2-mercaptoethanol)	It is low-cost use with the application on large scale. High yield. Effective. Lower degree of degradation in acidic degumming process	Sericin is highly degraded in alkaline degumming method. Recovery of sericin is difficult. Purifications cause effluent problems. EPS is needed to remove chemical contamination. Toxic to cells. This process is not environmentally friendly	[16, 17]

(continued)

Table 5.1 (continued)

Degumming methods	Approaches	Advantages	Disadvantages	References
Biological	Proteolytic enzymes	Effective. Completely removes sericin. Saves water, chemicals, and energy. Environmentally friendly, no effluent problem	Expensive. Time consuming. Degrades sericin extensively	[16, 17, 18]
Modern	Infrared heating, microwave, ultrasound, steam treatment, and carbon dioxide supercritical fluid	High yield and improved quality of silk sericin. Lower energy and water consumption. Short extraction time. Increased efficiency in lower temperature	Extra equipment needed. Limited commercial viability. Higher equipment cost. Some extraction techniques required chemicals, which are not well established in the industry	[1, 19, 20]

residues. Steam treatment offers chemical-free extraction with preserved physicochemical properties. Ultrasound enhances degumming at lower temperatures, and supercritical CO_2 provides high-purity sericin suitable for biomedical and pharmaceutical applications.

5.4.2 *Purification Methods*

Following degumming, purification and isolation are necessary to remove insoluble fibroin and other impurities from sericin, ensuring its suitability for medical applications. These steps also help evaluate extraction efficiency [9]. Initial separation is often achieved by paper filtration and centrifugation. Precipitation methods, including chemical coagulation, organic solvent precipitation, acidulation, and salting out, are also used to solidify and separate sericin. However, these approaches typically introduce chemicals (e.g., citric acid, sodium bicarbonate, urea), pose risks of secondary pollution, and yield relatively low recovery rates of about 40%. Dialysis can further purify chemically extracted sericin.

Advanced purification techniques include gel chromatography and membrane-based filtration (ultrafiltration, nanofiltration), which separate proteins by molecular size and achieve recovery and purity levels above 80%. Once isolated, sericin is dried into powder or solid form through lyophilization or spray drying. Overall, purification and isolation reduce sericin degradation while improving purity, yield, and functional efficiency, making the protein highly valuable for biomedical and pharmaceutical applications [1].

5.5 Functional Properties of Sericin

Sericin exhibits diverse biological and physicochemical properties that make it valuable in biomedicine, cosmetics, and textiles. Its richness in hydrophilic amino acids such as glycine, serine, and aspartic acid enhances its use in hydrogels, films, and sponges for biomedical and cosmetic applications. The biomedical applications of silk sericin, including its biocompatibility, antimicrobial, antioxidant, and moisture retention properties in different forms such as hydrogels, films, scaffolds, nanoparticles, microparticles, fibers, conduits, and devices, are presented in Fig. 5.3. The main functional properties are outlined below.

5.5.1 Biocompatibility and Non-toxicity

Biocompatibility is the most critical feature of sericin for medical use. It is immunologically inert, does not activate macrophages, and shows minimal cytotoxicity in cultured cells, confirming its safety for biomedical applications [1, 30, 31]. Studies with different silkworm strains reported that sericin decreased IFN-γ secretion without affecting TNFα or IL-10 release [32, 33]. Topical application of sericin-based creams in rat wound models reduced inflammatory cytokines such as IL-1β and TNFα, supporting its role in wound healing [32, 34]. Low-molecular-weight sericin (<10 kDa), obtained by high-temperature and pressure extraction, showed good compatibility with macrophages and human adipose stem cells. It enhanced wound repair by upregulating adhesion-related and wound healing genes [34]. Sericin is also biodegradable, allowing silk-based biomaterials to degrade naturally after use. This eliminates the need for removal from wound sites, reducing patient discomfort and protecting regenerating tissue. These qualities make sericin suitable for drug delivery, wound dressings, and tissue engineering [33, 35].

5.5.2 Antioxidant Properties

Sericin exhibits significant antioxidant activity, thereby protecting cells from oxidative stress and aging-related damage [36]. Its activity, measured by methods such as DPPH radical scavenging, chemiluminescence, and electron spin resonance, is higher in pigmented cocoon strains than in non-pigmented ones. Flavonoids and the abundance of serine and threonine residues enhance free radical scavenging by

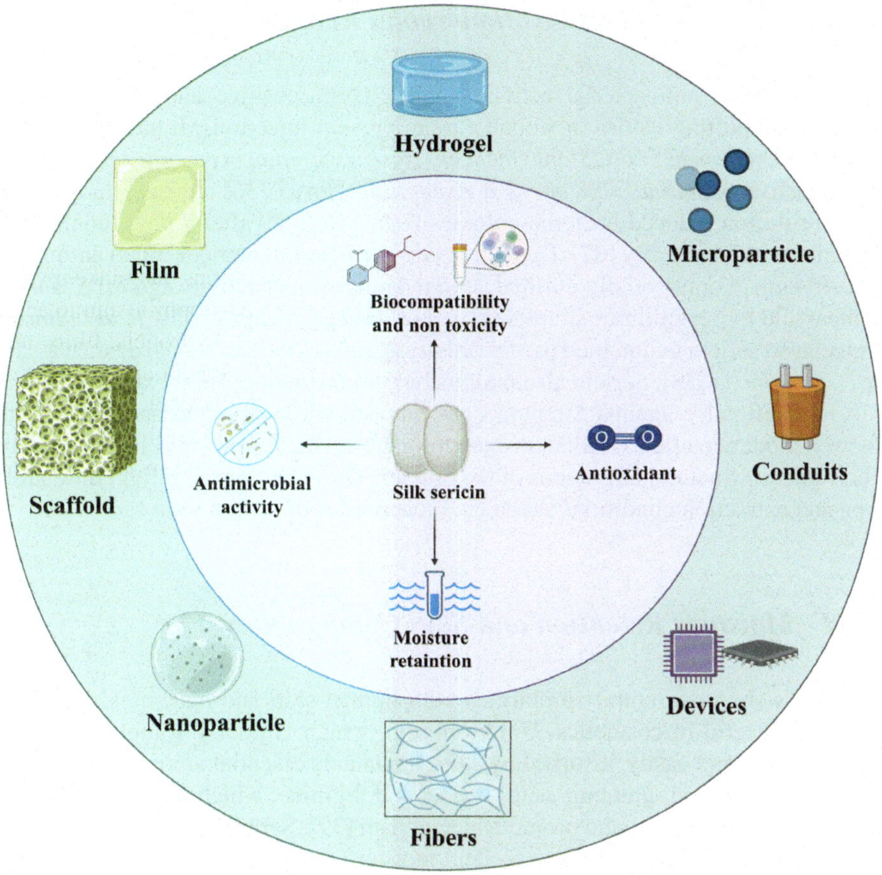

Fig. 5.3 Biomedical applications of silk sericin highlighting its biocompatibility, antimicrobial, antioxidant, and moisture retention properties in various forms such as hydrogels, films, scaffolds, nanoparticles, microparticles, fibers, conduits, and devices

chelating transition metal ions [1]. Extraction methods influence antioxidant potential; autoclaving produces sericin with higher activity than acidic extraction in *B. mori.* Non-mulberry silks, such as *A. assamensis* and *S. ricini,* also yield sericin with strong antioxidant properties using conventional methods. In vivo studies confirmed sericin's antioxidative effects in rat brain homogenates, while in vitro studies demonstrated cryoprotective roles in stem cells, hepatocytes, spermatozoa, and embryonic cells [13].

5.5.3 Antibacterial and Antimicrobial Activity

Sericin exhibits antibacterial activity against Gram-positive and Gram-negative bacteria, supporting its use in wound care to prevent infections. It has been shown to inhibit *Escherichia coli*, *Staphylococcus aureus*, *Streptococcus pneumoniae*, and fungi such as *Candida albicans* and *Aspergillus flavus* [36]. In one study, a 2% sericin solution reduced bacterial colonies from 146 to 29 after 7 days, confirming its antimicrobial potency [37]. Extraction and purification methods affect antimicrobial strength. Commercially purified sericin demonstrated activity against *S. aureus* comparable to penicillin or streptomycin, but weaker effects against *P. aeruginosa*. Autoclaved sericin maintained partial activity against *S. aureus* but not against *E. coli* or *P. aeruginosa* [38]. Sericin also inhibits biofilm formation. Urea-extracted sericin was most effective against Streptococcus mutans, while heat and acid extractions showed moderate effects. Alkaline extraction, however, lacked anti-biofilm activity, likely due to structural alterations of the protein. These findings highlight how strain type and extraction conditions shape the bioactivities of sericin [6].

5.5.4 Moisture Retention and Skin Compatibility

Silk fibers share structural similarities with human skin and hair, making sericin especially useful in cosmetics. Unlike fibroin, which is poorly soluble, sericin is water-soluble and easily absorbed by skin. It contains essential amino acids such as serine, aspartic acid, glutamic acid, lysine, and arginine, which nourish epidermal cells, enhance elasticity, and promote metabolism [39]. Sericin's high hydrophilicity and antioxidant capacity confer excellent moisture retention, comparable to natural moisturizing factors (NMF) in the stratum corneum [40]. Its amino acid composition closely resembles that of NMF, enabling it to bind water, improve hydration, and maintain skin softness. These properties also support wound healing, UV protection, and applications in hair care products such as shampoos and conditioners [39, 40].

5.6 Biomedical Applications of Silk Sericin

Silk sericin, once considered a by-product of the silk industry, has emerged as a valuable biomaterial due to its unique properties. It is biocompatible, biodegradable, and low immunogenic, demonstrating antioxidant and cell-regulating activities, which make it suitable for a wide range of biomedical applications. Recent studies have highlighted its applications in wound healing, tissue engineering, drug delivery, regenerative medicine, and cosmetics (Fig. 5.4). Notably, the use of sericin also supports sustainability by transforming silk industry waste into useful products.

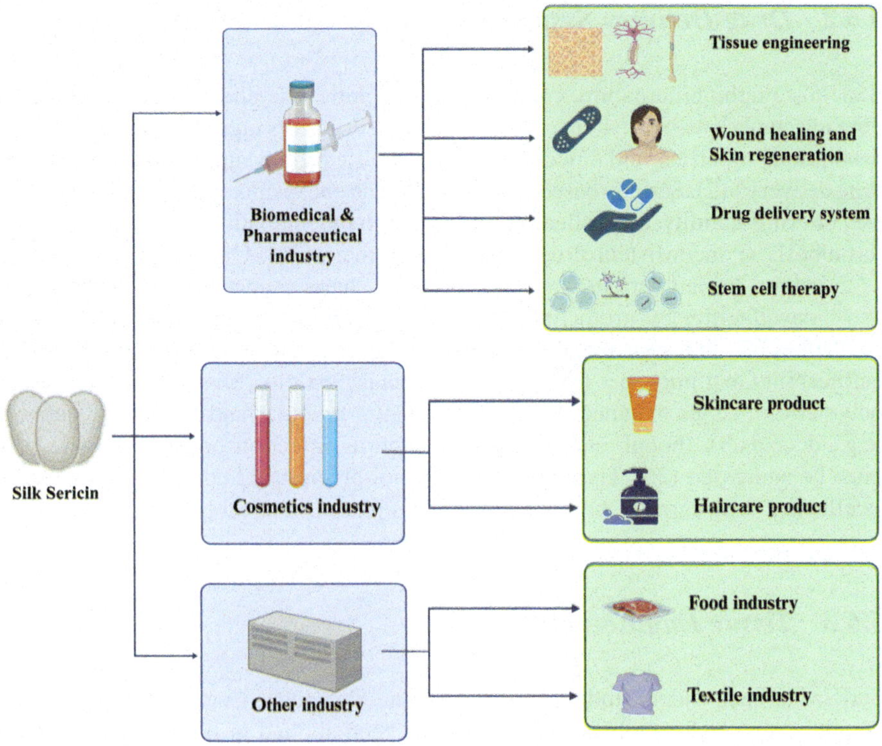

Fig. 5.4 Multifunctional uses of silk sericin across biomedical, cosmetic, and industrial fields

5.6.1 *Wound Healing and Skin Regeneration*

Wound healing is a complex biological process that involves hemostasis, inflammation, proliferation, and remodeling [5]. Sericin-based materials, including scaffolds, films, hydrogels, foams, fibers, and microspheres, have demonstrated effectiveness in treating both localized and systemic wounds. Methionine in sericin plays a key role in collagen synthesis, and sericin with higher molecular weights (e.g., 400 kDa) has been shown to enhance fibroblast growth by up to 250% within 72 h. This activity is strongly influenced by amino acid composition. Sericin also promotes fibroblast migration and collagen production, both critical in wound healing. In clinical trials, a standard burn cream containing 8% sericin accelerated epithelialization, reduced inflammation, and shortened hospitalization time [40]. Animal studies further confirmed reduced inflammatory cytokines and faster wound closure. Sericin hydrogels are particularly promising, as they enable scarless skin regeneration and recovery of skin appendages. Due to their safety, simplicity, and low cost, sericin-based dressings are widely considered an effective and sustainable option for wound care.

5.6.2 Drug Delivery Systems

Sericin's amphiphilic nature, combining hydrophilic side chains with hydrophobic domains, makes it an excellent carrier for diverse drugs. Its high moisture absorption, desorption properties, and moderate in vivo half-life further support its application in drug delivery [12]. Sericin-based nanoparticles can encapsulate drugs for intravenous use, offering stability, controlled release, and high biocompatibility, though risks like thrombosis or inconsistent drug distribution remain [36]. Oral formulations, such as curcumin-loaded sericin–lipid nanoparticles, have shown improved absorption and bioavailability by protecting drugs from gastrointestinal degradation, though challenges like first-pass metabolism persist. Transdermal sericin gels improve skin permeability, but improper formulation may cause toxicity. Similarly, sericin-based polymeric micelles enhance the efficacy of intra-articular and bone drug delivery (e.g., docetaxel), though safety concerns such as infection or injection site pain must be addressed [36]. Hydrogels and ethanol-precipitated composites also serve as effective drug carriers, broadening sericin's pharmaceutical potential.

5.6.3 Tissue Engineering

Sericin has been widely studied for tissue engineering due to its effects on fibroblasts, keratinocytes, and osteoblasts [1, 17]. It has been applied in bone, skin, cartilage, neural, and muscle tissue regeneration. In bone engineering, sericin promotes hydroxyapatite nucleation and enhances osseointegration when used as a coating for titanium implants. Sericin-gelatin scaffolds show superior cytocompatibility compared to gelatin alone [41]. Hydrogels combining sericin methacryloyl with graphene oxide provide stability, osteoinductive properties, and strong biocompatibility, accelerating bone regeneration in animal models [42, 43].

In skin tissue engineering, sericin-based scaffolds facilitate fibroblast adhesion, proliferation, and differentiation, supporting wound and burn treatment [44, 45]. Stem cell-based tissue engineering and 3D bioprinting using sericin hydrogels as bioinks show strong potential for creating functional tissues [46]. Sericin has also been tested in corneal tissue engineering, where it supports endothelial cell growth and transparency [36].

5.6.4 Cosmetics and Dermatology

Sericin is widely used in cosmetics for its moisturizing, anti-aging, and soothing properties. In creams and gels, sericin improves skin hydration and reduces irritation, even in patients with conditions such as chronic kidney disease [25]. In hair care, sericin-based shampoos enhance strength, elasticity, and shine. It also improves nail health, with sericin-based nail polish showing superior stability and sheen compared to conventional formulations [47]. Dietary intake of sericin has been linked to improved epidermal hydration and increased filaggrin production, suggesting benefits for dry skin and conditions like atopic dermatitis [48].

5.6.5 Regenerative Medicine

Sericin-based scaffolds have shown promise in nerve regeneration, supporting Schwann cell proliferation and nerve growth factor expression. In rat models, sericin conduits successfully bridged sciatic nerve gaps, with recovery comparable to autologous grafts [43]. Sericin is also being studied for stem cell culture, where it enhances adhesion, proliferation, and differentiation [49, 50]. It has even been proposed as a serum substitute for spermatogonial stem cell culture. Additionally, sericin promotes thyroglobulin production and improves angiogenesis, highlighting its potential in regenerative therapies [34, 49]. Due to its biocompatibility, biodegradability, oxygen permeability, and antimicrobial properties, sericin continues to gain attention as a biomaterial for wound healing, tissue repair, and organ regeneration [4, 51, 52].

5.6.6 Other Applications

Beyond medicine, sericin is being applied in the food and textile industries. As a protein-based edible coating, sericin and sericin-based nanoparticles help reduce respiration rate, oxidation, and water loss in fruits and vegetables, while providing a low-cost, sustainable alternative to conventional coatings [53, 54]. In textiles, sericin improves moisture retention and antimicrobial activity, enhancing the performance of natural and synthetic fabrics. Its incorporation promotes sustainability and reduces environmental waste from sericulture [55].

5.7 Formulation Development Using Sericin

Silk sericin has been widely applied in formulation development across the medical field through the formation of sericin-based hydrogels, nanogels, microspheres, membranes, nanoparticles, and related structures. These advances demonstrate its potential in drug delivery systems and tissue engineering. The following subsections highlight some of the key applications of sericin in formulation development.

5.7.1 Hydrogels and Nanogels

Silk sericin-based hydrogels and nanogels are engineered to achieve high efficiency and provide several advantages over conventional polymers. Through precise fabrication, researchers have improved their biodegradability, biocompatibility, processability, selective design, low toxicity, and cell encapsulation potential, making sericin an attractive material for biomedical use [56, 57]. Hydrogels fabricated from pure mulberry sericin by crosslinking with glutaraldehyde have been successfully applied in drug delivery systems [58]. When blended with natural polymers such as collagen, sericin hydrogels and membranes can be used for tissue engineering scaffolds, artificial cartilage, wound dressings, and cosmetic formulations [56]. Thus, sericin-based hydrogels represent promising biomaterials for biomedical applications.

Nanogels formulated with sericin also play a significant role in controlled drug delivery. They are characterized by their biodegradability, biocompatibility, and low toxicity, making them suitable for a range of therapeutic applications [40]. Sericin nanoformulations have been shown to retain important biological functions including anti-inflammatory, antioxidant, anticoagulant, antimicrobial, and anticancer properties. These features make them valuable in drug delivery, wound healing, disease therapy, tissue engineering, and diagnostic applications. In practice, they have been used to fabricate sponges, films, scaffolds, fibers, hydrogels, and nano-or micro-structured materials [9].

5.7.2 Nanoparticles and Microspheres

Sericin nanoparticles have shown potential in artificial skin production, articular cartilage repair, tissue regeneration scaffolds, and targeted small-molecule drug delivery systems. They are capable of releasing drugs under specific conditions such as changes in pH. Importantly, sericin nanoparticles have demonstrated the ability to inhibit cancer cell proliferation while remaining non-toxic to normal fibroblasts [9].

Sericin-based microspheres have also been developed for wound dressings. When incorporated into scaffolds, they provide antimicrobial activity against both Gram-positive and Gram-negative bacteria without causing tissue irritation. Furthermore, sericin nanoparticles can be used as carriers for exogenous DNA delivery through protein desolvation. In one study, green fluorescent protein (GFP)-encoding plasmid, pEGFP-C1 was successfully delivered into mouse fibroblast cells, as confirmed by fluorescent intensity, demonstrating the system's effectiveness [59]. Similarly, small interfering RNA (siRNA) delivery has been explored using albumin-sericin nanoparticles (Alb-Ser-NPs) for therapeutic gene silencing. In laryngeal cancer treatment, these nanoparticles effectively inhibited cancer cell growth and induced apoptosis, supporting their potential role in future siRNA delivery strategies [60].

Sericin microspheres fabricated by spray drying and glutaraldehyde crosslinking have been employed for the delivery of the anti-leukemic enzyme L-asparaginase.

Immobilization in sericin microspheres enhanced enzyme stability compared to the native form, providing superior resistance. Additionally, sericin-alginate-chitosan microcapsules created through electrohydrodynamic spraying and genipin crosslinking have been applied for live cell encapsulation. This system successfully maintained hepatocyte viability and established a model for cell encapsulation [61]. Given their diverse biological activities, sericin microspheres are considered highly valuable in bioactive drug delivery systems.

5.7.3 Fibers and Membranes for Biomedical Devices

Sericin-based fibers and membranes are widely used in biomedical applications owing to their unique physicochemical properties. Their high moisture retention makes them suitable for cosmetic formulations, and they can also serve as coatings for medical devices [40, 51]. To enhance mechanical strength, sericin membranes are often crosslinked and integrated into structures such as films, scaffolds, and hydrogels, which improve efficiency in biomedical applications [62]. These membranes support fibroblast and keratinocyte adhesion and promote skin tissue repair, maintaining stability for up to 4 weeks in aqueous environments [63]. Recent studies indicate that sericin membranes and fibers combined with hydroxyapatite exhibit osteogenic activity suitable for periodontal bone tissue regeneration. By supporting human periodontal ligament stem cells (hPDLSCs), these composite membranes induced osteogenic differentiation and activated osteoblast-related genes including ALP, OCN, Runx2, and OPN without the need for additional inducers [64].

5.7.4 Bioinks for 3D Bioprinting Applications

Three-dimensional bioprinting is an emerging technology aimed at creating functional tissues to replace damaged or diseased organs. In this approach, hydrogels incorporating desired cell types serve as bioinks, which are crosslinked after printing to stabilize and form the final construct. Bioinks can be derived from natural materials, synthetic materials, or their combinations. Recent research has demonstrated the application of sericin in bioinks for 3D bioprinting [65]. Sericin-based scaffolds can encapsulate cells within a microenvironment suitable for tissue engineering and wound healing applications. Studies also suggest that sericin extracted from cocoons through autoclaving enhances wound healing due to its high collagen content. Heat-degraded sericin possesses superior structural features that provide resistance under high temperature and pressure [49]. In addition, sericin has been used to prepare N-isopropylacrylamide hydrogels crosslinked with genipin, which support cell proliferation and allow temperature-sensitive detachment of cultured L929 fibroblasts [66]. Furthermore, sericin-polyvinyl alcohol (PVA) 3D scaffolds crosslinked with genipin have been developed as skin substitutes for tissue engineering applications [67].

5.8 Sericin–Polymer Composites for Biomedical Engineering

Sericin–polymer composites combine the unique properties of sericin, such as high biocompatibility and hydrophilicity, with synthetic polymers like polylactic acid (PLA) and polycaprolactone (PCL), or natural polymers such as collagen and chitosan. This blending results in hybrid materials with enhanced bioactivity, mechanical stability, and cell adhesion compared to individual components [68, 69]. Incorporating sericin with synthetic polymers like PCL and PLA improves bioactivity for tissue engineering, where cell interaction plays a crucial role. The mechanical properties of these composites can be fine-tuned to match targeted tissue characteristics [68, 69]. Similarly, blending with natural polymers such as collagen or chitosan enhances biological functions, promotes tissue regeneration, and mimics the extracellular matrix (ECM), supporting excellent cell integration. Sericin-based nanocomposites, due to their ECM-like composition, show promise as advanced biomaterials for skin tissue engineering [70, 71]. Further improvements in cell adhesion and tissue regeneration can be achieved by functionalizing composite surfaces with bioactive molecules or growth factors. Crosslinking techniques are also employed to increase the mechanical strength and stability of sericin-based composites [69]. These composites have applications in tissue engineering scaffolds for skin, bone, cartilage, and nerve regeneration; wound dressings due to their high moisture absorption, biocompatibility, and healing properties; and drug delivery systems that allow sustained release of therapeutic agents [12, 40, 70]. Beyond biomedical uses, sericin's hygroscopic and biocompatible properties make it a valuable surface modifier in textiles. For instance, sericin-modified cellulose fibers reduce skin discomfort and preserve skin flora, while sericin-TiO_2 nanocomposites coated on cotton fabrics provide antibacterial effects [67, 72]. In dentistry, sulfated sericin exhibits anticoagulant activity similar to fibroin and improves denture comfort and stability in dry oral cavities [73].

5.9 Advances in Nanotechnology Using Sericin

Sericin-based nanoparticles are synthesized from silk cocoon extracts using techniques such as precipitation, emulsification, and electrospraying to obtain particles with desirable sizes and properties. These are characterized by methods including TEM, FTIR, and DLS to assess morphology, size, and surface functional groups [74, 75]. Due to their hydrophilicity, low toxicity, and high protein content, sericin nanoparticles are well tolerated in the human body and can enable controlled drug release. They are especially useful in targeted drug delivery for tumors when conjugated with ligands, and in in vivo imaging when tagged with fluorescent dyes or metallic nanoparticles. Their versatile binding capacity makes sericin a suitable biorecognition element in biosensors for detecting antibiotics, antigens, and small

Table 5.2 Application, properties, and advantages of polymer combined with silk sericin

Polymer	Application	Properties	Advantages	References
Polycaprolactone (PCL)	Drug encapsulation and delivery	Non-Newtonian behavior, high absorption capacity, biodegradable	Enhanced biocompatibility, controlled drug release	[76]
Polyvinyl polymers	General biomedical applications	Enhanced mechanical strength	Improved stability and compatibility	[71]
Polymethyl methacrylate (PMMA) (with fibroin)	Robust biomedical applications	Improved mechanical strength	Enhanced durability and functionality	[77]
Other synthetic polymers	Various biomedical applications	Enhanced mechanical properties, biocompatibility	Improved functionality, stability	[78]

molecules. Upon binding of target molecules, sericin-based biosensors generate measurable signals [40, 74, 75]. At the bio-nano interface, sericin promotes cell adhesion and proliferation, while coating metal nanoparticles or polymers with sericin improves biocompatibility and reduces cytotoxicity [12]. Besides, combining sericin with synthetic polymers enhances its functional properties, expanding applications across biomedical and pharmaceutical industries [74]. Examples of polymer–sericin composites are summarized in Table 5.2.

5.10 Toxicological Studies and Safety Considerations

Toxicological evaluation is essential for the safe biomedical use of sericin. Cytotoxicity studies assess its ability to damage cells by disrupting membranes or interfering with signaling pathways. Evidence shows that purified sericin extracts have low cytotoxicity [79]. Genotoxicity assessments examine DNA damage, mutation, or chromosomal abnormalities. Current findings confirm that sericin is non-mutagenic and non-genotoxic in both in vitro and in vivo models, with no significant DNA damage detected [7]. In addition, sericin elicits negligible immune responses after implantation, reflecting high biocompatibility and anti-inflammatory potential [4, 7]. Its physicochemical properties, together with low toxicity, biodegradability, and biocompatibility, support applications in drug delivery, wound dressings, tissue engineering, and bone grafting. For clinical use, biomaterials such as sericin are expected to mimic the extracellular matrix (ECM), support scaffold formation, enable cell and drug transport, and provide adequate mechanical stability [7].

5.11 Challenges and Future Prospects

Despite its potential, sericin faces several challenges that limit its clinical applications. Extraction methods often lead to protein degradation, affecting stability, purity, and large-scale production. The processes are also expensive, with issues such as variable molecular weight, pH and temperature sensitivity, and inconsistent purity, which require additional purification steps during scale-up [9, 11, 17]. To address these challenges, more in vivo research and clinical trials are needed to assess the effectiveness of sericin-based nanoparticles in medical use. The molecular mechanisms underlying sericin's interactions with cells remain insufficiently understood, leaving gaps in knowledge about how to optimize its biomedical properties. Enhancing its mechanical strength and stability through both in vitro and in vivo studies is essential for its independent use in drug delivery systems and scaffolds. Currently, these limitations make sericin unsuitable for applications requiring strong structural support, such as bone and cartilage regeneration. However, combining sericin with other biomaterials can improve its mechanical properties and broaden its use in biomedical engineering.

5.12 Importance of Sericin in the Circular Economy

Sericin possesses multiple functional properties, such as biocompatibility, biodegradability, antioxidant activity, antimicrobial effects, and hydrophilic adhesion, that allow applications in tissue engineering, wound healing, drug delivery, cosmetics, textiles, and food packaging. Although traditionally treated as a sericulture by-product, its valorization contributes significantly to economic development and sustainability. Integrating sericin recovery into waste management aligns with circular economy principles, turning industrial waste into valuable resources. The challenge lies in balancing economic gain with environmental conservation. Effective policies, supportive regulations, and collaboration among stakeholders are needed to expand such practices. By advancing waste-to-resource technologies, sericin can strengthen the sustainability of the Indian sericulture industry while contributing to environmentally responsible economic growth.

5.13 Conclusion

Research on silk sericin has advanced rapidly, establishing its position as a promising biomaterial with diverse biomedical applications. Moving forward, solving practical challenges will matter more than finding new applications. Standardized extraction and purification methods, long-term stability studies, scalable production processes, and rigorous toxicological evaluations are urgently needed to bring sericin closer to clinical use. Without addressing these gaps, translation from laboratory studies to patient care will remain limited. At the same time, the multifunctional nature of sericin continues to inspire new directions. Its integration with advanced technologies such as nanomedicine, 3D bioprinting, and polymeric composites paves the way for a future where sericin-based systems support regenerative medicine, personalized therapies, and smart biomedical devices. Equally important is the broader perspective. Utilizing sericin within a circular bioeconomy not only reduces industrial waste but also adds value to healthcare innovation. Thus, sericin research represents a unique convergence of scientific opportunity and societal need. Focused efforts are needed to strengthen cross-disciplinary collaboration and innovation. Meeting these challenges could enable sericin to emerge as a next-generation biomaterial linking sustainable resource use with transformative healthcare solutions.

Declaration of AI Usage AI tools were utilized solely for language refinement and grammatical correction in the preparation of this chapter. The authors take full responsibility for the accuracy, interpretation, and originality of the final content presented herein.

References

1. Aad R, Dragojlov I, Vesentini S (2024) Sericin protein: structure, properties, and applications. J Funct Biomater 15:322. https://doi.org/10.3390/jfb15110322
2. Vijayakumar N, Sanjay AV, Al-Ghanim KA et al (2024) Development of biodegradable bioplastics with sericin and gelatin from silk cocoons and fish waste. Toxics 12(7):453. https://doi.org/10.3390/toxics12070453
3. Paladini F, D'Urso F, Panico A et al (2025) Comparative analysis of highly purified Sericin and waste-derived Sericin: implications for biomedical applications. Biomimetics 10:387. https://doi.org/10.3390/biomimetics10060387
4. Kim J-Y, Kim S-G, Garagiola U (2023) Relevant properties and potential applications of Sericin in bone regeneration. Curr Issues Mol Biol 45:6728–6742. https://doi.org/10.3390/cimb45080426
5. Borges A, Calvo MLM, Vaz JA et al (2024) Enhancing wound healing: a comprehensive review of Sericin and Chelidonium majus L. as potential dressings. Dent Mater 17:4199. https://doi.org/10.3390/ma17174199
6. Wang S-L, Zhuo J-J, Fang S-M et al (2024) Silk sericin and its composite materials with antibacterial properties to enhance wound healing: a review. Biomolecules 14:723. https://doi.org/10.3390/biom14060723
7. Liu J, Shi L, Deng Y et al (2022) Silk sericin-based materials for biomedical applications. Biomate 287:121638. https://doi.org/10.1016/j.biomaterials.2022.121638
8. Padamwar MN, Pawar AP (2004) Silk sericin and its applications: a review. J Sci Ind Res 63(4):323–329
9. Das G, Shin H-S, Campos EVR et al (2021) Sericin based nanoformulations: a comprehensive review on molecular mechanisms of interaction with organisms to biological applications. J Nanobiotechnol 19:30. https://doi.org/10.1186/s12951-021-00774-y
10. Mondal M, Trivedy K, Nirmal KS (2007) The silk proteins, sericin and fibroin in silkworm, Bombyx mori Linn.,-a review. Caspian J Environ Sci 5(2):63–76
11. Seo S-J, Das G, Shin H-S et al (2023) Silk sericin protein materials: characteristics and applications in food-sector industries. Int J Mol Sci 24:4951. https://doi.org/10.3390/ijms24054951

12. Silva AS, Costa EC, Reis S et al (2022) Silk sericin: a promising sustainable biomaterial for biomedical and pharmaceutical applications. Polymers 14(22):4931. https://doi.org/10.3390/polym14224931
13. Reddy N, Aramwit P (2021) Sustainable uses of byproducts from silk processing. John Wiley and Sons
14. Oh H, Lee JY, Kim MK et al (2011) Refining hot-water extracted silk sericin by ethanol-induced precipitation. Int J Biol Macromol 48:32–37. https://doi.org/10.1016/j.ijbiomac.2010.09.008
15. Yang Y, Lee SM, Lee HS et al (2013) Recovery of silk sericin from soap-alkaline degumming solution. Int J Ind Entomol 27:203–208. https://doi.org/10.7852/ijie.2013.27.1.203
16. Wang W, Pan Y, Gong K et al (2019) A comparative study of ultrasonic degumming of silk sericin using citric acid, sodium carbonate and papain. Color Technol 135:195–201. https://doi.org/10.1111/cote.12392
17. Lamboni L, Gauthier M, Yang G et al (2015) Silk sericin: a versatile material for tissue engineering and drug delivery. Biotechnol Adv 33:1855–1867. https://doi.org/10.1016/j.biotechadv.2015.10.014
18. More SV, Chavan S, Prabhune AA (2018) Silk degumming and utilization of silk Sericin by hydrolysis using alkaline protease from *Beauveria* Sp. (MTCC 5184): a green approach. J Nat Fibers 15:373–383. https://doi.org/10.1080/15440478.2017.1330718
19. Gupta D, Agrawal A, Chaudhary H et al (2013) Cleaner process for extraction of sericin using infrared. J Clean Prod 52:488–494. https://doi.org/10.1016/j.jclepro.2013.03.002
20. Wang R, Zhu Y, Shi Z et al (2018) Degumming of raw silk via steam treatment. J Clean Prod 203:492–497. https://doi.org/10.1016/j.jclepro.2018.08.286
21. Freddi G, Mossotti R, Innocenti R (2003) Degumming of silk fabric with several proteases. J Biotech 106:101–112. https://doi.org/10.1016/j.jbiotec.2003.09.006
22. Yun H, Oh H, Kim MK et al (2013) Extraction conditions of Antheraea mylitta sericin with high yields and minimum molecular weight degradation. Int J Biol Macromo 52:59–65. https://doi.org/10.1016/j.ijbiomac.2012.09.017
23. Çapar G, Aygün SS (2015) Characterization of sericin protein recovered from silk wastewaters. Turk Hij Den Biyol Derg 72(3):219–234. https://doi.org/10.5505/TurkHijyen.2015.47113
24. Lalit Jajpura AR (2015) The biopolymer Sericin: extraction and applications. J TextE Sci Eng 05. https://doi.org/10.4172/2165-8064.1000188
25. Aramwit P, Siritientong T, Srichana T (2012) Potential applications of silk sericin, a natural protein from textile industry by-products. Waste Manag Res 30:217–224. https://doi.org/10.1177/0734242X11404733
26. Devi R (2012) Biotechnological application of proteolytic enzymes in post cocoon technology. Int J Sci Nat 3:237–240
27. Suwannaphan S, Fufeungsombut E, Promboon A et al (2017) A serine protease from newly isolated bacillus sp. for efficient silk degumming, sericin degrading and colour bleaching activities. Int Biodeterior Biodegrad 117:141–149. https://doi.org/10.1016/j.ibiod.2016.12.009
28. Yakul K, Takenaka S, Nakamura K et al (2019) Characterization of thermostable alkaline protease from bacillus halodurans SE5 and its application in degumming coupled with sericin hydrolysate production from yellow cocoon. Process Biochemistry 78:63–70. https://doi.org/10.1016/j.procbio.2019.01.003
29. Krishnamurthy K, Khurana HK, Soojin J et al (2008) Infrared heating in food processing: an overview. Comp Rev Food Sci Food Safe 7:2–13. https://doi.org/10.1111/j.1541-4337.2007.00024.x
30. Terada S, Nishimura T, Sasaki M et al (2002) Sericin, a protein derived from silkworms, accelerates the proliferation of several mammalian cell lines including a hybridoma. Cytotechnology 40:3–12. https://doi.org/10.1023/A:1023993400608
31. Terada S, Sasaki M, Yanagihara K et al (2005) Preparation of silk protein sericin as mitogenic factor for better mammalian cell culture. J Biosci Bioeng 100:667–671. https://doi.org/10.1263/jbb.100.667
32. Aramwit P, Kanokpanont S, De-Eknamkul W et al (2009) Monitoring of inflammatory mediators induced by silk sericin. J Biosci Bioeng 107:556–561. https://doi.org/10.1016/j.jbiosc.2008.12.012

33. Mandal BB, Priya AS, Kundu SC (2009) Novel silk sericin/gelatin 3-D scaffolds and 2-D films: fabrication and characterization for potential tissue engineering applications. Acta Biomater 5:3007–3020. https://doi.org/10.1016/j.actbio.2009.03.026
34. Cherng J-H, Chang S-J, Chiu Y-K et al (2022) Low molecular weight Sericin enhances the in vitro of immunological modulation and cell migration. Front Bioeng Biotechnol 10:925197. https://doi.org/10.3389/fbioe.2022.925197
35. Dash BC, Mandal BB, Kundu SC (2009) Silk gland sericin protein membranes: fabrication and characterization for potential biotechnological applications. J Biotech 144:321–329. https://doi.org/10.1016/j.jbiotec.2009.09.019
36. Kabir M, Hasan MK, Rafi MN et al (2024) Biowaste transformation to functional materials: structural properties, extraction methods, applications, and challenges of silk Sericin. ChemistrySelect 9:e202403777. https://doi.org/10.1002/slct.202403777
37. Jassim KN, Al-Saree OJ (2010) Study of the antimicrobial activity of silk sericin from silkworm bombyx mori. Iraqi J Med Sci 23:130–133
38. Kanyora MG, Kegode TM, Kurgat J et al (2024) Evaluating antibacterial and antioxidant properties of sericin recovered from cocoons of Bombyx mori, Gonometa postica and Samia ricini in Kenya. PLoS One 19:e0316259. https://doi.org/10.1371/journal.pone.0316259
39. Sheng JY, Xu J, Zhuang Y et al (2013) Study on the application of Sericin in cosmetics. Adv Mater Res 796:416–423. https://doi.org/10.4028/www.scientific.net/AMR.796.416
40. Kunz RI, Brancalhão RMC, Ribeiro LDFC et al (2016) Silkworm Sericin: properties and biomedical applications. Biomed Res Int 2016:1–19. https://doi.org/10.1155/2016/8175701
41. Nayak S, Dey T, Naskar D et al (2013) The promotion of osseointegration of titanium surfaces by coating with silk protein sericin. Biomater 34:2855–2864. https://doi.org/10.1016/j.biomaterials.2013.01.019
42. Qi C, Liu J, Jin Y et al (2018) Photo-crosslinkable, injectable sericin hydrogel as 3D biomimetic extracellular matrix for minimally invasive repairing cartilage. Biomater 163:89–104. https://doi.org/10.1016/j.biomaterials.2018.02.016
43. Veiga A, Foster O, Kaplan DL et al (2024) Expanding the boundaries of silk sericin biomaterials in biomedical applications. J Mater Chem B 12:7020–7040. https://doi.org/10.1039/D4TB00386A
44. Hafeman AE, Zienkiewicz KJ, Zachman AL et al (2011) Characterization of the degradation mechanisms of lysine-derived aliphatic poly (ester urethane) scaffolds. Biomater 32:419–429. https://doi.org/10.1016/j.biomaterials.2010.08.108
45. Kundu B, Kurland NE, Bano S et al (2014) Silk proteins for biomedical applications: bioengineering perspectives. Prog Polym Sci 39:251–267. https://doi.org/10.1016/j.progpolymsci.2013.09.002
46. Hilmi ABM, Halim AS (2015) Vital roles of stem cells and biomaterials in skin tissue engineering. World J Stem Cells 7(2):428–436. https://doi.org/10.4252/wjsc.v7.i2.428
47. Giovannelli L, Milanesi A, Ugazio E et al (2021) Effect of methyl–β–Cyclodextrin and Trehalose on the freeze–drying and spray–drying of Sericin for cosmetic purposes. Pharmaceuticals 14:262. https://doi.org/10.3390/ph14030262
48. Kim H, Lim Y, Park J-H et al (2012) Dietary silk protein, sericin, improves epidermal hydration with increased levels of filaggrins and free amino acids in NC/Nga mice. Br J Nutr 108:1726–1735. https://doi.org/10.1017/S0007114511007306
49. Cao T-T, Zhang Y-Q (2017) The potential of silk sericin protein as a serum substitute or an additive in cell culture and cryopreservation. Amino Acids 49:1029–1039. https://doi.org/10.1007/s00726-017-2396-3
50. Ha S-J, Kim B-G, Lee Y-A et al (2016) Effect of antioxidants and apoptosis inhibitors on cryopreservation of murine germ cells enriched for Spermatogonial stem cells. PLoS One 11:e0161372. https://doi.org/10.1371/journal.pone.0161372
51. Arango MC, Montoya Y, Peresin MS et al (2021) Silk sericin as a biomaterial for tissue engineering: a review. Int J Polym Mater 70:1115–1129. https://doi.org/10.1080/00914037.2020.1785454

52. Jo Y-Y, Kweon H, Oh J-H (2020) Sericin for tissue engineering. Appl Sci 10:8457. https://doi.org/10.3390/app10238457
53. Shaw S, Mondal R, Dam P et al (2024) Synthesis, characterization and application of silk sericin-based silver nanocomposites for antibacterial and food coating solutions. RSC Adv 14:33068–33079. https://doi.org/10.1039/D4RA07056A
54. Tarangini K, Kavi P, Jagajjanani Rao K (2022) Application of sericin-based edible coating material for postharvest shelf-life extension and preservation of tomatoes. eFood 3:e36. https://doi.org/10.1002/efd2.36
55. Kongdee A, Okubayashi S, Tabata I et al (2007) Impregnation of silk sericin into polyester fibers using supercritical carbon dioxide. J Appl Polym Sci 105:2091–2097. https://doi.org/10.1002/app.26314
56. Qi C, Deng Y, Xu L et al (2020) A sericin/graphene oxide composite scaffold as a biomimetic extracellular matrix for structural and functional repair of calvarial bone. Theranostics 10:741–756. https://doi.org/10.7150/thno.39502
57. Zhang S, Shah SA-M, Basharat K et al (2022) Silk-based nano-hydrogels for futuristic biomedical applications. J Drug Deliv Sci Technol 72:103385. https://doi.org/10.1016/j.jddst.2022.103385
58. Wang Z, Zhang Y, Zhang J et al (2014) Exploring natural silk protein sericin for regenerative medicine: an injectable, photoluminescent, cell-adhesive 3D hydrogel. Sci Rep 4:7064. https://doi.org/10.1038/srep07064
59. Das SK, Dey T, Kundu SC (2014) Fabrication of sericin nanoparticles for controlled gene delivery. RSC Adv 4:2137–2142. https://doi.org/10.1039/C3RA44990D
60. Yalcin E, Kara G, Celik E et al (2019) Preparation and characterization of novel albumin-sericin nanoparticles as siRNA delivery vehicle for laryngeal cancer treatment. Prep Biochem Biotechnol 49:659–670. https://doi.org/10.1080/10826068.2019.1599395
61. Nayak S, Dey S, Kundu SC (2014) Silk sericin–alginate–chitosan microcapsules: hepatocytes encapsulation for enhanced cellular functions. Int J Biol Macromol 65:258–266. https://doi.org/10.1016/j.ijbiomac.2014.01.042
62. Nayak S, Talukdar S, Kundu SC (2012) Potential of 2D crosslinked sericin membranes with improved biostability for skin tissue engineering. Cell Tissue Res 347:783–794. https://doi.org/10.1007/s00441-011-1269-4
63. Akturk O, Tezcaner A, Bilgili H et al (2011) Evaluation of sericin/collagen membranes as prospective wound dressing biomaterial. J Biosci Bioeng 112:279–288. https://doi.org/10.1016/j.jbiosc.2011.05.014
64. Ming P, Rao P, Wu T et al (2022) Biomimetic design and fabrication of Sericin-hydroxyapatite based membranes with osteogenic activity for periodontal tissue regeneration. Front Bioeng Biotechnol 10:899293. https://doi.org/10.3389/fbioe.2022.899293
65. Gungor-Ozkerim PS, Inci I, Zhang YS et al (2018) Bioinks for 3D bioprinting: an overview. Biomater Sci 6:915–946. https://doi.org/10.1039/C7BM00765E
66. Zhang Q, Dong P, Chen L et al (2014) Genipin-cross-linked thermosensitive silk sericin/poly(*N*-isopropylacrylamide) hydrogels for cell proliferation and rapid detachment: Genipin-cross-linked thermosensitive SS/PNIPAm hydrogels. J Biomed Mater Res 102:76–83. https://doi.org/10.1002/jbm.a.34670
67. Cao T-T, Zhang Y-Q (2016) Processing and characterization of silk sericin from Bombyx mori and its application in biomaterials and biomedicines. Mater Sci Eng C 61:940–952. https://doi.org/10.1016/j.msec.2015.12.082
68. Cunha C, Panseri S, Antonini S (2011) Emerging nanotechnology approaches in tissue engineering for peripheral nerve regeneration. NBM 7:50–59. https://doi.org/10.1016/j.nano.2010.07.004
69. Stratton S, Shelke NB, Hoshino K et al (2016) Bioactive polymeric scaffolds for tissue engineering. Bioact Mater 1:93–108. https://doi.org/10.1016/j.bioactmat.2016.11.001
70. Hu D, Xu Z, Hu Z et al (2017) pH-triggered charge-reversal silk Sericin-based nanoparticles for enhanced cellular uptake and doxorubicin delivery. ACS Sustain Chem Eng 5:1638–1647. https://doi.org/10.1021/acssuschemeng.6b02392

71. Tao G, Wang Y, Cai R et al (2019) Design and performance of sericin/poly (vinyl alcohol) hydrogel as a drug delivery carrier for potential wound dressing application. Mater Sci Eng C 101:341–351. https://doi.org/10.1016/j.msec.2019.03.111
72. Doakhan S, Montazer M, Rashidi A et al (2013) Influence of sericin/TiO2 nanocomposite on cotton fabric: part 1. Enhanc Antibact Eff Carbohydr Polym 94:737–748. https://doi.org/10.1016/j.carbpol.2013.01.023
73. Tamada Y, Sano M, Niwa K et al (2004) Sulfation of silk sericin and anticoagulant activity of sulfated sericin. J Biomater Sci Polym Ed 15:971–980. https://doi.org/10.1163/1568562041526469
74. Das G, Seo S, Yang I-J et al (2023) Synthesis of biogenic gold nanoparticles by using Sericin protein from Bombyx mori silk cocoon and investigation of its wound healing, antioxidant, and antibacterial potentials. Int J Nanomedicine 18:17–34. https://doi.org/10.2147/IJN.S378806
75. Miguel GA, Álvarez-López C (2020) Extraction and antioxidant activity of sericin, a protein from silk. Braz J Food Technol 23:e2019058. https://doi.org/10.1590/1981-6723.05819
76. Deb D, Khatun B, BD M et al (2024) Utilizing silk Sericin as a biomaterial for drug encapsulation in a hydrogel matrix with Polycaprolactone: formulation and evaluation of antibacterial activity. ACS Omega acsomega.4c02453. doi:https://doi.org/10.1021/acsomega.4c02453
77. Meenakshi SI, Maheshkumar H, Dureja S et al (2024) A biomechanical reinforcement of PMMA denture base material with sericin and fibroin: a preliminary screening study. Technol Health Care 32:2643–2656. https://doi.org/10.3233/THC-231772
78. Reddy MSB, Ponnamma D, Choudhary R et al (2021) A comparative review of natural and synthetic biopolymer composite scaffolds. Polymers 13:1105. https://doi.org/10.3390/polym13071105
79. Vaishnav SR, Singh SA (2023) Sericin, a by-product of the silk industry: extraction and applications. In: Value-addition in Agri-food industry waste through enzyme technology. Elsevier, pp 199–208

Chapter 6
Recent Advances on Harnessing Sericin for Biomedical Applications

Vijay Kumar Guna and Sohini Chakraborty

6.1 Introduction

Silk is a naturally occurring animal fiber secreted by several organisms, including spiders, mollusks, scorpions, bees, ants, and silkworms [1]. Currently, around 90% of global commercial silk production comes from the mulberry silkworm (*Bombyx mori*, B. mori), a species of insect that undergoes complete metamorphosis and is classified in the Lepidoptera order and the Bombycidae family [2]. This silk is generally preferred owing to its high quality, short production time, and commercial feasibility. Silk has a rich history spanning thousands of years, primarily used in fabric manufacturing due to its lustrous appearance, superior mechanical strength, flexibility, and characteristics such as luxury, comfort, elegance, and purity [3, 4]. The initial advancement in the use of silk as a biomedical clinical suture, due to its exceptional biological properties, has led to the consideration of silk in a wide range of biomedical fields, including the development of tissue-engineered constructs, drug delivery systems, and other related applications [5].

The structure of silk fibers includes two primary components: silk fibroin (SF) and sericin (SS). Silk fibroin constitutes around 60–80% of the silk fiber and constitutes its structural semi-crystalline core. Due to its high mechanical strength and crystallinity, SF contributes to the load-bearing capacity of silk. SF has also been considered as a key material for the fabrication of tissue engineering scaffolds and hydrogels for wound healing, due to its versatility, low immunogenicity, and high mechanical performance. Sericin, a glue-like protein found in the outer layer of silk, serves to

V. K. Guna (✉)
Department of Mechanical Engineering, RNS Institute of Technology, Channasandra, Bengaluru 560098, Karnataka, India
e-mail: vijaykumar.g@rnsit.ac.in

S. Chakraborty
Department of Chemistry, Stella Maris College (Autonomous), University of Madras, Chennai, India

M. Shahid et al. (eds.), *Science and Technology of Silk*, Textile Science and Clothing Technology, https://doi.org/10.1007/978-3-032-26880-8_6

bind the fibrous fibroin (SF) within the cocoon and comprises about 15–35% the overall composition of silk. Furthermore, the proportion of SF and SS varies significantly across different species of silkworm. In textile manufacturing, SS is typically removed during degumming to obtain the fiber used in silk fabric production. Based on recent findings, approximately 50,000 tons of sericin are disposed of as waste from the global production of 400,000 tons of dried cocoons, thereby contributing to environmental pollution [1, 5, 6]. Thus, the utilization of sericin for therapeutic applications owing to its biocompatibility, low allergenicity, ability to degrade, and low immunogenicity has been promoted as an effective strategy for waste management and repurposing. Moreover, silk sericin protein has been investigated for its photoluminescent characteristics, antioxidant properties, and cell-function regulating activities, in addition to its basic biological applications [7, 8]. Additionally, sericin has also been used for the development of functional and controllable biomaterials through chemical modification, precipitation, physical blending, or cross-linking with other polymers. Thus, based on these aspects, the recovery of sericin has been extensively explored for its potential in diverse biological applications and effective waste utilization. This not only reduces environmental impact, but also provides significant value addition in the field of biomedical research [6, 8].

6.2 Structure of Silk Sericin

Many organisms produce silk, but silk filaments originating from domesticated *B.mori* silkworms are considered the most well-established source of silk. Among silk-producing insects, such as silkworms and spiders, silk fibers are secreted in specialized glands and consist of two protein subunits: a heavy chain with a molecular weight of approximately 350–400 kDa and a light chain of about 25–30 kDa. [9]. Sericin is an adhesive protein that forms an outer layer around the protein core by binding two parallel fibroin filaments together within the cocoons. Silk sericin is composed of polypeptides that is composed of 18 amino acids, with molecular weights of ~ 20 to 400 kDa, which primarily depend on the extraction method used. In contrast to SF, SS is a hydrophilic protein comprised of 42% polar amino acids (serine, lysine, and aspartic acid), 46% hydroxy amino acids (serine and threonine), and 12% nonpolar amino acids. Silk sericin exhibits the ability to form crosslinks, undergo co-polymerizations, and combine with different polymers [5, 10]. However, the stratified structure of sericin, composed of the inner, middle, and outer layers, significantly influences its water solubility and is further classified into Sericin A, Sericin B, and Sericin C. Sericin A, which forms the outermost layer of the silk fiber, is hydrophilic and exhibits solubility in boiling water due to the presence of a significant amount of polar amino acids such as serine, glycine, aspartic acid, and threonine. In the intermediate layer formed by Sericin B, the amino acids present in Sericin A are included, with tryptophan also being a key component. The inner layer of the fiber contains Sericin C, which consists of hydrophobic amino acids and

Table 6.1 Silk sericin protein types, production regions and instar stages of *B. mori* [6, 8, 12]

Type of SS protein	Region of production	Instar stages of *B. mori*
Ser 1	Posterior region of the MSG	V instar
Ser 2	Anterior region of the MSG	II to V instar
Ser 3	Middle of MSG	V instar
Ser 4	Anterior region of the MSG	II to V instar
Ser 5	Ser5 exhibits co-localization with Ser1 in the multilayer structure of sericin (SER)	I to IV instar

is insoluble in boiling water. Alkali or acid solutions at high temperatures can be utilized for the removal of the Sericin C from fibroin [11].

Structurally, sericin is a glycoprotein comprised of β-sheets and a high number of random coils, corresponding to both crystalline and amorphous regions, respectively. Some factors, such as mechanical stretching, moisture absorption, and temperature variations, can cause a structural transition from a random coil to a β-sheet structure, which facilitates the sol–gel process. The protein exhibits solubility in water at temperatures of 50–60 °C or higher. In contrast, lower temperatures lead to decreased solubility, and random coil structures are converted to β-sheets, which leads to the formation of a gel. Further, the chemical cross-linking can be employed for the gelation of sericin to form a stable β-sheet structure [5, 10].

In *Bombyx mori*, the silk gland is divided into three distinct regions: the anterior silk gland (ASG), which is responsible for excretion; the middle silk gland (MSG), involved in the production of various types of sericin; and the posterior silk gland (PSG), which secretes fibroin. However, the fibers are spun with the help of a spinneret that is located on top of the gland [10]. Currently, five different types of sericin have been reported, namely Ser1, Ser2, Ser3, Ser4, and Ser5, and characteristics of each type are represented in Table 6.1 [6, 8, 12].

6.3 Extraction Methods of Sericin

The degumming of silk, also known as silk sericin extraction, is a crucial process in the textile industry that involves the removal of the sericin layer from silk filaments. Figure 6.1 summarizes the various degumming methods studied on *B. mori,* including physical, chemical, and biological processes. The conventional process of degumming utilizes high temperature (~100 °C) in the presence of alkaline agents (sodium carbonate) and soap (Marseille soap) or boiling of silk cocoons in an alkaline solution [13].

However, the degumming solution contains sericin along with soap and salt, which can result in the degradation and decreased recovery of silk sericin. The sericin peptides and hydrolysates obtained exhibit limited compatibility with certain biomaterial applications, thereby limiting their potential uses [14]. In contrast to traditional

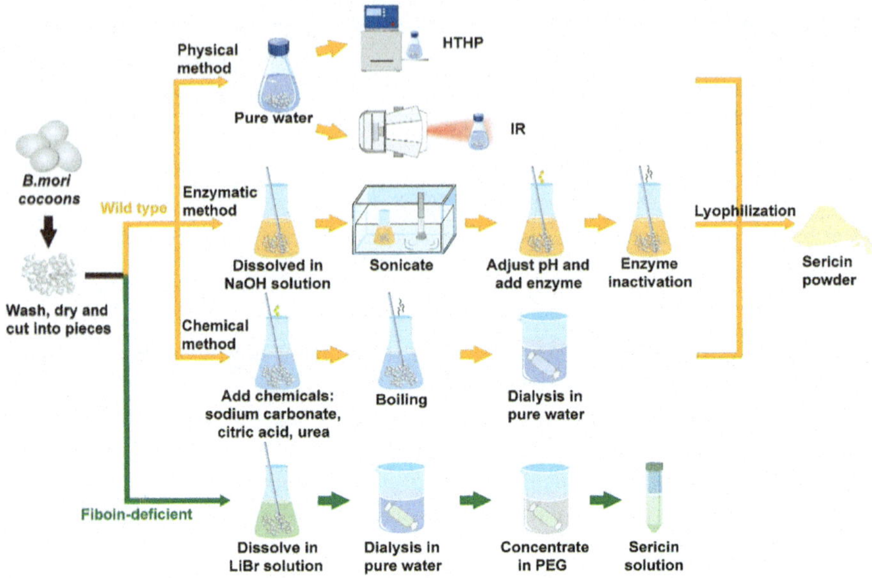

Fig. 6.1 An overview of various methods for the extraction of sericin. Adapted with permission from reference [5], copyright 2022, Elsevier

methods, autoclave degumming under high-temperature and high-pressure (HTHP) conditions have demonstrated considerable improvement in sericin recovery by up to 98% [15]. Nevertheless, the prolonged exposure to heat during HTHP extraction causes the denaturation and degradation of the protein [16]. Consequently, other alternative degumming techniques have been explored. Among various recent advancements, infrared (IR) heating-assisted extraction of sericin has gained increasing attention due to its faster heating rate and easy temperature control [17]. Similarly, the steam explosion method, which involves the removal of sericin under pressurized steam, has been considered as a feasible method for the degumming of silk. Optimized steam treatment at a reduced pressure of 0.14 MPa for 90 min has been shown to maintain the physicochemical properties of degummed silk fibers [18]. Furthermore, both steam explosion and IR heating are considered as more sustainable processes, as the degumming can be carried out without any chemicals and with lower water consumption compared to traditional methods. In addition, these methods have higher thermal efficiency, leading to lower processing costs and energy consumption while producing higher-quality silk sericin [17, 18]. Moreover, several other energy-efficient methods, such as microwave-assisted ultrasound, ultrasound combined with enzyme treatment, and deep eutectic solvent, have also been widely studied for their clean and efficient approach for degumming silk sericin [16]. Both temperature and extraction time have significant influence on the molecular weight of the obtained sericin [13]. Various chemical reagents, including salts (e.g., sodium chloride), alkaline agents (such as sodium carbonate, sodium hydroxide, potassium hydroxide, and urea), and acidic agents (like ammonium sulfate, citric acid, and tartaric acid), are used in combination with heating to enhance the efficiency of the degumming process [5, 19]. Sodium carbonate is the most commonly used alkali, which facilitates the extraction of sericin at higher temperatures. In contrast, ammonium sulfate is an acid that facilitates extraction at lower temperatures. Urea, a mild alkaline compound, facilitates the degumming of sericin without causing damage to the silk fibers. On the other hand, the enzymatic method has been promoted as it extracts sericin with high purity and yield [19]. The efficiency of enzymatic degumming is influenced by various factors, including enzyme concentration, treatment duration, pH levels, and temperature [13]. Enzymes such as proteases, lipases, and amylases have been used for breaking down the sericin protein [19]. Alkaline protease produced by *Beauveria sp.* (MTCC 5184) was effectively utilized in degumming of Chinese bivoltine raw silk fibers. The complete removal of sericin occurred within 45 min at an enzyme concentration of 75 International Units (IU) per gram of silk. The optimal degumming conditions were found to be at 50 °C and pH 9.0, resulting in lower molecular weight peptides of 1139.7 and 616.2 Da [20]. Further, the development of fibroin-deficient mutants such as *B. mori* 185 Nd-s and 140 Nd-s has facilitated the efficient isolation of intact sericin by employing a 6 M LiBr solution at 35 °C. This method of extraction under mild conditions provides a high sericin yield of more than 90%, while maintaining the structure of protein [3, 20]. Consequently, it is necessary to optimize the method of sericin extraction as per the end requirements and the type of silkworm cocoons.

6.4 Biological and Functional Characteristics of Sericin

6.4.1 Biocompatibility and Immunogenicity

One of the primary criteria that must be considered when selecting a material for biomedical applications is its interaction and compatibility with biological tissues. Several biopolymers have been used in this respect. Sericin, with its enhanced cell adhesion, migration, and proliferation, as well as its low immunogenicity, has also been extensively investigated for these applications. Although earlier reports have stated that sericin induces immune responses, particularly leading to T-cell-mediated allergic reactions in silk-based tissue-engineered constructs, the underlying mechanisms contributing to adverse immune responses induced by native silk still remain unclear. Recent investigations have revealed that silk-derived proteins, such as sericin and fibroin, exhibit low immunogenicity when cultured individually with macrophages under in vitro conditions. For instance, low immunogenicity with no significant immunological response has been reported for sericin extracted from the genetically modified fibroin-deficient silkworm mutant, 140 Nd-s. Additionally, numerous studies have presented the anti-inflammatory characteristics of sericin when added to cultured cells. In bone regeneration applications, the release of sericin from fabricated silk mats has been associated with increased expression of TNF-α, indicating a potential promotion of bone regeneration. In contrast, several other reports have revealed that immune responses induced by some silk-based substitutes are attributed to the presence of fibroin. Additionally, the physical interaction between fibroin and sericin has been proposed to trigger immunological responses [8, 10, 21, 22].

6.4.2 Photoluminescent Properties

Photoluminescent behavior of a material is characterized by the emission of light with a longer wavelength subsequent to absorption of high-energy excitation. Sericin proteins extracted using LiBr exhibit two distinct emission bands, particularly within the spectral ranges of 280–300 nm and 340–600 nm, when irradiated with excitation wavelengths in the range of 280 to 560 nm. The photoluminescent properties of sericin are primarily attributed to the presence of aromatic amino acids, such as phenylalanine and tyrosine. Degraded sericin, obtained via hydrothermal processes, retains its photoluminescent behavior, but the emission spectra display a slight shift toward longer wavelengths [8, 23].

6.4.3 Anti-tyrosinase and Antioxidant Activity

Excessive free radicals are highly reactive and can interact with various biomolecules, including nucleic acids, lipids, and proteins, resulting in oxidative stress that contributes to cellular and tissue damage and is further associated with the development of numerous diseases. To maintain optimal biological function, it is crucial to balance the levels of free radicals and ensure redox stability. Antioxidants, which may be synthetic or naturally derived, play a key role in neutralizing free radicals at low concentrations. Several studies have demonstrated the antioxidant behavior of SS, attributed to its ability to scavenge reactive oxygen species (ROS), prevent lipid peroxidation, and inhibit the function of tyrosinase. Sericin consists of abundant phenolic hydroxyl groups (tyrosine) and is considered to scavenge the free radicals produced by transferring hydrogen atoms or single electrons. In addition, the antioxidant mechanism is associated with its ability to increase the activities of several important antioxidant enzymes, such as superoxide dismutase (SOD), catalase (CAT), and glutathione peroxidase (GPx). Moreover, the accumulation of pigments, such as flavonoids and carotenoids in the sericin layer, may also contribute to its scavenging and anti-tyrosinase activity. Along with antioxidant activity, sericin has also exhibited anti-tyrosinase activity. Tyrosinase, a crucial rate-limiting enzyme in melanin synthesis, catalyzes the conversion of tyrosine to dihydroxyphenylalanine (DOPA) and the subsequent chemical modification of DOPA through oxidation. Inhibition of tyrosinase activity can reduce the excessive melanin production, thus providing a therapeutic approach for skin hyperpigmentation conditions such as age spots, solar lentigo, and other related diseases. Further, amino acids such as valine and arginine, which possess the ability to bind to tyrosinase, may further enhance the anti-tyrosinase activity exhibited by sericin [8, 11, 24, 25].

6.4.4 Cell Adhesion and Proliferation

Cell adhesion and proliferation are crucial processes that facilitate tissue regeneration during the repair phase. To evaluate the fundamental mechanism of cell adhesion and proliferation, the structures and physicochemical properties of sericin-based materials have been investigated. Based on amino acid sequence analysis, three fractions of sericin have been identified, namely Sericin A, Sericin M, and Sericin P. Studies have revealed that the M fraction of sericin influences cell adhesion. Further investigations indicated that the ≈170 kDa peptide fraction isolated from Sericin M might be responsible for the observed cell adhesion, due to the presence of serine-rich repetitive sequences. In addition, the actin cytoskeleton plays a pivotal role in cell attachment and migration by assembling into distinct structures that enable dynamic cellular processes. The presence of hydrophilic amino acids in the sericin also contributed to cell adhesion and proliferation. Nevertheless, the molecular mechanism underlying the biological activity of sericin remains unclear and requires further in-depth investigation [8].

6.5 Sericin in Biomedical Applications

6.5.1 Tissue Engineering Application

6.5.1.1 Wound Healing Application

The skin acts as a defense organ that protects our body from external threats. Trauma, infections, or various medical conditions that affect the skin, mucous membranes, or connective tissues can lead to the formation of wounds. Wounds are generally categorized into two primary types: acute and chronic. Diabetic ulcers and burns, which are common types of chronic wounds, are marked by a delayed healing process, leading to a substantial challenge for the healthcare system. This highlights the importance of developing advanced wound dressings to accelerate healing and facilitate the regeneration of skin tissue [26].

In this context, sericin-based materials with their multifunctional properties, promoting cell proliferation, antioxidant and anti-inflammatory activities, have been used for wound healing applications. Furthermore, additional characteristics, such as non-immunogenicity, anticoagulant and antibacterial properties, along with superior biocompatibility, also promote sericin as a promising candidate for wound repair. Studies have indicated that silk sericin influences the paracrine signals of mesenchymal stem cells that regulate the behavior of tissue-resident cells, thereby promoting wound regeneration. Moreover, it also possesses abundant polar amino acids (serine and threonine) and functional groups (hydroxyl, carboxylic, and amino), which provide binding sites for cross-linking and functional modifications. Consequently, numerous techniques and matrices have been utilized to enhance the ability of sericin to support healing in complex wound microenvironments [27, 28].

For instance, sericin/gelatin scaffolds reinforced with heparin-encapsulated poly (lactic-co-glycolic acid) (PLGA) nanoparticles have been developed using the electrospinning method to accelerate wound healing. Additionally, in vitro studies revealed the critical role of these scaffolds with the lowest degradation rate and marginal concentration of heparin in enhancing the cell proliferation of L929 fibroblast cells, thereby suggesting the suitability of the construct for tissue regeneration [29]. Among various wound dressing choices, hydrogels are widely considered due to their ability to maintain a moist environment, absorb the wound exudate, and act as a barrier against microbial penetration. Hydrogels have been developed by blending sericin with polyvinyl alcohol and carboxymethylcellulose to produce dressings with enhanced mechanical performance. Additionally, bioactive agents, such as natural plant extracts from onion and banyan leaves, were incorporated into the hydrogel matrix to further enhance wound healing activity. Wounds treated with the sericin/onion and banyan leaf extract hydrogel showed faster healing compared to the control. This hydrogel exhibited superior biocompatibility, enhanced infiltration of collagen fibers, fibroblasts, and keratinocytes at the wound site, and promoted angiogenesis [26].

In a similar study, sericin-modified methacrylic anhydride hydrogels were developed with the inclusion of phenylboronic acid and imidazole functional groups (SIP hydrogel) to enhance the antibacterial efficiency of the hydrogel, particularly for treating diabetic ulcers caused by methicillin-resistant *Staphylococcus aureus* (MRSA). The phenylboronic acid in the developed hydrogel dressing efficiently captured the bacteria, while the imidazole effectively eliminated them, further proving that the developed dressing could be highly effective for bacterial treatment. Notably, experimental analyses conducted both in vitro and in vivo using animal models demonstrated the significant role of the SIP hydrogel in promoting cell proliferation, epithelial regeneration, angiogenesis, and reducing inflammation at the injury site (Fig. 6.2a–c) [30]. Although porous hydrogels exhibit superior bioactive molecule loading efficiency, penetration through the skin remains a challenge, thereby reducing the overall potential of hydrogel dressings.

Recent advancements in microneedle patches address these limitations by facilitating the transdermal delivery of loaded drugs or bioactive agents into the outer epidermal layer and the superficial dermis of the skin. A novel oxygen-generating and glucose-responsive microneedle platform using boronate ester cross-linking of

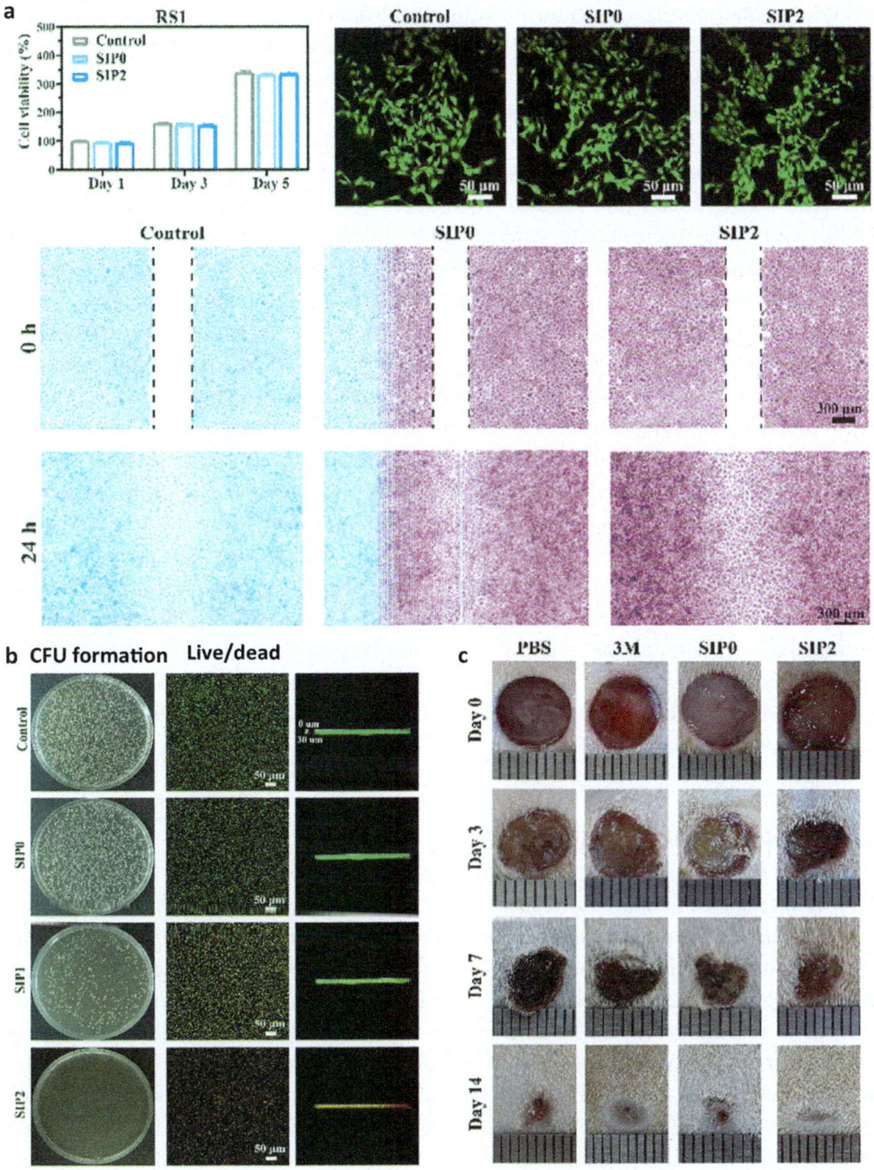

Fig. 6.2 **a** In vitro cellular studies on SIP hydrogels; **b** the anti-MRSA efficacy of SIP hydrogels was evaluated through agar plate assays and live/dead staining techniques; **c** in vivo evaluation of wound closure in models with full-thickness skin defects. Adapted and edited with permission from reference [30], copyright 2025, Elsevier

dopamine-functionalized sericin protein and 4-amino-3-fluorophenylboronic acid-functionalized hyaluronic acid, preloaded with verteporfin, was proposed for effective diabetic wound healing. The functionalized sericin protein, under oxidative stress conditions, efficiently eliminates excessive reactive oxygen species, which further generates oxygen and reduces the conditions of hypoxia. By activating the extracellular signal-regulated kinase and heme oxygenase-1 pathways, this process enhanced angiogenesis and reduced the expression of pro-inflammatory cytokines. Additionally, the microneedle patches under hyperglycemic conditions gradually disintegrate and trigger the controlled release of verteporfin, which promotes scarless re-epithelialization by inhibiting the yes-associated protein pathway [28].

Furthermore, the development of tissue adhesives that function as wound sealants and repair with instantaneous adhesion to wet biological tissue highlights a substantial improvement in wound healing processes and tissue regeneration. Herein, a study presents the fabrication of hydrophobic adhesive obtained by the organized assembly of sericin protein containing hydrolyzed disulfide bonds and tannic acid (rSer-TA), which simulated the interaction between hydrophobic amino acids and adhesive components present in marine organisms including mussels and barnacles. The wet adhesion of the developed adhesive is achieved through the self-aggregation of nonpolar amino acids, which results in water repulsion upon exposure to water and further promotes the formation of hydrogen bonds and electrostatic forces at the interface. This mechanism contributes to robust, repeatable, and durable underwater adhesion. In contrast to commercially available cyanoacrylate-based adhesives, the sericin-based adhesive demonstrated superior tissue sealing, enhanced adhesion, rapid hemostatic activity, and excellent healing properties [31].

6.5.1.2 Neural Tissue Engineering

After heart disease, cancer, and respiratory disorders, stroke is considered the fifth primary cause of death globally. Severe ischemic stroke leads to neuronal damage, resulting in extracellular fluid (ECF)-filled irregular-shaped cavities. This causes severe implications and substantially reduces the patient's overall well-being. Currently, the methods utilized in brain tissue regeneration remain ineffective. However, a promising alternative is the implantation of biomaterials that offer extracellular matrix and structural support for both endogenous and/or implanted stem cells to facilitate the development of the new neural networks [32, 33]. A hydrogel developed using sericin-conjugated genipin (Gen-SH) has exhibited promising result in treating ischemic stroke, facilitating the delivery of neuronal cells and providing protection against oxidative stress. The Gen-SH hydrogel was observed to facilitate efficient cell adherence, proliferation, and neuronal development in both in vitro and in vivo conditions. A 17-fold increase in the Bcl-2 to Bax ratio was observed in cells treated with sericin, compared to control groups treated with PBS and BSA, highlighting potential neuroprotective properties of sericin. Moreover, in hypoxic or glucose-deficient environments, the degradation products of Gen-SH support primary neuron survival and maintain cellular stability, further highlighting the role of Gen-SH in neuronal protection and prevention of cell death during brain tissue healing. This study emphasizes the impact of sericin in activating the Lkb1 and Nuak1 kinase pathways, which enhance axonal outgrowth. Together, sericin and degraded intermediates of Gen-SH demonstrated feasibility in applying the hydrogel for treatment and repair of ischemic stroke [32].

The incomplete gelation of conventional hydrogels is caused by the dilution with existing extracellular fluid and leakage into cerebral ventricles. This results in the formation of unstable and irregular macrostructures, which fail to provide structural support effectively. Injectable shape-memory scaffolds have been proposed to address these challenges. A carbon nanotube (CNT)-doped silkworm sericin scaffold (SS) with injectability, photoluminescence, and programmable shape-memory properties was fabricated and utilized for the repair of stroke-affected brain (Fig. 6.3a). The incorporation of CNTs into sericin conferred the CNT-SS composite with programmable shape-memory properties (Fig. 6.3b), where the recovery dynamics of the CNT-SS were mathematically calculated over a time scale of seconds by adjusting the concentration of CNT. Additionally, the hydrogels developed in this study can be designed to precisely fit any irregular cavities. The composite scaffold demonstrated the ability to recover its pre-designed shape after injection and precisely fit into the cavity, as demonstrated by a preclinical stroke model. Remarkably, the photoluminescence properties enabled the non-invasive approach for real-time tracking of the implanted cells in vivo (Fig. 6.3c). Furthermore, the fabricated CNT-SS scaffold effectively delivered stem cells into brain tissues and facilitated neuronal differentiation (Fig. 6.3d, e). Therefore, the injectable shape-memory scaffolds developed from silk sericin demonstrate promising potential for stroke repair (Fig. 6.3f) [33].

Sericin-based scaffolds have been employed for addressing chronic nerve compression (CNC), which is a condition that occurs when the nerve is compacted against surrounding tissues causing pain, numbness, and dysfunction. Similar to ischemic stroke, currently administered treatments for CNC have proven to be ineffective in some cases. Therefore, tissue engineering approaches have been proposed as efficient techniques for repairing CNC. A 3D scaffold, fabricated using chitosan and silk sericin crosslinked with genipin, was employed to deliver nerve growth factor (NGF) for CNC repair (Fig. 6.4a). The degraded derivatives of the sericin/chitosan scaffold have been shown to enhance the expression of nerve function recovery-related genes such as glial-derived neurotrophic factor (GDNF), early growth response 2 (EGR2), and neural cell adhesion molecule (NCAM) expressed in Schwann cells (Fig. 6.4b, c). However, the expression of inflammatory genes (TNF-α and IL-1β) in macrophages was downregulated. This fabricated scaffold has been identified as a significant candidate for treating CNC due to its ability to enhance nerve impulse conduction, accelerate tissue microstructure recovery, reduce nerve pain, and reduce gastrocnemius muscle degeneration, thereby leading to substantial nerve functional recovery, as demonstrated by studies in preclinical CNC animal models (Fig. 6.4d, e) [34].

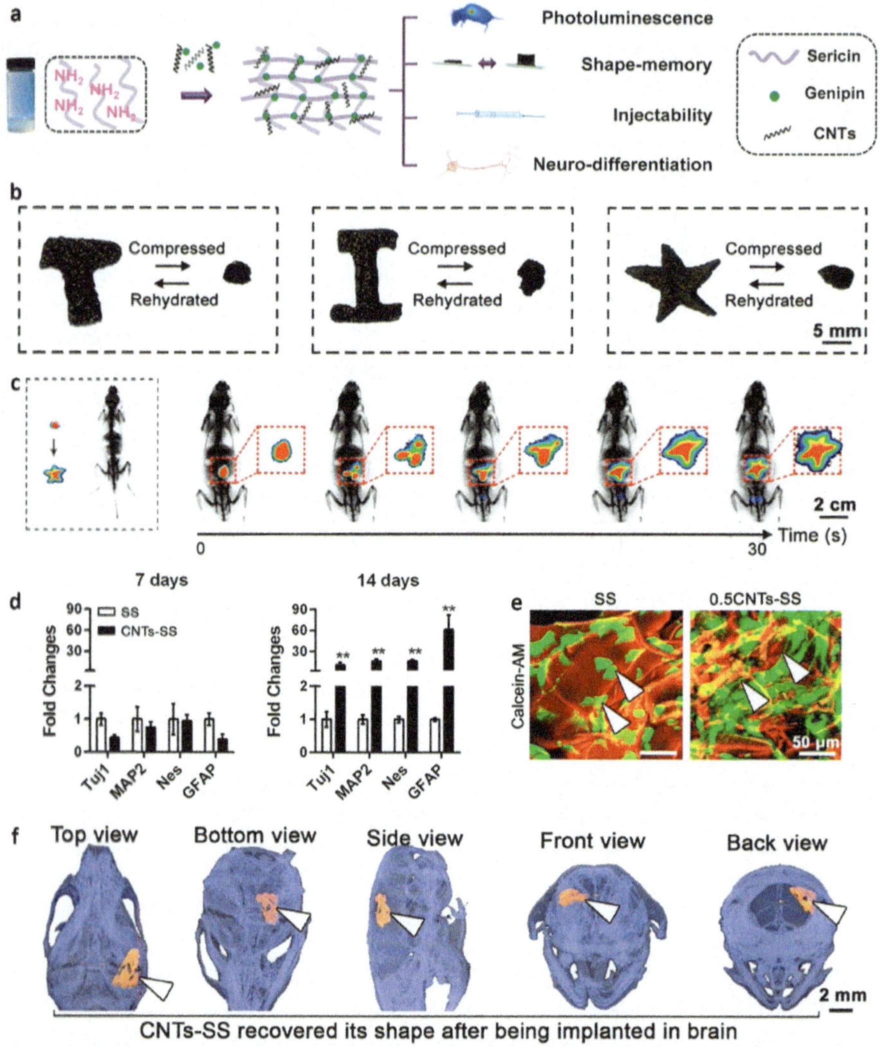

Fig. 6.3 **a** Schematics representing the design of CNTs-SS; **b** shape recovery response of 0.5CNTs-SS post-compression under hydrated conditions; **c** the compressed 0.5CNTs-SS demonstrated shape recovery to its original pentagram structure after in vivo intraperitoneal implantation; **d** the relative mRNA expressi levels of neuron-specific genes in Bone Marrow-derived Mesenchymal Stem Cells (BMSCs) cultured on 0.5CNTs-SS scaffolds for 7 and 14 days; **e** morphological assessment of live BMSCs usng Calcein-Acetoxymethyl staining; **f** micro-CT imaging illustrating the distribution of CNTs-SS following injection into the ischemic lesion of the brain [33]. Adapted and edited with permission from reference [33], copyright 2020, Elsevier

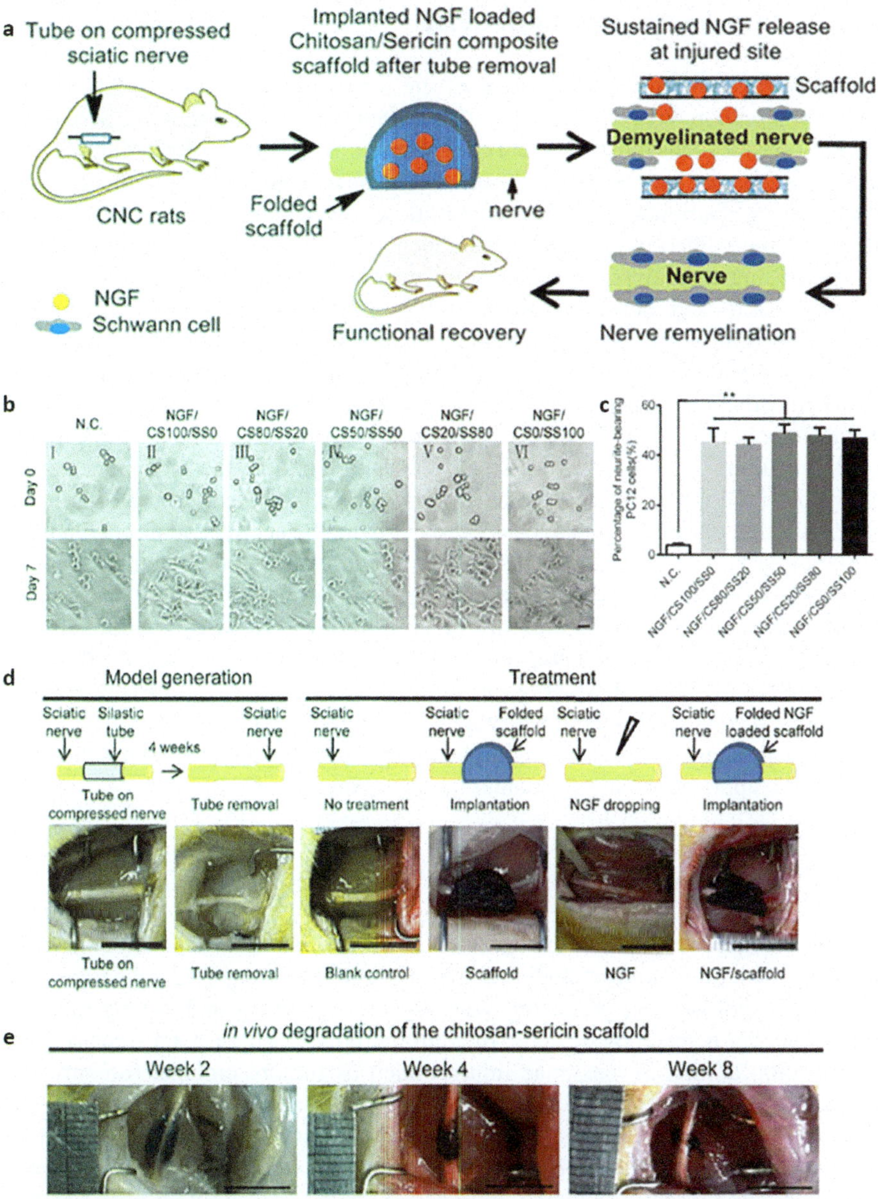

Fig. 6.4 **a** Schematic representation of utilization of chitosan/sericin scaffold to treat CNC; **b** microscopic phase contrast images of PC12 at days 0 and 7, cultured with scaffold fabricated; **c** assessment of the percentage of PC12 cells with neurites; **d** upper panel: Graphical illustration of the CNC model and the corresponding treatments. While corresponding surgical photographs to the upper panel are shown in the lower panel; **e** in vivo assessment of degradable chitosan-sericin scaffolds was conducted at 2, 4, and 8 weeks after implantation. Adapted and edited with permission from reference [34], copyright 2017, American Chemical Society

6.5.1.3 Periodontal Tissue Engineering

Periodontitis, a chronic inflammation of the periodontium, is mainly attributed to bacterial infection, which leads to the accumulation of plaque within the periodontal pocket. It is one of the most prevalent conditions, affecting nearly 50% of adults in the age group 30 and older. However, the effective management and repair of periodontal bone defects continues to present a considerable challenge in clinical treatment. Recently, the guided tissue regeneration technique has gained increasing attention as a promising alternative for periodontal regeneration therapy. For instance, sericin/hydroxyapatite and polyvinyl alcohol (PVA)/sericin/hydroxyapatite membranes have been developed and demonstrate remarkable biocompatibility. Further, the membranes promoted the osteogenic differentiation of human periodontal ligament stem cells (hPDLSCs). Moreover, PVA/sericin/hydroxyapatite membranes promoted the differentiation of hPDLSCs into osteoblasts via upregulation of osteogenic genes including ALP, Runx2, OCN, and OPN. In addition, the membranes exhibited substantial potential for treating periodontal tissue defects by facilitating the differentiation of hPDLSCs into osteoblasts without the need for additional inducers [35]. Similarly, sericin electrospun scaffolds combined with poly (lactic-co-glycolic acid) (PLGA) and ketoprofen were developed to provide promising environment for the adhesion and proliferation of human gingival fibroblasts. Additionally, the scaffolds exhibited sustained anti-inflammatory action, as revealed by the downregulation of the MMP-9 and MMP-3 (pro-inflammatory markers) and an upregulation of the IL-10 gene (anti-inflammatory gene) in lipopolysaccharide-stimulated RAW 264.7 macrophages. This study highlights the potential of sericin-based scaffolds in therapeutic applications in periodontal tissue engineering [36].

6.5.1.4 Bone Tissue Engineering

Among the various tissues in the human body, bone constitutes a major component with significant functions, including providing mechanical support and a framework for the skeletal muscle, protecting vital organs, depositing and storing minerals, and producing hematopoietic cells. In addition, bones also have self-repairing and regeneration ability. Severe bone injuries such as tumors, trauma, osteomyelitis, arthritis, and others lead to bone defects that require clinical treatment. Currently, the bone defects are treated using autografts and allografts, which are often limited due to their immunogenicity or shortage of donors. As a result, the tissue-engineered bone implants present a potential approach. Sericin, a silk protein obtained from silk cocoons, is widely explored in wide range of biomedical fields due to advantages including characteristics promoting cell adhesion and proliferation, excellent biocompatibility, and minimal immunogenic response. For instance, photo-crosslinked sericin methacryloyl scaffolds with graphene oxide were developed for functional bone repair. The implanted biomimetic scaffold in a rat calvarial bone defect model facilitated functional and structural repair within 12 weeks through the activation of MAPK, TNF, and chemokine-mediated signaling pathways, thereby promoting new bone regeneration through the migration and differentiation of BMSCs [37]. Similarly, poly (L-lactic acid) (PLLA)-sericin hybrid scaffolds loaded with gentamicin sulfate (GS) demonstrated antibacterial activity against *E. coli* and *S. aureus*, as well as a significant proliferative effect on osteoblast MC3T3-E1 cells, which highlights the prospective of sericin-based PLLA scaffolds for bone tissue engineering [38]. Although the bone defects caused by several bone-related injuries are often irregular in shape. The existing alternative approach of utilizing autologous grafts cannot be deemed as the effective therapeutic method for treating irregular bone defects due to increased complexity of graft harvest, adverse effects at donor sites, and additional operative time and cost. Nevertheless, artificial bone implants, particularly injectable hydrogels, can be used as an ideal candidate to fill irregular-shaped bone defects. The potential of an injectable alginate/sericin/graphene oxide (Alg/Ser/GO) hydrogel, developed using the Alg-Tyramine framework with enzymatic cross-linking, has been evaluated as an alternative for bone regeneration (Fig. 6.5a). The fabricated hydrogel revealed a controlled degradation behavior for sericin and GO. The incorporation of graphene oxide effectively promoted rat BMSC proliferation, osteogenic differentiation, and extracellular matrix mineralization. Furthermore, the release of sericin promoted M2 macrophage polarization and enhanced cell migration via activation of the NF-κB and MAPK pathways, which further contributed to osteogenic differentiation. The injected hydrogel with BMSCs effectively repaired distal femoral defects established in rats. Additionally, both in vitro and in vivo studies demonstrated macrophage infiltration into adjacent tissues, suppression of inflammatory responses, and fibrous capsule thickening (Fig. 6.5b, c) [39].

However, the design and development of the tissue-engineered substitutes primarily focused on promoting osteogenesis without considering the immune responses induced by the biomaterials. Prior studies have shown that host-immune rejection of the implanted substitute may result in fibrotic encapsulation, instability, or failure of the implantation. Hence, sericin combined with nano-hydroxyapatite and graphene oxide hydrogels has been considered for effective bone regeneration with both osteogenesis and an immunomodulatory effect. The rat cranial defect model study indicated reduced inflammatory responses and promoted bone regeneration mediated by osteogenic differentiation. In this study, sericin contributed to the immunomodulatory effect by stimulating macrophage M2-type polarization, which inhibited the immune response and resulted in the formation of a positive osteoimmune microenvironment, thereby promoting osteogenesis [40].

6.5.1.5 Cartilage Tissue Engineering

Cartilage defects, including cartilage injury and cartilage degradation due to aging or disease, affect millions of adults worldwide. Cartilage, particularly articular cartilage, is a simple-bearing tissue essential for the proper functional activity of joints. However, the repair of articular tissue after injury remains a significant clinical

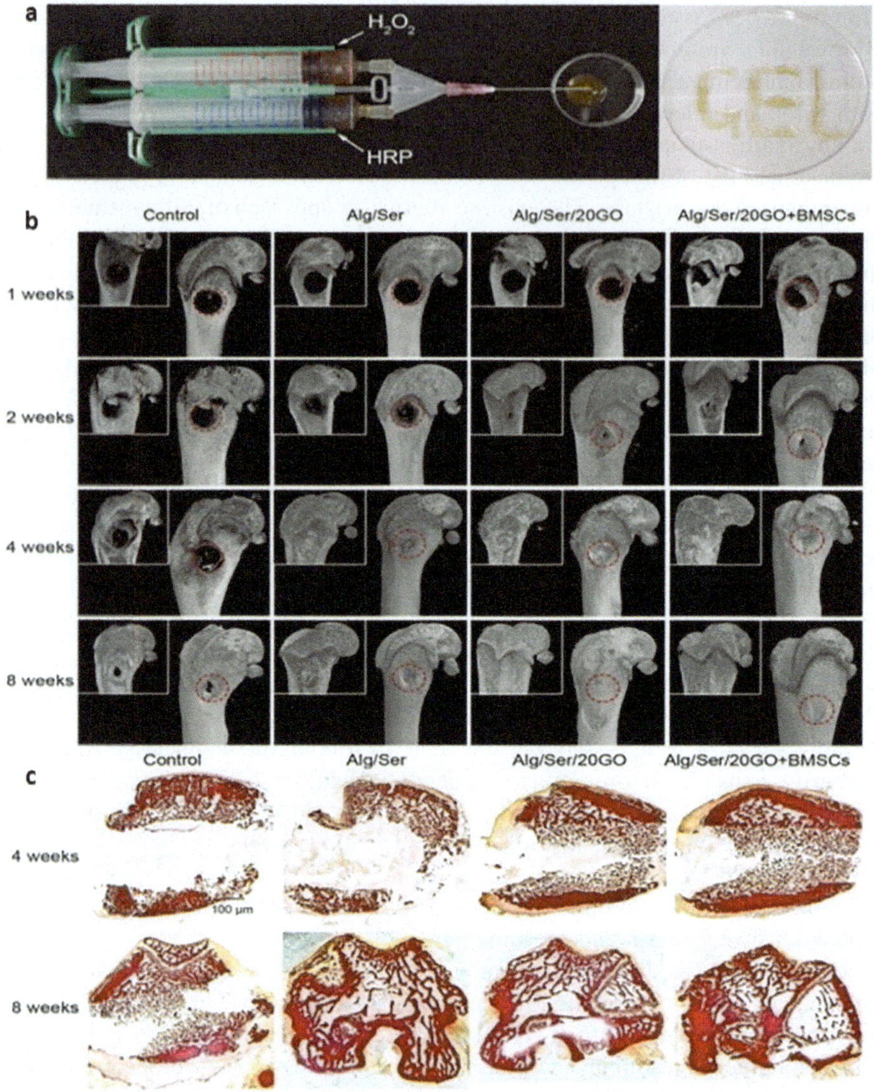

Fig. 6.5 **a** Photograph of rapid gelatinization properties following injection of developed Alg/Ser/GO gel; **b** coronal and 3D reconstructed micro-CT images of rat femurs after 1 to 8 weeks post-implantation; **c** histological Van Gieson's staining of axial distal femoral defects after hydrogel implantation at 4 and 8 weeks. Adapted and edited with permission from reference [39], copyright 2021, Elsevier

problem due to its poor self-repair and regeneration. At present, various treatment options are considered to treat articular cartilage defects, including arthroscopic debridement, implantation of chondrocytes, replacement of joints, allografts, autografts, and microfracture techniques. Although the current treatment methods employed are associated with severe complications such as the development of fibrocartilage and the overgrowth of subchondral bone due to microfracture, as well as delamination, hypertrophy, and graft failure associated with autograft and allograft procedures [41]. Advancements in regenerative therapeutics provide promising treatment alternatives for the treatment and repair of cartilage injuries, notably through the development of tissue-engineered substitutes, such as 3D scaffolds, hydrogels, and others. Consequently, researchers have investigated the potential of various biopolymer-based implants for effective cartilage repair. In particular, sericin with its inherent properties makes it a promising candidate for cartilage regeneration. In recent studies, a porous dual crosslinked hydrogel, containing sodium alginate (SA) and silk sericin (SS), was engineered for the in situ regeneration of cartilage defects (Fig. 6.6a). The hydrogel with 1% of SS facilitated chondrocyte adhesion and proliferation, while consequently reducing inflammatory cell infiltration (Fig. 6.6c, f). The hydrogel demonstrated systemic degradation behavior by varying the SS content that aligns with the rate of cartilage tissue repair. The hydrogel exhibited a systematic degradation behavior, regulated by varying the SS content, which aligned with the rate of cartilage tissue repair. Furthermore, the cartilage repair efficiency of the 3D sericin-based hydrogels was evaluated through in vivo studies using a rabbit patellar cartilage injury model (Fig. 6.6b, d, e). The study revealed the successful repair of defective cartilage within 4 weeks, with structural and functional properties similar to native cartilage [42].

In a subsequent study, injectable, photo-crosslinkable sericin hydrogels seeded with chondrocytes were proposed for cartilage repair. The functionalization of sericin into sericin methacryloyl (SerMA) facilitated in situ hydrogel formation through UV-initiated photopolymerization (Fig. 6.7a). Further, the photoluminescent properties of SerMA hydrogels facilitated continuous real-time monitoring. The hydrogels demonstrated superior characteristics, including excellent biocompatibility, promotion of chondrocyte adhesion, and enhanced cellular proliferation (Fig. 6.7b–f)0.6 Notably, the implantation of chondrocyte-laden SerMA hydrogels resulted in the formation of artificial cartilage within 8 weeks, exhibiting molecular and histological characteristics similar to native cartilage. Furthermore, the SerMA hydrogel further supported the expression of cartilage-related genes along with the accumulation of extracellular matrix component characteristics of cartilage [43].

Another study reported the synthesis of nano-gel soft agglomerates through metal ions complexation with silk sericin protein. The developed agglomerates were utilized as linkers to facilitate the assembly of mesenchymal stem cells (MSCs) into microspheres while promoting their differentiation into cartilage-like microtissues without the need for induction of exogenous growth factor. Moreover, the addition of sericin nanogels induced condensed micromass formation of MSCs, which subsequently developed into larger aggregates progressively over time. Further, in vitro studies demonstrated that various signaling pathways, such as calcium, FAK → Ras/

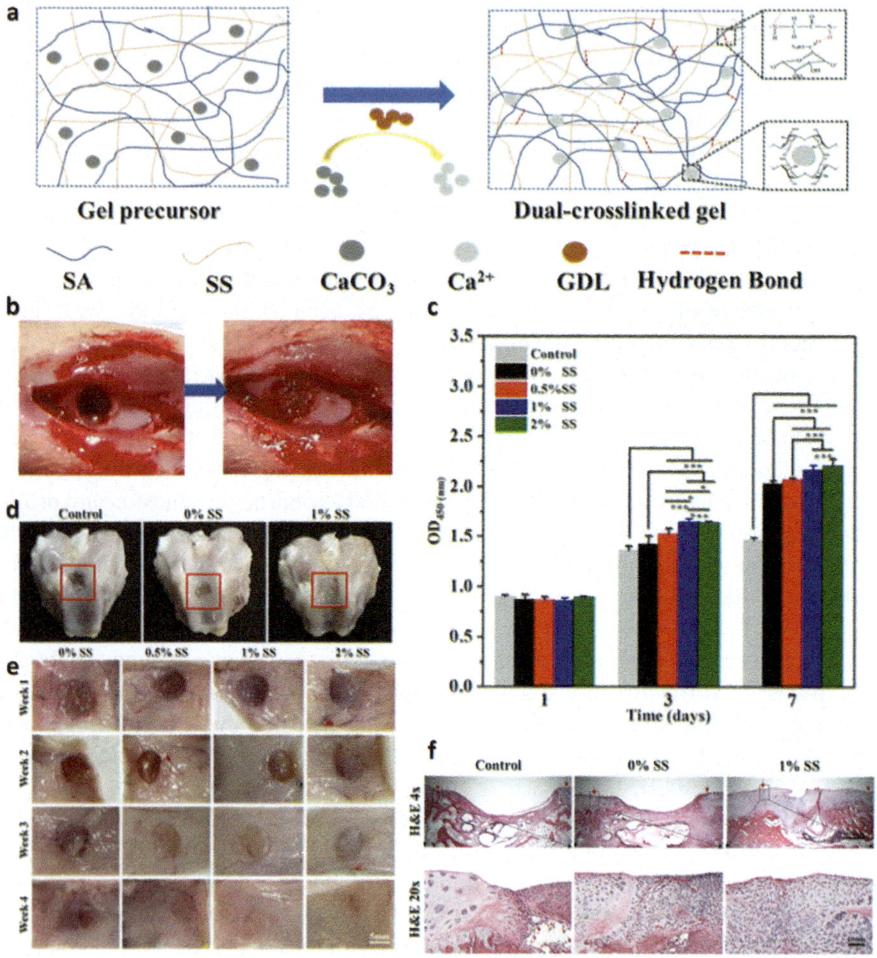

Fig. 6.6 **a** An illustrative representation of SA/SS dual crosslinked hydrogel; **b** in vivo studies to repair cartilage injury; **c** in vivo assessment of hydrogel degradation over time; **d** photographs of the cartilage repair; **e** observation of hydrogels post-subcutaneous implantation at 1, 2, 3, and 4 weeks; **f** histopathological examination of cartilage healing in rabbit patella using H&E-stained sections. Adapted and edited with permission from reference [42], copyright 2023, Elsevier

Rho, cAMP, hypoxia → PI3K–Akt/HIF-1α, and TGF-β → MAPK or ERK/MAPK, played a major contribution to the assembly of MSCs into micromasses and chondrogenic differentiation. [41]. Therefore, sericin-based implants exhibit potential for repairing defective cartilage in vivo, providing an effective approach toward the treatment of cartilage injury and related diseases.

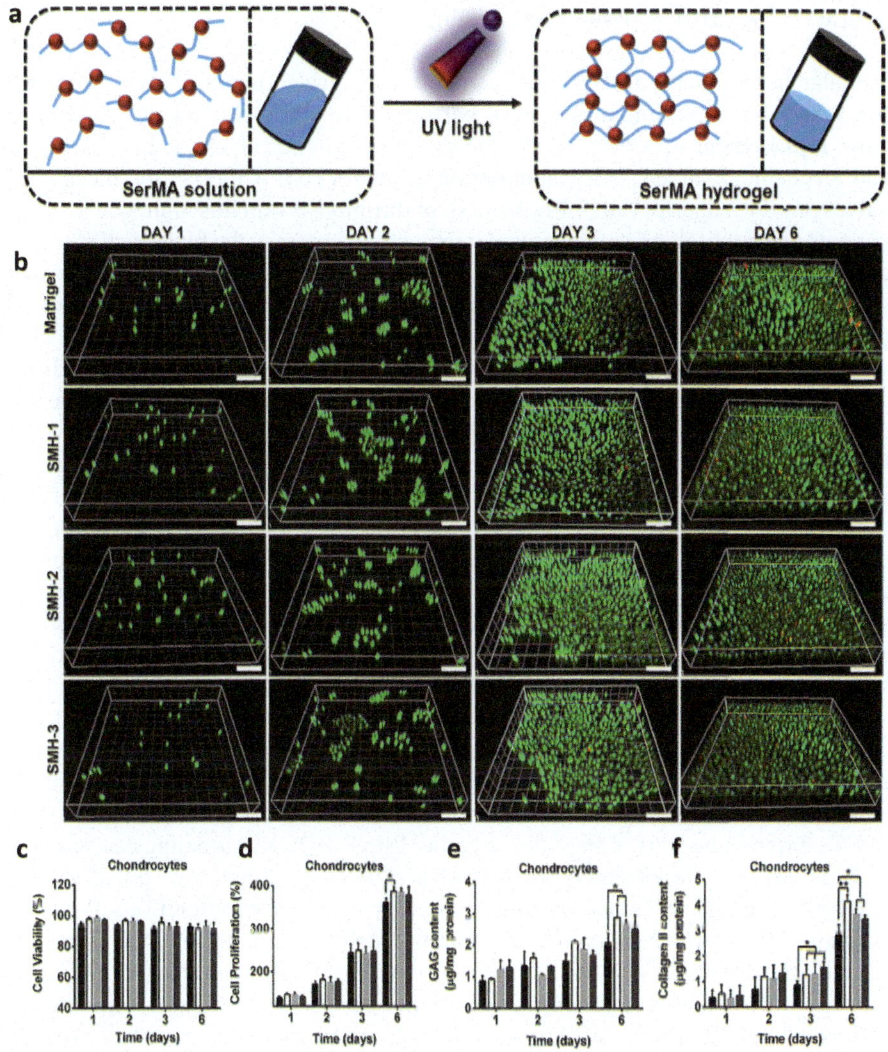

Fig. 6.7 **a** Preparation of SerMA hydrogel for cartilage repair and regeneration; **b** chondrocyte viability assessment via live (green) and dead (red) staining after 6 days of culture in the hydrogel. Quantification of **c** chondrocytes cell viability; **d** chondrocytes cell proliferation within 3D scaffolds at varying time points; **e** evaluation of glycosaminoglycan (GAG) accumulation in chondrocytes cultured within SerMA hydrogels; **f** quantification of collagen II content in 3D hydrogels containing chondrocytes. Adapted and edited with permission from reference [43], copyright 2018, Elsevier

6.5.1.6 Myocardial Tissue Engineering

Valvular heart disease (VHD) poses a significant risk to human health, with over 300,000 valve replacement surgeries performed annually worldwide. At present, valve replacement surgeries involve the use of prosthetic valves, such as biological and mechanical valves, as a viable option to treat VHD. Although the mechanical valves provide superior durability, the risk of thrombosis remains high, due to which patients require lifelong anticoagulant treatment. In contrast, the biological valve has a lower risk of thrombosis but is prone to valve calcification and degradation over a period of time. Thus, efforts have focused on engineering decellularized heart valves (DHVs) to overcome the limitations of the prosthetic valves. Decellularized heart valves (DHVs) are scaffolds that facilitate cell attachment, proliferation, and differentiation. However, DHVs are prone to rejection after implantation due to insufficient recellularization of the valve. Therefore, several surface engineering approaches have been employed to modify DHVs in order to promote re-endothelialization and improve biocompatibility [44, 45]. Sericin with abundant carboxyl, hydroxyl, and polar functional groups provides many reaction sites for the polymerization and cross-linking ability with other materials. Consequently, sericin was also considered to modify the surface of DHV to enhance endothelialization. For instance, in a study, Michael addition reaction was utilized to synthesize a covalently modified silk sericin valve (SCMV) by coupling maleylated silk sericin (MSS) with thiolated decellularized heart valves (DHVs). The results highlight that SCMV exhibits a smoother surface, enhanced hydrophilic properties, and a higher resistance to platelet adhesion than DHV. In addition, SCMV enhances the attachment of endothelial cells (ECs) in both stationary and dynamic flow conditions as well as promotes their survival in a subcutaneous mouse implantation model. Further, sericin facilitated the attachment of ECs through the activation of the integrins-RhoA-ROCK2 signaling cascade [44].

Similarly, the decellularized heart valve (DHV) modified with polydopamine (PDA) and sericin promoted the adhesion, migration, and proliferation of valvular interstitial cells, which are key to facilitating recellularization in DHVs. Moreover, in vivo studies revealed that the PDA/sericin-modified DHV exhibited a slower decay rate and a decrease in inflammatory cell infiltration, due to the anti-inflammatory effects inherent to sericin. Additionally, the fabricated valve demonstrated excellent biocompatibility with blood after transplantation into the rabbit carotid artery [45]. Besides, sericin-based tissue-engineered scaffolds have been identified as therapeutic biomaterials for the repair of other heart-related diseases, such as acute myocardial infarction (MI). A genipin-crosslinked, injectable sericin hydrogel was fabricated in situ for the repair of myocardial infarction (Fig. 6.8a). As a result of the application of the developed hydrogel to the area affected by myocardial infarction (MI), a reduction in scar tissue and infarct size, as well as an increase in wall thickness and neovascularization was observed (Fig. 6.8b–d). It also reduced inflammation and apoptosis induced by MI, which resulted in improvement in cardiac function. The results indicated that sericin reduces the levels of pro-inflammatory cytokines (TNF-α, IL-18) and chemokines (CCL2), while downregulating TNF-α expression by suppressing the TLR4-MAPK/NF-κB signaling pathways (Fig. 6.8e, f). Further, sericin induces

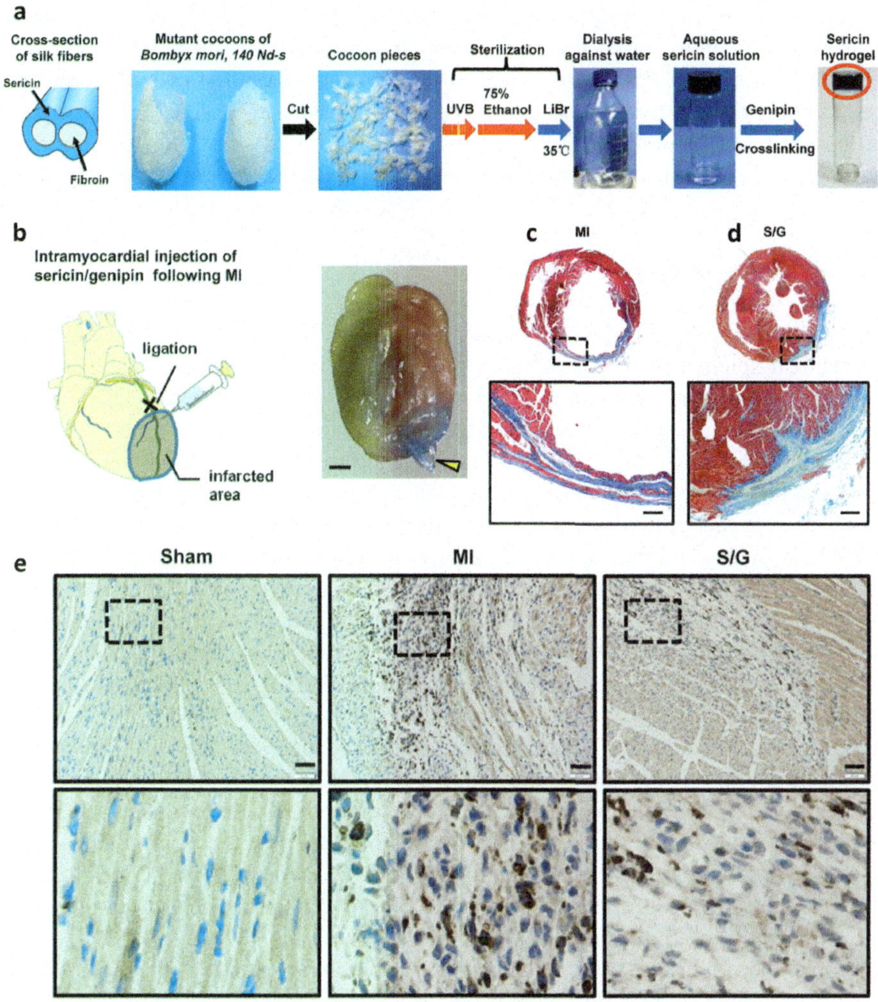

Fig. 6.8 **a** Fabrication of sericin-based hydrogel for ischemic myocardial repair; **b** a schematic representation and macroscopic view of a wild-type heart with the anatomical site covered with sericin/genipin hydrogel; assessment of morphological and functional cardiac recovery post-infarction using sericin hydrogels. Histological analysis of cardiac tissue using Masson's trichrome staining 6 weeks following MI, showing treatment with **c** PBS and **d** sericin hydrogels; **e** representative immunohistochemical images of CD68 + positive macrophage infiltration in infarcted hearts from the sham, MI, and S/G groups 1 week following MI. Adapted and edited with permission from reference [46], copyright 2016, Elsevier

angiogenesis by facilitating the migration and tubular network formation of human umbilical vein endothelial cells (HUVECs). In addition, sericin facilitates VEGF-A expression through ERK phosphorylation and prevents apoptosis in endothelial cells and cardiomyocytes by inhibiting the activation of caspase-3. Considering the properties and evidence indicating the ability of sericin-based substitutes in promoting angiogenesis, it represents a promising solution for cardiac tissue engineering [46].

6.5.1.7 Drug Delivery Constructs

Targeted drug delivery serves as a key role in the field of biomedical and pharmaceutical science which can enable prolonged and localized drug delivery. The effective design and development of targeted drug delivery systems (DDS) results in enhanced availability of drugs to target with minimal side effects on the healthy cells and/or body. Controlled and sustained drug release is a key characteristic of an optimal DDS [47]. Several natural and synthetic polymers have been extensively used in the fabrication of various polymeric scaffold forms of DDS, including hydrogels, nanogels, microgels, 3D-printed structures, and electrospun structures. However, natural polymers, particularly sericin, have obtained notable attention in recent studies for the fabrication of DDSs due to their hydrophilicity, biodegradability, and cytocompatibility [48]. Furthermore, the potential of sericin-coated nanolipid carriers (Se-NLCs) loaded with hesperidin (Se-He-NLCs) to treat gastric ulcers was evaluated in a study. Hesperidin release from Se-NLCs followed first-order kinetics, demonstrating a controlled and sustained delivery profile. The in vitro permeation and ex vivo mucoadhesion of Se-He-NLCs were increased, leading to higher bioavailability in vivo. Further, the developed DDS also produced a 2.24-fold increase in reduced glutathione and a 1.61-fold increase in glutathione-S-transferase in the gastric mucosa, while effectively regulating SOD and CAT activities to levels comparable to those in the control group. Moreover, the administration of Se-He-NLCs resulted in a considerable reduction in tissue lipid hydroperoxides (LOOH) and myeloperoxidase (MPO) activity. This approach presents the substantial ability of sericin-coated hesperidin nanocarriers in the treatment of gastric ulcers [49]. In another study, the potential of sericin-based nanocapsules in efficiently delivering the drug cisplatin (Cis-SNCs) was assessed as an advanced DDS for cancer therapy, particularly breast cancer (Fig. 6.9a). The fabricated nanocarriers, prepared with an effective concentration of 0.1 w/v % sericin, demonstrated an 84% drug encapsulation efficiency and maintained a negative net charge at the physiological pH of 7.4. Additionally, the developed DDC showed a controlled and pH-dependent drug release pattern, which becomes essential for sustaining a slightly acidic (pH $6.0-7.0$) tumor microenvironment. Cispt-SNCs demonstrated enhanced cellular uptake under mildly acidic conditions (pH 6), followed by DNA damage and nuclear condensation, leading to apoptotic cell death. Altogether, the results obtained represent the possibility of utilizing the SNCs as promising DDS in tumor-specific microenvironments. Although the antioxidant properties of sericin facilitated a remarkable protective effect on both normal and cancerous cells, the utilization of higher doses of cis-SNC

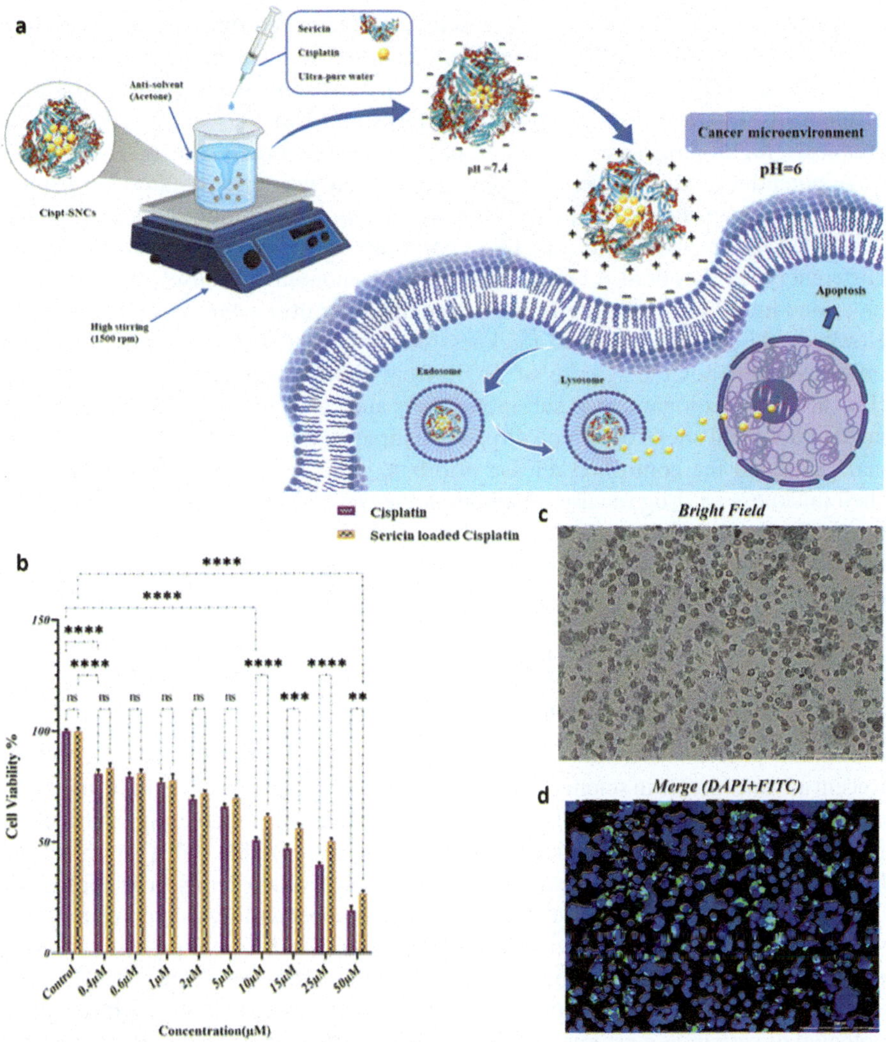

Fig. 6.9 **a** A graphical illustration of the synthesis and mechanism of sericin-based nanocapsules for cancer therapy via targeted drug delivery; **b** evaluation of cytotoxic effects of Cispt-SNCs and cisplatin on MCF-7 cells at varying concentrations; fluorescence imaging of FITC-labeled Cispt-SNCs uptake under pH 6 conditions; **c** bright field image; and **d** DAPI-stained fluorescent image. Adapted and edited with permission from reference [50], copyright 2024, American Chemical Society

at concentrations of 25 and 50 μM caused cytotoxicity for tumor cells. In addition, in vitro studies indicated the gradient reduction in Michigan Cancer Foundation-7 (MCF-7) cell viability (Fig. 6.9b–d), suggesting a sustained and controlled release of the drug. Interaction of the developed DDS with MCF-7 cells resulted in cell apoptosis, highlighting the potential of nanocarriers for breast cancer therapy [50].

A potential stimuli-responsive anti-cancer DDS was also synthesized using a sericin microcapsule as the matrix. Sericin and silica hybrid colloidosomes were developed and encapsulated with a hydrophilic anti-cancer drug, doxorubicin (DOX). Different types of colloidosomes, comprising sub-monolayer, densely packed monolayer, and multilayer SiO_2 shells, were developed and utilized in subsequent studies. However, the arrangement and thickness of the colloidal layer of silicon dioxide influenced the flexible release of DOX from the DDS developed in this study (Fig. 6.10a). In comparison to sub-monolayers and close-packed monolayers, multilayered colloidal shells of silicon dioxide maintained the structure of colloidosomes and facilitated the controlled release of drugs, indicative of the influence of SiO_2 shell thickness on drug release. In addition, the doxorubicin exhibited a toxic effect on the Human Bone Osteosarcoma cell line (MG-63) cells when released from the microcapsules, as a result of the degradation in the microenvironment of the cancer cells (Fig. 6.10b) [51].

6.5.2 Wearable Electronics

Recent advancements in science and technology have made wearable multifunctional electronic devices considerably essential in biomedical engineering, particularly for monitoring personal health and physical activities. Traditionally, strain sensors have been fabricated using metal or semiconductors, which limits their application due to their weight, reduced stretchability and compressibility, as well as a narrow sensing range. Therefore, polymeric substrates have been combined with a variety of conductive nanomaterials, including graphene, carbon nanotubes (CNTs), and carbon black, etc. for sensing applications. Researchers have recently highlighted the potential of sericin as a promising polymeric substrate, primarily due to its excellent biocompatibility and hydrophilic properties [52, 53]. Previous studies have shown that the biocompatibility of sericin significantly neutralizes the biotoxicity of CNTs and also facilitates enhanced dispersion of CNTs in water. Moreover, a melamine-based carbon foam (CMF) combined with sericin-functionalized carbon nanotubes and in situ-formed carbon nanofibers (CNFs) has been developed as a multifunctional wearable thermal therapy device. The resulting carbon foam consisted of a hierarchical porous micro-nanostructure that significantly improved the interaction between CNTs and CNFs. This improved interaction enabled the foam to withstand substantial deformation, with a compressive strain tolerance ranging from 0 to 80% and excellent cyclic stability at a 20% strain. Additionally, the sensors demonstrated remarkable piezoresistive performance, with a sensitivity of 0.11 k/Pa within a pressure range of 2–20 kPa, along with a quick response time (150 ms for response, 120 ms

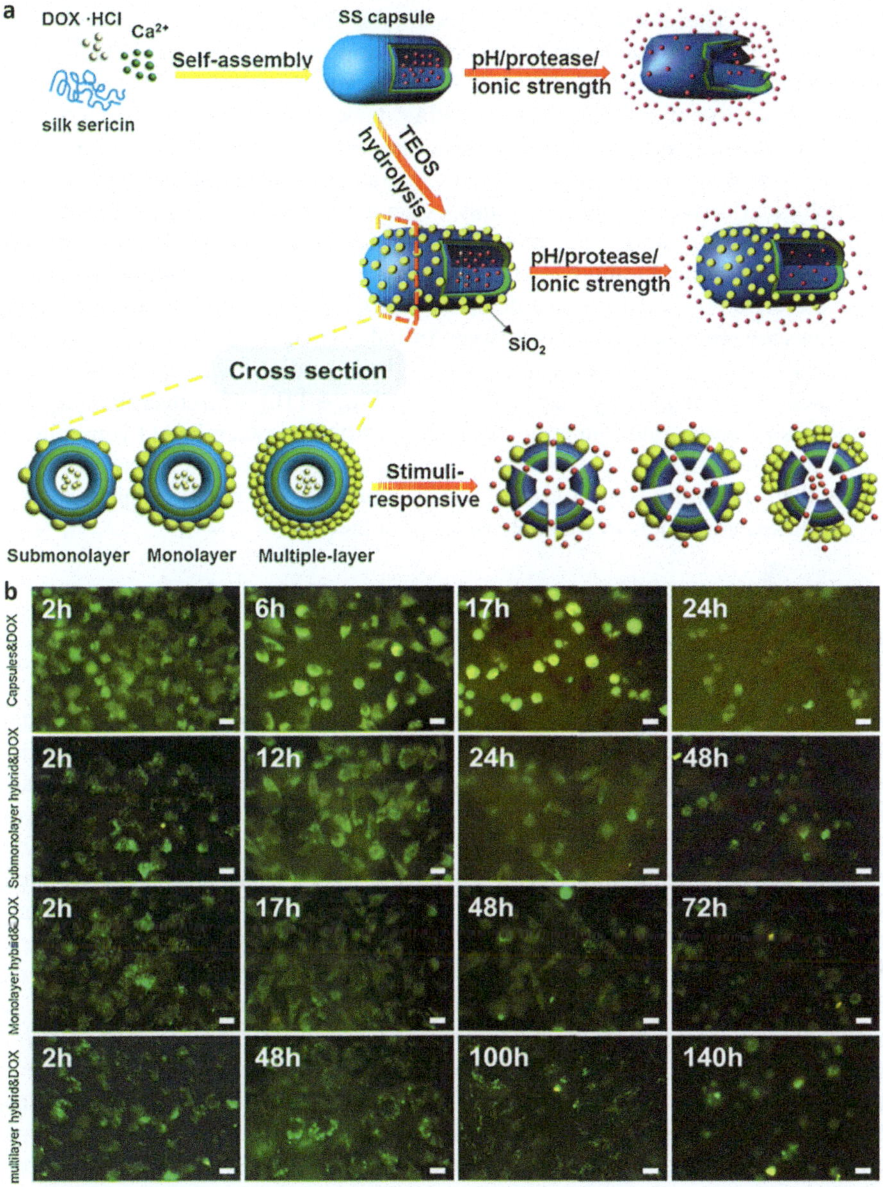

Fig. 6.10 **a** Schematics of colloidosome preparation, drug encapsulation, and release from sericin microcapsules and hybrid colloidosomes under varying pH levels, protease concentrations, and ionic strengths; **b** MG-63 cell morphology stained with DiO after co-culturing with DOX-loaded sericin microcapsules and colloidosomes bearing SiO2 shells in sub-monolayer, monolayer, and multilayer arrangements, evaluated at multiple time points. Adapted and edited with permission from reference [51], copyright 2017, Elsevier

for recovery). The material exhibited remarkable thermal insulation and effective Joule heating, with notable performance in a strain-dependent Joule heating mode at low voltages. Such characteristics of carbon foam are indicative of its utilization in the fabrication of energy-efficient heating devices in wearable medical devices [53]. Significant improvements have been made in the design and development of personalized wearable healthcare devices, particularly for real-time monitoring of physiological metabolites and biofluids. A screen-printed composite consisting of carbon black and recycled sericin on cotton fabrics (CB/rS@fabrics) has been fabricated for monitoring biofluids such as sweat (Fig. 6.11a). Sericin was incorporated as an effective polymer for sweat detection due to its hydrophilic properties, which facilitates the efficient collection of sweat. In addition, the fabric's sensing capability across different sweat pH conditions, ranging from acidic to neutral and alkaline, is attributed to the amphoteric nature of sericin (Fig. 6.11b). Moreover, the CB/rS@ fabrics also demonstrated superior characteristics such as exceptional hydrophilicity (86°), flexibility, resistivity (61.7 Ωcm^{-2}), and 99.1% retention after 30 washes, indicating their durability (Fig. 6.11c–e), further reiterating the practical application of sericin-based constructs in wearable bioelectronics [54].

6.5.3 Implantable Electronic Device

Advancements in implantable electronic devices have raised significant potential for improved diagnostic and therapeutic outcomes, as well as effective prevention and intervention of diseases, by offering real-time monitoring of physiological and metabolic signals. Nevertheless, increased research attention is focused on developing self-powered, biodegradable, and biocompatible implantable devices with reduced in vivo effects. Implantable systems designed using piezoelectric biomaterial have presented a considerable alternative to overcome the limitations of traditional implants [55, 56]. Among various biopolymers, silk sericin has been proposed for developing self-powdered systems due to its significant biological activities and abundant asymmetric amino acids, which facilitate efficient functionalization, as well as physical and chemical modifications. Furthermore, the presence of asymmetric amino acids in the sericin structure generates dipole moments, indicating its piezoelectric properties. Although the piezoelectricity of silk sericin remains unclear due to the lack of domain alignment and the absence of suitable methods for structurally ordered assembly, sericin has been reported to exhibit piezoelectric properties through structural engineering. An energy-generating device has been fabricated using a functionalized sericin-based (F-SS-based) piezoelectric film with a high longitudinal piezoelectric tensor (d33) of 12 $pC\ N^{-1}$, which converts mechanical force into electrical energy in both in vitro studies and in vivo animal models. Moreover, through repeated manual tapping, the EG device was able to store electrical energy in a commercial capacitor (1.1 μF), which was efficiently utilized to power light-emitting diodes or operate a timer. Additionally, the achieved instantaneous power density (218.5 $\mu W\ m^{-2}$) was sufficient to support cardiac pacing to revive a

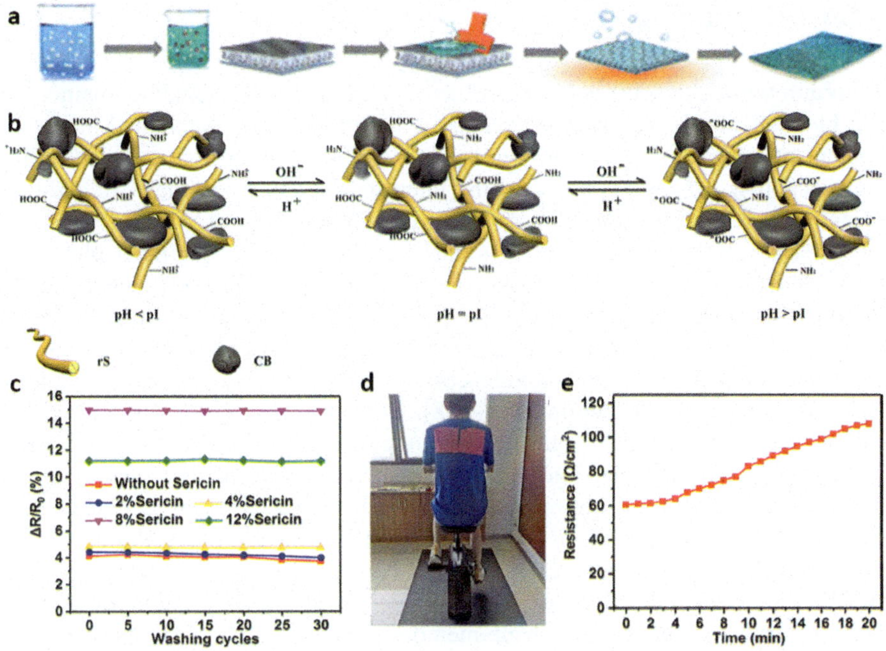

Fig. 6.11 **a** An illustrative representation of fabrication of recycled sericin-based wearable sensor; **b** mechanisms for sensing sweat with varying pH levels (neutral, acidic, and alkaline) using CB/rS-functionalized fabrics; **c** washability and durability of the developed wearable sensor assessed across multiple washing cycles; **d** monitoring of exercise activity using a sweat-monitoring system; **e** variation in the resistance measured by the sweat monitor as a function of exercise duration. Adapted and edited with permission from reference [54], copyright 2022, American Chemical Society

non-beating heart or restore atrioventricular conduction in a preclinical model [55]. Consequently, implantable hydrogel-based supercapacitors (SCs) have been developed using aminated graphene oxide (A-rGO) and sericin modified with methacrylic anhydride (SrMA), which were sequentially crosslinked with polyethylene glycol succinimidyl carbonate and polyethylene glycol acrylate. By incorporating a conductive multi-network into the SrMA/A-rGO-based SC implant, the resulting device exhibits an equivalent series resistance of 21 Ω cm^{-2}, a volumetric energy density of 26.0 μW cm^{-2}, and a specific capacitance retention of more than 76.4% after extended cyclic charging and discharging. Two SCs connected in series efficiently powered a light-emitting diode in both in vitro and in vivo experiments. In addition, in vivo studies demonstrated excellent biocompatibility and biodegradability of SCs. The developed implant showcased superior electrochemical performance and was further utilized as an output power source to electrically stimulate and revive the function of a stopped heart [56].

6.5.4 3D Printing

3D printing has extensive applications across various fields, including engineering and biomedical sciences. Advancements in 3D printing have facilitated the development of complex shapes and geometrical structures with high customizability and scalability, which has attracted the interest of many researchers. In addition, recent progress in machine learning, computer-aided design (CAD) software, and the exploration of diverse novel materials, ranging from conventional plastics to both synthetic and natural polymers, are further expanding the potential of 3D printing technologies. Biopolymers have garnered attention in artificial 3D organ printing as a promising area of research in biomedical engineering, owing to their superior biological properties, including biocompatibility, minimal immunogenicity, and diverse functional capabilities [57, 58]. Of particular interest, silk sericin (SS) exhibits superior biological characteristics, including biocompatibility and tunable gelling properties, as well as good hydrophilicity and ease of functional modification, making SS a promising biomaterial. Moreover, SS has been recognized as a suitable additive component in bioinks used for 3D extrusion bioprinting applications, for instance, when photocrosslinked with bipyridylruthenium (II) hexahydrate (Ru)/sodium persulfate (SPS) to develop 3D constructs using volumetric additive manufacturing techniques. The shape-memory and cytocompatible properties of the developed pristine SS 3D construct are also discussed [59]. Hydrogels composed of silk sericin (SS) and gelatin modified with methacrylic anhydride (GelMA) have been successfully produced through 3D printing for wound healing applications. The resulting 3D-printed sericin-based scaffolds exhibited a porous microstructure, excellent biocompatibility, and low immunogenicity, providing a favorable microenvironment for re-epithelialization and highlighting the potential of sericin-based bioinks in 3D bioprinting technologies [60].

6.5.5 Cell Culture Supplement

Cell and tissue culture techniques have been extensively utilized in the fields of biological science and biomedical engineering. Serum, a primary component of media used for these cultures, contains essential constituents such as lipids, hormones, proteins, growth factors, and adhesion factors. To date, fetal bovine serum (FBS) has been the most widely used animal-based serum in mammalian cell culture, due to its ability to promote cell proliferation and growth. However, preparation of FBS or animal-based sera presents various challenges such as poor batch consistency, increased risk of microbial contamination (particularly Mycoplasma), and ethical concerns about the use of a large number of cattle, all of which result in serious problems for both research and industries related to cell culture. Increasingly, concerns have been raised about the development of alternative substitutes for sera [61].

Sericin, an adhesive silk protein, has been shown to promote proliferation of various cell types, including human skin fibroblasts, HepG2 liver carcinoma cells, Balb3T3 fibroblast cells, HeLa cervical cancer cells, HEK-293 human kidney cells, and RC4 corneal epithelial cells. The incorporation of sericin into culture media was first documented in 2002 [62]. Sericin, composed of 18 distinct amino acids, is well-suited for application as a supplement in serum-free culture media. The presence of mitogenic and antioxidant properties promotes its effectiveness in serum-free medium formulation. In a recent systematic study, silk sericin used as an FBS substitute exhibited similar cellular morphology, improved comparable cell viability, shortened population doubling time, and increased S-phase cell fraction, with a G2/G1 ratio comparable to controls. Furthermore, at the transcriptomic level, genes regulating cell growth and proliferation were upregulated, similar to those observed in cells cultured in FBS-substituted media. Collectively, this study highlights sericin as a potential component for the development of an FBS-free medium [63]. In another study, different concentrations of sericin were added to the in vitro maturation medium (IVM). Incorporation of 0.5% sericin enhanced nuclear-cytoplasmic maturation and supported the expansion of cumulus cells. The proportion of blastocysts obtained from sericin-supplemented media was higher than that of the control media. In contrast, higher concentrations of sericin (1% and 2.5%) resulted in substantially lower percentages of cleavage and blastocyst rates than the control. Thus, this study emphasized the potential of using lower concentrations of sericin as supplements in IVM medium [64]. Besides, the potential use of sericin in cryopreservation media has been explored. For instance, a study developed a novel freezing medium containing sericin without serum to evaluate the effectiveness of SS in the cryopreservation of islet cells. This study demonstrated that the survival rate of islets and insulin production stimulated by the freezing media with 1% sericin were comparable to those observed in media containing 10% FBS. After transplantation, the islets effectively reversed hyperglycemia and sustained normal glucose levels. Further, the addition of 7% DMSO as a cryoprotectant with sericin demonstrated similar outcomes to those observed with higher DMSO concentrations in FBS-containing media [61].

6.6 Challenges and Future Perspectives

Silk sericin exhibits a broad range of beneficial properties; nevertheless, its application in biomedical fields remains limited due to inherent challenges. One of the primary concerns is selecting an appropriate extraction method to obtain high-quality silk sericin. Currently, conventional extraction methods involve the degradation and denaturation of protein, which limits comprehensive studies on the characteristics of sericin. Currently, due to the lack of quality-testing procedures and standards, it is not feasible to compare silk sericin proteins derived from different batches. Thus, there remains a significant opportunity to develop an optimized, scalable extraction process capable of producing high-quality silk sericin without altering the physicochemical characteristics. In addition, the underlying mechanisms responsible for the diverse biological functions of sericin remain unclear, primarily due to the involvement of multiple genes in its synthesis. Thus, the complex biological effects of sericin on cells can be examined using high-throughput sequencing techniques and bioinformatics. In addition, with increasing advances in artificial intelligence, the protein structure–activity of sericin can be further elucidated. Although several studies have reported the low immunogenicity of sericin, immune responses have been observed in certain cases. Thus, more detailed investigations need to be focused in this direction. In BALB/c mice, sericin extracted from both wild-type and fibroin-deficient B. mori cocoons showed low immunogenicity and reduced allergic responses. However, the biocompatibility of the sericin obtained from the different silkworm cocoon varieties, along with their immune responses in other animal models, needs to be evaluated. Sericin, an adhesive protein, contains numerous random coil structures, which leads to increased brittleness and a reduced molecular weight of the isolated peptides due to the utilization of harsh physical and chemical methods, resulting in poor mechanical properties. Consequently, scaffolds are often required to be combined with other biomaterials to improve their mechanical characteristics. Furthermore, with the increasing demand for personalized and responsive healthcare systems, there is a need for the design and development of intelligent devices with multiple functionalities. The integration of sericin-based biomaterials into intelligent electronic devices or smart micro/nanorobots may hold promise for a promising future perspective, enabling real-time monitoring of disease-related signals through specific bio-signal detection, transmission, and feedback. This provides a significant potential for advanced diagnostic and therapeutic applications [1, 8, 11, 23].

In summary, this chapter highlights the various extraction methodologies and biological activities of sericin. The increasing volume of research on sericin-based materials has demonstrated their utilization in 3D tissue-engineered constructs in various forms, such as hydrogels, scaffolds, and bioinks, within different biomedical fields, including tissue engineering, cell culture supplements, 3D bioprinting, and the development of wearable and implantable devices. Collectively, numerous sericin-based materials have been successfully developed, exhibiting great potential across various biomedical fields, despite the current challenges and limitations associated with sericin, which have also been discussed. Overall, this review provides recent advancements in sericin-related research in biomedicine, highlights the existing challenges, and explores its future prospects in biomedical applications.

Conflict of Interests The authors do not have any conflict of interests to declare.

Ethical Approval Not Applicable.

List of Abbreviations

ALP	Alkaline phosphatase
OCN	Osteocalcin

OPN	Osteopontin
Runx2	Runt-related transcription factor 2
TNF-α	Tumor necrosis factor alpha
Bcl-2	B-cell lymphoma 2
Bax	Bcl-2-associated X protein
PBS	Phosphate-buffered saline
BSA	Bovine serum albumin
Lkb1 kinase	Liver kinase B1
Nuak1 kinase	Nuak family kinase 1
IL-1β	Interleukin-1 beta
MMP-9	Matrix metalloproteinase-9
MMP-3	Matrix metalloproteinase-3
IL-10	Interleukin-10
MAPK	Mitogen-activated protein kinase
NF-κB	Nuclear factor kappa-light-chain-enhancer of activated B cells
M2-type	M2-type macrophage polarization
FAK	Focal adhesion kinase
Ras/Rho	Ras and Rho signaling proteins
CAMP	Cyclic adenosine monophosphate
PI3K	Phosphoinositide 3-kinase
Akt/HIF-1α	Protein kinase B/hypoxia-inducible factor 1-alpha
TGF-β	Transforming growth factor-beta
ERK	Extracellular signal-regulated kinase
Integrins–RhoA–ROCK2	Integrins–RhoA–Rho-associated protein kinase 2 axis
IL-18	Interleukin-18
TLR4	Toll-like receptor 4
VEGF-A	Vascular endothelial growth factor A
MCF-7	Michigan Cancer Foundation-7 cells
HepG2	Human hepatocellular carcinoma cells
Balb/3T3	Mouse embryonic fibroblast cells
HeLa	Human cervical cancer cells
HEK-293	Human embryonic kidney 293 cells
RC4	Rat corneal epithelial cells
BALB/c	Albino laboratory-bred strain of house mice

References

1. Veiga A, Foster O, Kaplan D, Oliveira AL (2024) Expanding the boundaries of silk-sericin biomaterials in biomedical applications. J Mater Chem B 12:7020–7040. https://doi.org/10.1039/D4TB00386A

2. Bitar L, Isella B, Bertella F, Vasconcelos CB, Harings J, Kopp A, van der Meer Y, Vaughan TJ, Bortesi L (2024) Sustainable *Bombyx mori* silk fibroin for biomedical applications as a molecular biotechnology challenge: a review. Int J Biol Macromol 264:130374. https://doi.org/10.1016/j.ijbiomac.2024.130374
3. Liu J, Ge X, Liu L, Xu W, Shao R (2022) Challenges and opportunities of silk protein hydrogels in biomedical applications. Mater Adv 3:2291–2308. https://doi.org/10.1039/D1MA00960E
4. Yao X, Zou S, Fan S, Niu Q, Zhang Y (2022) Bioinspired silk fibroin materials: from silk building blocks extraction and reconstruction to advanced biomedical applications. Mater Today Bio 16:100381. https://doi.org/10.1016/j.mtbio.2022.100381
5. Liu J, Shi L, Deng Y, Zou M, Cai B, Song Y, Wang Z, Wang L (2022) Silk sericin-based materials for biomedical applications. Biomaterials 287:121638. https://doi.org/10.1016/j.biomaterials.2022.121638
6. Kunz RI, Brancalhão RMC, Ribeiro LDFC, Natali MRM (2016) Silkworm sericin: properties and biomedical applications. Biomed Res Int 2016:8175701. https://doi.org/10.1155/2016/8175701
7. Jiao Z, Song Y, Jin Y, Zhang C, Peng D, Chen Z, Chang P, Kundu SC, Wang G, Wang Z, Wang L (2017) In vivo characterizations of the immune properties of sericin: An ancient material with emerging value in biomedical applications. Macromol Biosci 17:1700229. https://doi.org/10.1002/mabi.201700229
8. Wang J, Liu H, Shi X, Qin S, Liu J, Lv Q, Liu J, Li QS, Wang Z, Wang L (2024) Development and application of an advanced biomedical material–silk sericin. Adv Mater 36:2311593. https://doi.org/10.1002/adma.202311593
9. Zhang H, Xu D, Zhang Y, Li M, Chai R (2022) Silk fibroin hydrogels for biomedical applications. Smart Med 1:e20220011. https://doi.org/10.1002/SMMD.20220011
10. Silva AS, Costa EC, Reis S, Spencer C, Calhelha RC, Miguel SP, Ribeiro MP, Barros L, Vaz JA, Coutinho P (2022) Silk sericin: a promising sustainable biomaterial for biomedical and pharmaceutical applications. Polymers 14:4931. https://doi.org/10.3390/polym14224931
11. Aad R, Dragojlov I, Vesentini S (2024) Sericin protein: structure, properties, and applications. J Funct Biomater 15:322. https://doi.org/10.3390/jfb15110322
12. Saad M, El-Samad LM, Gomaa RA, Augustyniak M, Hassan MA (2023) A comprehensive review of recent advances in silk sericin: extraction approaches, structure, biochemical characterization, and biomedical applications. Int J Biol Macromol 250:126067. https://doi.org/10.1016/j.ijbiomac.2023.126067
13. Bascou R, Hardouin J, Mlouka MAB, Guénin E, Nesterenko A (2022) Detailed investigation on new chemical-free methods for silk sericin extraction. Mater Today Commun 33:104491. https://doi.org/10.1016/j.mtcomm.2022.104491
14. Cao TT, Zhang YQ (2016) Processing and characterization of silk sericin from *Bombyx mori* and its application in biomaterials and biomedicines. Mater Sci Eng C 61:940–952. https://doi.org/10.1016/j.msec.2015.12.082
15. Jaramillo-Quiceno N, Callone E, Dirè S, Álvarez-López C, Motta A (2021) Boosting sericin extraction through alternative silk sources. Polym J 53:1425–1437. https://doi.org/10.1038/s41428-021-00539-2
16. Pan M, Jin Y, Ye Y, Jiang W, Zhu L, Lu W (2024) An efficient and eco-friendly method for removing sericin using microwave-assisted steam degumming. Environ Technol Innov 35:103674. https://doi.org/10.1016/j.eti.2024.103674
17. Gupta D, Agrawal A, Chaudhary H, Gulrajani M, Gupta C (2013) Cleaner process for extraction of sericin using infrared. J Clean Prod 52:488–494. https://doi.org/10.1016/j.jclepro.2013.03.016
18. Wang R, Zhu Y, Shi Z, Jiang W, Liu X, Ni QQ (2018) Degumming of raw silk via steam treatment. J Clean Prod 203:492–497. https://doi.org/10.1016/j.jclepro.2018.08.286
19. Kabir M, Hasan MK, Rafi MN, Repon MR, Islam T, Saha J, Rahman M (2024) Biowaste transformation to functional materials: structural properties, extraction methods, applications, and challenges of silk sericin. ChemistrySelect 9:e202403777. https://doi.org/10.1002/slct.202403777

20. More SV, Chavan S, Prabhune AA (2018) Silk degumming and utilization of silk sericin by hydrolysis using alkaline protease from *Beauveria* sp. (MTCC 5184): a green approach. J Nat Fibers 15:373–383. https://doi.org/10.1080/15440478.2017.1330718
21. Karthick SA, Manjari K, Devi MG (2023) Biocompatible and bioactive PVA/sericin/chitosan nanofibrous wound dressing matrix. Appl Surf Sci Adv 13:100362. https://doi.org/10.1016/j.apsadv.2022.100362
22. Hu D, Li T, Xu Z, Liu D, Yang M, Zhu L (2018) Self-stabilized silk sericin-based nanoparticles: in vivo biocompatibility and reduced doxorubicin-induced toxicity. Acta Biomater 74:385–396. https://doi.org/10.1016/j.actbio.2018.05.024
23. Lei X, Li C, Jing Y, Yin Y, Li Q, Zhou Q, Yi L, Shen Y (2023) Evaluation of sericin protein photoluminescence: contribution attribution of amino acid units and insight into the emission mechanism and application exploration. ACS Sustain Chem Eng 11:7719–7729. https://doi.org/10.1021/acssuschemeng.2c07755
24. Joyjamras K, Chaotham C, Chanvorachote P (2022) Response surface optimization of enzymatic hydrolysis and ROS scavenging activity of silk sericin hydrolysates. Pharm Biol 60:308–318. https://doi.org/10.1080/13880209.2022.2032208
25. Chlapanidas T, Faragò S, Lucconi G, Perteghella S, Galuzzi M, Mantelli M, Avanzini MA, Tosca MC, Marazzi M, Vigo D, Torre ML, Faustini M (2013) Sericins exhibit ROS-scavenging, anti-tyrosinase, anti-elastase, and *in vitro* immunomodulatory activities. Int J Biol Macromol 58:47–56. https://doi.org/10.1016/j.ijbiomac.2013.03.054
26. Zahoor S, Tahir HM, Ali S, Ali A, Muzamil A, Murtaza Z, Zahoor N (2023) Diabetic wound healing potential of silk sericin protein based hydrogels enriched with plant extracts. Int J Biol Macromol 242:125184. https://doi.org/10.1016/j.ijbiomac.2023.125184
27. Baptista-Silva S, Borges S, Costa-Pinto AR, Costa R, Amorim M, Dias JR, Ramos Ó, Alves P, Granja PL, Soares R, Pintado M, Oliveira AL (2021) *In situ* forming silk sericin-based hydrogel: a novel wound healing biomaterial. ACS Biomater Sci Eng 7:1573–1586. https://doi.org/10.1021/acsbiomaterials.0c01745
28. Liu H, Qin S, Zhang H, Chen Z, Zhao Y, Liu J, Deng Y, Liu M, Chen W, Wang Z, Wang L (2025) Silk sericin-based ROS-responsive oxygen generating microneedle platform promotes angiogenesis and decreases inflammation for scarless diabetic wound healing. Adv Funct Mater 35:2404461. https://doi.org/10.1002/adfm.202404461
29. Başaran DDA, Gündüz U, Tezcaner A, Keskin D (2021) Topical delivery of heparin from PLGA nanoparticles entrapped in nanofibers of sericin/gelatin scaffolds for wound healing. Int J Pharm 597:120207. https://doi.org/10.1016/j.ijpharm.2021.120207
30. Wang J, Ge X, Xiang Y, Qi X, Li Y, Xu H, Cai E, Zhang C, Lan Y, Chen X, Shi Y, Li Z, Shen J (2025) An ionic liquid functionalized sericin hydrogel for drug-resistant bacteria-infected diabetic wound healing. Chin Chem Lett 36:109819. https://doi.org/10.1016/j.cclet.2024.109819
31. Liu H, Qin S, Liu J, Zhou C, Zhu Y, Yuan Y, Wang L (2022) Bio-inspired self-hydrophobized sericin adhesive with tough underwater adhesion enables wound healing and fluid leakage sealing. Adv Funct Mater 32:2201108. https://doi.org/10.1002/adfm.202201108
32. Zhao H, He L (2022) Fabrication of neuroprotective silk-sericin hydrogel: potential neuronal carrier for the treatment and care of ischemic stroke. J Exp Nanosci 17:362–376. https://doi.org/10.1080/17458080.2022.2075545
33. Wang J, Li X, Song Y, Su Q, Xiaohalati X, Yang W, Xu L, Cai B, Wang G, Wang Z, Wang L (2020) Injectable silk sericin scaffolds with programmable shape-memory property and neuro-differentiation-promoting activity for individualized brain repair of severe ischemic stroke. Bioact Mater 6:1988–2002. https://doi.org/10.1016/j.bioactmat.2020.12.017
34. Zhang L, Yang W, Tao K, Song Y, Xie H, Wang J, Li X, Shuai X, Gao J, Chang P, Wang G, Wang Z, Wang L (2017) Sustained local release of NGF from a chitosan–sericin composite scaffold for treating chronic nerve compression. ACS Appl Mater Interfaces 9:3432–3444. https://doi.org/10.1021/acsami.6b14691
35. Ming P, Rao P, Wu T, Yang J, Lu S, Yang B, Xiao J, Tao G (2022) Biomimetic design and fabrication of sericin-hydroxyapatite based membranes with osteogenic activity for periodontal

tissue regeneration. Front Bioeng Biotechnol 10:899293. https://doi.org/10.3389/fbioe.2022.899293
36. Chachlioutaki K, Karavasili C, Adamoudi E, Bouropoulos N, Tzetzis D, Bakopoulou A, Fatouros DG (2022) Silk sericin/PLGA electrospun scaffolds with anti-inflammatory drug-eluting properties for periodontal tissue engineering. Biomater Adv 133:112723. https://doi.org/10.1016/j.msec.2022.112723
37. Qi C, Deng Y, Xu L, Yang C, Zhu Y, Wang G, Wang Z, Wang L (2020) A sericin/graphene oxide composite scaffold as a biomimetic extracellular matrix for structural and functional repair of calvarial bone. Theranostics 10:741–759. https://doi.org/10.7150/thno.39502
38. Pankongadisak P, Jaikaew N, Kiti K, Chuenjitkuntaworn B, Supaphol P, Suwantong O (2019) The potential use of gentamicin sulfate-loaded poly(l-lactic acid)-sericin hybrid scaffolds for bone tissue engineering. Polym Bull 76:2867–2885. https://doi.org/10.1007/s00289-018-2520-x
39. Jiang LB, Ding SL, Ding W, Su DH, Zhang FX, Zhang TW, Yin XF, Xiao L, Li YL, Yuan FL, Dong J (2021) Injectable sericin-based nanocomposite hydrogel for multi-modal imaging-guided immunomodulatory bone regeneration. Chem Eng J 418:129323. https://doi.org/10.1016/j.cej.2021.129323
40. Fu M, Li J, Liu M, Yang C, Wang Q, Wang H, Cheng B, Fu Q, Sun G (2023) Sericin/nano-hydroxyapatite hydrogels based on graphene oxide for effective bone regeneration via immunomodulation and osteoinduction. Int J Nanomed 18:1875–1895. https://doi.org/10.2147/IJN.S399487
41. Zhang Q, Zhou W, Yang F, Shi J (2023) Sericin nano-gel agglomerates mimicking the pericellular matrix induce the condensation of mesenchymal stem cells and trigger cartilage micro-tissue formation without exogenous stimulation of growth factors in vitro. Biomater Sci 11:6480–6491. https://doi.org/10.1039/D3BM00501A
42. Han L, Wang W, Chen Z, Cai Y, Chen C, Chen G, Wang F (2023) Sericin-reinforced dual-crosslinked hydrogel for cartilage defect repair. Colloids Surf B Biointerfaces 222:113061. https://doi.org/10.1016/j.colsurfb.2022.113061
43. Qi C, Liu J, Jin Y, Xu L, Wang G, Wang Z, Wang L (2018) Photo-crosslinkable, injectable sericin hydrogel as 3D biomimetic extracellular matrix for minimally invasive repairing cartilage. Biomaterials 163:89–104. https://doi.org/10.1016/j.biomaterials.2018.02.016
44. Song Y, Wang H, Hu C, Liu J, Qiao W, Huang L, Dong N, Wang L (2024) Silk sericin is a potent, multifunctional biomaterial for endothelialization of tissue-engineered heart valves. Adv Funct Mater 34:2315826. https://doi.org/10.1002/adfm.202315826
45. Bai P, Kong G, Qiao W, Song Y, Wang Y, Shi J, Zhang N, Liu C, Chu C, Xiong T, Zhou Y, LuC WL, Dong N (2022) A novel method to improve the physical property and biocompatibility of decellularized heart valve scaffold with sericin and polydopamine. J Bionic Eng 19:1109–1123. https://doi.org/10.1007/s42235-022-00191-3
46. Song Y, Zhang C, Zhang J, Sun N, Huang K, Li H, Wang Z, Huang K, Wang L (2016) An injectable silk sericin hydrogel promotes cardiac functional recovery after ischemic myocardial infarction. Acta Biomater 41:210–223. https://doi.org/10.1016/j.actbio.2016.05.039
47. Fazal T, Murtaza BN, Shah M, Iqbal S, Rehman MU, Jaber F, Dera AA, Awwad NS, Ibrahium HA (2023) Recent developments in natural biopolymer based drug delivery systems. RSC Adv 13:23087–23121. https://doi.org/10.1039/D3RA03369D
48. Tuanchai A, Timjan S, Tuancharoensri N, Phimnuan P, Chartarrayawadee W, Worajittiphon P, Liu Y, Ross GM, Viennet C, Viyoch J, Chang HT, Unno M, Ross S (2024) Hybrid nanomaterials of carbon dots of silk sericin and folic acid embedded in bio-derived poly(lactic acid)/silk sericin nanogels and their incorporation with surface-porous electrospun fibers of poly(lactide-co-glycolide) for potential use as scaffolds in tissue engineering and drug delivery system. Adv Compos Hybrid Mater 7:1–19. https://doi.org/10.1007/s42114-024-01040-z
49. Khan NU, Razzaq A, Rui Z, Xie C, Khan ZU, Ullah A, Elbehairi SI, Shati AA, Alfaifi MY, Iqbal H, Jin ZM (2024) Bio-evaluations of sericin coated hesperidin nanoparticles for gastric ulcer management. Colloids Surf B Biointerfaces 234:113762. https://doi.org/10.1016/j.colsurfb.2024.113762

50. Bahremand K, Aghaz F, Bahrami K (2024) Enhancing cisplatin efficacy with low toxicity in solid breast cancer cells using pH-charge-reversal sericin-based nanocarriers: development, characterization, and in vitro biological assessment. ACS Omega 9:14017–14032. https://doi.org/10.1021/acsomega.3c09361
51. Yang Y, Cai Y, Sun N, Li R, Li W, Kundu SC, Kong X, Yao J (2017) Biomimetic synthesis of sericin and silica hybrid colloidosomes for stimuli-responsive anti-cancer drug delivery systems. Colloids Surf B Biointerfaces 151:102–111. https://doi.org/10.1016/j.colsurfb.2016.12.013
52. Vurro D, Liboà A, D'Onofrio I, De Giorgio G, Scaravonati S, Crepaldi M, Barcellona A, Sciancalepore C, Galstyan V, Milanese D, Riccò M, D'Angelo P, Tarabella G (2025) Sericin electrodes with self-adhesive properties for biosignaling. ACS Biomater Sci Eng 11:1776–1791. https://doi.org/10.1021/acsbiomaterials.4c02234
53. Su B, Wang B, Cao Y, Dou H, Zhang T, He X (2024) A multi-functional flexible piezoresistive sensor based on sericin-functionalized CNTs decorated carbon foams for wearable thermal-therapy devices. Ceram Int 50:10305–10313. https://doi.org/10.1016/j.ceramint.2023.12.3424
54. Ma H, Li J, Zhou J, Luo Q, Wu W, Mao Z, Ma W (2022) Screen-printed carbon black/recycled Sericin@Fabrics for wearable sensors to monitor sweat loss. ACS Appl Mater Interfaces 14:11813–11819. https://doi.org/10.1021/acsami.1c23341
55. Lv Q, Chen S, Luo D, Liu H, Song Y, Liu M, Xiao F, Wang Z, Wang L (2025) An implantable and degradable silk sericin protein film energy harvester for next-generation cardiovascular electronic devices. Adv Mater 37:2413610. https://doi.org/10.1002/adma.202413610
56. Lv Q, Li X, Tian X, Fu D-A, Liu H, Liu J, Song Y, Cai B, Wang J, Su Q, Chen W, Zou M, Xiao F, Wang S, Wang Z, Wang L (2023) A degradable and biocompatible supercapacitor implant based on functional sericin hydrogel electrode. Adv Energy Mater 13:2203814. https://doi.org/10.1002/aenm.202203814
57. Tetsuka H, Shin SR (2020) Materials and technical innovations in 3D printing in biomedical applications. J Mater Chem B 8:2930–2950. https://doi.org/10.1039/D0TB00034E
58. Bozkurt Y, Karayel E (2021) 3D printing technology; methods, biomedical applications, future opportunities and trends. J Mater Res Technol 14:1430–1450. https://doi.org/10.1016/j.jmrt.2021.07.050
59. Xie M, Lian L, Mu X, Luo Z, Garciamendez-Mijares CE, Zhang Z, López A, Manríquez J, Kuang X, Wu J, Sahoo JK, Zertuche González F, Li G, Tang G, Maharjan S, Guo J, Kaplan DL, Zhang YS (2023) Volumetric additive manufacturing of pristine silk-based (bio) inks. Nat Commun 14:210. https://doi.org/10.1038/s41467-023-35807-7
60. Chen CS, Zeng F, Xiao X, Wang Z, Li XL, Tan RW, Liu WQ, Zhang YS, She ZD, Li SJ (2018) Three-dimensionally printed silk-sericin-based hydrogel scaffold: a promising visualized dressing material for real-time monitoring of wounds. ACS Appl Mater Interfaces 10:33879–33890. https://doi.org/10.1021/acsami.8b10072
61. Cao TT, Zhang YQ (2017) The potential of silk sericin protein as a serum substitute or an additive in cell culture and cryopreservation. Amino Acids 49:1029–1039. https://doi.org/10.1007/s00726-017-2396-3
62. Cao TT, Zhang YQ (2015) Viability and proliferation of L929, tumour and hybridoma cells in the culture media containing sericin protein as a supplement or serum substitute. Appl Microbiol Biotechnol 99:7219–7228. https://doi.org/10.1007/s00253-015-6576-3
63. Liu L, Wang J, Duan S, Chen L, Xiang H, Dong Y, Wang W (2016) Systematic evaluation of sericin protein as a substitute for fetal bovine serum in cell culture. Sci Rep 6:31516. https://doi.org/10.1038/srep31516
64. Aghaz F, Hajarian H, Shabankareh HK, Abdolmohammadi A (2015) Effect of sericin supplementation in maturation medium on cumulus cell expansion, oocyte nuclear maturation, and subsequent embryo development in Sanjabi ewes during the breeding season. Theriogenology 84:1631–1635. https://doi.org/10.1016/j.theriogenology.2015.08.013

Chapter 7
Silk Biomaterials for Tissue Engineering, Implants, and Surgical Tools

Bantamlak Birlie Kassie, Tekalgn Mamay Daget, Dehenenet Flatie Tassaw, and Abay Mulusew Birhanu

7.1 Introduction

Silk-based biomaterials have garnered considerable scientific interest owing to their exceptional mechanical robustness, intrinsic biocompatibility, and multifunctional nature, rendering them highly advantageous for advanced biomedical and biotechnological applications. As a naturally occurring proteinaceous biopolymer, silk is biosynthesized by a diverse array of species, including *Bombyx mori* silkworms, spiders, scorpions, fleas, and mites [1, 2]. The material is secreted by epithelial cells, stored within specialized glandular lumens, and subsequently extruded into fibrous form. Notably, the hierarchical structure and mechanical performance of silk fibers are highly dependent on their phylogenetic origin, which critically governs their functional properties and end-use potential across various scientific and engineering domains. Among the different types of silk, glycine-rich varieties exhibit high elasticity and strength, while spider dragline silk surpasses other forms in mechanical robustness. Silkworm silk, widely studied for its biomedical applications, is particularly valued for its superior water resistance, accessibility, mechanical stability, and ability to withstand high temperatures [3, 4].

B. B. Kassie (✉) · T. M. Daget · D. F. Tassaw · A. M. Birhanu
Textile Faculty, Ethiopian Institute of Textile and Fashion Technology, Bahir Dar University, Bahir Dar, Ethiopia
e-mail: bantamlakbirlie@gmail.com

B. B. Kassie · T. M. Daget · A. M. Birhanu
Textile Science and Engineering Research Cener, Ethiopian Institute of Textile and Fashion Technology, Bahir Dar University, Bahir Dar, Ethiopia

D. F. Tassaw
Biorefinery Research Cener, Ethiopian Institute of Textile and Fashion Technology, Bahir Dar University, Bahir Dar, Ethiopia

Institute of Architecture of Textiles, Faculty of Textiles and Design, Lodz University of Technology, 116 Zeromskiego Str. 90-924 Lodz, Poland

M. Shahid et al. (eds.), *Science and Technology of Silk*, Textile Science and Clothing Technology, https://doi.org/10.1007/978-3-032-26880-8_7

The growing interest in silk-based biomaterials stems from their outstanding biocompatibility, making them promising candidates for advancing medical technology. Silk can be processed into a wide range of structural forms, such as hydrogels, films, scaffolds, and sponges, by employing advanced fabrication methods like electrospinning, freeze-drying, and 3D printing. These forms are specifically designed to meet the unique requirements of various biomedical uses [1]. The clinical application of silk has been well-established, particularly in tissue engineering and regenerative medicine. Tissue engineering integrates living cells with biochemical and physical stimuli such as growth factors and mechanical loading alongside biocompatible materials to repair, replace, or enhance biological tissues [5, 6]. An ideal biomaterial for tissue regeneration must meet several criteria, including biocompatibility, non-toxicity, controlled biodegradability, mechanical resilience, and antibacterial properties [7]. Although synthetic materials like metals, ceramics, bioactive glass, and polymers have been extensively utilized in tissue regeneration, natural polymers are often more biocompatible, a critical attribute for the success of artificial implants. Among these, extracellular matrix proteins are particularly valuable, as they promote cell attachment and movement, which are vital for effective tissue engineering [8]. Silk, in particular, stands out due to its distinctive molecular structure, which imparts exceptional mechanical properties, including impressive tensile strength (reaching up to 1 GPa), flexibility, and resilience under physiological conditions [9]. Silk fibroin offers a more supportive and stable environment for sensitive biological molecules than synthetic polymers such as poly(lactic-co-glycolic acid) (PLGA). It also combines strong mechanical properties with tunable degradation behavior, allowing researchers to tailor its performance for diverse biomedical purposes [10]. Silk fibroin exhibits generally low immunogenic responses in various studies, though its biocompatibility can be influenced by the presence of sericin and processing conditions [11]. Research into silk biomaterials has advanced significantly since *Bombyx mori* silk fibers were first introduced in clinical sutures [12]. Advancements in isolating silk proteins and refining processing methods have enabled the creation of a variety of two- and three-dimensional structures, including films, sponges, hydrogels, fibrous mats, and 3D-printed scaffolds. These matrices can be tailored for specific biomedical applications. While *Bombyx mori* silk has been widely studied, recent research also explores non-mulberry silk variants to broaden potential applications [13]. Despite its biocompatibility, prolonged exposure to silk-based products may elicit immune responses, necessitating further investigation to mitigate this challenge [14].

Silk-based biomaterials have demonstrated substantial efficacy in musculoskeletal tissue repair and regeneration due to their mechanical strength, chemical adaptability, tunable degradation rate, and excellent biocompatibility [15]. Research has demonstrated that silk supports bone tissue regeneration by creating a favorable environment for osteogenesis. When bioactive components like growth factors and stem cells are integrated into silk-based structures, their capacity to promote cell differentiation and aid in bone healing is significantly improved [16]. Additionally, silk hydrogels have shown promise in cartilage repair, mimicking the extracellular matrix with mechanical properties comparable to native tissue [17]. Advances in genetic engineering have further expanded the potential of silk biomaterials, enabling modifications that incorporate functional motifs to enhance biological performance [18]. Recent developments in multi-scale simulation techniques have accelerated the molecular-level design of bioactive silk materials [19]. Additionally, innovations in bio-fabrication techniques have enabled the replication of artificial cellular environments, optimizing silk-based materials for tissue regeneration [20].

The biomedical applications of silk extend beyond tissue engineering to include surgical tools, implants, and bioactive coatings as shown in Fig. 7.1. Due to its tunable degradation properties and mechanical resilience, silk has been widely used in sutures, meshes, and biodegradable surgical devices [21]. Silk-based vascular grafts have shown promising endothelialization and patency rates, indicating their potential in cardiovascular applications [22, 23]. Moreover, functionalized silk coatings for surgical tools and drug-eluting sutures have been developed to prevent infections and improve post-surgical outcomes [24–27]. Historically, silk sutures have been extensively used in clinical settings, with modern refinements minimizing inflammatory responses [28]. Silk meshes have also been employed in hernia repair, demonstrating reduced fibrosis compared to synthetic counterparts [29]. Emerging applications include biodegradable orthopedic fixtures [30], drug-eluting surgical coatings [31], and 3D-printed, patient-specific constructs using silk bioinks [32].

Despite significant advancements, various challenges continue to impede the widespread clinical translation of silk biomaterials. Large-scale production, batch-to-batch consistency, effective sterilization techniques, and regulatory approval remain critical obstacles [33–35]. Further optimization of sterilization methods, degradation profiles, and cost-effective manufacturing processes is required for broader clinical translation [17]. The integration of nanocomposite-enhanced silk materials, recombinant silk proteins, and bioactive silk-based systems represents a promising avenue for overcoming these limitations [36–38]. With genetic engineering, silk templates can be tailored to improve their bioactivity and functional properties for advanced medical applications [18]. Moreover, surface nanopatterning, chemical functionalization, and tunable mechanical properties play critical roles in modulating cellular responses, including adhesion, proliferation, migration, differentiation, and signaling pathways [20]. These innovations highlight the significant potential of silk biomaterials to revolutionize biomedical engineering, regenerative therapies, and advanced surgical technologies. This chapter offers an in-depth review of silk biomaterials, encompassing their fundamental characteristics, biomedical applications, and fabrication methods that govern their functional performance. Furthermore, emerging innovations in genetically engineered silk, silk composites, and nanotechnology-enhanced biomaterials were discussed. By addressing current challenges and highlighting recent breakthroughs, this chapter underscores silk's evolving role in advancing medical science and regenerative therapies.

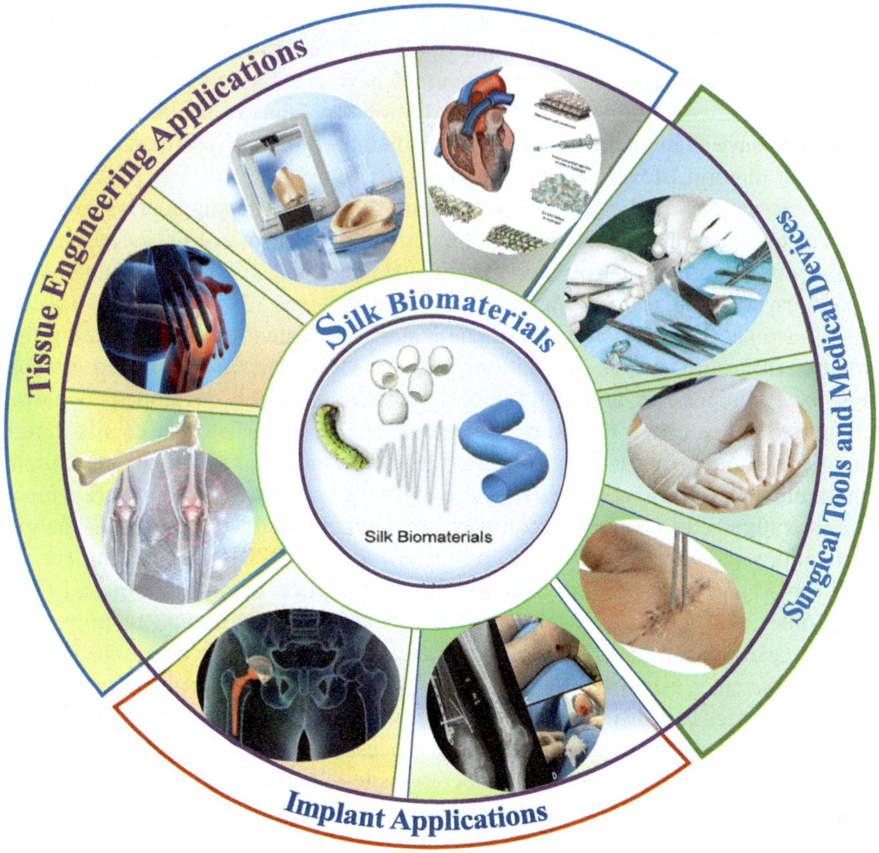

Fig. 7.1 Uses of silk-based biomaterials in tissue regeneration and implant technology

7.2 Properties of Silk Biomaterials

Silk, a naturally occurring polymer, has been a crucial material in medical applications for centuries, most notably in the form of sutures. The silk fibers secreted by insects or worms consist primarily of fibroin, a core filament protein. This protein is encapsulated by a sericin coating, which serves as an adhesive to preserve the structural integrity of the silk fibers. Over recent decades, silk fibroin has attracted significant interest in biomedical research due to its outstanding biocompatibility, controlled degradation profile, and superior mechanical strength. Furthermore, significant advancements in controlling its molecular structure and morphology, coupled with enhanced processing techniques and surface modifications, have expanded its potential applications, particularly in biomaterials and tissue engineering [39].

Silk fibroin (SF) is composed of two principal protein chains: a heavy chain (H) weighing approximately 390 kDa and a light chain (L) of about 26 kDa. These two chains are interconnected through disulfide bonds to form an H–L complex [40–43]. In addition to these two main chains, a glycoprotein known as P25 (25 kDa) is also a component of silk fibroin. P25 contains Asn-linked oligosaccharide chains and is hydrophobically associated with the H–L complex [44]. The three polypeptides H-chain, L-chain, and P25 are the primary constituents of the Bombyx mori silk cocoon, occurring in a molar ratio of 6:6:1, respectively. The H-chain's amino acid composition is largely glycine (45.9%), alanine (30.3%), serine (5.3%), valine (1.8%), with the remaining 4.5% comprising 15 other amino acids. Approximately 60–75% of the H-chain consists of a repeating Gly-X (GX) dipeptide motif, where hydrophobic residues within these repeats promote the formation of stable antiparallel β-sheet crystallites. Notably, two hexapeptide sequences, Gly-Ala-Gly-Ala-Gly-Ser and Gly-Ala-Gly-Ala-Gly-Tyr, account for around 70% of the GX motif region. [45–47].

Silk fibroin exists in two primary crystalline forms: silk I and silk II. Silk, I signify a metastable crystalline structure that contains bound water molecules, whereas silk II is a more stable crystalline configuration. The stability of silk II arises from strong hydrogen bonds formed among adjacent peptide blocks, which significantly enhance its mechanical properties, including tensile strength and rigidity. The secondary structure of regenerated silk fibroin (RSF) solutions comprises both crystalline and amorphous domains. Within the crystalline phase, silk features β-turns characteristic of silk I and insoluble folded β-sheet structures typical of silk II, as illustrated in Fig. 7.2.

Silk fibers exhibit remarkable mechanical properties, including toughness, ductility, biocompatibility, and biodegradability. These characteristics are attributed to the highly organized molecular structure of silk. Beyond its natural fiber form, silk has been incorporated into various advanced materials, such as regenerated silk, chemically modified variants, composite materials, and artificial silk-inspired substances. These materials are often synthesized through chemoenzymatic polymerization, offering further applications in diverse fields, particularly in the biomedical domain [48]. These remarkable properties position silk as an excellent candidate for diverse applications, ranging from medical sutures to advanced uses in tissue engineering, drug delivery systems, and beyond. The ongoing research into silk biomaterials continues to expand its utility, promising innovative solutions in the biomedical industry.

7.2.1 Mechanical Properties

The mechanical properties of silk fibroin textile constructs have been extensively studied to understand their potential in various applications, particularly in the biomedical and textile industries. These properties were evaluated through quasi-static tensile tests using an Instron 4505 Universal Testing. The primary parameters measured included tensile modulus, ultimate tensile strength, and strain at maximum

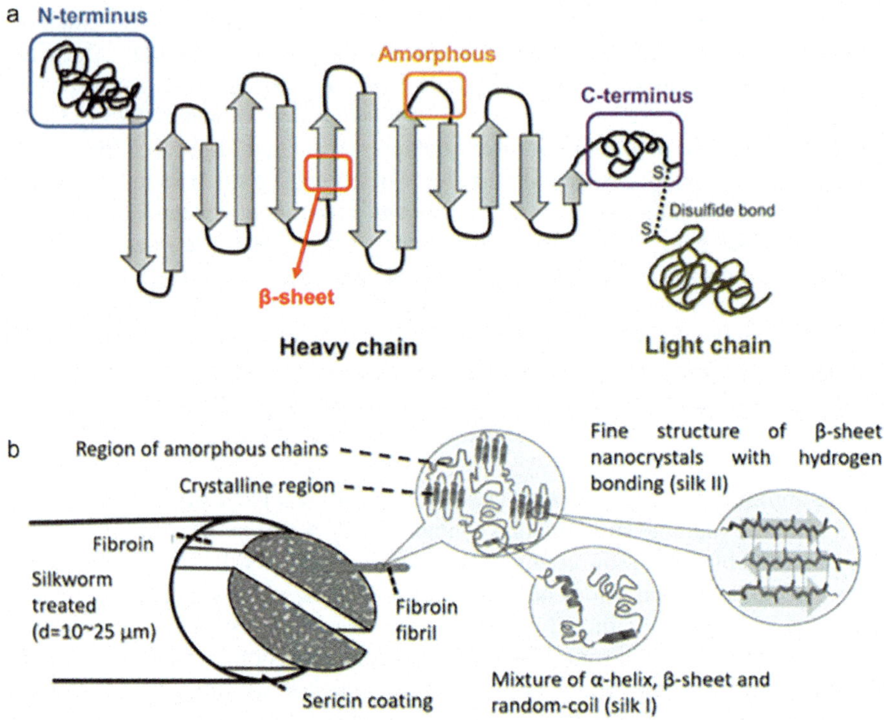

Fig. 7.2 Schematic representation of the silk structure **a** Heavy chain segments including N-terminus, β-sheets, amorphous regions, and C-terminus, along with the light chain connected via disulfide bonds [42], reproduced with permission from Copyright 2019. Adapted with permission from MDPI **b** Overview of the silkworm thread, fibril architecture, and silk fibroin polypeptide chains. Reproduced with permission from [41], Copyright 2017. Adapted with permission from MDPI

load, utilizing a 1 kN load cell and a crosshead speed of 5 mm/min. The tensile modulus was calculated from the most linear portion of the stress–strain curve using the secant method [49, 50]. Testing was carried out on both dry and hydrated samples. The dry samples were tested at 25 °C and 50% relative humidity. Hydrated samples, on the other hand, were immersed in a phosphate-buffered saline (PBS) solution at pH 7.4 for three days prior to testing. For each condition, five samples measuring 15 $\times$ 40 mm^2 were analyzed to ensure the reliability of the results.

Silk, an ancient biomaterial, is known for its exceptional strength, extensibility, and toughness. Made from simple protein building blocks, silk is a remarkable natural fiber with diverse applications. The molecular architecture of dragline spider silk, closely related to silkworm silk, is composed of semi-amorphous regions and nanocrystalline β-sheet protein domains [51]. This unique molecular arrangement contributes to the remarkable mechanical properties of silk fibers, including high break strain (4–26%), tensile strength (300–740 MPa), and toughness (70–78 $MJ \cdot m^{-3}$) [52]. Silk fibers exhibit a balanced combination of ductility and toughness, offering excellent elongation, modulus, and breaking strength. In comparison to synthetic and natural fibers, such as nylon and aramid, spider silk exhibits superior strength and elasticity. Both silkworm and spider silks demonstrate similar strain-hardening behaviors, as evidenced by their stress–strain curves [53]. This similarity in mechanical characteristics suggests that silk is suitable for high-load-bearing applications, such as tissue grafts or scaffolds, where mechanical strength is crucial [54, 55]. However, despite its potential, many current silk-based biomedical tissue-engineered constructs (BTFIs) do not match the mechanical performance or material-tissue interface interaction of native tissues. This discrepancy suggests that further research is needed to optimize silk processing methods, thereby better aligning its mechanical properties with the specific requirements of BTFIs. To evaluate the stiffness of silk fibers, the relative bending rigidity can be calculated using Eq. (7.1) [53].

$$\textit{Relative bending rigidity} = \left[\frac{1}{4\pi\beta_0}\right]\frac{\varepsilon E d^2}{\rho} \tag{7.1}$$

where E is the tensile modulus, d is the linear density, ρ is the density, ε is the fiber cross-section ($\varepsilon = 1$ for circular, <1 for flat or elliptical, >1 for hollow fiber), β_0 is a constant, and the units are shown in Table 7.1. Silk stands out as a versatile biomaterial due to its unique mechanical properties, especially a moderate bending rigidity of 5.2, which lies between the flexibility of nylon and the stiffness of polyester. This balance allows it to serve in braided sutures, woven or knitted scaffolds for tendon repair, and non-woven mats for wound healing, providing both pliability and structural support. Additionally, silk can be processed through braiding and knitting to produce vascular grafts and durable implants. The chapter emphasizes silk's ability to merge mechanical performance with biological functionality across tissue engineering, surgical tools, and implant applications.

7.2.2 Biocompatibility of Silk Biomaterials

Silk fibroin, a structural protein derived primarily from the *Bombyx mori* silkworm, has garnered significant attention in biomedical research due to its unique combination of mechanical strength, biodegradability, and biocompatibility [56]. Unlike synthetic polymers, SF exhibits remarkable versatility, allowing it to be processed into various forms, such as films, hydrogels, nanofibers, and porous scaffolds, making it suitable for diverse regenerative medicine applications [57]. Recent advancements in silk-based biomaterials have highlighted their ability to mimic the extracellular matrix (ECM), providing an optimal microenvironment for cellular activities [58]. Furthermore, functional enhancements have been achieved through modifications such as blending with bioactive molecules (e.g., growth factors, peptides) and hybridizing with other natural or synthetic polymers [59].

Table 7.1 Comparative analysis of the physical and mechanical properties of silk versus other fibers [53]

Fiber	Density (g/cm^3)	Tenacity cN/denier (MPa)	Breaking strain (%)	Initial modulus N/tex (GPa)	Relative bending rigidity
Raw silk	1.34	4.2 (507)	20–24	7.02 (9–10)	5.2
De-gummed silk	1.27	(610–690)	4–16	(15–17)	-
Spider silk	1.31	(875–972)	17–18	(11–13)	-
Cellulose, rayon	1.50	2.6 (351)	18	6.12 (9.2)	4.1
Nylon	1.14	7.5 (770)	25	3.15 (3.6)	2.8
Polyester	1.38	6.3 (782)	16	11.7 (16.2)	8.5
Polypropylene	0.91	7.0 (573)	18	7.20 (6.6)	7.9
Polyurethane	1.20	0.4 (43)	500	0.01 (0.012)	0.01
Aramid (para)	1.44	27.5 (3564)	2.4	88.2 (127)	61.3
Steel cord	7.85	3.0 (2120)	8	24.5 (192)	3.1
Spectra 1000	0.97	36 (3142)	3.3	112 (109)	115.5

7.2.2.1 Interaction with Cells, Tissues, and the Immune Response

Silk fibroin influences cell behavior by enhancing the attachment, growth, and specialization of stem cells as well as mature cells like osteoblasts, chondrocytes, and fibroblasts. It also triggers only a mild immune reaction, with minimal neutrophil presence and macrophages favoring an anti-inflammatory state, which helps tissue repair without strongly activating adaptive immunity [60]. The presence of integrin-binding motifs, including RGD sequences, in SF promotes enhanced cell adhesion, while its adjustable mechanical properties enable the tailoring of scaffold stiffness to closely mimic that of native tissues [61]. A critical aspect of SF biocompatibility is its low immunogenicity after sericin removal. Recent in vivo studies have confirmed that degummed SF elicits minimal inflammatory responses, comparable to clinically approved collagen-based materials [28]. Macrophage polarization studies indicate that SF predominantly induces an M2 (anti-inflammatory) phenotype, promoting tissue remodeling rather than chronic inflammation [62]. Moreover, SF's slow degradation rate ensures structural support during tissue regeneration while avoiding abrupt immune reactions associated with rapid polymer breakdown.

7.2.3 Host Immune Response and Organ Regeneration

The host immune response plays a decisive role in the integration and long-term success of implanted biomaterials. SF's ability to modulate immune reactions makes it particularly advantageous for applications requiring prolonged in vivo stability [63]. For instance, silk-based wound dressings functionalized with antimicrobial agents have shown accelerated healing in chronic wounds by reducing bacterial load while maintaining a favorable immune microenvironment [64]. In organ regeneration, SF has been explored for liver, cardiac, and neural tissue engineering due to its ability to support vascularization and ECM deposition [65]. Recent studies using 3D-printed SF scaffolds for cartilage repair demonstrated enhanced chondrogenesis with reduced fibrotic scarring, attributed to SF's immune-tolerant properties [66]. Additionally, SF hydrogels loaded with immunomodulatory cytokines have shown promise in reducing post-implant fibrosis, further highlighting its potential in complex tissue regeneration.

7.2.4 Degradation Behaviors of Silk Biomaterials

Silk fibroin (SF), classified as a non-degradable material by the U.S. Pharmacopeia, exhibits biodegradability under physiological conditions due to enzymatic and foreign body-mediated degradation processes [67]. Unlike synthetic polymers, silk fibroin degradation does not typically provoke significant immunogenic reactions, making it a favorable candidate for biomedical applications [68]. Biodegradation, defined as the breakdown of polymeric materials into smaller compounds via physical, chemical, or biological mechanisms, occurs in silk primarily through enzymatic action [69]. Proteolytic enzymes, such as protease XIV, play a pivotal role in hydrolyzing SF, classifying it as an enzymatically degradable polymer. The degradation process involves two key stages: (1) enzyme adsorption onto the substrate surface via a binding domain and (2) subsequent hydrolysis of peptide bonds [70].

7.2.4.1 Biodegradation Properties of Silk Fibroin

Silk fibroin has been widely studied for use in tissue engineering scaffolds and controlled drug delivery systems owing to its adjustable degradation rates [71]. To evaluate the biodegradation behavior, freeze-dried SF gels have been subjected to protease XIV solutions, revealing distinct degradation rates between SF C-gels (low crystallinity) and P-gels (higher crystallinity) [72]. Studies indicate that SF C-gels degrade completely within two weeks in 1.0 U/mL protease XIV at 37 °C, whereas P-gels exhibit slower degradation due to their higher crystallinity and lower porosity [73]. The accelerated degradation of C-gels is attributed to enhanced enzyme penetration facilitated by their porous structure [74]. Moreover, degradation kinetics are influenced by SF concentration. Increasing the SF concentration from 2.3 to 7.9% (w/v) extends the time required for 50% mass loss from 4 to 15 days [75]. Control samples incubated in phosphate-buffered saline (PBS) without protease exhibit minimal degradation, confirming the enzyme-dependent nature of SF breakdown. The rapid degradation of SF C-gels makes them suitable for short-term applications, such as temporary tissue scaffolds or burst-release drug carriers [71].

Structural analysis via scanning electron microscopy (SEM) reveals that pore architecture in 7.9% SF C-gels remains intact for up to 28 days before eventual collapse due to progressive enzymatic erosion. This degradation behavior underscores the importance of tailoring SF crystallinity and porosity for specific biomedical applications.

7.2.5 Functionalization Potential

Silk fibroin-based biotextile structures have shown considerable potential in tissue engineering (TE) owing to their strong mechanical strength, interconnected porous networks, and regulated biodegradability. Various surface modification techniques, such as sodium hydroxide (NaOH) treatment, ultraviolet/ozone (UV/O_3) exposure, and oxygen (O_2) plasma treatment, followed by immersion in acrylic acid, vinyl phosphonic acid, or vinyl sulfonic acid, have been explored to enhance SF scaffold performance [50]. These modifications influence mechanical strength, surface morphology, wettability, and elemental composition. In vitro studies with L929 fibroblasts confirm cell adhesion and proliferation on both untreated and modified silk constructs, underscoring the adaptability of these treatments for specific biomedical applications [76].

Silk fibroin fiber (SFF) surface modifications can be categorized based on the technique used and their intended applications. The primary surface modification approaches include the following:

- Physical Modification.
- Chemical Modification.
- Nanomaterial-Based Functionalization.

7.2.5.1 Surface Modification by Physical Techniques

Surface modification of silk biomaterials using physical techniques provides a sustainable approach to enhancing material properties without the use of chemical additives. These methods improve wettability, adhesion, and biocompatibility while preserving the bulk characteristics of silk. The following sections discuss key physical surface modification techniques, including UV treatment, gas treatment, and plasma treatment, along with their mechanisms and effects on silk fibroin.

(a) ***UV Treatment***

Ultraviolet (UV) irradiation, particularly within the vacuum ultraviolet (VUV) spectrum, triggers the photochemical cleavage of silk fibroin chains, leading to the creation of micropores and the formation of free radicals. These radicals interact with atmospheric oxygen to produce reactive species, such as singlet oxygen ($O(^1D)$), which introduce polar functional groups (e.g., hydroxyl and carbonyl) onto the silk surface. This surface modification enhances the wettability and adhesion properties of silk while minimally affecting its weight, crystallinity, and tensile strength. However, prolonged UV exposure may degrade textile properties, including hand feel and air permeability, due to surface oxidation [77]. Despite this limitation, UV treatment remains a cost-effective and scalable method suitable for industrial applications.

(b) ***Gas Treatment***

Gas-phase surface treatments, such as plasma exposure, physical vapor deposition (PVD), and ozone oxidation, modify silk's adhesion, wettability, and chemical resistance. Among these, ozone (O_3) treatment a strong oxidizing agent has been widely studied for cotton bleaching but remains less explored for silk processing. Recent studies indicate that ozone treatment increases silk's pliability while also inducing yellowness due to the oxidation of amino acids such as tyrosine, glycine, and alanine [78]. Notably, tyrosine content decreases by over 40%, forming chromophores like 3,4-dihydroxyphenylalanine (DOPA) and carbonyl derivatives, which contribute to the observed color changes.

(c) ***Plasma Treatment***

Silk biomaterials have attracted considerable interest in biomedical applications owing to their biocompatibility, mechanical robustness, and biodegradability. Nonetheless, surface modifications like plasma treatment can further improve their functionality in tissue engineering, implants, and surgical instruments. Plasma treatment is an eco-friendly surface modification technique that uses ionized gases to functionalize silk surfaces without altering bulk properties. Plasma treatment introduces functional groups (e.g., –OH, –COOH, $–NH_2$) on silk surfaces, improving wettability, cell adhesion, and bioactivity [79]. In tissue engineering, plasma-treated silk scaffolds promote better cell proliferation and differentiation, crucial for regenerative medicine [80]. For implants, plasma modification reduces immune rejection by enhancing surface biocompatibility. Additionally, plasma-treated silk surgical tools exhibit improved antimicrobial properties, reducing infection risks [81]. Recent studies highlight atmospheric pressure plasma as an efficient, eco-friendly method for silk modification without altering bulk properties [82]. This progress facilitates the advancement of next-generation silk-based medical devices.

7.2.5.2 Surface Modification Through Chemical Approaches

Silk biomaterials exhibit exceptional biocompatibility and mechanical strength, rendering them highly suitable for applications in tissue engineering, implants, and surgical tools. However, chemical surface modifications further enhance their functionality [53]. Methods such as plasma treatment, covalent conjugation, and chemical grafting introduce reactive groups (e.g., carboxyl, amine, or hydroxyl) that improve cell adhesion, antibacterial properties, and overall bioactivity. For example, carbodiimide chemistry facilitates the attachment of growth factors or peptides, promoting tissue regeneration, while silane coupling agents enhance hydrophilicity, essential for implant integration. Additionally, chemical crosslinking with agents like glutaraldehyde or genipin reinforces silk scaffolds without compromising biodegradability. These modifications expand silk's biomedical potential, improving its performance in wound healing, bone regeneration, and surgical sutures. Despite these advancements, the chemical modification of silk fibroin (SF) is challenging due to the limited reactivity of most amino acids, which are often embedded in crystalline regions. To address this, mild reactive agents are used to directly modify SF surfaces, enabling controlled alterations in both structure and function.

7.2.5.3 Surface Functionalization with Nanomaterials

Nanomaterial immobilization enhances the performance of silk by introducing novel properties. Quantum dots, metal nanoparticles, and inorganic nanostructures have been incorporated into SFFs to enhance functionality [83]. The design of silk-based nanomaterials relies on the organization of components, interfacial interactions, and hierarchical structuring to optimize performance [84]. Initially, gold nanoparticles, carbon nanotubes, and graphene oxide were introduced through physical adsorption, covalent bonding, or electrostatic interactions to enhance conductivity, mechanical strength, and bioactivity. Subsequently, silver nanoparticles were utilized for their antimicrobial properties, thereby reducing the risk of infection in implants and surgical instruments. Later, hydroxyapatite nanoparticles were integrated to enhance osteoconductivity, supporting bone tissue engineering. More recently, quantum dots have been utilized for bioimaging, enabling real-time monitoring of tissue regeneration. Advancements in functionalization techniques, such as layer-by-layer assembly, now ensure uniform nanomaterial distribution while maintaining silk's biodegradability. These innovations have made silk-based composites highly effective for regenerative medicine, implant coatings, and smart surgical devices, bridging the gap between natural and synthetic biomaterials.

7.3 Silk in Tissue Engineering

7.3.1 Silk Scaffolds for Regenerative Medicine

Regenerative medicine focuses on enhancing cell performance to bridge the gap between in vitro and in vivo drug release studies [85]. Tissue engineering combines cells, scaffold biomaterials, and growth factors to promote organ regeneration. Among natural biopolymers, silk fibroin is distinguished by its superior mechanical strength, biocompatibility, hemocompatibility, and biodegradability, positioning it as a highly promising scaffold material in regenerative medicine. The increasing focus on silk-based biomaterials underscores their significant potential in tissue engineering applications. [86].

7.3.2 Silk-Based Biomaterial Scaffolds

Silk scaffolds are widely utilized in tissue engineering due to their biocompatibility and diverse functional properties. These scaffolds can be categorized into 2D films, 3D sponges, and hydrogels, each serving distinct functions in regenerative medicine [13, 87].

7.3.2.1 Silk Hydrogels

Silk hydrogels are formed when aqueous silk fibroin solutions experience sol–gel transitions induced by dehydrating agents, acids, or sonication [88]. These hydrogels are versatile, serving as both injectable and non-injectable delivery systems. Their strong mechanical characteristics make them especially suitable for load-bearing tissue engineering uses, like cartilage regeneration, positioning them as an ideal option for scaffold-based tissue repair [89, 90].

7.3.2.2 3D Porous Silk Scaffolds

Three-dimensional silk scaffolds, fabricated via freeze-drying [91], porogen leaching [92], or solid free-form [93] techniques, offer precise control over pore architecture (60–250 μm), critical for cell infiltration and nutrient diffusion in bone and cartilage regeneration [94]. To augment mechanical strength, these scaffolds are often reinforced with silk microfibers or hybrid composites (e.g., hydroxyapatite for bone or chondrogenic factors for cartilage). While organic/inorganic fillers improve bioactivity and load-bearing capacity, challenges persist in optimizing interfacial compatibility between components [95]. The adaptability of these scaffolds makes them a versatile platform for structurally demanding tissue engineering applications.

7.3.2.3 Silk in 3D Bioprinting

Silk fibroin has emerged as a promising bioink for 3D bioprinting due to its biocompatibility, tunable mechanical properties, controllable biodegradation, and cost-effectiveness, all of which are critical for tissue regeneration [17]. Its resemblance to the extracellular matrix improves cell attachment and growth, making it well-suited for creating biomimetic tissue constructs [96]. Over the last thirty years, silk fibroin-based bioinks have become increasingly important in precision medicine, allowing for the precise layering of cells, biomaterials, and growth factors to engineer functional tissues and organs. This technology bridges the gap between research and clinical applications, offering potential for personalized therapies, organ implantation, and bone regeneration [97].

3D bioprinting uses additive manufacturing to construct complex, biocompatible structures using computer-aided design. Key techniques include the following:

- Inkjet bioprinting (high-speed, droplet-based deposition).
- Extrusion bioprinting (continuous filament deposition for high-viscosity bioinks).
- Photocuring-based bioprinting (light-induced crosslinking for high-resolution structures).

These methods enable precise control over scaffold architecture, pore size, and mechanical strength, which are essential for mimicking native tissues [98, 99]. Silk fibroin scaffolds, whether as hydrogels, porous sponges, or bioprinted matrices, play a pivotal role in tissue engineering. Their versatility supports diverse applications, from cartilage and bone regeneration to soft tissue repair, underscoring their potential in regenerative medicine [100, 101].

7.3.3 Applications in Bone, Cartilage, and Skin Tissue Engineering

Tissue engineering is a multidisciplinary field that merges concepts from biology, engineering, and medicine to create functional replacements for injured tissues and organs [102]. This strategy combines cells, biocompatible materials, and bioactive agents, such as growth factors, to repair, maintain, or enhance tissue function. Central to tissue engineering is the use of scaffolds that facilitate cell adhesion and proliferation while delivering essential biochemical and mechanical signals to direct tissue regeneration [103]. Through these strategies, tissue engineering offers promising solutions for repairing or replacing compromised tissues (Fig. 7.3). Significant progress has been made in specialized areas such as skin tissue engineering for wound healing and burns, bone tissue engineering for fracture repair and defect reconstruction, and cartilage tissue engineering for joint and musculoskeletal repair. These advances highlight the potential of tissue engineering to revolutionize regenerative medicine, providing clinically viable alternatives to traditional transplantation and surgical interventions [52].

7.3.3.1 3D Bone Tissue Engineering

Bone is a unique type of connective tissue that plays a vital role in providing structural support to the human body. It plays several critical physiological roles, including facilitating movement through mechanical support and protecting vital organs from

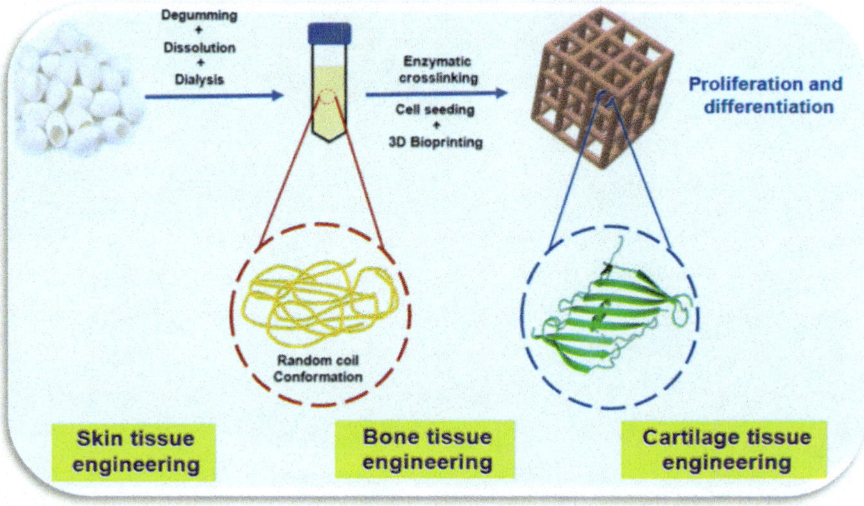

Fig. 7.3 Application of silk fibroin protein [104]. Copyright 2022. Adapted with permission from Elsevier Ltd

injury [105]. Bone is structurally made up of both organic and inorganic components. The organic matrix, mainly composed of collagen, imparts flexibility and tensile strength, while the inorganic portion provides stiffness and compressive strength [106]. Bone tissue engineering plays a crucial role when extensive defects resulting from cancer surgeries, congenital disorders, or trauma exceed the bone's natural ability to heal and regenerate properly [107]. Traditional treatments often struggle to address large-scale defects, necessitating the use of advanced regenerative solutions. In this context, 3D bioprinting has emerged as a transformative approach in tissue engineering, enabling the fabrication of patient-specific, customized constructs that precisely match anatomical and biological requirements [108]. This approach holds great potential for restoring complex tissue and organ damage, ultimately contributing to improved clinical outcomes (Fig. 7.4).

7.3.3.2 3D Cartilage Tissue Engineering

Cartilage is a distinct type of connective tissue that provides a smooth, lubricated surface for joint movement and supports the distribution of low-friction loads. However, its lack of blood vessels significantly limits its ability to self-repair, making it prone to gradual deterioration following injury or disease [110]. This inherent challenge has driven extensive research in cartilage tissue engineering, intending

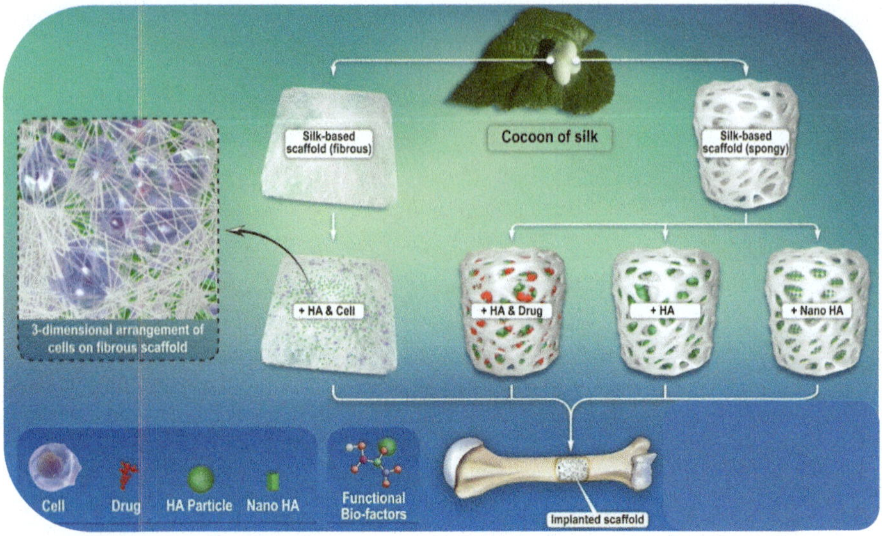

Fig. 7.4 Schematic representation of bone tissue regeneration [109], copyright 2017. Adapted with permission from Elsevier Inc

to develop functional substitutes for damaged tissue. Despite significant advancements, the clinical translation of engineered cartilage remains a complex and time-consuming endeavor, requiring further refinement of biomaterials and fabrication techniques [111, 112].

A key challenge in 3D bioprinting cartilage is designing scaffolds that replicate the mechanical resilience and load-bearing properties of native tissue [113]. Among the biomaterials under investigation, silk fibroin has emerged as a leading candidate due to its exceptional mechanical strength, biocompatibility, and tunable degradation kinetics. The structural integrity of silk fibroin (Fig. 7.5) arises from its β-sheet crystalline domains, which can be modulated to enhance durability while maintaining flexibility. Furthermore, silk fibroin's ability to retain hydration and support cell viability makes it particularly suitable for cartilage tissue engineering, where maintaining a hydrated extracellular matrix is critical for functionality [114]. By harnessing advanced bioprinting techniques, researchers can create patient-specific 3D constructs replicating the complex structure of natural cartilage. These innovations offer a promising solution to the shortcomings of conventional repair methods, bringing engineered cartilage closer to clinical viability [5, 6].

Silk fibroin is often combined with other biomaterials, such as gelatin or decellularized extracellular matrix, to enhance printability, maintain cell viability, and promote chondrogenic differentiation. As a bioink, silk fibroin provides a versatile platform for 3D bioprinting of cartilage, offering an exceptional balance of mechanical strength, tunable degradation kinetics, and inherent chondroinductive capabilities [116, 117]. To advance the field, future research must prioritize the following:

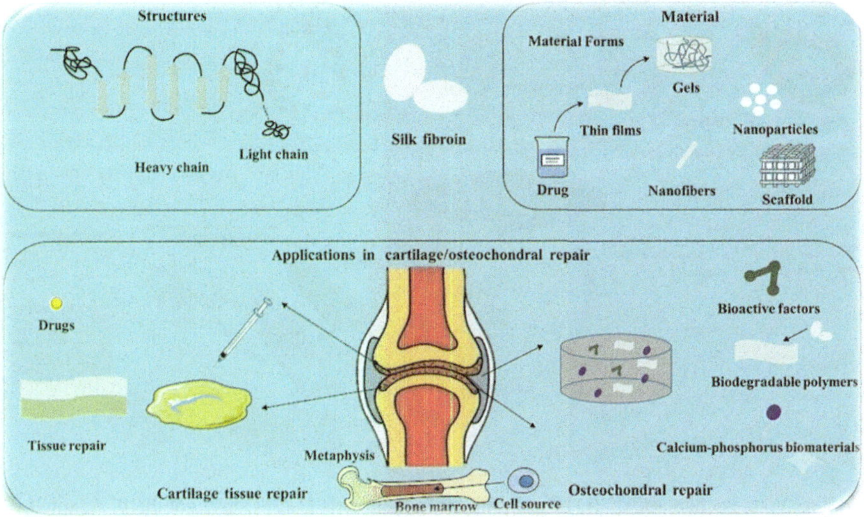

Fig. 7.5 Schematic representation of silk fibroin for 3D Cartilage tissue engineering [115]. Copyright 2023, Adapted with permission from MDPI

- Vascularization strategies to ensure nutrient diffusion in thick constructs.
- Long-term biomechanical performance under physiological loading.
- Scalable manufacturing and preclinical validation to bridge the gap toward clinical adoption.

Addressing these challenges will be crucial for developing functional cartilage grafts that can repair complex joint defects [118].

7.3.3.3 3D Skin Tissue Engineering

The skin, the main organ in the human body, is a multilayered structure comprising three primary layers: the epidermis, dermis, and hypodermis [119, 120]. Despite its remarkable regenerative capacity, severe skin damage, such as deep burns or chronic wounds, poses significant clinical challenges [121]. Conventional treatments such as autografts and allografts are limited by several challenges, including donor site complications, scarce tissue availability, poor vascularization, and incomplete recovery of epidermal functionality. These constraints underscore the urgent need for advanced tissue engineering strategies that can replicate the skin's native architecture while promoting rapid vascular integration and long-term functional recovery [122, 123].

To address these challenges, 3D bioprinting has arisen as a transformative approach for engineering biomimetic skin substitutes that recapitulate native tissue architecture (Fig. 7.6). Among potential bioinks, silk fibroin has gained significant

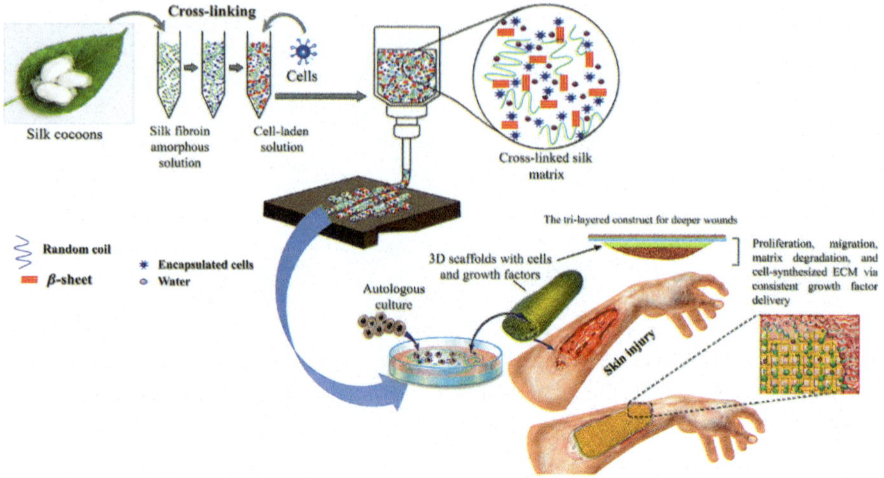

Fig. 7.6 Visual schematic of 3D bioprinting for skin models [126] Copyright 2019. Adapted with permission from Hindawi

attention due to its exceptional biocompatibility, tunable mechanical properties, and inherent ability to promote cell adhesion and proliferation, critical attributes for skin regeneration [124, 125].

Recent advances highlight silk fibroin's versatility in skin bioprinting. In one study, digital light processing-based 3D printing was used to fabricate a multilayered skin scaffold from a photo-cross-linkable silk fibroin/four-arm polyethylene glycol (PEG) composite. The resulting construct not only supported robust keratinocyte proliferation but also facilitated the development of a stratified keratin layer, closely resembling the epidermal structure in vivo [127].

Alternative strategies have incorporated functionalized silk fibroin blends to enhance regenerative outcomes. For instance, a gelatin-sulfonated silk fibroin bioink, combined with bioactive growth factors, demonstrated accelerated wound healing in preclinical models. The bioprinted constructs promoted organized extracellular matrix deposition and skin appendage formation, suggesting improved functional recovery compared to conventional grafts [128, 129].

Collectively, these studies underscore silk fibroin-based bioprinting as a promising avenue for developing next-generation skin substitutes. Nonetheless, significant challenges persist, including the need to scale up production, maintain long-term stability, and achieve full-thickness skin regeneration with integrated appendages and vascular networks—essential steps toward successful clinical translation [130].

Although silk fibroin-based bioinks exhibit considerable potential for skin tissue engineering, this field remains in its promising stages. Key challenges that require further investigation include the following:

- Enhancing vascularization within bioprinted skin constructs.
- Improving mechanical resilience under dynamic physiological stresses.

- Ensuring long-term functional integration in clinical settings.

Overcoming these challenges is essential to enhance the translational potential of silk fibroin-based bioprinted skin grafts for clinical use [86, 131].

7.3.3.4 Cell-Adhesive Silk Fibroin Scaffolds for Neural Repair

The first synthetic nerve canal material, silicone, has limitations, including low degradation and extension capacity, which can lead to nerve compression and fibrosis, thereby restricting its long-term use [132]. To address these issues, a range of natural polymers, such as chitosan, alginate, collagen, gelatin, and hyaluronic acid as well as synthetic polymers like polylactide (PLA), polyglycolide (PGA), poly(lactic-co-glycolic acid) (PLGA), and polycaprolactone (PCL) have been developed [133, 134]. Natural polymers are preferred for peripheral nerve regeneration because of their biocompatibility, lower toxicity, and ability to promote cell proliferation and migration [135]. The remarkable biocompatibility of silk fibroin has positioned it as a highly promising biomaterial for peripheral nerve repair applications. Recent advances have demonstrated its particular utility in the development of nerve guidance conduits, where its structural and biological properties offer significant advantages [136]. Current methodologies employ carbodiimide chemistry to covalently conjugate biologically active molecules, specifically collagen and laminin derived from human placenta, onto silk fibroin matrices. This biofunctionalization approach has been shown to enhance the performance of nerve regeneration scaffolds by:

- Improving initial Schwann cell adhesion.
- Promoting long-term cellular viability.
- Supporting robust cell proliferation.

The modified scaffolds exhibit consistent and reproducible performance, with laminin-functionalized surfaces demonstrating particularly rapid cellular attachment kinetics. These technical developments, combined with silk fibroin's inherent mechanical strength and tunable degradation profile, have accelerated its adoption in neural tissue engineering applications [137, 138].

Cell adhesion assays confirm that functionalizing silk fibroin with either laminin or collagen enhances Schwann cell attachment compared to unmodified controls. Notably, laminin-functionalized substrates exhibit a particularly rapid acceleration in cell adhesion kinetics. These findings highlight the potential of biofunctionalized silk fibroin scaffolds to enhance cellular interactions, supporting their application in advanced peripheral nerve repair strategies [139]. The evaluation of cell proliferation on three-dimensional silk fibroin scaffolds has provided important insights into scaffold design parameters. Studies demonstrate that scaffold fabrication temperature significantly influences pore architecture, with 80 °C processed scaffolds exhibiting superior performance despite having smaller pore sizes compared to 20 °C counterparts. Specifically, scaffolds with 50–75 μm pores (80 °C) consistently support greater cell proliferation than those with 75–100 μm pores (20 °C), highlighting that optimal cell growth depends on multiple architectural factors beyond simple pore dimensions [140].

Analysis of scaffold microstructure reveals that enhanced porosity and interconnectivity facilitate more efficient cell migration throughout the three-dimensional matrix. This structural characteristic prevents localized cell overcrowding during initial growth phases, thereby improving overall cell survival rates. Migration studies confirm this mechanism, showing active cellular movement from central seeding points to peripheral regions of the scaffolds [141].

The freeze-drying fabrication method enables precise control over these critical microstructural features through manipulation of processing temperatures. This tunability, combined with silk fibroin's inherent biocompatibility and mechanical strength, establishes these scaffolds as particularly suitable for bone and hard tissue engineering applications. The technology offers a reliable platform for creating functional tissue constructs where controlled pore architecture is essential for successful cell proliferation and tissue integration [142].

7.4 Silk as an Implant Material

Implantable biomedical textiles and fiber-based implants are extensively used in medical applications, for example wound closure (sutures) and replacement surgeries (e.g., soft-tissue implants, vascular prostheses, and artificial tendons/ligaments). A key requirement for these materials is biocompatibility, as they are designed for in vivo use and must interact safely with the body's biological systems. Ensuring compatibility with living tissues is crucial to prevent adverse reactions and support successful integration and long-term functionality within the body [53].

7.4.1 Types of Silk Implants

7.4.1.1 Sutures and Meshes

The use of non-bio-incompatible mesh materials in surgical implants often leads to significant post-operative complications such as delayed wound healing, seroma formation, infections, inadequate tissue infiltration, and chronic pain. These complications stem from adverse host responses, including excessive inflammation and foreign body reactions. When persistent inflammatory reactions occur, they can trigger fibrotic responses that increase the risk of bacterial infections and fluid accumulation (e.g., hematomas) [143, 144]. Additionally, excessive scar tissue formation may bridge across the mesh pores, obstructing cellular infiltration and further impairing wound healing [145]. Prolonged inflammation also disrupts proper collagen deposition, leading to abnormal scar formation that stiffens the abdominal wall. Another critical concern is mesh shrinkage, which can compromise implant functionality and, if left unaddressed, contribute to erosion of surrounding tissues [146].

In contrast, silk has emerged as a promising bioabsorbable polymer due to its enzymatic degradability and minimal immunogenicity. Unlike synthetic meshes, silk degrades into biocompatible by-products such as peptides and amino acids, minimizing adverse immune responses [147]. The degradation process typically begins within eight weeks, with complete absorption occurring over one year. Interestingly, implants with larger pore sizes degrade faster than those with smaller pores, influencing their integration and remodeling in vivo. This favorable degradation profile, combined with silk's mechanical strength and biocompatibility, makes it an attractive alternative for soft tissue repair, reducing the risks of chronic inflammation and long-term complications associated with conventional synthetic meshes.

7.4.1.2 Dental Implants

Dental implants are artificial structures implanted into the jawbone to serve as replacements for missing tooth roots. Despite their widespread use, implant failures often occur due to insufficient osseointegration, the critical biological bonding between bone and implant surface, resulting in poor peri-implant bone formation [148, 149]. Improving osseointegration is essential for maximizing bone-implant contact, particularly through optimized surface characteristics that promote bone nucleation. Recent advances demonstrate that nanoscale surface modifications, including bioactive biomolecular coatings, can significantly enhance peri-implant bone regeneration and increase bone-to-implant contact ratios [150, 151]. These surface engineering approaches offer promising strategies to improve implant success rates and long-term stability.

Two innovative techniques have been employed to enhance dental implant performance: depositing biodegradable silk fibroin coatings that can withstand implantation stresses. The first approach utilized electrodeposition to directly coat titanium substrates, while the second employed melted electrogel deposition. Importantly, the coatings maintained structural integrity during simulated implantation. These silk fibroin-based coatings offer significant advantages, including mechanical robustness, biocompatibility, and biodegradability, while avoiding harsh chemical processing. The successful application to various titanium implant geometries suggests promising potential for clinical translation, providing a foundation for functionalized dental implants with improved osseointegration capabilities [152].

7.4.2 *In Vivo Performance and Comparative Analysis of Silk Biomaterials*

The growing environmental burden of medical waste has become a critical global concern. According to World Health Organization (WHO) statistics, healthcare facilities in high-income countries generate substantial amounts of hazardous waste, averaging 0.5 kg per hospital bed per day [150]. This alarming figure, combined with the well-documented ecological and health impacts of synthetic medical materials, has driven significant research efforts toward developing sustainable alternatives in healthcare applications.

Among various natural materials being investigated, silk fibroin has emerged as a particularly promising biomaterial due to its exceptional combination of properties [153]:

- **Mechanical Superiority**: Silk fibroin exhibits impressive tensile strength, ranging from 47.2 to 67.7 MPa, and toughness between 60,000 and 160,000 J/kg, levels comparable to those of many synthetic polymers.
- **Tunable Physical Properties**: a Young's modulus ranging from 1.9–3.9 GPa; silk fibroin can be engineered for specific medical applications.
- **Biocompatibility and Safety**: Excellent tissue compatibility with minimal immune response.
- **Environmental Advantages**: Natural biodegradability and reduced ecological footprint compared to petroleum-based materials.

The integration of silk fibroin with other natural compounds presents a viable pathway toward developing medical solutions that address both clinical performance and environmental sustainability. This approach is particularly relevant for reducing dependence on conventional chemical-based treatments and medical devices [28]. Recent research trends indicate an increasing interest in silk fibroin-based biomaterials, with studies exploring new applications in tissue engineering, drug delivery, and medical device coatings [101]. The unique characteristics of silk fibroin, combined with modern material processing technologies, are opening new possibilities for sustainable healthcare solutions that meet both medical and environmental requirements [68, 154].

7.5 Processing and Fabrication of Silk Biomaterials

Silk biomaterials, renowned for their biocompatibility, biodegradability, and mechanical strength, are highly suitable for various biomedical applications, including tissue engineering, wound healing, and drug delivery [155]. The process begins with extracting fibroin from silkworms, which is then purified and reassembled into various forms, including films, scaffolds, and fibers. Techniques such as electrospinning, solvent casting, and freeze-drying enable the fabrication of structures that mimic the extracellular matrix, thereby supporting cell adhesion and growth. Electrospinning produces nanofibrous mats, solvent casting creates flexible films, and freeze-drying forms porous sponges for 3D culture. Chemical crosslinking enhances material stability, while genetic modifications enable the functionalization of silk. These versatile silk biomaterials show significant promise in regenerative medicine and other biomedical applications.

7.5.1 Extraction and Purification

Silk fibroin-based scaffolds have garnered significant attention in tissue engineering due to their excellent biocompatibility, natural degradability, and robust mechanical properties. One of the critical steps in utilizing silk fibroin for medical applications is the extraction and purification process, which ensures the removal of sericin, a glue-like protein that can cause immune reactions while preserving the mechanical and biological properties of fibroin. Figure 7.7 illustrates the process of preparing silk fibroin solution from Bombyx mori cocoons, which includes essential steps such as cocoon cutting, boiling, fiber washing, incubation in lithium bromide (LiBr), dialysis, centrifugation, and subsequent storage.

7.5.1.1 Degumming Process

The preparation of silk for biomedical applications begins with the degumming process, which removes sericin, a sticky protein coating the silk fibers. Sericin can trigger immune responses when used in the human body, so its removal is crucial for making silk suitable for biomedical use. Degumming involves treating raw silk with a mild alkali solution to dissolve sericin without damaging fibroin, the protein responsible for silk's strength [156]. A widely used approach involves boiling silk in a 0.02 M sodium carbonate ($Na2CO_3$) solution for 30 min, repeated three times to thoroughly eliminate sericin [157]. Sodium carbonate is favored due to its efficiency and its capacity to enhance the crystallinity index, a key factor influencing silk's suitability for biomedical uses. While sodium carbonate remains the most common agent, alternative substances such as Marseille soap, ethylenediamine, and tartaric acid have also been utilized for this purpose [158]. However, sodium carbonate consistently provides a higher crystallinity index, making it the preferred choice for most applications. Enzymatic degumming, using cocoonase, offers a promising alternative. This method gently removes sericin while preserving the quality of fibroin. Recent studies show that neutral proteases, such as cocoonase, effectively remove sericin and maintain the molecular weight of fibroin, resulting in silk with better texture, gloss, and color retention compared to alkali treatments [159]. Regulating the degumming process is crucial to achieving an optimal balance between removing sericin and maintaining the structural integrity of fibroin. This balance is vital to preserving silk's biocompatibility, mechanical strength, and functionality, which are essential

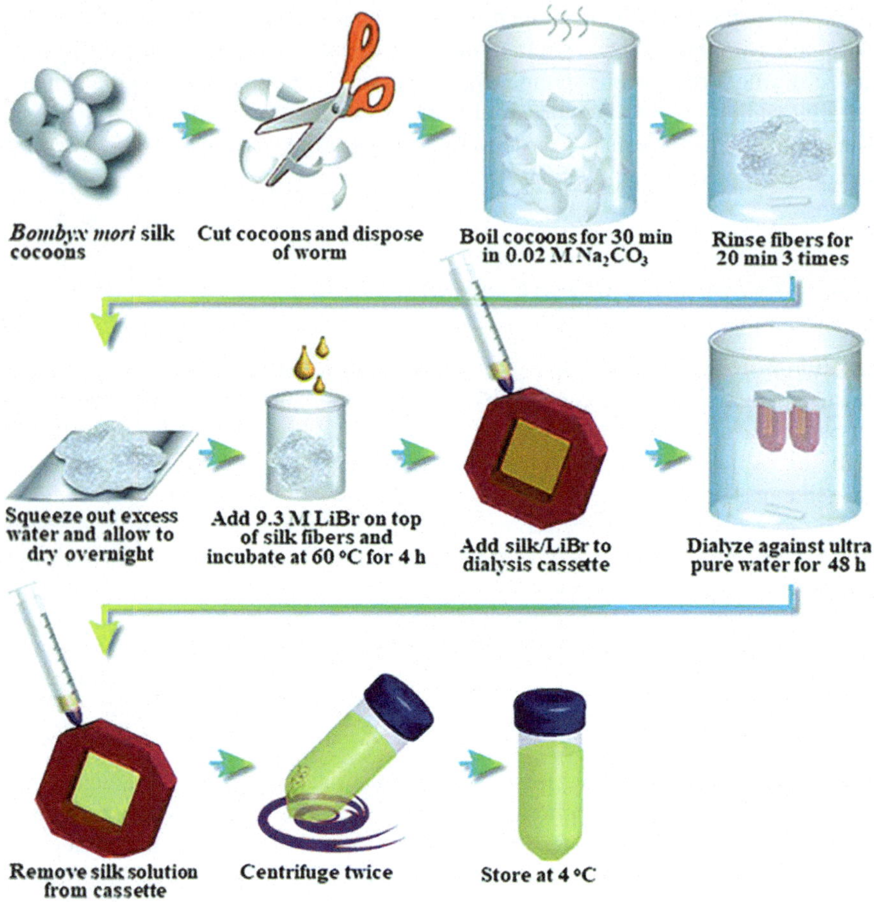

Fig. 7.7 Silk fibroin solution from Bombyx mori cocoons is prepared by cutting, boiling, rinsing, LiBr incubation, dialysis, centrifugation, and storage [6], Copyright 2024. Adapted with permission from frontiers

for various biomedical applications, including sutures, tissue engineering scaffolds, and drug delivery platforms.

7.5.1.2 Optimization for Medical-Grade Silk

Silk fibroin is a valuable biomaterial due to its biocompatibility, strength, and biodegradability [160]. To meet medical standards, it must be purified to remove impurities while preserving its structural integrity. The first step is degumming, which removes sericin, a glycoprotein that can trigger immune responses. This process must strike a balance between sericin removal and fibroin preservation. Factors such as temperature, time, and the use of degumming agents influence efficiency. Higher temperatures accelerate sericin removal but can weaken fibroin, while prolonged treatment may compromise its structure. Chemical agents, such as sodium carbonate, or enzymes affect purity and stability, with enzymatic and ultrasonication-assisted methods offering improved control [109]. Further purification removes residual chemicals and contaminants through repeated washing, dialysis, centrifugation, and filtration. The purified silk fibroin obtained is highly suitable for tissue engineering applications, including the regeneration of bone, cartilage, and skin. Additionally, it facilitates wound healing by enhancing cell adhesion and serves as an effective material for controlled drug release in delivery systems. Optimizing purification ensures silk fibroin meets medical standards, with ongoing research enhancing its biomedical potential [28]. Careful control of processing ensures high purity and functionality, with ongoing research enhancing its biomedical potential.

7.5.2 Fabrication Methods for Silk-Based Biomaterials

Silk fibroin scaffolds for tissue engineering are produced through diverse fabrication techniques, each influencing key material properties including porosity, mechanical strength, and biocompatibility. Commonly employed methods comprise electrospinning, casting, and freeze-drying. These approaches enable the tailoring of silk-based biomaterials to meet specific requirements, thereby expanding their suitability for various tissue engineering applications based on the scaffold's intended characteristics and functions.

7.5.2.1 Electrospinning

Electrospinning is a highly efficient method for fabricating nanofibrous silk scaffolds, playing a crucial role in tissue engineering. This process involves applying a high-voltage electric field to a silk fibroin solution, causing the solution to stretch into fine fibers [161]. These fibers, collected onto a grounded collector, have diameters similar to those of the natural extracellular matrix, which is crucial for supporting cell adhesion and growth. The versatility of electrospinning enables the arrangement of fibers in various patterns and structures, making it adaptable to diverse tissue engineering applications. As a result, electrospun scaffolds typically exhibit excellent mechanical properties and a large surface area, which are beneficial for tissue regeneration [162]. Electrospinning is a fiber fabrication technique that utilizes electrical forces to transform polymer solutions into fine fibers. This method enables the production of non-woven mats with nanoscale features, particularly in terms of pore size and structural characteristics, which are challenging to achieve through conventional fiber manufacturing techniques [163]. The resulting nanofibrous mats closely replicate the architecture of the natural extracellular matrix [164], which is essential for applications in tissue engineering. In the electrospinning process, a silk fibroin solution is placed in a syringe connected to a spinning machine, set at a specific distance and angle. An adjustable voltage of 10–35 kV and a flow rate of 0.2–4.0 mL/h create Coulombic forces, forming a Taylor cone. The charged jet extends, solidifies, and is collected by a grounded collector [165]. The morphology and diameter of silk fibroin (SF) fibers are influenced by several factors, including the properties of the spinning solution, such as its nanostructures, concentration, viscosity, and solvent as well as the processing parameters, such as voltage, flow rate, temperature, and the distance between the capillary tip and the collector [166]. These parameters provide precise control over the fiber properties, enabling customization to meet the specific requirements of various tissue engineering applications.

7.5.2.2 Casting

Casting is a widely employed technique for fabricating silk fibroin scaffolds due to its ability to create scaffolds with specific shapes and structures. In this process, a silk solution is poured into a mold and left to solidify, resulting in a scaffold with a defined geometry. This method is particularly advantageous when creating scaffolds that replicate the structure of organs or tissues, offering preF7cise control over the shape and architecture of the final product. One of the main benefits of casting is its ability to adjust the pore size and density within the scaffold, factors critical to promoting cell infiltration, nutrient exchange, and the overall biological performance of the scaffold [167]. By modifying the silk solution's concentration and altering curing conditions, the mechanical properties of the scaffold can be optimized to meet the needs of various tissue types, ensuring compatibility with the intended application.

An advanced variation of casting, solvent casting with particle leaching, is utilized to produce porous scaffolds. In this method, the silk solution is mixed with a volatile organic solvent and water-soluble salt particles. This mixture is then cast into a three-dimensional scaffold mold. After casting, the scaffold is immersed in a solvent that dissolves the salt particles while leaving the polymer matrix intact. This process creates pores where the salt particles were initially present [168]. The addition of various pore-forming agents, such as salt particles, sugars, glucose, or gelatin, enables fine control over the scaffold's pore morphology [169]. The choice of casting solvent significantly impacts the silk fibroin scaffold's properties, including β-sheet content, mechanical strength, stability, and degradation rates [170]. In general, casting techniques, including solvent casting with particle leaching, provide precise control over structure, pore morphology, and mechanical properties, making silk fibroin scaffolds adaptable for tissue engineering.

7.5.2.3 Freeze-Drying

Freeze-drying, also known as lyophilization, is a widely utilized technique for fabricating porous silk fibroin scaffolds, essential in various biomedical applications. The process begins with an aqueous silk fibroin solution, which is poured into a mold and subsequently frozen in an ultralow-temperature freezer below its triple point [5]. This freezing process prepares the material for sublimation. In the first phase of freeze-drying, the frozen scaffold is transferred to a freeze dryer, where approximately 95% of the water is removed by sublimation, transitioning directly from ice to vapor. The second phase eliminates the remaining bound water by applying a higher temperature, which helps to break the bonds between the water molecules and the silk fibroin material. This results in the formation of a highly porous scaffold.

The porosity of the scaffold is crucial for promoting cell migration, nutrient exchange, and tissue ingrowth, making freeze-dried scaffolds ideal for applications like wound healing and bone regeneration. Factors such as polymer concentration, viscosity, and the amount of aqueous phase influence the porosity and pore size. Recent advancements include a two-step freeze-drying process, which creates micro/nanofibrous networks, enhancing cell attachment and proliferation [171]. These advancements in freeze-drying techniques continue to enhance the functionality and applicability of silk fibroin scaffolds in regenerative medicine and tissue engineering.

7.5.2.4 Gas Foaming

Gas foaming is an innovative fabrication technique used to produce porous silk-based biomaterials in a solvent-free manner, avoiding the need for high temperatures. The process incorporates a gas-forming agent, such as ammonium bicarbonate or sodium bicarbonate, into a silk solution or scaffold. When exposed to heat or an acidic environment, these agents decompose and release gas, creating a network of pores within the silk matrix. The size and distribution of the pores can be controlled by adjusting the concentration of the gas-forming agent and processing conditions. This method offers several advantages for biomedical applications, especially in the creation of scaffolds for tissue regeneration. The interconnected porous networks created through gas foaming improve cell infiltration, nutrient transport, and tissue integration. This feature is especially advantageous for bone and cartilage regeneration, where elevated porosity is critical to support cell adhesion and facilitate tissue ingrowth [172]. Additionally, gas foaming is more biocompatible and environmentally friendly compared to other porogen-based methods, as it does not require harsh chemicals.

Despite these advantages, one limitation of gas foaming is the lack of precise control over pore uniformity, which may influence mechanical properties and cell behavior. To address this, researchers often combine gas foaming with other techniques such as freeze-drying or particulate leaching. For example, Maniglio et al. [173] used N_2O to create fibroin scaffolds, where pore size was influenced by gas pressure and protein concentration, while Rao et al. [174] combined electrospinning and gas foaming to develop 3D nanofiber-sponge scaffolds.

7.5.2.5 3D Printing

3D printing is an advanced fabrication method that allows for the precise design and construction of silk-based biomaterials with tailored architectures. The process involves the layer-by-layer deposition of silk-based bioinks, which consist of silk fibroin solutions or composites mixed with other biocompatible materials. This technique enables the creation of scaffolds with complex structures, making it suitable for tissue engineering, patient-specific implants, drug delivery systems, and organ models. The ability to control scaffold architecture at the microscale is a key advantage of 3D printing, particularly for creating scaffolds that closely mimic the extracellular matrix, enhancing cell adhesion, proliferation, and differentiation [175]. The process begins with preparing a silk-based bioink with suitable viscosity, stability, and printability. Once the structures are printed, they are stabilized using post-processing techniques such as crosslinking, drying, or curing, which help enhance their mechanical strength and maintain shape. Despite its advantages, challenges remain in optimizing print resolution, ensuring long-term stability, and improving mechanical properties for clinical use. Silk fibroin is an ideal material for 3D printing due to its tunable mechanical properties and processability. However, its low viscosity and unsuitable rheology for printing often require modifications, such as purification or concentration [37].

To improve the printability and mechanical properties of silk fibroin bioinks, researchers often combine them with other biomaterials like gelatin or hydroxypropyl methylcellulose. These combinations balance mechanical strength and degradation rates, supporting the growth of mesenchymal progenitor cells. Moreover, a two-step method integrating silk fibroin with hydroxypropyl methylcellulose (HPMC) or calcium alginate has demonstrated improvements in printability and mechanical strength, thereby enhancing the suitability of silk fibroin-based scaffolds for prolonged cell culture applications [176]. These innovations position silk fibroin bioinks as a promising material for 3D bioprinting in tissue engineering.

7.5.3 Chemical Modifications of Silk Fibroin

Chemical modifications of silk biomaterials are commonly used to improve their properties and expand their use in medical applications. Techniques such as crosslinking, grafting, and functionalization enable the modification of silk's surface characteristics, enhancing its mechanical strength, biocompatibility, and biodegradability [177]. Crosslinking, for example, boosts the material's resistance to enzymatic degradation, which is beneficial for creating silk-based scaffolds in tissue engineering. Grafting bioactive molecules or peptides on the silk surface promotes targeted cellular interactions, fostering cell adhesion, proliferation, and differentiation. These modifications make silk biomaterials adaptable for various medical uses, including controlled drug delivery, wound healing, and tissue regeneration [178]. The integration of bioactive molecules, including growth factors, significantly improves cell compatibility and biological activity; while crosslinking processes reinforce scaffold stability and prevent early degradation in vivo. Additionally, surface functionalization can be used to regulate how the material interacts with the biological environment, optimizing its performance in regenerative medicine and tissue engineering.

7.5.3.1 Crosslinking of Silk Biomaterials

Crosslinking is a widely used chemical modification to enhance the mechanical stability and durability of silk biomaterials. By forming covalent bonds between silk molecules, this process strengthens the material's structure, making it more rigid and resilient [179]. Crosslinking enhances the performance of silk in biomedical and textile applications, providing better resistance to degradation and mechanical stress. Various crosslinking agents and methods are employed to optimize silk's properties, ensuring its suitability for diverse functional uses in advanced material science.

(a) **Improvement of Mechanical Properties**: Crosslinking enhances the mechanical strength of silk biomaterials, increasing their resistance to deformation under mechanical stress. This reinforcement is essential for applications requiring structural stability, such as tissue scaffolds and implants.
(b) **Resistance to Enzymatic Degradation**: Silk biomaterials are prone to enzymatic degradation, which can be accelerated by biological fluids. Crosslinking increases the resistance of silk to enzymatic breakdown, ensuring that the material maintains its stability over extended periods in vivo.
(c) **Application in Tissue Engineering**: Crosslinked silk scaffolds are used to create a temporary matrix that supports tissue growth and regeneration. The modified silk scaffold remains intact long enough for new tissues to form and integrate with the body.

7.5.3.2 Grafting Bioactive Molecules and Peptides

Grafting refers to the process of attaching bioactive molecules, such as growth factors, peptides, or signaling molecules, onto the surface of silk materials. This modification enhances the biological functionality of silk by enabling it to interact more effectively with cells [180]. The incorporation of these bioactive molecules can trigger specific cellular responses, including enhanced cell adhesion, proliferation, and differentiation, which are essential for numerous biomedical applications. Precise grafting of such molecules onto the silk surface improves the material's functionality and effectiveness in tissue engineering, wound healing, and other medical domains [180].

By altering the surface chemistry of silk, it becomes a more versatile and functional material, capable of supporting complex biological processes. This strategy not only enhances the biocompatibility of silk but also enables the controlled release of bioactive substances, providing a platform for tailored therapeutic interventions.

(a) **Promotion of Cell Adhesion and Proliferation**: By grafting adhesion peptides or growth factors onto silk, the material facilitates enhanced cell attachment, spreading, and proliferation. This is particularly beneficial for tissue regeneration applications, where controlling cell behavior is necessary.
(b) **Induction of Specific Cellular Responses**: Grafted molecules can direct cell differentiation, migration, or other desired cellular responses, thereby improving the functional integration of the silk biomaterial with the surrounding tissues.
(c) **Applications in Wound Healing and Drug Delivery**: Grafted silk biomaterials can be used for wound healing, where enhanced cell interactions are necessary. Additionally, they are valuable for controlled drug delivery systems, enabling the sustained release of therapeutic agents.

7.5.3.3 Surface Functionalization of Silk Biomaterials

Surface functionalization involves chemically modifying the surface of silk to introduce specific functional groups or molecules. This process is essential for tailoring the silk's interactions with biological environments, enhancing its compatibility and functionality in various applications [181]. By attaching specific functional groups, such as amino, carboxyl, or hydroxyl groups, the surface characteristics of silk can be altered, which affects its hydrophilicity, charge, and overall reactivity [182]. These modifications can improve the silk's ability to support cell adhesion, proliferation, and differentiation [183], making it a valuable material for tissue engineering and other biomedical applications. Additionally, surface functionalization enables the incorporation of bioactive molecules, including growth factors or drugs, allowing for their controlled release over time. This versatility enables silk to be customized for a wide range of clinical and industrial uses, contributing to its growing role in advanced biomaterial design.

(a) **Modification of Biological Interactions**: Surface functionalization enhances the biocompatibility of silk materials by promoting favorable interactions with cells and tissues. By incorporating functional groups, the material's ability to integrate with the body can be significantly improved.
(b) **Improvement of Cell Compatibility**: Functionalization allows the modification of the silk surface to enhance cell adhesion, migration, and proliferation, thereby ensuring the material effectively supports the biological processes required for tissue regeneration.
(c) **Control Over Drug Release**: Functionalization can also influence the release of therapeutic agents from silk-based drug delivery systems. By modifying the surface properties, it is possible to control the rate and duration of drug release, enhancing the material's utility in controlled drug delivery applications.

Surface functionalization enhances the biological performance of silk biomaterials in tissue engineering and implant applications. Although silk fibroin is inherently biocompatible, modifying its surface allows better control over cell attachment, proliferation, and tissue integration. Chemical approaches enable the binding of bioactive molecules such as peptides or polymers, improving cellular responses and stability. Physical techniques, including plasma treatment and surface adsorption, adjust wettability and surface energy without affecting bulk properties. In addition, antimicrobial coatings reduce infection risks in surgical and wound-care uses. These strategies transform silk from a passive structural material into an active interface that supports healing and long-term functionality.

7.6 Clinical Applications and Challenges

Silk biomaterials have emerged as promising candidates for various clinical applications due to their exceptional biocompatibility, mechanical strength, and controlled biodegradability. These characteristics render silk-based materials highly suitable for applications in wound healing, tissue engineering, and drug delivery. However, the transition from laboratory research to clinical application involves numerous challenges, including regulatory compliance, variability in in vivo performance, and the need for patient-specific customization.

7.6.1 Regulatory Hurdles

The clinical translation of silk-based biomaterials faces stringent regulatory challenges that must be meticulously addressed to ensure safety and efficacy. As a naturally derived polymer, silk fibroin presents complexities in standardization compared to synthetic biomaterials. Regulatory agencies, including the U.S. Food and Drug Administration (FDA) and the European Medicines Agency (EMA), enforce stringent criteria for verifying consistency in material properties, validating sterilization processes, and establishing long-term safety profiles [12]. One major concern lies in the inherent variability of silk's molecular structure, which is influenced by its source (e.g., Bombyx mori versus wild silk species) and the extraction methodologies used. These differences can significantly impact mechanical properties and degradation rates, potentially affecting clinical outcomes [56]. Emerging sterilization techniques, including supercritical CO_2 processing, have demonstrated promise but require further validation to gain regulatory approval.

The classification of silk-based medical products further complicates regulatory pathways. Depending on their intended clinical application, whether as surgical meshes, drug delivery systems, or tissue scaffolds, these materials may be categorized under different regulatory frameworks, each with specific preclinical testing and clinical trial requirements. Addressing these regulatory hurdles is vital for ensuring the safe and effective integration of silk biomaterials into medical practice.

7.6.2 In Vivo Performance

The clinical effectiveness of silk biomaterials is largely determined by their interaction with host tissues over time. Although silk fibroin is generally considered biocompatible, its in vivo behavior is influenced by processing methods, implantation sites, and degradation kinetics. A primary advantage of silk-based biomaterials is the ability to tailor their degradation rate, ranging from weeks to several years, through adjustments in processing techniques such as alcohol treatment or water annealing [184]. However, aligning degradation rates with tissue regeneration timelines remains a clinical challenge. For instance, slow-degrading formulations are typically preferred for bone regeneration, whereas soft tissue applications may require more rapid resorption [185]. Despite its generally low immunogenicity, silk fibroin must be carefully purified to eliminate residual sericin, which can elicit inflammatory responses. While advanced purification techniques have significantly reduced this risk, further studies are essential to fully understand the immunomodulatory effects of silk, particularly in immunocompromised patients. Maintaining mechanical integrity throughout the degradation process is another crucial consideration, especially for applications involving load-bearing functions. Although silk exhibits a high strength-to-weight ratio, its gradual hydrolysis in vivo may lead to premature mechanical failure in some cases. To mitigate this limitation, composite approaches incorporating silk with synthetic polymers or ceramics are being actively explored [160]. These hybrid materials aim to enhance mechanical performance while preserving the favorable biological properties of silk.

7.6.3 Customization for Patients

The advent of personalized medicine has increased the demand for biomaterials that can be custom-made to individual patient needs. Silk's versatility positions it as a strong candidate for customized medical solutions, particularly in tissue engineering and regenerative medicine. Advancements in 3D bioprinting technologies now enable the fabrication of patient-specific silk scaffolds with intricate geometries that closely match anatomical defects [186]. By carefully controlling parameters such as pore size, porosity, and mechanical stiffness, researchers can design silk-based constructs that are optimized for specific tissue types and functional requirements. One challenge associated with silk-based implants is their natural radiolucency, which can hinder postoperative monitoring. To address this limitation, researchers have incorporated contrast agents such as barium sulfate or iodine into silk matrices, enabling radiographic visualization without compromising the material's mechanical and biological properties [187]. While laboratory-scale production of customized silk implants is feasible, scaling up these techniques for widespread clinical application remains a significant challenge. Efforts are currently focused on developing automated fabrication systems and establishing stringent quality control protocols to ensure consistency in large-scale manufacturing [188]. Overcoming these challenges will be key to unlocking the full potential of silk biomaterials in precision medicine.

7.7 Future Directions of Silk Biomaterials

The field of silk biomaterials is rapidly evolving, offering exciting possibilities for tissue engineering, medical implants, and surgical applications. Recent advances highlight silk's unique advantages, including its superior mechanical properties, biocompatibility, and tunable degradation rates (Attri et al.). Looking ahead, researchers are focusing on several key areas that promise to expand the clinical utility of silk. One major direction involves enhancing silk's bioactivity through advanced functionalization techniques. Scientists are incorporating growth factors, antimicrobial peptides, and conductive nanomaterials into silk matrices to create scaffolds that actively promote tissue regeneration [84]. These modifications could revolutionize treatments for complex wounds, neural injuries, and cardiac tissue repair. Fabrication technologies are also undergoing significant innovation. Cutting-edge approaches, such as 3D bioprinting and microfluidic spinning, enable precise control over silk scaffold architecture at multiple length scales [189]. These methods allow customization of pore size, mechanical strength, and degradation profiles to match specific tissue requirements. An emerging area of interest is the creation of "smart" silk composites that respond to physiological stimuli. These engineered silk-based materials can deliver therapeutic agents in response to variations in pH, temperature, or enzymatic activity [190]. Such systems could enable targeted drug delivery and real-time monitoring of healing processes. In surgical applications, next-generation silk sutures with built-in antimicrobial properties and enhanced tensile strength are showing promise for reducing infection rates and improving wound closure [191]. Similarly, silk-derived surgical adhesives are being developed as alternatives to traditional staples and sutures. Despite significant progress, challenges persist in scaling up production while ensuring material consistency and compliance with regulatory requirements. Future advancements will rely on interdisciplinary collaboration among materials scientists, biologists, and clinicians.

7.8 Conclusion

Silk-based biomaterials have attracted considerable interest in biomedical fields owing to their exceptional mechanical strength, biocompatibility, biodegradability, and capacity for functionalization. These properties make silk a promising alternative to synthetic and other natural materials in areas such as tissue engineering, medical implants, and surgical tools. Various silk-based scaffolds, including hydrogels, porous sponges, and 3D-printed structures, have shown exceptional potential in supporting cell adhesion, proliferation, and differentiation, making them particularly valuable in regenerative treatments for bone, cartilage, skin, and neural tissues. As an implant material, silk is extensively used in sutures, surgical meshes, and dental applications, offering durability and low immunogenicity. Additionally, silk's versatility extends to surgical tools, where it serves as a base for antimicrobial coatings and controlled drug delivery systems. However, despite these advantages, challenges remain in large-scale production, maintaining consistency between batches, improving sterilization methods, and meeting regulatory requirements.

Emerging advancements in genetically engineered silk proteins, nanocomposite materials, and smart responsive biomaterials offer promising solutions to these challenges, potentially expanding silk's applications in medicine. Continued interdisciplinary research and collaboration will be crucial in overcoming these barriers and unlocking silk's full potential in clinical settings. In essence, silk biomaterials serve as a bridge between conventional medical solutions and innovative regenerative therapies. By addressing existing limitations and integrating emerging technologies, silk-based innovations have the potential to reshape biomedical engineering, providing sustainable and patient-specific solutions for tissue repair, implantable medical devices, and advanced surgical applications. While the journey from research to widespread clinical use requires sustained effort, the potential impact of silk biomaterials in healthcare is immense.

Declaration of AI-Usage AI tools were used solely for language editing, grammar correction, and improving clarity of the manuscript. The authors take full responsibility for the originality, accuracy, and integrity of the content.

References

1. Babu PJ, Suamte L (2024) Applications of silk-based biomaterials in biomedicine and biotechnology. Eng Regen 5(1):56–69
2. Jauzein V, Colomban P (2009) Types, structure and mechanical properties of silk. In: Bunsell AR (ed) Handbook of Tensile Properties of Textile and Technical Fibres. Woodhead Publishing, pp. 144–178
3. Foo CWP, Kaplan DL (2002) Genetic engineering of fibrous proteins: spider dragline silk and collagen. Adv Drug Deliv Rev 54(8):1131–1143

4. Hakimi O, Knight DP, Vollrath F, Vadgama P (2007) Spider and mulberry silkworm silks as compatible biomaterials. Compos B Eng 38(3):324–337
5. Ma L, Dong W, Lai E, Wang J (2024) Silk fibroin-based scaffolds for tissue engineering. Front Bioeng Biotechnol 12:1381838
6. Ma Y, Deng B, He R, Huang P (2024) Advancements of 3D bioprinting in regenerative medicine: exploring cell sources for organ fabrication. Heliyon 10(3):e24593
7. Xie C, Ye J, Liang R, Yao X, Wu X, Koh Y, Wei W, Zhang X, Ouyang H (2021) Advanced strategies of biomimetic tissue-engineered grafts for bone regeneration. Adv Healthcare Mater 10(14):2100408
8. Łuczak JW, Palusińska M, Matak D, Pietrzak D, Nakielski P, Lewicki S, Grodzik M, Szymański Ł (2024) The future of bone repair: emerging technologies and biomaterials in bone regeneration. Int J Mol Sci 25(23):12766
9. Omenetto FG, Kaplan DL (2010) New opportunities for an ancient material. Science 329(5991):528–531
10. Saeed TN, Al-Hussainy AF, Sanghvi G, Ballal S, Singh A, Sudhakar T, Mishra S, Kubaev A, Taher SG, Alwan M (2025) Tunable silk fibroin platforms for controlled drug delivery: a biocompatible approach. J Drug Deliv Sci Technol 107649
11. Holland C, Numata K, Rnjak-Kovacina J, Seib FP (2019) The biomedical use of silk: past, present, future. Adv Healthcare Mater 8(1):1800465
12. Bucciarelli A, Motta A (2022) Use of Bombyx mori silk fibroin in tissue engineering: from cocoons to medical devices, challenges, and future perspectives. Biomaterials Advances 139:212982
13. Janani G, Kumar M, Chouhan D, Moses JC, Gangrade A, Bhattacharjee S, Mandal BB (2019) Insight into silk-based biomaterials: from physicochemical attributes to recent biomedical applications. ACS Appl Bio Mater 2(12):5460–5491
14. Baburaj A, Banerjee S, Aliyana AK, Shee C, Banakar M, Bairagi S, Kumar SN, Ali SW (2024) Biodegradable based TENGs for self-sustaining implantable medical devices. Nano Energy 127:109785
15. Lv Q, Li Q, Cao P, Wei C, Li Y, Wang Z, Wang L (2024) Designing silk biomaterials toward better future healthcare: the development and application of silk-based implantable electronic devices in clinical diagnosis and therapy. Adv Mater 2411946
16. Ding Z, Cheng W, Mia MS, Lu Q (2021) Silk biomaterials for bone tissue engineering. Macromol Biosci 21(8):2100153
17. Lin D, Li M, Wang L, Cheng J, Yang Y, Wang H, Ye J, Liu Y (2024) Multifunctional hydrogel based on silk fibroin promotes tissue repair and regeneration. Adv Func Mater 34(39):2405255
18. Zhou S, Huang W, Belton DJ, Simmons LO, Perry CC, Wang X, Kaplan DL (2015) Control of silicification by genetically engineered fusion proteins: Silk–silica binding peptides. Acta Biomater 15:173–180
19. Huang W, Ebrahimi D, Dinjaski N, Tarakanova A, Buehler MJ, Wong JY, Kaplan DL (2017) Synergistic integration of experimental and simulation approaches for the de novo design of silk-based materials. Acc Chem Res 50(4):866–876
20. Nie K, Zhou S, Li H, Tian J, Shen W, Huang W (2023) Advanced silk materials for musculoskeletal tissue regeneration. Front Bioeng Biotechnol 11:1199507
21. Sofia S, McCarthy MB, Gronowicz G, Kaplan DL (2001) Functionalized silk-based biomaterials for bone formation. J Biomed Mater Res: Off J Soc Biomater Jpn Soc Biomater 54(1):139–148
22. Enomoto S, Sumi M, Kajimoto K, Nakazawa Y, Takahashi R, Takabayashi C, Asakura T, Sata M (2010) Long-term patency of small-diameter vascular graft made from fibroin, a silk-based biodegradable material. J Vasc Surg 51(1):155–164
23. Meinel L, Hofmann S, Karageorgiou V, Kirker-Head C, McCool J, Gronowicz G, Zichner L, Langer R, Vunjak-Novakovic G, Kaplan DL (2005) The inflammatory responses to silk films in vitro and in vivo. Biomaterials 26(2):147–155
24. Bhumiratana S, Bernhard JC, Alfi DM, Yeager K, Eton RE, Bova J, Shah F, Gimble JM, Lopez MJ, Eisig SB (2016) Tissue-engineered autologous grafts for facial bone reconstruction. Sci Transl Med 8(343):343–383

25. Marelli B, Ghezzi CE, Barralet JE, Boccaccini AR, Nazhat SN (2010) Three-dimensional mineralization of dense nanofibrillar collagen−bioglass hybrid scaffolds. Biomacromol 11(6):1470–1479
26. Pritchard EM, Valentin T, Panilaitis B, Omenetto F, Kaplan DL (2013) Antibiotic-releasing silk biomaterials for infection prevention and treatment. Adv Func Mater 23(7):854–861
27. Schneider A, Wang X, Kaplan D, Garlick J, Egles C (2009) Biofunctionalized electrospun silk mats as a topical bioactive dressing for accelerated wound healing. Acta Biomater 5(7):2570–2578
28. De Giorgio G, Matera B, Vurro D, Manfredi E, Galstyan V, Tarabella G, Ghezzi B, D'Angelo P (2024) Silk fibroin materials: biomedical applications and perspectives. Bioengineering 11(2):167
29. Najm A, Niculescu A-G, Gaspar BS, Grumezescu AM, Beuran M (2023) A review of abdominal meshes for hernia repair—current status and emerging solutions. Materials 16(22):7124
30. Dharadhar S, Majumdar A (2019) Biomaterials and its medical applications. In: Application of Biomedical Engineering in Neuroscience. Springer, Singapore, pp. 355–380
31. Partlow BP, Bagheri M, Harden JL, Kaplan DL (2016) Tyrosine templating in the self-assembly and crystallization of silk fibroin. Biomacromol 17(11):3570–3579
32. Deshpande VA, Kore A, Kandasubramanian B (2023) Silk based bio–inks for medical applications. Eur Polymer J 196:112255
33. Bari E (2020) From bench to bedside: the long way towards GMP scale-up, preclinical and clinical trials for silk-based drug delivery systems. In: Silk-based Drug Delivery Systems. Royal Society of Chemistry, pp. 179–204
34. Seib FP (2017) Silk nanoparticles-an emerging anticancer nanomedicine. AIMS Bioengineering 42(2):239–258
35. Wang Y, Rudym DD, Walsh A, Abrahamsen L, Kim HJ, Kim HS, Kirker-Head C, Kaplan DL (2008) In vivo degradation of three-dimensional silk fibroin scaffolds. Biomaterials 29(24–25):3415–3428
36. Chantada-Vazquez MDP, Castro Lopez A, García-Vence M, Acea-Nebril B, Bravo SB, Núñez C (2020) Protein corona gold nanoparticles fingerprinting reveals a profile of blood coagulation proteins in the serum of Her2-overexpressing breast cancer patients. Int J Mol Sci 21(22):8449
37. Rodriguez MJ, Brown J, Giordano J, Lin SJ, Omenetto FG, Kaplan DL (2017) Silk based bioinks for soft tissue reconstruction using 3-dimensional (3D) printing with in vitro and in vivo assessments. Biomaterials 117:105–115
38. Zhang Y, Xia L, Zhai D, Shi M, Luo Y, Feng C, Fang B, Yin J, Chang J, Wu C (2015) Mesoporous bioactive glass nanolayer-functionalized 3D-printed scaffolds for accelerating osteogenesis and angiogenesis. Nanoscale 7(45):19207–19221
39. Wang Y, Kim HJ, Vunjak-Novakovic G, Kaplan DL (2006) Stem cell-based tissue engineering with silk biomaterials. Biomaterials 27(36):6064–6082
40. Mori K, Tanaka K, Kikuchi Y, Waga M, Waga S, Mizuno S (1995) Production of a chimeric fibroin light-chain polypeptide in a fibroin secretion-deficient naked pupa mutant of the silkwormbombyx mori. J Mol Biol 251(2):217–228
41. Qi Y, Wang H, Wei K, Yang Y, Zheng R-Y, Kim IS, Zhang K-Q (2017) A review of structure construction of silk fibroin biomaterials from single structures to multi-level structures. Int J Mol Sci 18(3):237
42. Tomeh MA, Hadianamrei R, Zhao X (2019) Silk fibroin as a functional biomaterial for drug and gene delivery. Pharmaceutics 11(10):494
43. Wongpinyochit T, Johnston BF, Seib FP (2018) Degradation behavior of silk nanoparticles—enzyme responsiveness. ACS Biomater Sci Eng 4(3):942–951
44. Tanaka K, Inoue S, Mizuno S (1999) Hydrophobic interaction of P25, containing Asn-linked oligosaccharide chains, with the HL complex of silk fibroin produced by Bombyx mori. Insect Biochem Mol Biol 29(3):269–276

45. Asakura T, Ito T, Okudaira M, Kameda T (1999) Structure of alanine and glycine residues of samia c ynthia r icini silk fibers studied with solid-state 15N and 13C NMR. Macromolecules 32(15):4940–4946
46. Gage LP, Manning R (1980) Internal structure of the silk fibroin gene of Bombyx mori. I The fibroin gene consists of a homogeneous alternating array of repetitious crystalline and amorphous coding sequences. J Biol Chem 255(19):9444–9450
47. Zhou CZ, Confalonieri F, Jacquet M, Perasso R, Li ZG, Janin J (2001) Silk fibroin: structural implications of a remarkable amino acid sequence. Proteins: Struct, Funct, Bioinform 44(2):119–122
48. Brown BN, Badylak SF (2014) Chapter 11 – Biocompatibility and immune response to biomaterials. In: Regenerative Medicine Applications in Organ Transplantation. Academic Press, pp. 151–162
49. Almeida LR, Martins AR, Fernandes EM, Oliveira MB, Correlo VM, Pashkuleva I, Marques AP, Ribeiro AS, Durães NF, Silva CJ (2013) New biotextiles for tissue engineering: development, characterization and in vitro cellular viability. Acta Biomater 9(9):8167–8181
50. Ribeiro VP, Almeida LR, Martins AR, Pashkuleva I, Marques AP, Ribeiro AS, Silva CJ, Bonifacio G, Sousa RA, Reis RL (2016) Influence of different surface modification treatments on silk biotextiles for tissue engineering applications. J Biomed Mater Res B Appl Biomater 104(3):496–507
51. Giesa T, Arslan M, Pugno N, Buehler MJ (2011) Nanoconfinement of spider silk fibrils begets superior strength, extensibility and toughness. Nano Letters 11(11):5038–5046
52. Li G, Sun S (2022) Silk fibroin-based biomaterials for tissue engineering applications. Molecules 27(9):2757
53. Li G, Li Y, Chen G, He J, Han Y, Wang X, Kaplan DL (2015) Silk-based biomaterials in biomedical textiles and fiber-based implants. Adv Healthcare Mater 4(8):1134–1151
54. Rajkhowa R, Gupta V, Kothari V (2000) Tensile stress–strain and recovery behavior of Indian silk fibers and their structural dependence. J Appl Polym Sci 77(11):2418–2429
55. Zhang Y, Yang H, Shao H, Hu X (2010) Antheraea pernyi silk fiber: a potential resource for artificially biospinning spider dragline silk. Biomed Res Int 2010(1):683962
56. Bitar L, Isella B, Bertella F, Vasconcelos CB, Harings J, Kopp A, van der Meer Y, Vaughan TJ, Bortesi L (2024) Sustainable Bombyx mori's silk fibroin for biomedical applications as a molecular biotechnology challenge: a review. Int J Biol Macromol 264(1):130374
57. Gan N, Wei M, Zhou Q, Dong K, Gao B (2024) Silk fibroin-based triboelectric nanogenerators for energy harvesting and biomedical applications. ACS Appl Nano Mater 7(8):8407–8423
58. Kumar Sahi A, Gundu S, Kumari P, Klepka T, Sionkowska A (2023) Silk-based biomaterials for designing bioinspired microarchitecture for various biomedical applications. Biomimetics 8(1):55
59. Klimek K, Ginalska G (2020) Proteins and peptides as important modifiers of the polymer scaffolds for tissue engineering applications—a review. Polymers 12(4):844
60. Zhu J, Du Y, Backman LJ, Chen J, Ouyang H, Zhang W (2025) Cellular interactions and biological effects of silk fibroin: implications for tissue engineering and regenerative medicine. Small 21(4):2409739
61. Kumar VB, Tiwari OS, Finkelstein-Zuta G, Rencus-Lazar S, Gazit E (2023) Design of functional RGD peptide-based biomaterials for tissue engineering. Pharmaceutics 15(2):345
62. Ode Boni BO, Bakadia BM, Osi AR, Shi Z, Chen H, Gauthier M, Yang G (2022) Immune response to silk sericin–fibroin composites: potential immunogenic elements and alternatives for immunomodulation. Macromol Biosci 22(1):2100292
63. Mariani E, Lisignoli G, Borzì RM, Pulsatelli L (2019) Biomaterials: foreign bodies or tuners for the immune response? Int J Mol Sci 20(3):636
64. Zou P, Li X, Zhao H, Qu S, Jiang Z, Lei X, Wu C, Liu L (2025) Recent advances of silk-fibroin-based hydrogel in the field of antibacterial application. Polym Int 74(3):183–198
65. Hosseini V, Maroufi NF, Saghati S, Asadi N, Darabi M, Ahmad SNS, Hosseinkhani H, Rahbarghazi R (2019) Current progress in hepatic tissue regeneration by tissue engineering. J Transl Med 17:1–24

66. Han Y, Li X, Zhang Y, Han Y, Chang F, Ding J (2019) Mesenchymal stem cells for regenerative medicine. Cells 8(8):886
67. Rockwood DN, Preda RC, Yücel T, Wang X, Lovett ML, Kaplan DL (2011) Materials fabrication from Bombyx mori silk fibroin. Nat Protoc 6(10):1612–1631
68. Nguyen TP, Nguyen QV, Nguyen V-H, Le T-H, Huynh VQN, Vo D-VN, Trinh QT, Kim SY, Le QV (2019) Silk fibroin-based biomaterials for biomedical applications: a review. Polymers 11(12):1933
69. Guo C, Li C, Kaplan DL (2020) Enzymatic degradation of Bombyx mori silk materials: a review. Biomacromol 21(5):1678–1686
70. Lu Q, Zhang B, Li M, Zuo B, Kaplan DL, Huang Y, Zhu H (2011) Degradation mechanism and control of silk fibroin. Biomacromol 12(4):1080–1086
71. Umuhoza D, Yang F, Long D, Hao Z, Dai J, Zhao A (2020) Strategies for tuning the biodegradation of silk fibroin-based materials for tissue engineering applications. ACS Biomater Sci Eng 6(3):1290–1310
72. Chiasson R, Hasan M, Al Nazer Q, Farokhzad OC, Kamaly N (2016) The use of silk in nanomedicine applications. In: Nanomedicine. Springer, New York, pp. 245–278
73. Kim MH, Park WH (2016) Chemically cross-linked silk fibroin hydrogel with enhanced elastic properties, biodegradability, and biocompatibility. Int J Nanomedicine 11:2967–2978
74. Singh H, Hassan S, Nabi SU, Mishra NC, Dhanka M, Purohit SD, Ganai NA, Bhaskar R, Han SS, Qurashi AUH (2024) Multicomponent decellularized extracellular matrix of caprine small intestine submucosa based bioactive hydrogel promoting full-thickness burn wound healing in rabbits. Int J Biol Macromol 255:127810
75. Wang H-Y, Zhang Y-Q, Wei Z-G (2021) Characterization of undegraded and degraded silk fibroin and its significant impact on the properties of the resulting silk biomaterials. Int J Biol Macromol 176:578–588
76. Rahman M, Dip TM, Nur MG, Padhye R, Houshyar S (2024) Fabrication of silk fibroin-derived fibrous scaffold for biomedical frontiers. Macromol Mater Eng 309(5):2300422
77. Zhang Z, Li M, Zhang X, Zhou F (2024) Novel strategies for tumor treatment: harnessing ROS-inducing active ingredients from traditional Chinese medicine through multifunctional nanoformulations. Int J Nanomedicine 19:9659–9688
78. Balcı N, Ömeroğulları Z, Kut D, Eren H (2015) Effects of plasma and ozone treatments on tensile and whiteness properties of 100% silk. Uludağ Üniversitesi Mühendislik Fakültesi Dergisi 20(2):43–56
79. Abrar S, Kiran S, Iqbal S, Munir B, Rasool A (2024) Recent advances in plasma modification of silk. In: Advances in Plasma Treatment of Textile Surfaces. Woodhead Publishing, pp. 37–56
80. Intranuovo F, Gristina R, Fracassi L, Lacitignola L, Crovace A, Favia P (2016) Plasma processing of scaffolds for tissue engineering and regenerative medicine. Plasma Chem Plasma Process 36:269–280
81. Thakur A, Kumar A, Kaya S, Marzouki R, Zhang F, Guo L (2022) Recent advancements in surface modification, characterization and functionalization for enhancing the biocompatibility and corrosion resistance of biomedical implants. Coatings 12(10):1459
82. Rida Galaly A, Dawood N (2024) Enhancing the production of eco-friendly silk fabrics through the application of nonthermal plasma wettability techniques. ACS Omega 9(19):20791–20806
83. Li G, Liu H, Li T, Wang J (2012) Surface modification and functionalization of silk fibroin fibers/fabric toward high performance applications. Mater Sci Eng: C 32(4):627–636
84. Wang Y, Guo J, Zhou L, Ye C, Omenetto FG, Kaplan DL, Ling S (2018) Design, fabrication, and function of silk-based nanomaterials. Adv Func Mater 28(52):1805305
85. Jain P, Kathuria H, Dubey N (2022) Advances in 3D bioprinting of tissues/organs for regenerative medicine and in-vitro models. Biomaterials 287:121639
86. Shabbirahmed AM, Sekar R, Gomez LA, Sekhar MR, Hiruthyaswamy SP, Basavegowda N, Somu P (2023) Recent developments of silk-based scaffolds for tissue engineering and regenerative medicine applications: a special focus on the advancement of 3D printing. Biomimetics 8(1):16

87. Ansari AI, Sheikh NA (2023) Biocompatible scaffold based on silk fibroin for tissue engineering applications. J Inst Eng (India): Ser C 104(1):201–217
88. Yin Z, Wu F, Xing T, Yadavalli VK, Kundu SC, Lu S (2017) A silk fibroin hydrogel with reversible sol–gel transition. RSC Adv 7(39):24085–24096
89. Okuducu C (2024) Optimization and quality assessment of silk fibroin sponges for scaffold-based 3D cell culture model. Master's thesis, University of Vienna, Austria
90. Parisi-Amon A (2013) Bones, bones, bones: using protein-engineered biomaterials to improve bone regeneration and implant osseointegration. Stanford University
91. Lv Q, Feng Q (2006) Preparation of 3-D regenerated fibroin scaffolds with freeze drying method and freeze drying/foaming technique. J Mater Sci - Mater Med 17(12):1349–1356
92. Burger D, Beaumont M, Rosenau T, Tamada Y (2020) Porous silk fibroin/cellulose hydrogels for bone tissue engineering via a novel combined process based on sequential regeneration and porogen leaching. Molecules 25(21):5097
93. Kang H-W, Seol Y-J, Cho D-W (2008) Development of an indirect solid freeform fabrication process based on microstereolithography for 3D porous scaffolds. J Micromech Microeng 19(1):015011
94. Adel IM, ElMeligy MF, Elkasabgy NA (2022) Conventional and recent trends of scaffolds fabrication: a superior mode for tissue engineering. Pharmaceutics 14(2):306
95. Zhang L, Zhang W, Hu Y, Fei Y, Liu H, Huang Z, Wang C, Ruan D, Heng BC, Chen W (2021) Systematic review of silk scaffolds in musculoskeletal tissue engineering applications in the recent decade. ACS Biomater Sci Eng 7(3):817–840
96. Kasoju N, Bora U (2012) Silk fibroin based biomimetic artificial extracellular matrix for hepatic tissue engineering applications. Biomed Mater 7(4):045004
97. Yao X, Zou S, Fan S, Niu Q, Zhang Y (2022) Bioinspired silk fibroin materials: from silk building blocks extraction and reconstruction to advanced biomedical applications. Materials Today Bio 16:100381
98. He Y, Gu Z, Xie M, Fu J, Lin H (2020) Why choose 3D bioprinting? Part II: methods and bioprinters. Bio-Des Manuf 3(1):1–4
99. Vanaei S, Parizi M, Salemizadehparizi F, Vanaei H (2021) An overview on materials and techniques in 3D bioprinting toward biomedical application. Engineered Regeneration 2:1–18
100. Montaseri Z, Abolmaali SS, Tamaddon AM, Farvadi F (2023) Composite silk fibroin hydrogel scaffolds for cartilage tissue regeneration. J Drug Deliv Sci Technol 79:104018
101. Sun W, Gregory DA, Tomeh MA, Zhao X (2021) Silk fibroin as a functional biomaterial for tissue engineering. Int J Mol Sci 22(3):1499
102. Hommadova B, Poliykov D (2025) Methodology of teaching cell and tissue engineering. ОБРАЗОВАНИЕ И НАУКА В XXI ВЕКЕ(59–4 (том 1))
103. Han F, Wang J, Ding L, Hu Y, Li W, Yuan Z, Guo Q, Zhu C, Yu L, Wang H (2020) Tissue engineering and regenerative medicine: achievements, future, and sustainability in Asia. Frontiers in bioengineering and biotechnology 8:83
104. Chakraborty J, Mu X, Pramanick A, Kaplan DL, Ghosh S (2022) Recent advances in bioprinting using silk protein-based bioinks. Biomaterials 287:121672
105. Florencio-Silva R, Sasso GRDS, Sasso-Cerri E, Simões MJ, Cerri PS (2015) Biology of bone tissue: structure, function, and factors that influence bone cells. BioMed Res Int 2015(1):421746
106. Olszta MJ, Cheng X, Jee SS, Kumar R, Kim Y-Y, Kaufman MJ, Douglas EP, Gower LB (2007) Bone structure and formation: a new perspective. Mater Sci Eng R Rep 58(3–5):77–116
107. Henkel J, Woodruff MA, Epari DR, Steck R, Glatt V, Dickinson IC, Choong PF, Schuetz MA, Hutmacher DW (2013) Bone regeneration based on tissue engineering conceptions—a 21st century perspective. Bone Res 1(1):216–248
108. Turnbull G, Clarke J, Picard F, Riches P, Jia L, Han F, Li B, Shu W (2018) 3D bioactive composite scaffolds for bone tissue engineering. Bioactive Materials 3(3):278–314
109. Farokhi M, Mottaghitalab F, Samani S, Shokrgozar MA, Kundu SC, Reis RL, Fatahi Y, Kaplan DL (2018) Silk fibroin/hydroxyapatite composites for bone tissue engineering. Biotechnol Adv 36(1):68–91

110. Krakowski P, Rejniak A, Sobczyk J, Karpiński R (2024) Cartilage integrity: a review of mechanical and frictional properties and repair approaches in osteoarthritis. Healthcare 12(16):1648
111. Jeyaraman M, Jeyaraman N, Nallakumarasamy A, Ramasubramanian S, Yadav S (2024) Critical challenges and frontiers in cartilage tissue engineering. Cureus 16(1):e53095
112. Zhang L, Hu J, Athanasiou KA (2009) The role of tissue engineering in articular cartilage repair and regeneration. Crit Rev Biomed Eng 37:1–57
113. Cipollaro L, Ciardulli MC, Porta GD, Peretti GM, Maffulli N (2019) Biomechanical issues of tissue-engineered constructs for articular cartilage regeneration: in vitro and in vivo approaches. Br Med Bull 132(1):53–80
114. Cheng G, Davoudi Z, Xing X, Yu X, Cheng X, Li Z, Deng H, Wang Q (2018) Advanced silk fibroin biomaterials for cartilage regeneration. ACS Biomater Sci Eng 4(8):2704–2715
115. Su X, Wei L, Xu Z, Qin L, Yang J, Zou Y, Zhao C, Chen L, Hu N (2023) Evaluation and application of silk fibroin based biomaterials to promote cartilage regeneration in osteoarthritis therapy. Biomedicines 11(8):2244
116. Veiga A, Silva IV, Duarte MM, Oliveira AL (2021) Current trends on protein driven bioinks for 3D printing. Pharmaceutics 13(9):1444
117. Yan K, Zhang X, Liu Y, Cheng J, Zhai C, Shen K, Liang W, Fan W (2023) 3D-bioprinted silk fibroin-hydroxypropyl cellulose methacrylate porous scaffold with optimized performance for repairing articular cartilage defects. Mater Des 225:111531
118. Lovett M, Lee K, Edwards A, Kaplan DL (2009) Vascularization strategies for tissue engineering. Tissue Eng Part B Rev 15(3):353–370
119. Kumar MSA (2024) The skin. In: Techniques in Small Animal Wound Management. Wiley-Blackwell, pp. 1–36
120. Yadav N, Parveen S, Chakravarty S, Banerjee M (2019) Skin anatomy and morphology. In: Dwivedi A, Agarwal N, Ray L, Tripathi A (eds) Skin Aging & Cancer. Springer, Singapore, pp. 1–10
121. Girard D, Laverdet B, Buhé V, Trouillas M, Ghazi K, Alexaline MM, Egles C, Misery L, Coulomb B, Lataillade J-J (2017) Biotechnological management of skin burn injuries: challenges and perspectives in wound healing and sensory recovery. Tissue Eng Part B Rev 23(1):59–82
122. Jorgensen AM, Mahajan N, Atala A, Murphy SV (2023) Advances in skin tissue engineering and regenerative medicine. J Burn Care Res 44(Supplement_1):S33-S41
123. Kolimi P, Narala S, Nyavanandi D, Youssef AAA, Dudhipala N (2022) Innovative treatment strategies to accelerate wound healing: trajectory and recent advancements. Cells 11(15):2439
124. Skardal A, Mack D, Kapetanovic E, Atala A, Jackson JD, Yoo J, Soker S (2012) Bioprinted amniotic fluid-derived stem cells accelerate healing of large skin wounds. Stem Cells Transl Med 1(11):792–802
125. Xiong S, Zhang X, Lu P, Wu Y, Wang Q, Sun H, Heng BC, Bunpetch V, Zhang S, Ouyang H (2017) A gelatin-sulfonated silk composite scaffold based on 3D printing technology enhances skin regeneration by stimulating epidermal growth and dermal neovascularization. Sci Rep 7(1):4288
126. Kaur A, Midha S, Giri S, Mohanty S (2019) Functional skin grafts: where biomaterials meet stem cells. Stem cells Int 2019(1):1286054
127. Sultan MT, Lee OJ, Lee JS, Park CH (2022) Three-dimensional digital light-processing bioprinting using silk fibroin-based bio-ink: recent advancements in biomedical applications. Biomedicines 10(12):3224
128. Mathur V, Agarwal P, Kasturi M, Srinivasan V, Seetharam RN, Vasanthan KS (2025) Innovative bioinks for 3D bioprinting: exploring technological potential and regulatory challenges. J Tissue Eng 16:20417314241308024
129. Muñoz-Castiblanco T, Moreno-Marín JP, Osorio M (2025) Natural macromolecule-based bioinks for 3D bioprinting: A systematic review of composition, physicochemical characterization, and biomedical applications. Bioprinting 48:e00407

130. Ramakrishnan R, Chouhan D, Vijayakumar Sreelatha H, Arumugam S, Mandal BB, Krishnan LK (2022) Silk fibroin-based bioengineered scaffold for enabling hemostasis and skin regeneration of critical-size full-thickness heat-induced burn wounds. ACS Biomater Sci Eng 8(9):3856–3870
131. Bhar B, Das E, Manikumar K, Mandal BB (2024) 3D bioprinted human skin model recapitulating native-like tissue maturation and immunocompetence as an advanced platform for skin sensitization assessment. Adv Healthcare Mater 13(15):2303312
132. Cunha C, Panseri S, Antonini S (2011) Emerging nanotechnology approaches in tissue engineering for peripheral nerve regeneration. Nanomedicine: Nanotechnol, Biol Med 7(1):50–59
133. Borschel GH, Kia KF, Kuzon WM Jr, Dennis RG (2003) Mechanical properties of acellular peripheral nerve. J Surg Res 114(2):133–139
134. Tan A, Rajadas J, Seifalian AM (2012) Biochemical engineering nerve conduits using peptide amphiphiles. J Control Release 163(3):342–352
135. Gu X, Ding F, Yang Y, Liu J (2011) Construction of tissue engineered nerve grafts and their application in peripheral nerve regeneration. Prog Neurobiol 93(2):204–230
136. Farokhi M, Mottaghitalab F, Shokrgozar MA, Kaplan DL, Kim H-W, Kundu SC (2017) Prospects of peripheral nerve tissue engineering using nerve guide conduits based on silk fibroin protein and other biopolymers. Int Mater Rev 62(7):367–391
137. Burdick JA, Mauck RL (eds) (2010) Biomaterials for Tissue Engineering Applications: A Review of the Past and Future Trends. Springer, Vienna
138. Teuschl AH, Neutsch L, Monforte X, Rünzler D, Van Griensven M, Gabor F, Redl H (2014) Enhanced cell adhesion on silk fibroin via lectin surface modification. Acta Biomater 10(6):2506–2517
139. Schuh CM, Monforte X, Hackethal J, Redl H, Teuschl AH (2016) Covalent binding of placental derived proteins to silk fibroin improves schwann cell adhesion and proliferation. J Mater Sci Mater Med 27:1–8
140. Mandal BB, Kundu SC (2009) Cell proliferation and migration in silk fibroin 3D scaffolds. Biomaterials 30(15):2956–2965
141. Kim U-J, Park J, Kim HJ, Wada M, Kaplan DL (2005) Three-dimensional aqueous-derived biomaterial scaffolds from silk fibroin. Biomaterials 26(15):2775–2785
142. Kim HJ, Kim U-J, Vunjak-Novakovic G, Min B-H, Kaplan DL (2005) Influence of macroporous protein scaffolds on bone tissue engineering from bone marrow stem cells. Biomaterials 26(21):4442–4452
143. Mangir N, Roman S, Chapple CR, MacNeil S (2020) Complications related to use of mesh implants in surgical treatment of stress urinary incontinence and pelvic organ prolapse: infection or inflammation? World J Urol 38:73–80
144. See CW, Kim T, Zhu D (2020) Hernia mesh and hernia repair: a review. Eng Regen 1:19–33
145. Meshkin DH, Firriolo JM, Karp NS, Salibian AA (2023) Management of complications following implant-based breast reconstruction: a narrative review. Ann Transl Med 11(12):416
146. Zhu L-M, Schuster P, Klinge U (2015) Mesh implants: An overview of crucial mesh parameters. World J Gastrointest Surg 7(10):226
147. Bandyopadhyay A, Chowdhury SK, Dey S, Moses JC, Mandal BB (2019) Silk: a promising biomaterial opening new vistas towards affordable healthcare solutions. J Indian Inst Sci 99(3):445–487
148. Chrcanovic B, Albrektsson T, Wennerberg A (2014) Reasons for failures of oral implants. J Oral Rehabil 41(6):443–476
149. Govindharajulu JP, Chen X, Li Y, Rodriguez-Cabello JC, Battacharya M, Aparicio C (2017) Chitosan-recombinamer layer-by-layer coatings for multifunctional implants. Int J Mol Sci 18(2):369
150. Organization WH (2024) Health-care waste. https://www.who.int/news-room/fact-sheets/detail/health-care-waste. Accessed 28 Mar
151. Pachauri P, Bathala LR, Sangur R (2014) Techniques for dental implant nanosurface modifications. J Adv Prosthodont 6(6):498

152. Elia R, Michelson CD, Perera AL, Harsono M, Leisk GG, Kugel G, Kaplan DL (2015) Silk electrogel coatings for titanium dental implants. J Biomater Appl 29(9):1247–1255
153. Jaya Prakash N, Wang X, Kandasubramanian B (2023) Regenerated silk fibroin loaded with natural additives: a sustainable approach towards health care. J Biomater Sci Polym Ed 34(10):1453–1490
154. Paladini F, Pollini M (2022) Novel approaches and biomaterials for bone tissue engineering: a focus on silk fibroin. Materials 15(19):6952
155. Attri K, Shruthi G, Gulabrao DP, Teja KSS, Garai I, Pandey AK, Kumar GA, Jeevitha P, Bharathi BKM (2024). Silk biomaterials: applications and future prospects in biomedical engineering. Biomed Eng 45(16):205–216
156. Feng Y, Lin J, Niu L, Wang Y, Cheng Z, Sun X, Li M (2020) High molecular weight silk fibroin prepared by papain degumming. Polymers 12(9):2105
157. Anand P, Pandey JP, Pandey DM (2021) Study on cocoonase, sericin, and degumming of silk cocoon: computational and experimental. J Genet Eng Biotechnol 19(1):32
158. Gai T, Tong X, Han M, Li C, Fang C, Zou Y, Hu H, Xiang H, Xiang Z, Lu C (2020) Cocoonase is indispensable for Lepidoptera insects breaking the sealed cocoon. PLoS Genet 16(9):e1009004
159. Liu X, Huang Q, Pan P, Fang M, Zhang Y, Yang S, Li M, Liu Y (2023) Comparative study of the preparation of high-molecular-weight fibroin by degumming silk with several neutral proteases. Polymers 15(16):3383
160. Farokhi M, Mottaghitalab F, Reis RL, Ramakrishna S, Kundu SC (2020) Functionalized silk fibroin nanofibers as drug carriers: Advantages and challenges. J Control Release 321:324–347
161. Dai S, Liang H, Zhu M, Zhang Y (2024) Electrospun silk for biomedical applications. Med-X 2(1):22
162. Khorshidi S, Solouk A, Mirzadeh H, Mazinani S, Lagaron JM, Sharifi S, Ramakrishna S (2016) A review of key challenges of electrospun scaffolds for tissue-engineering applications. J Tissue Eng Regen Med 10(9):715–738
163. Belbéoch C, Lejeune J, Vroman P, Salaün F (2021) Silkworm and spider silk electrospinning: a review. Environ Chem Lett 19:1737–1763
164. Chen K, Li Y, Li Y, Pan W, Tan G (2023) Silk fibroin combined with electrospinning as a promising strategy for tissue regeneration. Macromol Biosci 23(2):2200380
165. Humenik M, Lang G, Scheibel T (2018) Silk nanofibril self-assembly versus electrospinning. Wiley Interdiscip. Rev.: Nanomedicine Nanobiotechnology 10(4):e1509
166. Mu Y (2020) Electrospinning of Silk Fibroin Fibres. PhD thesis, The University of Manchester, United Kingdom
167. Khademolqorani S, Tavanai H, Chronakis IS, Boisen A, Ajalloueian F (2021) The determinant role of fabrication technique in final characteristics of scaffolds for tissue engineering applications: a focus on silk fibroin-based scaffolds. Mater Sci Eng, C 122:111867
168. Loh QL, Choong C (2013) Three-dimensional scaffolds for tissue engineering applications: role of porosity and pore size. Tissue Eng Part B: Rev 19(6):485–502
169. Santos-Rosales V, Iglesias-Mejuto A, García-González CA (2020) Solvent-free approaches for the processing of scaffolds in regenerative medicine. Polymers 12(3):533
170. Chimerad M, Barazesh A, Zandi M, Zarkesh I, Moghaddam A, Borjian P, Chimehrad R, Asghari A, Akbarnejad Z, Khonakdar HA (2023) Tissue engineered scaffold fabrication methods for medical applications. Int J Polym Mater Polym Biomater 72(18):1455–1479
171. Li X, You R, Luo Z, Chen G, Li M (2016) Silk fibroin scaffolds with a micro-/nano-fibrous architecture for dermal regeneration. J Mater Chem B 4(17):2903–2912
172. Costantini M, Barbetta A (2018) Gas foaming technologies for 3D scaffold engineering. In: Deng Y, Kuiper J (eds) Functional 3D Tissue Engineering Scaffolds: Materials, Technologies, and Applications. Woodhead Publishing, pp. 127–149
173. Maniglio D, Bonani W, Migliaresi C, Motta A (2018) Silk fibroin porous scaffolds by N2O foaming. J Biomater Sci Polym Ed 29(5):491–506
174. Rao F, Yuan Z, Li M, Yu F, Fang X, Jiang B, Wen Y, Zhang P (2019) Expanded 3D nanofibre sponge scaffolds by gas-foaming technique enhance peripheral nerve regeneration. Artif Cells, Nanomedicine, Biotechnol 47(1):491–500

175. Wang Q, Han G, Yan S, Zhang Q (2019) 3D printing of silk fibroin for biomedical applications. Materials 12(3):504
176. Das S, Pati F, Choi Y-J, Rijal G, Shim J-H, Kim SW, Ray AR, Cho D-W, Ghosh S (2015) Bioprintable, cell-laden silk fibroin–gelatin hydrogel supporting multilineage differentiation of stem cells for fabrication of three-dimensional tissue constructs. Acta Biomater 11:233–246
177. Lau K, Akhavan B, Lord MS, Bilek MM, Rnjak-Kovacina J (2020) Dry surface treatments of silk biomaterials and their utility in biomedical applications. ACS Biomater Sci Eng 6(10):5431–5452
178. Radulescu DM, Andronescu E, Vasile OR, Ficai A, Vasile BS (2024) Silk fibroin-based scaffolds for wound healing applications with metal oxide nanoparticles. J Drug Deliv Sci Technol 96:105689
179. Tran HA, Hoang TT, Maraldo A, Do TN, Kaplan DL, Lim KS, Rnjak-Kovacina J (2023) Emerging silk fibroin materials and their applications: new functionality arising from innovations in silk crosslinking. Mater Today 65:244–259
180. Ren X, Feng Y, Guo J, Wang H, Li Q, Yang J, Hao X, Lv J, Ma N, Li W (2015) Surface modification and endothelialization of biomaterials as potential scaffolds for vascular tissue engineering applications. Chem Soc Rev 44(15):5680–5742
181. Sahoo JK, Hasturk O, Falcucci T, Kaplan DL (2023) Silk chemistry and biomedical material designs. Nat Rev Chem 7(5):302–318
182. Murphy AR, Kaplan DL (2009) Biomedical applications of chemically-modified silk fibroin. J Mater Chem 19(36):6443–6450
183. Murphy AR, John PS, Kaplan DL (2008) Modification of silk fibroin using diazonium coupling chemistry and the effects on HMSC proliferation and differentiation. Biomaterials 29(19):2829–2838
184. Hu X, Shmelev K, Sun L, Gil E-S, Park S-H, Cebe P, Kaplan DL (2011) Regulation of silk material structure by temperature-controlled water vapor annealing. Biomacromol 12(5):1686–1696
185. Motta A, Migliaresi C, Faccioni F, Torricelli P, Fini M, Giardino R (2004) Fibroin hydrogels for biomedical applications: preparation, characterization and in vitro cell culture studies. J Biomater Sci Polym Ed 15(7):851–864
186. Akilbekova D, Turlybekuly A (2023) Patient-specific 3D bioprinting for in situ tissue engineering and regenerative medicine. In: 3D Printing in Medicine (2nd ed.). Woodhead Publishing, pp. 149–178
187. Cuijpers VMJI (2016) Imaging strategies in bone tissue engineering. PhD thesis, Radboud University, Nijmegen, The Netherlands
188. Mehta PP (2024) Miniaturized devices: a silk route for innovative commercial prospects. In: Lab-on-a-chip Devices for Advanced Biomedicines: Laboratory Scale Engineering to Clinical Ecosystem. Royal Society of Chemistry, pp. 615–648
189. Serpe F, Casciola CM, Ruocco G, Cidonio G, Scognamiglio C (2024) Microfluidic fiber spinning for 3D bioprinting: Harnessing microchannels to build macrotissues. Int J Bioprinting 10(1):47–74
190. Reizabal A, Costa C, Pérez-Álvarez L, Vilas-Vilela J, Lanceros-Méndez S (2023) The new silk road: silk fibroin blends and composites for next generation functional and multifunctional materials design. Polym Rev 63(4):1014–1077
191. Baptista-Silva S, Borges S, Brassesco ME, Coscueta ER, Oliveira AL, Pintado M (2022) Research, development and future trends for medical textile products. In: Medical Textiles from Natural Resources. Woodhead Publishing, pp. 795–828

Chapter 8
Silk Fibroin: A Multifunctional Natural Biomaterial for Emerging Biomedical Applications

Poulami Bhadra, Barsha Roy, Shubhajit Shaw, and Debnirmalya Gangopadhyay

8.1 Introduction

8.1.1 Overview and Sources of Silk Fibroin

Silk is a natural fibrous protein secreted by various arthropods. However, the domesticated silkworm, *Bombyx mori* remains the most widely studied source for biomedical applications because it produces high-quality silk and is easy to rear. The fibroin extracted from *B. mori* is especially valued for its high β-sheet content, strong mechanical properties, and excellent biocompatibility. These features have established it as the foundation of silk-based biomedical research and development. In contrast, non-mulberry silks, derived from species such as *Antheraea pernyi*, *Antheraea mylitta*, *Antheraea assamensis*, and *Samia ricini*, offer their own unique advantages. For instance, *A. pernyi* fibroin contains RGD (Arginine-Glycine-Aspartic) motifs that enhance cell adhesion, while *S. ricini* fibroin is distinguished by its high hydrophilicity. Despite their advantages, these silks remain underexplored in clinical settings due to limited research and a lack of standardized processing methods. Spider silks, known for their extraordinary tensile strength and inherent biofunctionality, present another attractive alternative, but they are difficult to produce in large amounts, which limits their wider use [1, 2]. This review focuses primarily on *B. mori* silk fibroin due to its availability, reproducibility, and established clinical relevance, while briefly highlighting the emerging potential of alternative silk sources.

P. Bhadra · B. Roy · S. Shaw · D. Gangopadhyay (✉)
Department of Sericulture, Raiganj University, Raiganj, West Bengal, India
e-mail: dng@raiganjuniversity.ac.in

M. Shahid et al. (eds.), *Science and Technology of Silk*, Textile Science and Clothing Technology, https://doi.org/10.1007/978-3-032-26880-8_8

8.1.2 Biomedical Significance and Historical Perspective

Biomaterials are engineered substances designed to interact with biological systems for therapeutic or diagnostic purposes. An ideal biomaterial must exhibit biocompatibility, controlled biodegradability, bioactivity, and minimal immunogenicity. Silk fibroin inherently satisfies these criteria and has demonstrated compatibility in both in vitro and in vivo environments, with minimal foreign body reactions and low levels of chronic inflammation [3]. Silk fibroin is highly versatile and can be processed into various formats, including films, sponges, hydrogels, and three-dimensional scaffolds. These configurations support its use in various biomedical fields, including tissue engineering, wound healing, and regenerative medicine [4]. Besides, the material can also serve as a carrier for bioactive molecules, enhancing its utility in drug delivery and cancer therapy. Historically, silk has been utilized in medicine for thousands of years. Ancient Greek and Roman physicians employed silk sutures, and traditional Chinese medicine utilized silk bandages for wound care. Its tensile strength, flexibility, and biocompatibility made it superior to early alternatives [5]. In contemporary applications, silk-based scaffolds facilitate cell adhesion and proliferation, while modern wound dressings enhance tissue regeneration and antimicrobial properties. Due to its low immunogenicity, silk fibroin has received approval from the United States Food and Drug Administration (FDA) for clinical use [6]. Further, silk exhibits hemostatic and structural properties that support its use in surgical sutures, wound dressings, and drug-eluting systems [7].

8.2 Structural and Functional Characteristics of Silk Fibroin for Biomedical Applications

8.2.1 Molecular Architecture of Silk Fibroin

Silk fibroin, derived from *B. mori*, is composed of microfibrils formed from highly ordered amino acid sequences. The predominant amino acids include glycine (43%), alanine (30%), and serine (12%) [8]. Its primary structure is characterized by repeating hexapeptide motifs (Glycine-Serine-Glycine-Alanine-Glycine-Alanine)$_n$, which self-assemble into anti-parallel β-sheet crystalline structures. These β-sheets organize into nanofibrils measuring 3–5 nm, which further aggregate into microfibrils ranging from 20 to 200 nm in diameter, forming a highly aligned, mesh-like architecture that imparts significant mechanical strength and durability [9]. Silk fibroin exists in three polymorphic forms, namely Silk I (amorphous, hydrophilic, and α-helical), typically present in the aqueous silk dope; Silk II (crystalline, β-sheet-rich, and hydrophobic), found in the spun silk fiber; and Silk III, a metastable structure observed at air-water interfaces [10]. Structurally, silk fibroin comprises both crystalline β-sheet domains, which provide tensile strength and stability, and amorphous regions, which contribute elasticity and extensibility [11]. The high glycine content

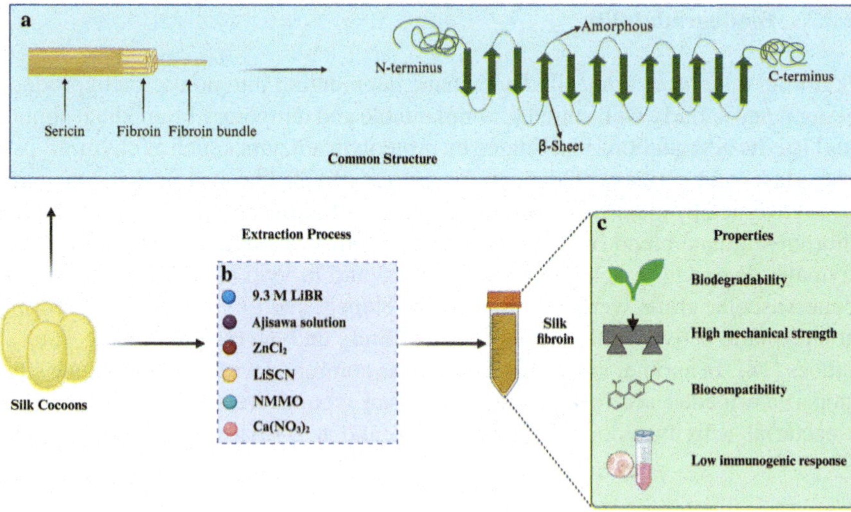

Fig. 8.1 Silk fibroin: structure, extraction, and biomedical functions. **a** Silk fiber composition with sericin and β-sheet-rich fibroin; **b** extraction methods using various solvents; **c** biomedical features of silk fibroin

facilitates conformational transitions between α-helices and β-sheets, enhancing its adaptability under physiological conditions [12]. Figure 8.1 provides a schematic overview of silk fibroin's structural features, extraction processes, and key biomedical properties.

8.2.2 *Physicochemical and Biological Properties*

8.2.2.1 Biocompatibility

Silk fibroin exhibits intrinsic biocompatibility with minimal immunogenicity because it degrades into non-toxic amino acids that are safely metabolized by the body. These features make silk fibroin suitable for a wide range of biomedical applications, from conventional sutures to advanced regenerative therapies [13]. It promotes cellular adhesion, proliferation, and differentiation, and is especially effective in wound healing, nerve regeneration, and three-dimensional (3D) bioprinting [14]. Silk fibroin-based hydrogels loaded with antimicrobial compounds such as rhein have demonstrated enhanced tissue repair and infection control [15]. The protein also supports the growth and function of neurons and Schwann cells, reinforcing its utility in neural repair and dermal regeneration. In 3D bioprinting, silk fibroin bioinks have enabled the creation of complex anatomical constructs such as tracheal and lung tissues [14], further affirming its regenerative potential.

8.2.2.2 Biodegradability

Silk fibroin undergoes controlled enzymatic degradation into non-toxic byproducts, making it particularly well suited for implantable and temporary biomedical applications [16]. Its degradation is facilitated by proteolytic enzymes such as chymotrypsin, carboxylase, and actinase. The degradation rate of silk fibroin-based porous materials is governed by material composition and crosslinking efficiency, with variations in fibroin–alginate blend ratio and structural stability, leading to markedly different enzymatic degradation behaviors both in vitro and in vivo [17]. Silk fibroin-based tubular vascular grafts were successfully developed, and in vitro studies confirmed that enzymatic activity influenced the crystallinity and degradation behavior of the SF tubes [18]. In another study, silk fibroin was combined with chitin to create silver nanoparticle-loaded scaffolds for antimicrobial wound dressings, showing strong antibacterial activity, good biodegradability, and non-toxic degradation products [19].

8.2.2.3 Mechanical Properties

Silk fibroin is among the strongest known natural biomaterials, exhibiting tensile strengths between 300 and 740 megapascals (MPa). It surpasses steel of equivalent diameter and outperforms many synthetic and natural biomaterials, including cotton and polylactic acid (PLA). Its elastic modulus ranges from 10 to 17 gigapascals (GPa), with elongation at break varying from 4 to 26%, offering a rare combination of strength and flexibility [7, 16]. These mechanical attributes are primarily derived from its β-sheet crystallinity and nanofibrillar organization. Processing techniques have a significant impact on mechanical performance; for example, thermal treatment increases β-sheet content and stiffness, while sericin removal tends to reduce tensile strength [16, 20]. Blending silk fibroin with other biocompatible polymers improves flexibility, toughness, and fatigue resistance. Its interconnected fibrillar structure allows for uniform stress distribution, enhancing its performance under mechanical load [21].

8.2.2.4 Biological Activity and Functional Performance

In addition to its structural and mechanical roles, silk fibroin exhibits essential biological activity that supports its use in tissue repair and wound healing. It contributes to all phases of the wound healing cascade, including hemostasis, inflammation regulation, cellular proliferation, and tissue remodeling. Silk fibroin enhances cell adhesion and migration, critical for efficient tissue regeneration, and activates intracellular signaling pathways such as extracellular signal-regulated kinase (ERK) and c-Jun N-terminal kinase (JNK), which promote cell motility and accelerate wound closure [22]. Nanofibrous silk fibroin scaffolds have shown superior cellular adhesion compared to many traditional materials, further validating their application in tissue engineering and regenerative medicine. Beyond wound healing, silk fibroin serves as an effective drug delivery carrier, enhancing probiotic stability, supporting oral formulations, and enabling sustained therapeutic release [16, 23]. A comparative summary of silk fibroin's biologically active properties and its performance relative to traditional biomaterials is presented in Table 8.1, highlighting its multifunctional role in biomedical applications.

8.3 Silk Biomaterial Fabrication Techniques

8.3.1 Extraction and Processing of Fibroin

Silk-based biomaterial development starts with extracting and purifying silk fibroin from silk cocoons through degumming, dissolution, dialysis, and concentration. Solvent choice, temperature, and purification steps influence silk fibroin's molecular weight and functional properties. An overview of these methods is shown in Fig. 8.2.

8.3.1.1 Degumming and Purification Techniques

Silk fibroin is naturally surrounded by sericin, a glue-like protein that must be removed prior to biomedical applications. The degumming process, typically the first step in silk fibroin extraction, involves boiling or autoclaving silk cocoons in an aqueous sodium carbonate (Na_2CO_3) solution for 30 to 60 mins. This effectively eliminates sericin, leaving behind purified fibroin filament. The filaments are then thoroughly rinsed with distilled water to remove residual chemicals and impurities, followed by air-drying [4, 9].

8.3.1.2 Solubilization Methods

After degumming, the fibroin must be solubilized by disrupting its crystalline β-sheet structure. This is typically done using strong solvents such as 9.3 M lithium bromide (LiBr), Ajisawa's reagent (a 1:8:2 molar ratio of calcium chloride, ethanol, and water), zinc chloride ($ZnCl_2$), lithium thiocyanate (LiSCN), N-methylmorpholine-N-oxide (NMMO), hexafluoroisopropanol (HFIP), or calcium nitrate [$Ca(NO_3)_2$] [4, 24, 25]. Among these, LiBr is commonly preferred due to its efficiency and consistency. The dissolved fibroin solution contains residual salts that must be removed to ensure biocompatibility. This is achieved through dialysis using a semi-permeable membrane placed in ultrapure water for 24 to 48 h, with frequent water changes to remove salts such as LiBr or $CaCl_2$. Additional purification steps such as centrifugation or filtration may be used to eliminate aggregates or insoluble debris. The

Table 8.1 Physicochemical properties of silk

Property	Description	References
Biocompatibility	Highly biocompatible with minimal immunogenic response	[13]; [16]
	Used in sutures, wound healing, neurological regeneration, and 3D bioprinting	
	Promotes cell development, nerve grafting, and tissue restoration	
	Silk fibroin-based hydrogels with antibacterial agents aid tissue repair and prevent infections	
Biodegradability	Naturally degrades with minimal toxicity, making it ideal for biomedical applications	[17]; [18]
	Degradation facilitated by enzymes like chymotrypsin, carboxylase, and actinase	
	Factors like pore size, silk fibroin concentration, and implantation site influence degradation rate	
	Observed gradual scaffold disintegration and vascularization in vivo	
Mechanical properties	Strong natural fiber with tensile strength of 300–740 MPa	[7]; [21]
	Elastic modulus ranges from 10 to 17 GPa	
	β-sheet crystallites provide stability and stiffness	
	Processing methods (e.g., heat treatment, polymer integration) alter mechanical properties	
	Enhanced elasticity and durability through polymer blending	
Biologically functional properties	Supports wound healing via hemostasis, inflammation, cell proliferation, and tissue remodeling	[16]; [23]
	Enhances cell adhesion and migration, crucial for tissue regeneration	
	Silk fibroin-based materials improve drug delivery efficiency and probiotic survival	
	Enables sustained drug release	
	Promotes cell adhesion better than synthetic polymers like PLA	

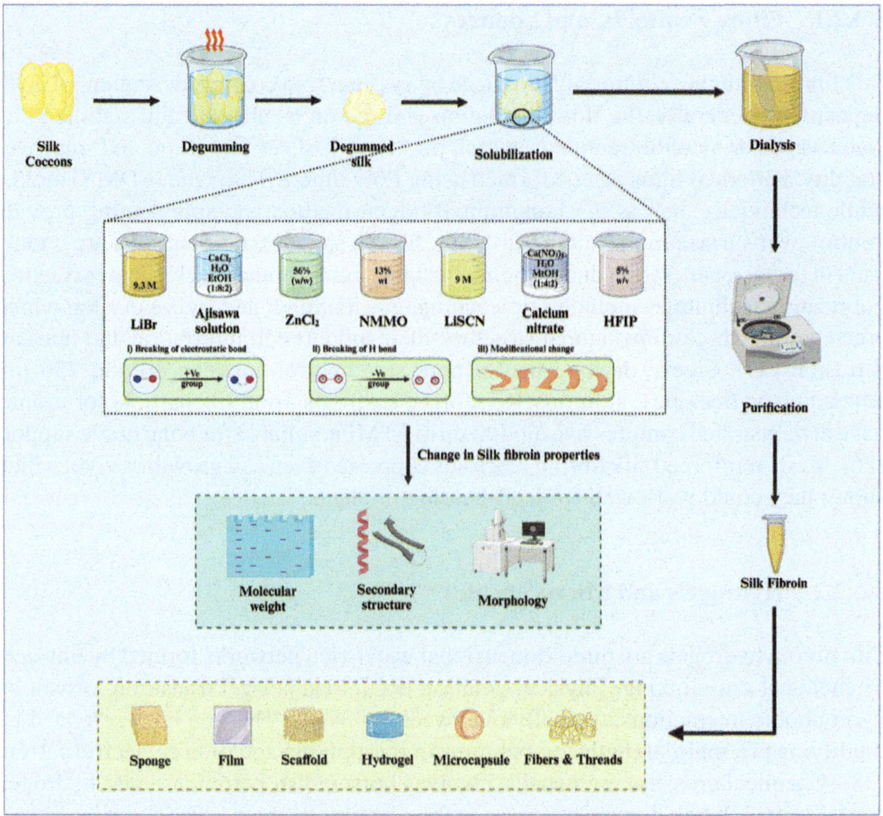

Fig. 8.2 Silk fibroin extraction steps and their influence on structure and biomedical formats

final clear and stable fibroin solution is suitable for biomedical use; storable at room temperature or 4 °C; and ready for fabrication into films, hydrogels, scaffolds, or sponges.

8.3.2 Forms of Silk Biomaterials

After extraction, silk fibroin is processed into diverse forms such as fibers, films, hydrogels, sponges, and nanoparticles. Each format is tailored for specific biomedical applications. Its structural and functional properties can be adjusted to control strength, porosity, degradation rate, and biocompatibility, making silk fibroin an adaptable material for drug delivery, wound healing, and tissue engineering.

8.3.2.1 Films, Scaffolds, and Sponges

Silk fibroin films are commonly produced by casting, spin coating, or sequential layer deposition. Generally, the fibroin solution is dried on a substrate and stabilized by water vapor or alcohol treatment, which promotes β-sheet formation and improves stability. Patterned films can be formed using Polydimethylsiloxane (PDMS) molds, while techniques such as the Langmuir Blodgett method and spin coating provide control over surface morphology [26]. Silk fibroin sponges and scaffolds are widely used in tissue regeneration due to their adjustable porosity and mechanical resilience. Fabrication techniques include salt leaching, gas foaming, and freeze drying, which create porous three-dimensional structures that support cell infiltration and nutrient diffusion [27]. Freeze drying enables pore size control between 60 and 250 μm depending on freezing conditions. Reinforced scaffolds using silk particles or strands have demonstrated compressive moduli up to 13 MPa, suitable for bone tissue support [26]. Mesh-reinforced silk fibroin scaffolds supported even cell growth in vivo, which shows they could work well for load-bearing implants.

8.3.2.2 Hydrogels and Microcapsules

Silk fibroin hydrogels are three-dimensional water-rich networks formed by physical or chemical crosslinking. Physical gelation occurs via sol-gel transition, driven by hydrophobic interactions, typically over weeks at 37 °C. This can be accelerated by modifying pH, ionic strength, temperature, or by applying agitation or electric current [28]. Chemical crosslinking usually involves horseradish peroxidase and hydrogen peroxide, which link tyrosine residues within the fibroin chain [29]. These hydrogels show adequate mechanical strength for cartilage regeneration and can be used in injectable or implantable drug delivery platforms. Silk fibroin microspheres and nanoparticles are effective carriers for targeted delivery of drugs, proteins, and genes. Self-assembled particles are typically formed by ethanol precipitation followed by rapid freezing [30]. Lipid-based microspheres are fabricated by emulsifying silk fibroin with lipid films such as dioleoyl phosphocholine, followed by freeze-thaw cycles, lyophilization, and methanol treatment to remove lipids and stabilize the structure [31].

8.3.2.3 Fibers and Threads for Sutures

Silk fibroin fibers are produced using methods such as wet spinning, dry jet spinning, and electrospinning [32]. Electrospinning generates nanofibers with a large surface area, suitable for incorporating bioactive molecules, while wet and dry spinning yield microfibers with controllable structure. In concentrated silk fibroin solutions, micelles form through hydrogen bonding and hydrophobic interactions, and fiber alignment occurs via shear force and dehydration. During electrospinning, an electric field directs the silk fibroin solution into a fiber stream that solidifies upon reaching a collector. Methanol treatment converts silk fibroin from a random coil or α-helix (Silk I) to a β-sheet (Silk II) structure. This conformational transition enhances crystallinity, thereby improving the material's mechanical strength [33]. Electrospun silk fibroin mats exhibit high porosity and support cell adhesion, making them well suited for tissue engineering applications such as nerve guidance conduits and vascular grafts [34]. Regenerated silk fibroin fibers created through wet spinning and microfluidic spinning further enable precise control over fiber diameter, alignment, and mechanical characteristics.

8.4 Biomedical Applications of Silk Fibroin: A Multifunctional Natural Biomaterial

Silk fibroin is increasingly recognized as a high-performance biomaterial owing to its unique combination of biocompatibility, controlled biodegradability, mechanical robustness, and ease of processing. These inherent properties enable its broad application across diverse biomedical domains, including tissue engineering, drug delivery, wound healing, and implantable devices [34]. Unlike many synthetic polymers, silk fibroin supports natural cellular responses and degrades into non-toxic byproducts, making it suitable for both short-term and long-term therapeutic use. An effective biomaterial must not only maintain structural integrity but also integrate seamlessly with host tissues while minimizing immune responses [9]. Conventional materials often fall short in achieving both biocompatibility and mechanical reliability, leading to poor clinical outcomes. In contrast, silk fibroin-based materials exhibit strong biological interactions and structural versatility, overcoming common limitations such as poor biointegration or inflammatory reactions. These advantages position silk fibroin as a highly adaptable and clinically relevant platform for next-generation biomedical technologies. The wide-ranging applications and advantages of silk fibroin are schematically depicted in Fig. 8.3, highlighting its multifunctional biomedical potential.

8.4.1 Bone Tissue Regeneration

Silk fibroin scaffolds are effective for bone regeneration due to their biocompatibility, non-immunogenic nature, and ability to support cell adhesion and proliferation. Incorporating bioactive molecules such as stromal cell-derived factor 1 (SDF-1) and bone morphogenetic protein 2 (BMP-2) enhances stem cell recruitment and promotes new bone formation, accelerating the healing process [35, 36]. Silk fibroin composites with hydroxyapatite and trace metals like magnesium, zinc, and strontium have

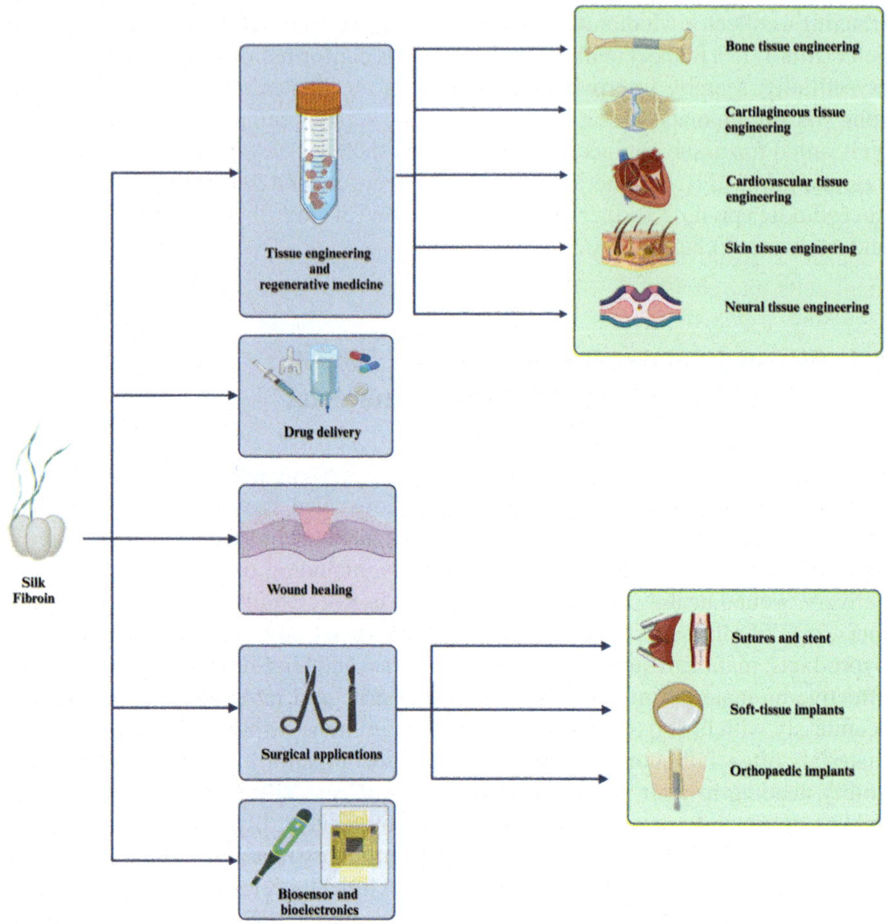

Fig. 8.3 Overview of major biomedical applications of silk fibroin, including tissue engineering, drug delivery, surgical implants, and bioelectronic devices

demonstrated improved osteoconductivity and mineralization. Among these, strontium plays a key role in osteoporotic conditions by promoting stem cell differentiation into osteoblasts, thereby enhancing bone healing [37]. Wu et al. [38] reported that silk fibroin-based scaffold, which resembles the structure and function of the native extracellular matrix, provides a three-dimensional framework that supports cell growth and facilitates nutrient diffusion, making them suitable for bone repair.

8.4.2 Cartilage Tissue Regeneration

Cartilage has limited self-repair ability due to its avascular structure; therefore, facilitating the delivery of nutrients or stem cells from surrounding tissues can promote cartilage regeneration [39]. Silk fibroin-based three-dimensional scaffolds with optimized pore sizes support chondrocyte infiltration and matrix deposition, thereby promoting cartilage regeneration [40, 41]. Combining silk fibroin with collagen enhances scaffold biocompatibility and promotes the differentiation of mesenchymal stem cells into chondrocytes through transforming growth factor beta 1 (TGF-β1) signaling, a pathway essential for cartilage formation [42]. Curcumin-loaded silk fibroin scaffolds offer additional therapeutic benefits, including anti-inflammatory and antioxidant effects, which help in cartilage repair [43]. Recent advances also include integrating gene delivery with silk fibroin platforms. For instance, lysyl oxidase gene-loaded GelMA hydrogels in silk fibroin matrices have shown potential in promoting osteochondral regeneration, representing an emerging approach for treating complex cartilage injuries [42].

8.4.3 Skin Tissue Regeneration

The skin, frequently exposed to injuries from burns, trauma, and surgical procedures, requires effective materials for functional repair. Silk fibroin is extensively studied in skin regeneration owing to its low immunogenicity and ability to stimulate cellular migration through NF-κB, MEK, and JNK signaling pathways. These biological responses promote wound healing and enhance local coagulation [44]. Silk fibroin-based biomaterials facilitate the development of three-layered in vitro skin models that mimic the epidermis, dermis, and hypodermis. These models provide a more accurate simulation of natural skin behavior compared to traditional two-layered constructs [45]. Among various fabrication methods, electrospinning is particularly effective for generating extracellular matrix-like nanofibers. These structures improve cell migration, vascularization, and overall biocompatibility [46, 47]. Other approaches such as salt leaching, freeze drying, and three-dimensional printing are also employed to produce hydrogels, films, and porous scaffolds suited for skin repair. The source of fibroin plays a critical role. For instance, fibroin from *Antheraea pernyi* contains Arg-Gly-Asp (RGD) motifs, which enhance cell adhesion through integrin binding. Moreover, silk fibroin films functionalized with polydopamine nanoparticles exhibit improved hydrophilicity and protein adsorption, promoting collagen deposition and rapid wound closure in animal studies [48].

8.4.4 Neural Tissue Regeneration

Injury or disease of the nervous system often leads to serious functional impairments because of its complex organization and limited capacity for regeneration [49]. Silk fibroin has emerged as a promising scaffold material for neural repair, offering flexibility, mechanical strength, and biocompatibility. However, the lack of inherent electrical conductivity limits its standalone use in neural tissue engineering. To overcome this, silk fibroin is often combined with conductive polymers such as polypyrrole and carbon-based nanomaterials. These composites improve electrical signaling, reduce inflammatory responses, and support neural cell adhesion and growth [50]. The enhanced scaffolds guide axonal extension, facilitate stem cell differentiation into neuronal lineages, and maintain structural integrity under physiological stress. In central nervous system disorders such as Parkinson's and Alzheimer's disease, silk fibroin-based materials have demonstrated neuroprotective effects by mitigating toxic protein aggregation and improving cognitive function [35, 51]. Additionally, silk fibroin hydrogels serve as delivery platforms for neurotrophins like neurotrophin 3, which support spinal cord healing and nerve regrowth [52]. For peripheral nerve repair, silk fibroin fibers are effective in reconnecting severed nerves and directing axonal regeneration. Semmler et al. [53] observed that fibroin from diverse silkworm species has superior nerve guidance and functional outcomes compared to conventional autografts.

8.4.5 Drug Delivery

Conventional cancer treatments such as chemotherapy often face limitations including low tumor specificity and significant systemic toxicity [54]. To address these challenges, silk fibroin nanoparticles have been investigated as drug carriers due to their biocompatibility, biodegradability, and capacity for controlled drug release. Genetically engineered silk elastin-like protein polymers (SELPs) have demonstrated effectiveness in delivering hydrophobic drugs such as doxorubicin. Modifying the silk to elastin ratio can enhance drug loading and release properties. Notably, doxorubicin-loaded silk fibroin nanoparticles exhibited nearly twice (1.8) the cytotoxic effect compared to the free drug, indicating improved therapeutic efficiency [55]. In addition, pH-responsive silk fibroin nanoparticles enable selective drug release in the acidic tumor microenvironment. This targeted release improves drug accumulation at the tumor site and reduces systemic side effects, enhancing the overall efficacy of cancer therapy [56].

8.5 Silk Fibroin in Implantable Devices

Biomedical textiles and fiber-based implants (BTFIs) are increasingly important in modern clinical practice due to their adaptability, biocompatibility, and compatibility with minimally invasive procedures. Compared to traditional rigid materials, BTFIs offer greater design flexibility, making them suitable for soft implants, fixation systems, and tissue repair applications [35]. This reflects a growing preference for soft, dynamic materials that better integrate with biological tissues and reduce postoperative complications. Silk fibroin can be made into fibers or yarns and then woven, knitted, or braided into custom structures for bone grafts, spinal repair, and soft tissue engineering. Its combination of mechanical strength, controlled biodegradability, and excellent tissue compatibility makes it ideal for implants requiring integration with host tissues. Silk fibroin-based textiles, either alone or blended with synthetic polymers such as medical grade polyester (PET), are used in a variety of implants including surgical sutures, vascular grafts, and artificial ligaments. These implants can be customized as mono- or multifilament constructs to optimize healing outcomes while minimizing immune response [57]. By selecting appropriate polymers and leveraging advanced fabrication techniques, silk fibroin continues to advance the performance of textile-based implants. This progress supports new directions in personalized medicine and regenerative therapies.

8.5.1 Biodegradable Implants

8.5.1.1 Sutures

Silk fibroin, derived from silkworms and spiders, has a long-standing role as a surgical suture material. Despite its classification as non-biodegradable due to its slow degradation rate (~2 years), silk fibroin offers high biocompatibility, affordability, and minimal inflammatory response, making it suitable for applications like vascular ligation and mucosal wound closure [35]. However, traditional braided silk sutures are susceptible to bacterial colonization because bacteria readily adhere to the proteinaceous surface of silk and can become trapped within the interstices of the braided structure, thereby promoting microbial adhesion and biofilm formation [58]. To overcome this, silk fibroin-based fibers inspired by the core-shell architecture of spider silk have been developed, offering improved mechanical strength and inherent antimicrobial properties [59]. The integration of antimicrobial peptides further reduces inflammation and bacterial adhesion, as demonstrated in both in vitro and in vivo models, positioning them as promising alternatives to conventional sutures [60].

8.5.2 Soft Tissue Artificial Implants

8.5.2.1 Artificial Skin

Silk-based biomaterials have been extensively explored for use in artificial skin due to their low immunogenicity and ability to support the growth of keratinocytes and fibroblasts [57]. Techniques such as electrospinning and nonwoven scaffold fabrication have yielded skin-like structures, including 3D silk fibroin scaffolds crosslinked with formic acid [61], nanofiber matrices treated with water vapor [62], and electrospun silk fibroin fibers ranging from 30 to 120 nm in diameter [63]. These constructs have shown favorable support for skin cell behavior in vitro. For enhanced regeneration in animal models, silk fibroin has been combined with natural polymers like collagen, chitin, and alginate, improving cell adhesion and matrix remodeling [64]. Bilayered silk fibroin constructs, which mimic native skin architecture, have further demonstrated effective reepithelialization and dermal repair [65]. This shows silk fibroin's potential in the development of functional artificial skin.

8.5.2.2 Artificial Nerve Guides

Although the peripheral nervous system can recover from minor injuries, severe damage often requires the use of artificial scaffolds to support regeneration. Silk fibers offer a suitable platform for nerve repair due to their biocompatibility, mechanical strength, and ability to support nerve cell adhesion and growth [57]. Spider silk incorporated into decellularized vein grafts has been shown to guide axonal regrowth and restore function in preclinical models. Silk-based scaffolds, especially when fabricated as nonwoven mats or reinforced fiber conduits, have shown improved outcomes in nerve regeneration. Li et al. [57] demonstrated that composites combining silk with chitosan or polylactic glycolic acid (PLGA) can effectively bridge nerve gaps in rat models and support aligned axonal growth and functional recovery.

8.5.2.3 Artificial Ligament and Tendon

Scaffold materials designed for ligament and tendon repair must exhibit elasticity, sufficient mechanical strength, and controlled biodegradability while supporting tissue regeneration. Silk-based biomaterials meet these requirements and have shown strong potential in ligament and tendon engineering. Spider silk fibers are particularly well suited for anterior cruciate ligament repair due to their impressive tensile strength, which can reach up to 4.8 gigapascals, and high flexibility, with elongation at break of up to 35% [66]. These properties allow the fabrication of scaffolds that closely resemble the mechanical behavior of native ligament tissue. Incorporating bioactive factors such as transforming growth factor beta enhances the regenerative potential of silk-based scaffolds, promoting tissue remodeling and integration.

Preclinical studies in animal models have demonstrated positive outcomes in tendon and ligament regeneration using silk-derived constructs [57].

8.5.3 Orthopedic Implants

Traditional orthopedic implants often rely on metals or synthetic polymers to stabilize and repair fractured bones. However, silk-based biomaterials are being increasingly explored as more biocompatible and flexible alternatives for skeletal repair [16]. Electrospun silk scaffolds incorporated with hydroxyapatite nanoparticles and bone morphogenetic protein 2 have shown the ability to support osteogenic differentiation of human mesenchymal stem cells, promoting effective bone regeneration [26]. This suggests a valuable role for silk constructs in orthopedic tissue engineering. In addition, silk ligatures have been successfully used in canine models of periodontitis. These silk-based tools offered a less invasive and more effective approach compared to conventional synthetic materials, highlighting the therapeutic promise of silk in dental and orthopedic contexts [67].

8.5.4 Fibroin-Coated Devices

8.5.4.1 Anti-fouling and Antimicrobial Coatings

Bacterial colonization and biofilm formation remain significant challenges in biomedical implants. Conventional antibiotic-based coatings risk promoting antimicrobial resistance. Silk fibroin offers a promising alternative due to its biocompatibility and ability to be surface modified. Nanoscale-patterned silk fibroin films have been shown to reduce bacterial attachment by up to 66%, primarily due to their hydrophobicity and surface topographies that disrupt microbial adhesion. Functionalization with antimicrobial agents further enhances their efficacy, enabling both antifouling and antimicrobial properties. Silk fibroin's biodegradability and compatibility with tissue make it well suited for safe, customizable coatings. Recent advances include silk fibroin films integrated with organic semiconductors, which improve microbial resistance while adding photonic features for imaging and light-activated therapies [68].

8.5.4.2 Stent

Vascular stents are widely used to treat occluded arteries but are still linked with complications such as thrombosis, inflammation, and restenosis. To reduce neointimal growth, drug-eluting stents incorporating biodegradable polymer coatings like

PLA or PGA have been developed to control local drug release, though polymer-related inflammatory reactions and issues with healing and late thrombosis remain significant challenges in clinical use [69]. Silk fibroin offers a promising alternative for stent coatings due to its biocompatibility, mechanical stability, and adaptable material forms. It can be processed into films, gels, nanofibers, or porous scaffolds, enabling sustained drug delivery while minimizing immune reactions. Silk fibroin-coated stents loaded with agents such as heparin, paclitaxel, or clopidogrel have been shown to inhibit smooth muscle cell proliferation and support endothelial regeneration. Its hemocompatibility also reduces platelet adhesion and thrombus formation. Preclinical studies in porcine models have demonstrated the functional stability and therapeutic efficacy of silk-based stent systems, indicating strong potential for clinical application [69]. Various applications of silk fibroin in the biomedical field, including tissue regeneration, drug delivery, and implantable devices, are summarized in Table 8.2.

8.6 Silk-Based Surgical Tools and Sutures

8.6.1 Fibroin Sutures: Characteristics and Applications

Silk fibroin sutures are widely used in surgery due to their excellent biocompatibility, tensile strength, and flexibility. Their clinical performance relies on mechanical properties like tensile strength, which are critical for maintaining wound stability during healing [80]. Mechanical behavior is influenced by factors such as gauge length and extension rate. Longer gauge lengths localize stress at filament defects, reducing tenacity. Knotting also introduces stress concentration points, often leading to structural failure. The viscoelastic nature of silk fibroin affects its tensile response; higher extension rates help preserve tenacity by minimizing molecular relaxation, while slower rates increase stress relaxation and reduce strength [81]. These insights highlight the need for precise control over manufacturing and application conditions to optimize suture performance.

8.6.2 Silk-Based Adhesives and Sealants

Effective tissue sealing in fluid-rich surgical environments remains challenging, as conventional methods like sutures and staples often fail to provide watertight closure. Commercial adhesives such as cyanoacrylates and fibrin sealants face limitations including cytotoxicity, poor wet adhesion, and limited mechanical strength [82, 83]. Silk fibroin has emerged as a promising alternative due to its biocompatibility, biodegradability and adjustable mechanical properties. However, the silk fibroin-tannic acid hydrogel effectively enhances β-sheet formation to achieve strong

Table 8.2 Applications of silk fibroin in the biomedical field

Target tissue / application	Scaffold type / condition	Cell type	In vitro / in vivo (animal)	Reference(s)
Bone tissue regeneration	3D-printed SF composites loaded with hydroxyapatite (HAp), SF-HT-3GEN, β-TCP, or collagen	Osteoblasts / bone precursor cells	In vitro / in vivo (biocompatible and osteoconductive)	[38]; [70]; [71]
Cartilage tissue regeneration	SF hydrogels (soft/ tough, γ-ray crosslinked for mechanical tunability)	Chondrocytes / MSCs	In vitro	[72]
	SF–chitosan / SF–PVA composites (enhanced flexibility and strength)	–	–	[73]
Skin tissue regeneration	SF films (flexible, transparent; suitable for dermal dressings)	–	In vivo (dermal sites)	[74]
	Nonwoven 3D SF scaffolds (formic acid crosslinked)	Human epidermal keratinocytes (HEK), dermal fibroblasts (HDF)	In vitro (co-culture, 75–95 days)	[61]
Neural tissue regeneration	SF–PEDOT:PSS / PVA composite (electroconductive film)	Corneal epithelial cells, potentially neurons	In vitro	[75]
Drug delivery	Porous SF particles (salting out, DOX and FA loaded; pH-responsive)	Tumor cells	In vitro	[76]
	SF nanoparticles (reinforcement in scaffolds for drug release)	–	–	[77]
Implantable devices / biomedical scaffolds	SF–PU composite fibers (wet-spun, enhanced flexibility)	Vascular endothelial cells	In vivo (rat model; vascular graft)	[78]; [79]

SF—Silk Fibroin; **HAp**—Hydroxyapatite; **β-TCP**—Beta-Tricalcium Phosphate; **MSC**—Mesenchymal Stem Cell; **PVA**—Polyvinyl Alcohol; **HEK**—Human Epidermal Keratinocytes; **HDF**—Human Dermal Fibroblasts; **PEDOT: PSS**—Poly (3, 4-ethylenedioxythiophene):Poly (styrenesulfonate); **DOX**—Doxorubicin; **FA**—Folic Acid; **PU**—Polyurethane; **SF-HT-3GEN**—SF Hardystonite-Gentamicin.

adhesion under wet and dynamic conditions [84]. This engineered silk-based sealant combines strong adhesion with rapid hemostasis and enables suture-free sealing of ruptured tissues in wet and dynamic biological environments while addressing critical surgical challenges [85].

8.7 Advances in Silk Biomaterials for Functional Implants

Silk fibroin has enabled the development of flexible implantable electronic systems that integrate closely with soft biological tissues. Its mechanical compliance minimizes tissue-device mismatch and improves signal fidelity in wearable sensors and neural interfaces. A key advancement is the emergence of transient electronics which are devices designed to degrade safely after use, eliminating the need for surgical removal. Silk fibroin serves as a biocompatible substrate for constructing resorbable antennas, circuits, and electrodes [86]. Its programmable degradation supports applications such as time-sensitive diagnostics and therapeutic monitoring. Silk-based sensors can also detect mechanical strain, infection, or inflammation in implants such as pacemakers or orthopedic prostheses. Unlike conventional rigid and nondegradable materials, silk fibroin offers a biodegradable platform better suited for dynamic biological environments. This shift enables implantable electronics that function reliably and disappear without intervention. One example is a wirelessly powered therapeutic device that combines magnesium-based heaters and drug-loaded silk fibroin layers. Upon activation, localized heating triggers targeted drug release for antimicrobial treatment [86].

8.8 Challenges in Clinical Translation

Silk fibroin offers significant potential for biomedical applications due to its biocompatibility, biodegradability, and favorable mechanical properties. However, several challenges continue to hinder its clinical translation. One of the main issues is the absence of standardized and scalable production methods. Although silk fibroin can be extracted and processed reliably in laboratory settings, no universally accepted protocol exists for industrial-scale production. This leads to inconsistencies in material quality and performance, as differences in silk source, amino acid composition, and processing techniques can significantly influence the final product [87]. Another major limitation is the limited availability of safety and efficacy data from large animal models and human clinical trials. Despite promising preclinical outcomes in areas such as wound healing and tissue engineering, clinical validation remains insufficient [9]. Furthermore, replicating the complex tissue architecture and maintaining the functional properties of silk fibroin-based constructs under physiological conditions remains technically demanding. This is particularly true for composite hydrogels, which require a finely tuned chemical and mechanical environment that mimics native tissue behavior. Addressing this challenge is essential for advancing silk fibroin from research to routine clinical practice.

8.9 Future Perspectives

Silk biomaterials continue to redefine possibilities in biomedical engineering. Advancements in fabrication techniques such as electrospinning, freeze drying, and three-dimensional printing have enabled the development of highly customized silk fibroin-based scaffolds tailored for regenerative medicine and tissue engineering. Recent progress in material science, genetic engineering, and synthetic biology allows for the design of silk constructs with adjustable mechanical strength, degradation rates, and functional properties, making them suitable for a wide range of medical applications. The integration of silk with polymers, nanoparticles, and other bioactive compounds has resulted in multifunctional composites that support enhanced tissue regeneration and targeted drug delivery [88]. These materials show promise in clinical applications including wound healing, cartilage and nerve repair, and orthopedic reconstruction. In addition, the compatibility of silk with bioelectronic systems presents new opportunities in wearable diagnostics and personalized therapies. This includes platforms for real-time monitoring and responsive treatment strategies, further expanding the role of silk in future medical technologies.

8.10 Conclusion

Silk fibroin has emerged as a highly promising biomaterial in the biomedical field due to its long history of medical use, excellent mechanical properties, and strong biocompatibility. Its unique β-sheet crystallinity contributes to outstanding tensile strength, thermal stability, and biodegradability. These features distinguish it from many conventional biomaterials. Silk fibroin can be processed into various forms including films, sponges, hydrogels, and three-dimensional scaffolds, enabling its use in tissue engineering, regenerative medicine, and drug delivery. Its ability to promote cell adhesion, proliferation, and differentiation supports the regeneration of skin, bone, cartilage, and nerve tissues. Moreover, silk fibroin is being explored for vascular grafts, nerve conduits, cardiac patches, orthopedic implants, sutures, and drug releasing stents. Despite these advantages, challenges remain in scaling production, standardizing processing methods, and ensuring consistent clinical performance. Ongoing research focused on material optimization and biological functionalization will be essential for translating silk fibroin-based technologies into broad clinical use.

Declaration of AI-Usage AI tools were used solely for language editing, grammar correction, and improving clarity of the manuscript. The authors take full responsibility for the originality, accuracy, and integrity of the content.

References

1. Kang Z, Wang Y, Xu J et al (2018) An RGD-containing peptide derived from wild silkworm silk fibroin promotes cell adhesion and spreading. Polymers 10:1193. https://doi.org/10.3390/polym10111193
2. Pereira RFP, Silva MM, de Zea BV (2015) *Bombyx mori* silk fibers: an outstanding family of materials. Macromol Mater Eng 300:1171–1198. https://doi.org/10.1002/mame.201400276
3. Li X, Cui R, Liu W et al (2013) The use of Nanoscaled fibers or tubes to improve biocompatibility and bioactivity of biomedical materials. J Nanomater 2013:728130. https://doi.org/10.1155/2013/728130
4. Rockwood DN, Preda RC, Yücel T et al (2011) Materials fabrication from *Bombyx mori* silk fibroin. Nat Protoc 6:1612–1631. https://doi.org/10.1038/nprot.2011.379
5. Marin E, Boschetto F, Pezzotti G (2020) Biomaterials and biocompatibility: an historical overview. J Biomed Mater Res Part A 108:1617–1633. https://doi.org/10.1002/jbm.a.36930
6. Holland C, Numata K, Rnjak-Kovacina J, Seib FP (2019) The biomedical use of silk: past, present, future. Adv Healthc Mater 8:1800465. https://doi.org/10.1002/adhm.201800465
7. Koh L-D, Cheng Y, Teng C-P et al (2015) Structures, mechanical properties and applications of silk fibroin materials. Prog Polym Sci 46:86–110. https://doi.org/10.1016/j.progpolymsci.2015.02.001
8. Vepari C, Kaplan DL (2007) Silk as a biomaterial. Prog Polym Sci 32:991–1007. https://doi.org/10.1016/j.progpolymsci.2007.05.013
9. Nguyen TP, Nguyen QV, Nguyen V-H et al (2019) Silk fibroin-based biomaterials for biomedical applications: a review. Polymers 11:1933. https://doi.org/10.3390/polym11121933
10. Pugina RS, Caiut JMA (2021) Silk: history, obtaining, structure and properties of an old material in the development of new technologies. IJAMB 4:27–45. https://doi.org/10.52466/ijamb.v4i1.88
11. Qiu W, Patil A, Hu F, Liu XY (2019) Hierarchical structure of silk materials versus mechanical performance and mesoscopic engineering principles. Small 15:1903948. https://doi.org/10.1002/smll.201903948
12. Wilson D, Valluzzi R, Kaplan D (2000) Conformational transitions in model silk peptides. Biophys J 78:2690–2701. https://doi.org/10.1016/S0006-3495(00)76813-5
13. Choi Y-J, Cho D-W, Lee H (2021) Development of silk fibroin scaffolds by using indirect 3D-bioprinting technology. Micromachines 13:43. https://doi.org/10.3390/mi13010043
14. Kim SH, Yeon YK, Lee JM et al (2018) Precisely printable and biocompatible silk fibroin bioink for digital light processing 3D printing. Nat Commun 9:1620. https://doi.org/10.1038/s41467-018-03759-y
15. Yin C, Han X, Lu Q et al (2022) Rhein incorporated silk fibroin hydrogels with antibacterial and anti-inflammatory efficacy to promote healing of bacteria-infected burn wounds. Int J Biol Macromol 201:14–19. https://doi.org/10.1016/j.ijbiomac.2021.12.156
16. Lujerdean C, Baci G-M, Cucu A-A, Dezmirean DS (2022) The contribution of silk fibroin in biomedical engineering. Insects 13:286. https://doi.org/10.3390/insects13030286
17. Wang Y, Fan S, Li Y et al (2020b) Silk fibroin/sodium alginate composite porous materials with controllable degradation. Int J Biol Macromol 150:1314–1322. https://doi.org/10.1016/j.ijbiomac.2019.10.141
18. Catto V, Farè S, Cattaneo I et al (2015) Small diameter electrospun silk fibroin vascular grafts: mechanical properties, in vitro biodegradability, and in vivo biocompatibility. Mater Sci Eng C 54:101–111. https://doi.org/10.1016/j.msec.2015.05.003

19. Mehrabani MG, Karimian R, Mehramouz B et al (2018) Preparation of biocompatible and biodegradable silk fibroin/chitin/silver nanoparticles 3D scaffolds as a bandage for antimicrobial wound dressing. Int J Biol Macromol 114:961–971. https://doi.org/10.1016/j.ijbiomac.2018.03.128
20. Vidya M, Rajagopal S (2021) Silk fibroin: a promising tool for wound healing and skin regeneration. Int J Polym Sci 2021:1–10. https://doi.org/10.1155/2021/9069924
21. Zhang H, Cao L, Li J et al (2023) Dual physically crosslinked silk fibroin Ionoelastomer with ultrahigh Stretchability and low hysteresis. Chem Mater 35:1752–1761. https://doi.org/10.1021/acs.chemmater.2c03536
22. Martínez-Mora C, Mrowiec A, García-Vizcaíno EM et al (2012) Fibroin and Sericin from *Bombyx mori* silk stimulate cell migration through upregulation and phosphorylation of c-Jun. PLoS One 7:e42271. https://doi.org/10.1371/journal.pone.0042271
23. Kwon G, Heo B, Kwon MJ et al (2021) Effect of silk fibroin biomaterial coating on cell viability and intestinal adhesion of probiotic bacteria. J Microbiol Biotechnol 31:592–600. https://doi.org/10.4014/jmb.2103.03031
24. Ajisawa A (1998) Dissolution of silk fibroin with calciumchloride/ethanol aqueous solution. J Seric Sci Jpn 67(2):91–94. https://doi.org/10.11416/kontyushigen1930.67.91
25. Wöltje M, Kölbel A, Aibibu D, Cherif C (2021) A fast and reliable process to fabricate regenerated silk fibroin solution from degummed silk in 4 hours. Int J Mol Sci 22:10565. https://doi.org/10.3390/ijms221910565
26. Kundu B, Rajkhowa R, Kundu SC, Wang X (2013) Silk fibroin biomaterials for tissue regenerations. Adv Drug Deliv Rev 65:457–470. https://doi.org/10.1016/j.addr.2012.09.043
27. Nazarov R, Jin H-J, Kaplan DL (2004) Porous 3-D scaffolds from regenerated silk fibroin. Biomacromolecules 5:718–726. https://doi.org/10.1021/bm034327e
28. Galateanu B, Hudita A, Zaharia C, Bunea MC, Vasile E, Buga MR, Costache M (2018) Silk-based hydrogels for biomedical applications. In: Cellulose-Based Superabsorbent Hydrogels. Springer, Cham, pp 1–26. https://doi.org/10.1007/978-3-319-76573-0_59-1
29. Chirila TV, Suzuki S, Papolla C (2017) A comparative investigation of *Bombyx mori* silk fibroin hydrogels generated by chemical and enzymatic cross-linking. Biotechnol Appl Biochem 64:771–781. https://doi.org/10.1002/bab.1552
30. Cao Z, Chen X, Yao J et al (2007) The preparation of regenerated silk fibroin microspheres. Soft Matter 3:910. https://doi.org/10.1039/b703139d
31. Wang X, Yucel T, Lu Q et al (2010a) Silk nanospheres and microspheres from silk/pva blend films for drug delivery. Biomaterials 31:1025–1035. https://doi.org/10.1016/j.biomaterials.2009.11.002
32. Pérez-Rigueiro J, Madurga R, Gañán-Calvo AM et al (2018) Straining flow spinning of artificial silk fibers: a review. Biomimetics 3:29. https://doi.org/10.3390/biomimetics3040029
33. Zhang KH, Mo XM (2011) Influence of post-treatment with methanol vapor on the properties of SF/P(LLA-CL) Nanofibrous scaffolds. Adv Mater Res 236–238:2221–2224. https://doi.org/10.4028/www.scientific.net/AMR.236-238.2221
34. Flores-Rojas GG, Gómez-Lazaro B, López-Saucedo F et al (2023) Electrospun scaffolds for tissue engineering: a review. Macromol 3:524–553. https://doi.org/10.3390/macromol3030031
35. De Giorgio G, Matera B, Vurro D et al (2024) Silk fibroin materials: biomedical applications and perspectives. Bioengineering 11:167. https://doi.org/10.3390/bioengineering11020167
36. Shen X, Zhang Y, Gu Y et al (2016) Sequential and sustained release of SDF-1 and BMP-2 from silk fibroin-nanohydroxyapatite scaffold for the enhancement of bone regeneration. Biomaterials 106:205–216. https://doi.org/10.1016/j.biomaterials.2016.08.023
37. Wang L, Pathak JL, Liang D et al (2020a) Fabrication and characterization of strontium-hydroxyapatite/silk fibroin biocomposite nanospheres for bone-tissue engineering applications. Int J Biol Macromol 142:366–375. https://doi.org/10.1016/j.ijbiomac.2019.09.107
38. Wu Z, Meng Z, Wu Q et al (2020b) Biomimetic and osteogenic 3D silk fibroin composite scaffolds with nano MgO and mineralized hydroxyapatite for bone regeneration. J Tissue Eng 11:2041731420967791. https://doi.org/10.1177/2041731420967791

39. Ribeiro VP, Da Silva MA, Maia FR et al (2018) Combinatory approach for developing silk fibroin scaffolds for cartilage regeneration. Acta Biomater 72:167–181. https://doi.org/10.1016/j.actbio.2018.03.047
40. Wang J, Yang Q, Cheng N et al (2016) Collagen/silk fibroin composite scaffold incorporated with PLGA microsphere for cartilage repair. Mater Sci Eng C 61:705–711. https://doi.org/10.1016/j.msec.2015.12.097
41. Wang Y, Bella E, Lee CSD et al (2010b) The synergistic effects of 3-D porous silk fibroin matrix scaffold properties and hydrodynamic environment in cartilage tissue regeneration. Biomaterials 31:4672–4681. https://doi.org/10.1016/j.biomaterials.2010.02.006
42. He S, Fu X, Wang L et al (2023) Self-assemble silk fibroin microcapsules for cartilage regeneration through gene delivery and immune regulation. Small 19:2302799. https://doi.org/10.1002/smll.202302799
43. Kim DK, In Kim J, Sim BR, Khang G (2017) Bioengineered porous composite curcumin/silk scaffolds for cartilage regeneration. Mater Sci Eng C 78:571–578. https://doi.org/10.1016/j.msec.2017.02.067
44. Sun F, Xiao D, Su H et al (2023) Highly stretchable porous regenerated silk fibroin film for enhanced wound healing. J Mater Chem B 11:1486–1494. https://doi.org/10.1039/D2TB01896A
45. Abbott RD, Kimmerling EP, Cairns DM, Kaplan DL (2016) Silk as a biomaterial to support long-term three-dimensional tissue cultures. ACS Appl Mater Interfaces 8:21861–21868. https://doi.org/10.1021/acsami.5b12114
46. Lee JM, Chae T, Sheikh FA et al (2016) Three dimensional poly(ε-caprolactone) and silk fibroin nanocomposite fibrous matrix for artificial dermis. Mater Sci Eng C 68:758–767. https://doi.org/10.1016/j.msec.2016.06.019
47. Lu G, Ding Z, Wei Y et al (2018) Anisotropic biomimetic silk scaffolds for improved cell migration and healing of skin wounds. ACS Appl Mater Interfaces 10:44314–44323. https://doi.org/10.1021/acsami.8b18626
48. Wang J, Chen Y, Zhou G et al (2019) Polydopamine-coated *Antheraea pernyi* (*a. pernyi*) silk fibroin films promote cell adhesion and wound healing in skin tissue repair. ACS Appl Mater Interfaces 11:34736–34743. https://doi.org/10.1021/acsami.9b12643
49. Boni R, Ali A, Giteru SG et al (2020) Silk fibroin nanoscaffolds for neural tissue engineering. J Mater Sci Mater Med 31:81. https://doi.org/10.1007/s10856-020-06422-5
50. Zhao Y, Liang Y, Ding S et al (2020) Application of conductive PPy/SF composite scaffold and electrical stimulation for neural tissue engineering. Biomaterials 255:120164. https://doi.org/10.1016/j.biomaterials.2020.120164
51. Chae H-S, Kang YK, Shin YK, et al (2004) The role of BF-7 on neuroprotection and enhancement of cognitive function. Korean J Physiol Pharmacol 8(4):173–179
52. Sha Q, Wang Y, Zhu Z et al (2023) A hyaluronic acid/silk fibroin/poly-dopamine-coated biomimetic hydrogel scaffold with incorporated neurotrophin-3 for spinal cord injury repair. Acta Biomater 167:219–233. https://doi.org/10.1016/j.actbio.2023.05.044
53. Semmler L, Naghilou A, Millesi F et al (2023) Silk-in-silk nerve guidance conduits enhance regeneration in a rat sciatic nerve injury model. Adv Healthc Mater 12:2203237. https://doi.org/10.1002/adhm.202203237
54. Hoang CNM, Nguyen SH, Tran MT (2025) Nanoparticles in cancer therapy: strategies to penetrate and modulate the tumor microenvironment—a review. Smart Mater Med 6:270–284. https://doi.org/10.1016/j.smaim.2025.07.004
55. Mottaghitalab F, Farokhi M, Shokrgozar MA et al (2015) Silk fibroin nanoparticle as a novel drug delivery system. J Control Release 206:161–176. https://doi.org/10.1016/j.jconrel.2015.03.020
56. Zou X, Jiang Z, Li L, Huang Z (2021) Selenium nanoparticles coated with pH responsive silk fibroin complex for fingolimod release and enhanced targeting in thyroid cancer. Artif Cells Nanomedicine Biotechnol 49:83–95. https://doi.org/10.1080/21691401.2021.1871620
57. Li G, Li Y, Chen G et al (2015) Silk-based biomaterials in biomedical textiles and fiber-based implants. Adv Healthc Mater 4:1134–1151. https://doi.org/10.1002/adhm.201500002

58. Chen X, Hou D, Tang X, Wang L (2015) Quantitative physical and handling characteristics of novel antibacterial braided silk suture materials. J Mech Behav Biomed Mater 50:160–170. https://doi.org/10.1016/j.jmbbm.2015.06.013
59. Liu M, Zhang Y, Liu K et al (2021) Biomimicking antibacterial Opto-electro sensing sutures made of regenerated silk proteins. Adv Mater 33:2004733. https://doi.org/10.1002/adma.202004733
60. Franco AR, Pirraco R, Fernandes EM et al (2022) Untangling the biological and inflammatory behavior of silk-like sutures in vivo. Biomaterials 290:121829. https://doi.org/10.1016/j.biomaterials.2022.121829
61. Dal Pra I, Chiarini A, Boschi A et al (2006) Novel dermo-epidermal equivalents on silk fibroin-based formic acid-crosslinked three-dimensional nonwoven devices with prospective applications in human tissue engineering/regeneration/repair. Int J Mol Med 18:241–247. https://doi.org/10.3892/ijmm.18.2.241
62. Min B, Jeong L, Lee KY, Park WH (2006) Regenerated silk fibroin nanofibers: water vapor-induced structural changes and their effects on the behavior of Normal human cells. Macromol Biosci 6:285–292. https://doi.org/10.1002/mabi.200500246
63. Min B-M, Lee G, Kim SH et al (2004) Electrospinning of silk fibroin nanofibers and its effect on the adhesion and spreading of normal human keratinocytes and fibroblasts in vitro. Biomaterials 25:1289–1297. https://doi.org/10.1016/j.biomaterials.2003.08.045
64. Wang G, Hu X, Lin W et al (2011) Electrospun PLGA–silk fibroin–collagen nanofibrous scaffolds for nerve tissue engineering. In Vitr Cell Dev Biol Anim 47:234–240. https://doi.org/10.1007/s11626-010-9381-4
65. Wendt H, Hillmer A, Reimers K et al (2011) Artificial skin—culturing of different skin cell lines for generating an artificial skin substitute on cross-weaved spider silk Fibres. PLoS One 6:e21833. https://doi.org/10.1371/journal.pone.0021833
66. Chen X, Qi Y-Y, Wang L-L et al (2008) Ligament regeneration using a knitted silk scaffold combined with collagen matrix. Biomaterials 29:3683–3692. https://doi.org/10.1016/j.biomaterials.2008.05.017
67. Kim SE, Lee ER, Lee Y et al (2012) A modified method for inducing periodontitis in dogs using a silk-wire twisted ligature. J Vet Sci 13:193. https://doi.org/10.4142/jvs.2012.13.2.193
68. Tullii G, Donini S, Bossio C et al (2020) Micro- and Nanopatterned silk substrates for antifouling applications. ACS Appl Mater Interfaces 12:5437–5446. https://doi.org/10.1021/acsami.9b18187
69. Wang X, Zhang X, Castellot J et al (2008) Controlled release from multilayer silk biomaterial coatings to modulate vascular cell responses. Biomaterials 29:894–903. https://doi.org/10.1016/j.biomaterials.2007.10.055
70. Huang T, Fan C, Zhu M et al (2019) 3D-printed scaffolds of biomineralized hydroxyapatite nanocomposite on silk fibroin for improving bone regeneration. Appl Surf Sci 467–468:345–353. https://doi.org/10.1016/j.apsusc.2018.10.166
71. Hadisi Z, Bakhsheshi-Rad HR, Walsh T et al (2020) In vitro and in vivo evaluation of silk fibroin-hardystonite-gentamicin nanofibrous scaffold for tissue engineering applications. Polym Test 91:106698. https://doi.org/10.1016/j.polymertesting.2020.106698
72. Wu N, Yu H, Sun M et al (2020a) Investigation on the structure and mechanical properties of highly tunable elastomeric silk fibroin hydrogels cross-linked by γ-ray radiation. ACS Appl Bio Mater 3:721–734. https://doi.org/10.1021/acsabm.9b01062
73. Xing X, Han Y, Cheng H (2023) Biomedical applications of chitosan/silk fibroin composites: a review. Int J Biol Macromol 240:124407. https://doi.org/10.1016/j.ijbiomac.2023.124407
74. Srivastava CM, Purwar R, Kannaujia R, Sharma D (2015) Flexible silk fibroin films for wound dressing. Fibers Polym 16:1020–1030. https://doi.org/10.1007/s12221-015-1020-y
75. Bhattacharjee P, Ahearne M (2020) Fabrication and biocompatibility of Electroconductive silk fibroin/PEDOT: PSS composites for corneal epithelial regeneration. Polymers 12:3028. https://doi.org/10.3390/polym12123028
76. Sun N, Lei R, Xu J et al (2019) Fabricated porous silk fibroin particles for pH-responsive drug delivery and targeting of tumor cells. J Mater Sci 54:3319–3330. https://doi.org/10.1007/s10853-018-3022-9

77. Shen Y, Wang X, Li B et al (2022) Development of silk fibroin-sodium alginate scaffold loaded silk fibroin nanoparticles for hemostasis and cell adhesion. Int J Biol Macromol 211:514–523. https://doi.org/10.1016/j.ijbiomac.2022.05.064
78. Tanaka T, Ibe Y, Jono T et al (2021) Characterization of a water-dispersed biodegradable polyurethane-silk composite sponge using 13C solid-state nuclear magnetic resonance as coating material for silk vascular grafts with small diameters. Molecules 26:4649. https://doi.org/10.3390/molecules26154649
79. Suganuma K, Matsuda H, Asakura T (2022) Characterization of polyurethane and a silk fibroin-polyurethane composite fiber studied with NMR spectroscopies. Polym J 54:803–813. https://doi.org/10.1038/s41428-022-00629-9
80. Kim J-C, Lee Y-K, Lim B-S et al (2007) Comparison of tensile and knot security properties of surgical sutures. J Mater Sci Mater Med 18:2363–2369. https://doi.org/10.1007/s10856-007-3114-6
81. Thilagavathi G, Viju S (2015) Silk as a suture material. In: Advances in Silk Science and Technology. Woodhead Publishing, pp 219–232. https://doi.org/10.1016/B978-1-78242-311-9.00011-2
82. Annabi N, Tamayol A, Shin SR et al (2014) Surgical materials: current challenges and nano-enabled solutions. Nano Today 9:574–589. https://doi.org/10.1016/j.nantod.2014.09.006
83. Annabi N, Zhang Y-N, Assmann A et al (2017) Engineering a highly elastic human protein–based sealant for surgical applications. Sci Transl Med 9:eaai7466. https://doi.org/10.1126/scitranslmed.aai7466
84. De Brito VP, De Souza Ribeiro MM, Viganó J et al (2022) Silk fibroin hydrogels incorporated with the antioxidant extract of Stryphnodendron adstringens bark. Polymers 14:4806. https://doi.org/10.3390/polym14224806
85. Bai S, Zhang X, Cai P et al (2019) A silk-based sealant with tough adhesion for instant hemostasis of bleeding tissues. Nanoscale Horiz 4:1333–1341. https://doi.org/10.1039/C9NH00317G
86. Patil AC, Xiong Z, Thakor NV (2020) Toward nontransient silk bioelectronics: engineering silk fibroin for bionic links. Small Methods 4:2000274. https://doi.org/10.1002/smtd.202000274
87. Li M, You J, Qin Q et al (2023) A comprehensive review on silk fibroin as a persuasive biomaterial for bone tissue engineering. IJMS 24:2660. https://doi.org/10.3390/ijms24032660
88. Attri K, Shruthi GH, Gulabrao DP et al (2024) Silk biomaterials: applications and future prospects in biomedical engineering. UPJOZ 45:205–216. https://doi.org/10.56557/upjoz/2024/v45i164301

Chapter 9
Silk Enhanced Polymer Composites

K. R. Sangeetha Gowda, Deity S. Anand, Patel Disha Chiragkumar, and Sunil S. More

9.1 Introduction

Natural fibers have been considered a promising material for fiber-reinforced polymer composites for the past few decades due to their ability to biodegrade, thereby reducing environmental pollution. When natural fibers are utilized as a reinforcing agent in composites, they enhance the high molding flexibility and can serve as a viable alternative to traditional fillers such as calcium, mica, glass, and carbonate. Natural fiber reinforcements in polymer matrices can replace the use of inorganic fillers, leading to cost-effectiveness, environmental friendliness, biodegradability, and improved insulation and thermal properties. These would lead to a wide range of applications, including food packaging materials, bulletproof vests, high-performance textiles, protective gear, aerospace, and sports equipment [1]. Among natural fibers, silk has been widely explored as a reinforcing agent in polymer matrix biocomposites. Silk is a naturally occurring fibrous protein sourced from silkworms (*Bombyx mori*) or spiders (*Nephila madagascariensis*). Silk fiber contributes to good mechanical strength, biocompatibility, and biodegradability properties. The integration of silk fibers into polymer matrices has a wide range of applications in biomedical fields [2]. A wide range of both natural and synthetic polymers have been investigated as matrices for silk-based composites [3]. Studies showed that the integration of silk in these materials results in regulated degradation rates, enhanced elasticity,

K. R. Sangeetha Gowda · D. S. Anand · P. D. Chiragkumar · S. S. More (✉)
School of Basic and Applied Sciences, Dayananda Sagar University, Bangalore, India
e-mail: sunilsmorey@gmail.com

K. R. Sangeetha Gowda
e-mail: sangeethagowdakr@gmail.com

D. S. Anand
e-mail: deityanand.s-rs-sbas@dsu.edu.in

P. D. Chiragkumar
e-mail: pateldisha.chiragbhai-rs-sbas@dsu.edu.in

M. Shahid et al. (eds.), *Science and Technology of Silk*, Textile Science and Clothing Technology, https://doi.org/10.1007/978-3-032-26880-8_9

increased strength, and improved resistance to environmental factors [4]. Certain techniques are involved in establishing a silk-reinforced polymer composite, one of which is electrospinning, which facilitates the formation of fine fibers with a high surface area ratio, making them suitable for biomedical applications such as scaffolds, wound dressings, and controlled drug-delivery systems [5]. Solvent casting is another method that involves mixing the polymer components with the silk in a solvent, followed by solvent evaporation, to yield composite films. The film obtained from this method is employed for its flexible membranes and biodegradable packaging materials [6]. Melt blending is a technique in which silk fibers are integrated with thermoplastic polymers at elevated temperatures, thereby enhancing the durability and structural integrity for various industrial applications [7]. Additionally, 3D printing has been utilized in the field of tissue engineering, which involves the development of precise, functionalized medical implants [8]. In biomedical engineering, tissue regeneration is a prominent field where silk has been utilized in scaffolds facilitating cell adhesion, proliferation, and differentiation for the regeneration of skin, cartilage, and bone [9]. In the domain of drug delivery, silk-reinforced polymers play a vital role in the controlled and sustained release of drugs, with silk nanoparticles being incorporated into the polymer that serve as nanocarriers for the effective delivery of proteins, antitumor efficacy, and anti-cancer drugs [10]. Thus, silk-based composites offer immense potential across various fields, driven by their unique combination of biological functionality and environmental sustainability.

9.2 Fundamentals of Silk as a Reinforcement Material

As a natural fiber, silk is renowned for its excellent properties, including elasticity, mechanical strength, biocompatibility, and biodegradability. These properties contribute to silk being an excellent option for use as a reinforcing agent in polymer composites, thereby enhancing the structural and functional integrity of the polymer matrices [11]. In order to use silk as a reinforcement to polymer, it is important to understand the silk's function in the polymer matrix and the structural inherent properties utilized in composite production.

9.2.1 Structure and Composition of Silk

Silk is a natural protein-based biopolymer that is initially in a liquid state and eventually solidifies to form solid fibers. Silk consists of large proteins, approximately 250–350 kDa in size, which comprise non-repetitive hydrophilic amino- and carboxy-terminal domains and a repetitive hydrophobic core. The backbone of silk proteins is composed of a polyamide chain and is rich in amino acids, which facilitate the assembly of β-strands into secondary crystalline β-sheet structures. These crystalline domains are widely scattered throughout the amorphous area formed by the helical

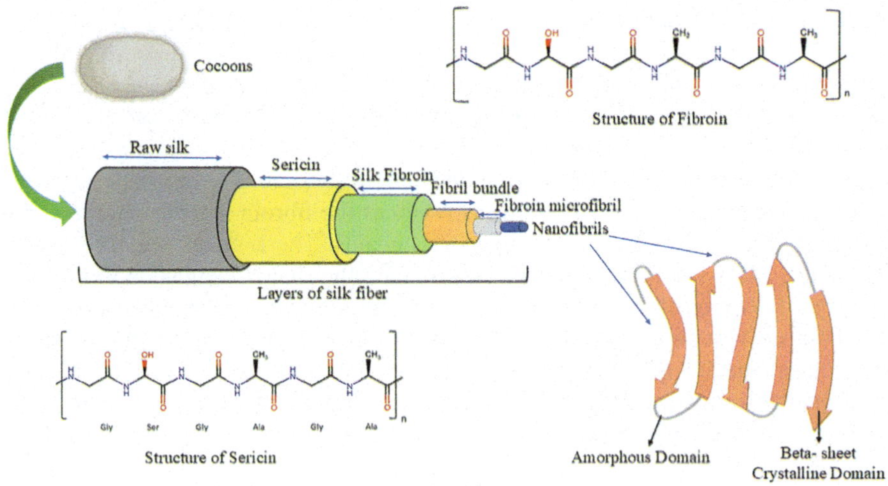

Fig. 9.1 Structure of silk fiber from *Bombyx mori* [15, 16]

structures and other secondary structural components of the proteins present in the silk. The mechanical strength of the silk fiber is determined by the quantity of β-sheet nanocrystallites and non-crystalline amorphous areas within the silk proteins [12].

The silk fiber obtained from the silkworm consists of two key proteins: fibroin and sericin, each of which plays a distinct role in determining silk's structural and functional properties. Fibroin consists of 40–60% of β-sheets and its crystalline secondary structures are formed by repetitive amino acid sequences which include glycine (Gly), alanine (Ala) and serine (Ser). Fibroin exhibits a semi-crystalline structure in which the amorphous segments increase flexibility and the crystalline β-sheets provide strength resulting in optimum ratio of elasticity to stiffness ensuring resistance to considerable stress and strain without fracturing. The predominant protein of the silk is sericin. Sericin serves as an adhesive and protective layer to fibroin by its hydrophilic glycoprotein. It contributes to the silk's thermal stability and water retention capabilities. Sericin is generally eliminated from the silk by the process called degumming to improve the biocompatibility and mechanical properties of fibroin [12, 13]. Despite sericin being removed, sericin retains significant bioactive properties and has been widely researched for its potential applications in wound healing, antioxidant roles, and biomaterial coatings [14]. Figure 9.1 depicts the structure of fibroin and sericin.

9.2.1.1 Hierarchical Structural Organization of Silk

The silks spun by *Nephila pilipes* spiders and *Bombyx mori* silkworms consist of a five-level hierarchical network structure [17].

9.2.1.1.1 Level One: Primary Structure

The primary structure of silk protein at the level one consists of an amino acid sequence. The composition of silk from the silkworm and the spider is different from each other. The composition of *Bombyx mori* silk fibers is mainly composed of two structural proteins, i.e., fibroin and sericin. Sericin is hydrophilic in nature and acts as glue located on the fibroin. Whereas the fibroin is hydrophobic and is further divided into light (L) and heavy (H) chains. The L-chain weighs ~25 kDa and plays a less significant role. The H, on the other hand, weighs ~350 kDa and is primarily responsible for excellent mechanical performance as it self-assembles into β-sheet crystallites. The H-chain consists of glycine (G), alanine (A) and serine (S) with about 5% tyrosine (Y) [18]. The silk fibroin H-chain contains 12 large hydrophobic repetitive domains, which are interrupted by 11 smaller hydrophilic non-repetitive amorphous domains. These are the key units for forming β-sheet and into β-β-crystallites.

The composition of *Nephila pilipes* silk fibers has different amino acid sequences. Major ampullate (MA) dragline spider silk proteins include polyalanine repeats, which are responsible for β-sheet structures, glycine-alanine (GA) repeats, glycine-rich (GGX) sequences, and GPGXX sequences. GPGXX sequences form β-turn spirals, eventually contributing to the fine elasticity and flexibility of spider silk.

9.2.1.1.2 Level Two: Secondary Structure

Level two consists of symmetrical structures formed by folding peptide chains, and bonds between amine and carboxyl groups are stabilized by hydrogen bonds (H-bonds). The most common elements present are α-helices and β-sheets. α-helices are characterized by H-bonds that form within the same strand, leading to better extensibility in silk materials. β-sheets are characterized by forming H-bonds between adjacent strands. β-sheet naturally forms an antiparallel arrangement, making it more stable due to shorter and linear H-bonds. Intramolecular β-sheets contribute a strain-stiffening effect, which is observed in spider dragline and *A. pernyi* silk fibers [19].

9.2.1.1.3 Level Three: β-crystallites

B-crystallites are formed by the stacking of several adjacent β-sheets from different molecules into an organized and highly compact 3D protein structure. β-crystallites are stabilized via hydrogen bonds, hydrophobic interactions, and van der Waals interactions between β-sheets; these interactions are collectively referred to as crystalline binding interactions. β-crystallites contribute to the high toughness, strain hardening, and stiffness of the silk material [20, 17].

9.2.1.1.4 Level Four: Nanofibrils

Nanofibrils are the basic mesoscopic structural units of hierarchical structure. A nanofibril consists of a crystal network in which β-crystallites are linked by amorphous chains. When β-crystallites are cross-linked, the silk materials adopt a nano-fishnet topology. This contributes to high level of toughness [21]. The mechanical properties are determined by the orientation of crystallites (orientation function f). For example, spider silk with $f \approx 0.93$, when compared to silkworm silk with $f \approx 0.84$, leads to higher breaking stress. Additionally, a higher density results in a stronger network due to the presence of more β-crystallites, which can contribute to load sharing. When compared to the silkworm silk, spider silk has double the β-crystallites density [17].

9.2.1.1.5 Nanofibrils Networks

Nanofibril networks are formed by the interconnectivity of numerous individual nanofibrils. This level determines the macroscopic performance of silk materials. The interconnecting nanofibrils exhibit strong interactions due to their highly oriented, twisted nature, which results in interlocked structures that prevent free movement. Their helically twisted morphology prevents slips and forms a non-slip fibril bundle (N-SFB) structure, which significantly toughens the fiber. Non-fibrous silk materials exhibit weak inter-nanofibril interactions, as they are patterned in a disorderly manner, for example, silk used in hydrogels and films [22].

9.2.2 Properties of Silk

9.2.2.1 Spider Silk Properties

Orb-weaving spiders such as *Araneidae* spiders are well-researched models because they produce different types of silk, each with unique amino acid composition and material properties [23]. In accordance with mechanical properties, the major ampullate (MA) silk from *Araneus* has a maximum strength of 1.1 GPa [24]. Dragline silk is known to be stronger than steel per unit weight in terms of high tensile strength and is more resilient than the finest synthetic fibers [25]. Viscid silk from *Araenus* can approximately stretch up to 270% before breaking, whereas MA silk from *Araenus* has an extensibility of 0.27. Vicid silk is considered to stretch 200% before breaking [24]. Compared to Kevlar fibers, some of the spider silks are more resilient. The toughness of silk is linked to its nanofibrillar composition and the strong interactions it exhibits [26]. Super-contraction is a unique occurrence, when exposed to water, majorly ampullate spider silks tend to shorten to roughly half their original length and double in diameter. This can be reversed by using a solvent, resulting in an increase in fiber extensibility and a decrease in stiffness. Supercontractions are caused by water molecules altering or destroying the hydrogen bonds, leading to molecular chain motion and disorientation [27]. Chemical composition, as mentioned earlier, includes a major structural protein in spider silk known as spidroins, which is rich in Gly, Ala, and Glu with significant amounts of Tyr and Leu residues. Dragline spider silk degrades very rapidly when implanted onto the skin, which limits the sustained release of any drug [24].

9.2.2.2 Silkworm Silks Properties

Although the silkworm silk is stiff, it is brittle when compared to spider silk due to the higher crystallinity compared to spider silk. The mechanical strength of the *Bombyx mori* silk is 0.6 GPa and the extensibility of the silk is 0.18. The toughness of silk is ~70 MJ m^{-3}. Unlike spider silk, the silkworm silk exhibits minimal change in length or shape when submerged in water [24]. *B. mori* silk fibroin is generally considered slow-degradable material as it takes more than 60 days to break down in vivo. Hence its slow and predictable degradability makes it useful for applications requiring support and load transfer for the developing tissues [28]. The chemical composition of *B. mori* silk fiber is primarily composed of fibroin and is covered with sericin. Two chains, known as heavy and light fibroin are joined by a disulfide bond to form fibroin linked to P25 (glycoprotein) in the silk fiber. Sericins are gummy-like proteins that assist in holding the fibroin together making the fiber more stable [29]. Sericins have properties like UV resistance, moisture absorption, oxidation resistance, and antibacterial activity. It has been reported that sericin protein coating stimulates immune response. However degummed silk fibers are essentially inert and might prevent macrophage activation [30].

9.2.3 Types of Silk Used in Polymer Composites

Silk exhibits a vivid range of characteristic properties based on the source of the silk. There are several types of silk that are used in polymer composites which include natural silk, regenerated silk fibroin, and recombinant silk proteins.

9.2.3.1 Natural Silk

Natural silk fibers are secreted from silkworms and are a popular material across textile industry. There are many different kinds of silk, such as mulberry silk secreted from *Bombyx mori* and wild silk which are non-mulberry silks. Wild silks are derived from the silkworm species that are not domesticated, primarily inhabiting the forested areas. Wild silks include tasar silk secreted by *Anthheraea mylitta*, eri silk from *Samia ricini* and muga silk from *Antheraea assamensis* [31].

9.2.3.1.1 Mulberry Silk

Bombyx mori is a species of mulberry silkworm and has been domesticated certain parts of India, Chaina and Japan. It is known for production of large number of silk cocoons. Apart from textile industry *Bombyx mori* is widely used in biomedicine due to its excellent biocompatibility and mechanical strength [32]. Poly lactic acid (PLA) composites with silk fiber exhibits good thermal stability and crystallization. Silk fiber has the ability to improve PLA's ductility and mechanical properties such as its modulus, strain percentage at break, tensile strength and modulus of elasticity. Silk fiber acts as a barriers for heat insulations, increasing the glass temperature and improving the thermal stability of PLA-silk bio composites. Additionally silk reinforcements increased the weight of residues and the PLA matrix capacity for enzymatic breakdown. Silk fiber can improve dimension stability and increase the PLA matrix's stiffness at higher temperatures (70–160 °C) [33]. The mechanical properties of Poly (butylene succinate) (PBS) are enhanced by silk fibers and the tensile strength is improved noticeably by increasing the fiber content upto 50%. Depending on the length and fiber content, silk fibers can significantly enhance PBS's static and dynamic mechanical qualities [34]. Gelatin's tensile strength and tensile modulus are increased when blend with silk fiber. Additionally, it also raises the hardness and impact strength of gelatin. Hence can be utilized in potential tissue engineering [35]. When compared to poly (lactide-co-glycolide) (PLGA) scaffolds, electrospun-knitted silk composite scaffolds exhibit better cell adherence, improved cell proliferation, and higher collagen content [36].

9.2.3.1.2 Tasar Silk

Tasar silk, also known as Chinese oak tasar is derived from the wild non-mulberry silkworm species *Antheraea mylitta*. Tasar silk fibroin is a promising and alluring biopolymer for producing biomaterials due to the presence of RGD sequences (Arg-Gly-Asp) enhancing self-assembling capability, adaptable structure, chemical resistance, affordability and biocompatibility [37]. Studies showed that tasar silk and chitosan have combined to generate polyelectrolyte complexes (PECs). These composites seek to combine the mechanical stability of chitosan with the biocompatibility and biodegradability of tasar silk [38]. In comparison to calcium polyphosphate (CPP) alone, tasar silk can be associated with CPP to produce composite scaffolds that exhibit potential for bone repair, and have better compressive strength, degradation behavior and biocompatibility [39]. PCL and tasar silk have been combined to generate nanofiber using electrospinning technique. Due to the RGD tripeptides and PCL's enhanced stability, this blend showed adhesion and growth of PC12 neuron-like cells [40]. Blend of PCL and gelatin with tasar silk has established a novel tasar silk fibroin-based growing medium for retinal pigment epithelial (RPE) cells which is excellent in in vivo and in vitro cytocompatibility for potential application as prosthetic Bruce's membrane in RPE transplantation [41]. Due to the propensity for RGD binding, tasar silk fibroin has been conjugated with poly (ethylenimine)

(PEI) to supply growth factors simultaneously and act as a coating later for PEI/DNA complexes to decrease the PEI's cytotoxicity while increasing transfection effectiveness [42].

9.2.3.1.3 Muga Silk

Muga silk is produced from a non- mulberry silkworm known as *Antheraea assamensis*, which is mainly found in the northeastern region, particularly in Assam, India. Agarose and *Antheraea assama* silk fibroin have been blend to form blended porous hydrogels. These hydrogels are considered to be utilized for cartilaginous tissue. This blend offers adequate elasticity and sustain high quantities of collagen and sulphated glycosaminoglycans and upregulating genes that are specific to cartilage tissues such as collagen type II, aggrecan and sox-8 [43]. Muga silk has been subjected to plasma polypropylene grafting to form a sterilized suture material. The modified muga silk exhibited excellent mechanical properties and wound healing activity [44]. Muga silk fibroin has been modified through plasma-induced grafting of chitosan, followed by penicillin fortification. This blend severed as a good antibacterial and antithrombogenic material [45].

9.2.3.1.4 Eri Silk

Eri silk is derived from silkworm known as *Samia ricini*. It is particularly found in the northeastern states of India including Assam and Meghalaya. These are non-violent silk alternatives which means the silk of the eri cocoon is reeled without harming the pupa. Eri silk is known for excellent moisture absorption capabilities [46]. Wound-dressing mats were prepared from eri silk fibroin blended with PLA by electrospinning techniques [47]. Eri silk fibroin and gelatin are blended to prepare a film using as a model for hydrophilic drug release. Highest drug release rate was observed when compared to native gelatin [48].

9.2.3.2 Regenerated Silk Fibroin

Regenerated silk fibroin (RSF) is a natural silk fibroin that is obtained through a series of chemical and physical alterations. RSF is achieved by separating sericin and water-soluble glycoprotein leading to modification in the silk fiber, thereby increasing its versatility and compatibility with the polymer composites. The process of converting silk fibroin to RSF involves four main steps namely, degumming, dissolution, dialysis, and centrifugation. Degumming is an essential step that separates sericin from fibroin by heating the raw silk in aqueous solution. Dissolution converts insoluble silk fibroin to soluble RSF; this is achieved by either acid dissolution, salt dissolution, or alkaline dissolution methods. Dialysis is done to ensure all the salt ions are completely removed. Centrifugation is performed to separate the RSF [49].

Research showed that concentrated RSF liquid maintains its naive conformation due to surface engineering with PEI and PS which contribute to the viscoelastic properties with controllable phase transition, in order to form new fabrication technique with tunable mechanical properties. Applications of RSF contribute to injection-based writing, compression molding, and shaping RSF bioconjugates into various geometries [50].

9.2.3.3 Recombinant Silk Protein

Recombinant DNA technology is significantly enhancing biomaterial sciences by altering the design and production of next-generation biomaterial domains. The mechanism of generating recombinant silk protein is by genetic engineering that involves inducing artificial or natural genes into the plasmid DNA, followed by introduction into the host cell. This leads to expression of genes that were induced and production of the desired recombinant protein. The customization of the properties includes incorporating non-natural amino acids, functionalization of the sequence, producing hybrid designs, and dynamic stimuli-responsive systems [51]. For example, silk-elastin-like polypeptides (SELPs). SELPs are formed as a type of recombinant blend; these are engineered to combine properties of both elastin and silk. These contain repetitive sequences or elastin and silk domains. Another example is a recombinant protein blend: silk-bacterial collagens (SBCs); they are formed by combining silk with bacterial collagen-like domain [52]. A specific example of SBCs includes Scl2 bacterial collagen-like domain fused with fibronectin binding sites, MMP binding sites, heparin binding sites, or integrin binding sites. Silk and collagen proteins are fused to provide robust structure and help avoid scaffold shrinkage which provide sustained structures for biomedical applications that include artificial cornea, wound dressings, and tissue engineering [53]. Table 9.1 summarizes the types of silk and their blend with different polymers.

9.3 Polymer Matrices Used in Silk Composites

Silk fibers are great reinforcements in composite materials due to their excellent tensile strength, toughness, and biocompatibility. The mechanical strength and biodegradability are frequently integrated into the qualities of a polymer matrix to produce a composite with controllable performance [54].

The matrix phase of reinforced silk composites influences properties such as stiffness, toughness, degradation rate, and sustainability. In general, polymer matrices can be divided into three major categories: synthetic, biodegradable, and natural (bio-based) polymers. Each category has advantages and disadvantages, particularly in terms of strength, processability, and biological activity. Additionally, natural

Table 9.1 Types of silk and their use in polymer composites

Type of silk	Source	Key properties	Polymer composite applications
Mulberry silk (*Bombyx mori*)	Domesticated silkworm (India, China, Japan)	High biocompatibility, mechanical strength, thermal stability	• PLA/silk: tensile strength, thermal stability, biodegradation [33] • PBS/silk: enhanced quasistatic and dynamic mechanical strength up to 50% fiber content [34] • Gelatin/silk: increased tensile strength, hardness, impact strength (tissue engineering) [35] • PLGA/silk: better cell adhesion, proliferation, collagen and cartilage regeneration [36]
Tasar silk (*Antheraea mylitta*)	Wild, non-mulberry silkworm (India, China)	Contains RGD sequences, biocompatible, self-assembling, chemical resistance	• Tasar silk + chitosan: PECs with mechanical stability and biodegradability [38] • Tasar silk + CPP: bone repair scaffolds with better compressive strength and biocompatibility [39] • Tasar silk + PCL: nanofibers for neural growth (PC12 cells) [40] • Tasar silk + gelatin/PCL: scaffold for retinal pigment epithelial (RPE) cells [41] • Tasar silk + PEI: reduced cytotoxicity, improved transfection [42]
Muga silk (*Antheraea assamensis*)	Wild, non-mulberry silkworm (Northeast India, Assam)	Strong, durable, good elasticity	• Agarose + muga silk: porous hydrogels for cartilage tissue [43] • Plasma-grafted muga silk: sterilized suture with wound healing activity [44] • Muga silk + chitosan + penicillin: antibacterial and antithrombogenic material [45]
Eri silk (*Samia ricini*)	Non-violent silk (Northeast India: Assam, Meghalaya)	High moisture absorption, eco-friendly	• Eri silk + PLA: electrospun wound dressing mats [47] • Eri silk + gelatin: hydrophilic drug release films with higher release rate [48]

(continued)

Table 9.1 (continued)

Type of silk	Source	Key properties	Polymer composite applications
Regenerated silk fibroin (RSF)	Chemically and physically processed natural silk	Versatile, tunable mechanical properties, biocompatible	• RSF + PEI/PS: enhanced viscoelastic properties with phase transition control and Applications in injection-based writing, compression molding, bioconjugates for biomedical devices [50]
Recombinant silk proteins	Genetically engineered (host cells with silk genes)	Customizable, hybrid designs, stimuli-responsive	• Silk-Elastin Like Polypeptides (SELPs): combines elasticity and strength and Silk-Bacterial Collagens (SBCs): prevents scaffold shrinkage, supports tissue engineering (cornea, wound dressings, bone, cartilage) [52]

matrices are biocompatible but mechanically weaker; synthetic plastics offer toughness and durability, while biodegradable polymers provide renewability, biocompatibility, and sustainability [55]. The choice of a suitable matrix is also critical for the intended application, such as biomedical implants, structural parts, and green packaging, because each requires different matrix properties [56, 57].

9.3.1 Natural Polymers

Natural polymers derived from organic sources such as plants, microbes, algae, and animals provide excellent biocompatibility and biodegradability, rendering them suitable for biomedical and eco-friendly applications, such as tissue regeneration scaffolds, drug delivery, and imaging (magnetic resonance imaging (MRI), fluorescence imaging, and ultrasound) [58, 59]. The group of natural polymers includes polyesters, proteins, polysaccharides, cellulose, and natural rubber, among others, which are derived from various organic sources [60]. The composite often exhibits enhanced bioactivity and controlled degradation when silk fibroin is incorporated into a natural polymer matrix. A common biopolymer matrix is chitosan, which is a deacetylated form of chitin. It is naturally antibacterial, biocompatible, and helps in blood coagulation. The free amino (–NH_2) group in chitosan facilitates high interfacial adhesion through hydrogen bonding with silk fibroin strands. The strength and stability of the composite are enhanced by this strong bonding. Due to considerable silk fibroin (SF) and chitosan (CS) bonding, CS-SF blend films provide a higher tensile strength and flexibility than silk alone. In tissue scaffolds and wound dressings silk-chitosan composites are used because they promote cell development while preventing bacterial infection [61, 62].

Another well-known natural matrix is Gelatin, sometimes known as denatured collagen. Gelatin creates flexible gels and offers cell-adhesion sequences, such as RGD motifs, which encourage cell adhesion and proliferation. It possesses low immunogenicity and good biocompatibility. Gelatin and silk can be used to increase mechanical compliance and scaffold printability. Gelatin serves as a spacer between silk chains in composite hydrogels, thereby preventing excessive β-sheet formation and maintaining open pores. This improves mechanical toughness and cell infiltration. According to research, 3D-printed silk and gelatin photo-crosslinked hydrogel enhances fibroblast proliferation and wound healing without the use of harsh chemicals [63].

Furthermore, starch-based polymers derived from plants (found in maize, potatoes, etc.) provide biodegradable, affordable, and renewable materials when integrated with silk fibers. These starch-silk composites have been investigated for temporary devices and biodegradable packaging. Silk provides strength, while starch's hydrophilic properties can enhance cell adherence and the rate of deterioration in biomedical scaffolds by creating a moist environment. For instance, calcium phosphate-filled porous starch-silk fibroin scaffolds have been suggested for bone tissue regeneration, exhibiting slow resorption and high cell survival [64]. In conclusion, natural matrices accumulate silk composites, providing bioactive and environmentally beneficial qualities by allowing them to interact with cells and break down gradually after usage [65].

9.3.2 Synthetic Polymers

Synthetic polymer matrices offer superior durability, thermal stability, and process control, rendering them ideal for engineering, industrial, and high-performance applications. In contrast to natural polymers, synthetic variants provide enhanced control over the material properties of composites, resulting in improved toughness, stiffness, and resistance to environmental factors. Commonly synthetic materials include in silk composites are polypropylene (PP), polyethylene (PE), and polyurethane (PU) [66].

PE, a prevalent thermoplastic polymer, is recognized for its excellent impact resistance, lightweight characteristics, and chemical stability. When reinforced with silk fibers, PE composites demonstrate superior uniform resistance and mechanical strength, making them suitable for applications in the automotive, aerospace, and protective sectors. Likewise, PP is another commonly employed thermoplastic from silk reinforcement, which enhances its toughness, stiffness, and fatigue resistance, while preserving its lightweight nature. PP-silk composites are utilized in automotive components, consumer products, and medical prosthetics due to their cost-effectiveness and remarkable durability [67]. PU is a versatile elastic polymer and is frequently combined with silk fibers to produce flexible, high-strength composites. PU-silk composites are widely applied in smart textiles, biomedical implants, and soft robotics, where elasticity, wear resistance, and biocompatibility are essential.

The capacity of synthetic polymers to improve structural reinforcement, long-term stability, and high-impact performance positions makes them an optimal choice for various advanced applications [68].

9.3.3 Biodegradable Polymers

Silk is increasingly being combined with biodegradable, bio-sourced polymers to create genuinely environmentally friendly composites. These polymers decompose in the environment into non-toxic byproducts, rendering them particularly suitable for applications such as biodegradable implants, drug delivery, and eco-friendly packaging solutions. Polycaprolactone (PCL), polylactic acid (PLA), and polyhydroxyalkanoates (PHAs) are notable examples of biodegradable polymers, which are frequently utilized in silk composites [69].

PLA is a translucent, rigid polyester derived from renewable agricultural materials, including sugarcane or corn. It is widely employed in silk composites and also FAD-approved for implants and biodegradable applications. PLA is commonly combined with silk to overcome its hydrophobicity and slow hydrolysis, while preserving mechanical integrity due to its good rigidity, optical clarity, and biocompatibility. Silk-fibroin (SF) enhances wettability, cell adhesion, and tunable degradation in electrospun PLA-SF scaffolds, which are characteristics utilized in meniscus tissue-engineering scaffolds. Additionally, PLA-SF has been confirmed in vivo as a resorbable hernia-repair mesh that lowers adhesions and inflammation in comparison to PLA alone [70].

PCL is a flexible, slowly resorbing aliphatic polyester that is valued for drug administration compatibility and long-term implant stability, and it has a slower deterioration rate than the other resorbable. For tissue engineering and wound care, blending PCL and SF used for produces scaffolds with increased bioactivity and toughness. SF-PCL has been developed into drug-eluting, biodegradable stents that combine local delivery and mechanical support, further helping in wound dressings and regenerative medicine [71, 72].

Microbial polyesters known as PHAs produce completely bio-based, highly biodegradable, and sustainable alternatives to conventional petroleum-based plastics. Beyond medical applications, PHAs are also used in green packaging, highlighting their sustainability profile. PHA when combined with SF typically with osteoconductive fillers like nano-hydroxyapatite shows strong processability and load-bearing potential for tissue scaffolds while maintaining high biodegradability [73, 74].

In general, the incorporation of silk fibers or fibroin into PLA, PCL, and PHA matrices provides composites that better balance stiffness, toughness, cell friendliness, and controlled bioresorption, enabling applications ranging from wound dressings and regenerative scaffolds to resorbable meshes and stents.

9.4 Processing Techniques for Silk-Polymer Composites

9.4.1 Solution Casting and Film Formation

The solution casting technique is widely used in silk-polymers. This technique involves three types of materials: silk fibroin, polymers, and solvents. Silk fibroin was extracted from spider silk after removing sericin from the fibroin. Thus obtained, fibroin and polymers are dissolved in suitable solvents such as water, formic acid, ethanol, and ionic liquids. Followed by the process of filtration to remove the impurities from the solution containing fibroin and polymer. Consequently, the two solutions are mixed in a specific ratio to achieve a homogeneous mixture. The mixture is poured into a flat mold or petri plates and allowed to dry at 60 °C for 48 h (Fig. 9.2). This process produces flexible, thin films that can be used for coatings, sensors, packaging, or biomedical devices while maintaining biocompatibility. To reinforce the materials, researchers have incorporated nano-additives into these films. The result indicated the cellulose nanofibrils with silk fibroin with a tensile strength that is approximately 44% higher than clean silk films, which was performed by the solution casting method. A mixture of silk and another polymer (such as PVA or PLA) is often dialyzed or filtered to eliminate impurities, then cast and allowed to dry gradually. However, it is difficult to control the thickness and completely remove the remaining solvent. These composite films are well-suited for biomedical applications and packaging, particularly when solvents are used to mitigate toxicity.

The solvent casting method has certain limitations, including the presence of solvent residue in the film, the thickness of the films affecting the drying process, and brittleness. To overcome this problem, designing hybrid composites with the

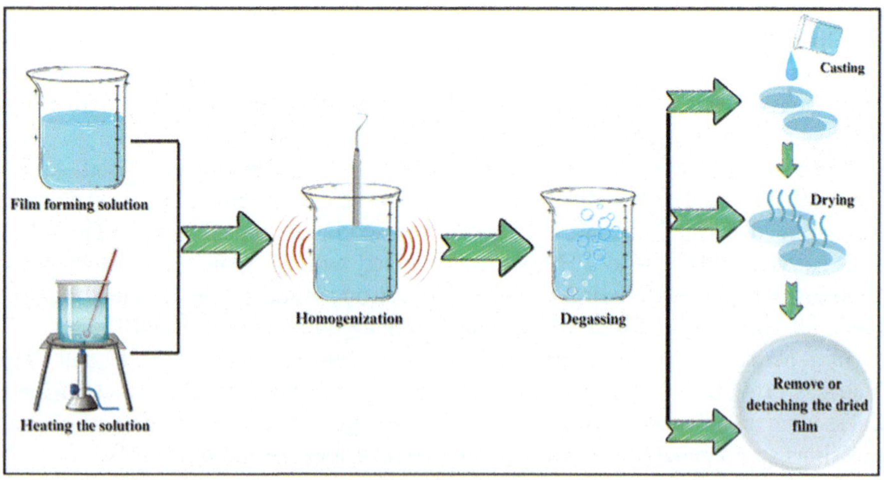

Fig. 9.2 Solution casting and film formation technique (*Source* Canva imaging)

aid of nanoparticles (NPs) enhances mechanical properties and utilizes eco-friendly solvents to mitigate toxicity [75, 76].

9.4.2 *Electrospinning for Nanofiber Composites*

Electrospinning composites are a versatile type of material that offers a range of properties and applications. Numerous materials, including metals, carbon nanotubes, polymers, composite ceramics, bacteria, and viruses, can currently be created or integrated into micro- or nano-fibers either directly through electrospinning or via a post-spinning procedure. The method, characteristics, and applications of electrospinning and electrospun nanofibers have been the subject of numerous publications, including reviews and research studies. The ultra-fine diameter, high surface area to volume ratio, large porosity, high aspect ratio of length to diameter, high mechanical properties due to the high molecular orientation along the fiber axis, and easy-tail nanofiber products are some of the special qualities of electrospun nanofibers and nanofiber materials [77]. Due to their exceptional qualities, electrospun nanofibers and nanofiber materials have been used in a wide range of applications such as pH and temperature responsive materials and sensors, electrospun special and functional polymers for multipurpose protective clothing, high-performance electrospun polymers for creating nanofiber reinforced composites, short electrospun fibers for sponges, and many more.

Silk-polymer-based nanofibers are an ideal product that combines the superior mechanical properties of silk fibers with the potential for enhanced biocompatibility. The electrospinning process involves generating a powerful electric field between a polymer solution held in a container, such as a syringe equipped with a needle of various sizes, which leads to the formation of fibers that are gathered on a metal plate called a collector [78]

Steps of the electrospinning process are as follows:

- Droplet Formation: Surface tension causes the polymer liquid to first form a small ring droplet at the needle tip. A steady droplet or meniscus hangs from the spinneret due to the syringe pump steadily feeding the solution [79].
- Charge Accumulation and Taylor Cone: Charge accumulates on the droplet, and as the voltage rises, the coulombic forces within the external electric field lead to the formation of a sharp cone known as the Taylor cone.
- Jet Ejection and Stretching: A fine charged jet of polymer solution emerges from the Taylor cone apex at a critical voltage. The jet initially moves in a straight line, but small disturbances cause it to bend and whip (a phenomenon known as bending instability). The jet is rapidly stretched and thinned out as it speeds in the direction of the collector. The solvent quickly evaporates, and the polymer hardens during this phase, causing the jet to narrow into a continuous solid fiber.
- Fiber Collection: Finally, the grounded metal collection plate (either a flat plate or revolving drum) is covered with the solidified fiber. The fiber randomly settles on

the collectors as a result of the jet's uncontrolled whipping; in the end, a multilayer mat of nanofibers develops. Electrospun fibers are typically incredibly thin, with diameters ranging from thousands to nanometers to a few micrometers [80].

9.4.2.1 Advantages of the Electrospinning Technique

The main advantages of electrospinning are an extremely high surface-to-volume ratio, a network of interconnected pores for the transportation of nutrients, and adjustable fibre diameter and position orientation [81]. In tissue scaffolds, the high porosity and surface area promote waste/nutrient exchange, providing a large number of binding sites for sensors or catalysts. In the case of aligned electrospun fibres, the modulus difference was 33 times, at high collector speeds, and the fibre direction can be over 30× stiffer than the transverse direction. Due to its directional control and compatibility with a wide range of natural and synthetic polymers, electrospinning is perfect for high-performance filter and regenerative medicine applications, like nerve conduits and vascular grafts. All together due to its exceptional advantages, electrospun silk composites will continue to expand in textile and biological applications in present scenario [82].

9.4.3 Melt Blending and Extrusion

Melt processing can be used to combine silk with thermoplastic polymers for large-scale production. Silk is sheared into powder or short fibers and added to the polymer by heating it above its melting point, and temperatures are maintained below silk deterioration (about 250–300 °C) [83]. After the molten mixture has been homogenized, it is extruded and then shaped using typical techniques (extrusion, spinneret, injection molding) into final forms. This method utilizes existing plastics machinery and can generate kilograms of materials per hour. The toughness and strength of the composite are increased when the silk fibers in the polymer melt align and fuse with the matrix [84].

9.4.3.1 Advantages of Melt Blending and Extrusion

Extrusion and melting mixing are excellent in terms of speed and scalability. Silk can be incorporated into current plastics extrusion lines without requiring any specific equipment modifications, enabling high volume production [85]. The heat can even improve polymer crystallinity and silk-matrix bonding, and the technique usually results in evenly distributed silk within the matrix. As a result, mechanical properties are stable and frequently better than those of the raw polymers. Their versatility ensures compatibility with a variety of applications by enabling the incorporation of a broad range of polymers, including thermoplastics like PLA, PCL, and PEG.

Melt blending and extrusion are crucial techniques for creating high performance, biodegradable, and sustainable polymer-based materials because of their versatility. This leads silk-polymer composites to have applications in different fields, like packaging, textiles, and biomedical [86].

9.4.4 3D Printing Techniques for Silk-Polymer Composites

For silk-polymer composites, 3D printing offers new creative freedom. It is possible to print intricate, specific shapes and structures, which makes it perfect for use in tissue engineering, flexible electronics, and biomedical engineering. Table 9.2 provides a summary of various 3D printing technologies. According to research, bone-mimetic scaffolds have been printed using silk Gelatin-hydroxyapatite bioinks, and unique bioresorbable implants have been printed using silk-PCL filaments [87, 88]. Cell growth can be guided by 3D-printed silk scaffolds that have region-specific porosity or resemble organ morphologies (e.g., stiffer "struts" where load-bearing, more porous sections for vascularization). Due to its rapid digital prototyping and precise architectural control, 3D printing is appealing for on-demand packaging prototypes, flexible electronics, and personalized medicine. All things considered, the potential geometries and functionalities of silk-polymer composites are significantly increased by additive manufacturing [89].

9.4.4.1 Advantages of 3D Printing Technologies

3D printing offers unparalleled precision and intricate design. It can create devices based on a patient's anatomy, for example, using CT (computed tomography) data to create cartilage scaffolds or a skull. It is possible to adjust the local porosity and stiffness of a scaffold by changing the infill and print patterns. According to Wei et al. [97], a silk-gelatin-HA-TCP bioink printed scaffold significantly increased osteogenic gene expression and stem cell proliferation, particularly when coated with growth factors. Additionally, variation is accelerated by digital design, as prototypes swiftly transition from CAD (Computer-Aided Design) to actual parts. These kinds of advantages put 3D printing in a position to transform silk-polymer biomaterials, to generate from smart textiles and customized implants to innovative biodegradable product designs [97]. These composites are essential components of net generation engineering and healthcare solutions because of their exceptional performance, which can be achieved by improving them at the microscopic level [98].

9.4.4.2 Applications of 3D Printing Techniques

3D-printed silk-polymer scaffolds can closely resemble the natural structure of tissues, providing an ideal environment for cell adhesion, growth, and regeneration,

Table 9.2 Different types of 3D printing techniques

Technique	Description	Materials used	Applications	References
Fused deposition modeling (FDM)	In FDM, extrudes a molten thermoplastic filament with a heated nozzle, layer by layer	Thermoplastic filaments (PLA, ABS, and PP)	Progressive methods for producing prototypes, toy industries, automobile parts, and moulds for casting implants in medical fields	[90, 91]
Vat photopolymerization (VP)	VP uses controlled light exposure to solidify liquid resin layer by layer, and creation of 3D objects with high resolution	Resin (Standard, flexible, castable waxes, and biocompatible resins)	Dental models, jewelry, low-run injection molding, and automotive fields	[92, 93]
Powder bed fusion (PBF)	PBF, use thermal energy to create items layer-by-layer through a combination of increased temperature and energy sources	It uses, powdered feedstock material (metal or alloyed powder)	Electronics, automotive, aviation industries, medical implants, and surgical tooling	[94]
Material jetting (MJ)	Process of applying liquid photopolymer droplets to specific region and curing them with UV light to form layers	Photopolymer resin, waxes, and come metals, ceramics	Electronics, solar cells and battery, pacemakers for heart and brain stimulation, and cost-effective wearable sensors	[95, 96]

making them highly suitable for application in regenerative medicine. Additionally, these composites are being integrated into drug-delivery systems, where 3D-printed structures enable controlled and sustained drug release, providing unique treatment solutions tailored to individual patients. The use of advanced manufacturing techniques also extends to the development of specialized medical implants, prosthetics, and surgical tools, thereby improving precision and enabling patient-specific adaptations.

Beyond biomedicine, silk-polymer composites are gaining popularity as sustainable packaging materials. Biodegradable films are produced using an advanced production process that offers excellent barrier properties, high strength, and environmentally friendly degradation. These films help reduce environmental waste while providing an alternative to traditional plastic packaging. Additionally, 3D printing technology is being used to create personalized, environmentally friendly containers, enabling creative and sustainable packaging solutions that reduce waste and its negative effects on the environment [99, 100]. The relationship between sustainability, biomedicine, and silk-polymer composites is expected to drive groundbreaking developments in environmental preservation and healthcare as this field of research progresses [101].

9.5 Characterization and Analysis of Silk-Polymer Composites

Silk-polymer composites are characterized and tested using a variety of sophisticated methods to evaluate their mechanical, structural, thermal, and biological characteristics. The microstructure, crystallinity, and chemical composition of the composite are examined using structural analytical methods such as Fourier-transform infrared spectroscopy (FTIR), X-ray diffraction (XRD), and scanning electron microscopy (SEM) to ensure appropriate material formation. To determine the composites' sustainability and performance under various stress conditions, mechanical attributes like tensile strength, compression resistance, and flexural strength are assessed through extensive testing.

Thermal analysis, such as thermogravimetric analysis (TGA) and differential scanning calorimetry (DSC), offers information about the composites phase transitions, decomposition patterns, and thermal stability, all of which are crucial for applications in high-temperature settings. Silk-polymer composites are optimized and safely applied in a variety of industries through these thorough characterization techniques [102, 103].

9.5.1 Mechanical Properties of Silk-Polymer Composites

Silk is distinguished from other natural fibers due to its extraordinary toughness and high mechanical strength. Numerous factors, including the type of silk protein used, the polymer matrix selected, and the processing technique employed during manufacturing, significantly impact the mechanical performance of silk-polymer composites [104]. Furthermore, the effectiveness of stress transfer in the composite is greatly influenced by the interfacial compatibility of the silk fibers and the polymer matrix. The overall mechanical behavior is directly impacted by the strength of the fiber-matrix interfacial zone, which controls the capacity to distribute and sustain applied stress. As a result, silk-based composites can have a broad range of mechanical properties, from increased tensile strength and modulus to better toughness and elasticity, due to the special fusion of silk's inherent molecular structure with customized polymer matrices [105]. Silk-polymer composites mechanical properties can be assessed through these parameters as mentioned below.

(i) Tensile Strength

A material's tensile strength indicates the ability to resist deformation, when under stress condition. It is an essential parameter for the materials used in load-bearing applications, like shields, biomedical implants, and structural components. In silk-polymer composites, the interfacial bonding between the silk fibers and the polymer matrix affects the tensile strength [106].

(ii) Toughness

Toughness is a significant mechanical characteristic for many applications, which is the capacity of a material to absorb energy and absorb plastic deformation without breaking. Toughness in silk-polymer composites depends on a number of variables, such as the micro structure of composite, the interfacial bonding between the two phases, and the fundamental properties of the silk and polymers [107].

(iii) Elasticity

The molecular structure of silk, which is made up of semi-crystalline protein chains, is largely responsible for its elastic properties. The material's amorphous region offers flexibility, while the crystalline regions, made up of beta-sheets, contribute to its strength and stiffness. This elastic behavior can differ based on the type of silk and the relative proportion of these (amorphous and crystalline) regions [108, 109].

9.5.2 Factors Affecting the Properties of Silk-Polymer Composites

To attain the intended quality, a number of variables that affect the characteristics of silk-polymer composites must be carefully managed. The main elements influencing the characteristics of silk-polymer composites are covered in detail below.

(i) Material Composition

The type of silk (such as *Bombyx mori* or wild silk) and polymer have a significant impact on composition. The native strength of *Bombyx mori* silk is quite high (3.8 GPa) and its biodegradation is predictable, but wild silks (*Antheraea mylitta*) may differ in their amino acid sequences or thermal stability. Similarly, a soft elastomeric polymer (like PCL or elastane) will produce a different composite behavior than a stiff polymer (like nylon or polystyrene) [110–112].

The selection of the polymer matrix, which states the mechanical, chemical, and biodegradation properties of the composite, is crucial; due to their adjustable crystallinity, hydrophobicity, and rates of degradation, synthetic polymers, like PLA,

PCL, and PEG, are frequently utilized in silk-polymer composites. While PCL offers greater flexibility and slower degradation, making it appropriate for long-term applications like soft tissue engineering, PLA offers high strength and biodegradability, making it perfect for medical implants [113].

(ii) Biological Properties

Biocompatibility and degradation rate are important biological factors for biomedical implants. Although the polymer matrix (such as PLA, PCL, or PVA) must be non-toxic and ideally medical-grade to ensure safety. Silk when incorporated along with the polymer must be biocompatible and suitable for implantable devices [114]. Their combined rate of degradation must be controlled due to silk's enzymatic degradation property which is typically slow. The overall degradation profile can be adjusted by selecting a polymer and processing parameters. In order to add additional function (such as osteoinductive or antibacterial), researchers also include bioactive compounds (drugs, growth factors) into the silk matrix [115]. Prior to any medical applications, compatibility and safety tests (such as hemocompatibility, cytotoxicity, and animal studies) are usually carried out [116, 117].

(iii) Environmental Factors

Environment has a significant impact on the properties and stability of silk-polymer composites. Silk is hydrophilic in nature and is hardened when exposed to air; the quality of the silk composite is influenced by moisture and humidity. Silk is frequently chemically crosslinked or added with less hygroscopic polymers to stabilize the structure. Another main environmental factor is temperature, since high temperatures can break down (glass transition around 170–180 °C thermal degradation around 230–265 °C) silk proteins and change the polymer matrix structural integrity; therefore, lower temperatures are needed for thermal processing [118, 119]. In order to increase the heat resistance of silk-based materials, careful polymer selection and thermal stabilization techniques are frequently used [120, 121]. Effectively controlling these environmental variables ensures that silk-polymer composites will continue to function and have structural integrity in a variety of applications.

(iv) Thermal Properties

Silk-polymer composites performed in different applications are greatly influenced by their thermal characteristics. A crucial component is thermal stability, which is affected by the polymer matrix and silk degradation temperatures. While synthetic polymers have varying thermal stabilities based on their composition, silk usually starts to break down at 250–300 °C. Another important consideration is the glass transition temperature (Tg), which affects the composites' mechanical behavior at different temperatures. A greater Tg improves dimensional stability and increases the composites' resistance to heat-induced deformation [122].

Furthermore, the arrangement and distribution of silk fibers within the polymer matrix have a major impact on the composite's thermal conductivity. While a random distribution may result in different thermal properties, proper fibre alignment can enhance heat dissipation. Optimising silk-polymer composites for uses requiring controlled heat transfer, thermal insulation, or high-temperature resistance requires an understanding of these thermal properties.

(v) Morphology and Microstructure

The mechanical, thermal, and biological properties of silk polymer composites are significantly influenced by their morphology and microstructure. Fibre introduction is one of the crucial components. While randomly organised silk fibers offer isotropic qualities that more evenly distributed in mechanical stress, aligned silk fibers improves directional strength and load-bearing capacity, making them perfect for structural applications. The permeability is another crucial factor that influences the composites surface area, permeability, and ability to interact with biological tissues [123].

In tissue engineering scaffolds and filtration systems, controlled porosity is important where a balance between fluid permeability and mechanical strength is maintained. Furthermore, the strength, thermal stability, and rate of degradation of the composite are all strongly influenced by the crystallinity of the silk and polymer phases. While greater crystallinity increases stiffness and durability, it may also decrease flexibility, which is important for biomedical applications, like drug-delivery systems and biodegradable implants. Modifying silk-polymer composites according to particular functional needs requires an understanding and attention to these microstructural parameters [124, 125].

9.6 Applications of Silk-Polymer Composites in Various Fields

The unique combination of high strength, flexibility, biocompatibility, and biodegradability has led to the widespread use of silk-enhanced polymer composites in various industries. Silk composites being environmentally sustainable find a wide range of applications in biomedical engineering, textiles, and high-performance structures.

9.6.1 Biomedical Applications

Biomedical engineering has been transformed by silk-enhanced polymer composites because of their superior mechanical qualities, controlled degradation, and biocompatibility. They are used in healthcare and medicine for the following applications.

9.6.1.1 Tissue Regeneration

Tissue Regeneration Scaffolds: Silk fibroin scaffolds are used in tissue engineering to stimulate cell development by simulating the extracellular matrix. In particular, the bone tissue scaffolds promote cell adhesion, proliferation, and differentiation by being highly porous and biocompatible. Osteogenic (bone-forming) responses are further enhanced by silk-collagen or silk-hydroxyapatite composites. According to studies, stem cell-containing silk-based scaffolds significantly boost bone mineralization and alkaline phosphatase activity in in vitro and rapid bone healing in animal models [126]. These results highlight the applicability of silk fibroin for implants that promote bone repair. Additionally, silk scaffolds can be designed into absorbable screws and plates or doped with growth hormones (such as BMP-2), these devices are biodegradable and gradually transmit load to healing tissue, to prevent the need for removal surgery [127]. Silk fibroin with imidazole framework-8 gave the excellent antibacterial activity, making it a promising component for many applications such as tissue engineering and other medical treatment [128].

Silk Regeneration: Silk-polymer dressing promotes wound healing and skin renewal without causing inflammation. Silk fibroin promotes angiogenesis and the creation of new tissue and is extremely biocompatible and biodegradable. Because of their pro-angiogenic and anti-inflammatory properties, silk fibroin dressings have been demonstrated to fast wound closure [129]. Thus, the FAD-approved SERI Surgical Scaffold from Allergan is a silk-based mesh used for breast reconstruction and other soft tissue repairs. SERI offers temporary structural support while the patient's tissue develops. Over the months, the silk fibroin slowly breaks down and is replaced by the patient's own tissue [130].

9.6.1.2 Drug-Delivery Systems

Silk composites allow for the regulated release of medicines in drug-delivery systems. Drugs can be encapsulated in silk fibroin and released over time as films, hydrogels, microspheres, and nanoparticles [131]. In comparison to many polymer carriers, these systems can be adjusted for size and rate of degradation, resulting in sustained release profiles. Silk carriers have been investigated in oncology to administer chemotherapy directly to tumors, minimizing systemic damage. In this regard, silk-based delivery systems have been developed for chemotherapeutics and other cancer medications, offering precise and regulated administration [132]. Transdermal patches to treat insomnia, the Thai biotech start-up EngineLife (Chulalongkorn University) has created a silk fibroin "Keep On Sleep" patch that delivers cannabidiol (CBD) is another application of silk composites. Thai authorities have approved this patch, which maintains therapeutic medication levels over hours by utilizing silk slow-release qualities. In general, silk is a desirable platform for vaccines, hormones, and sensitive biologics due to its durability and adjustable breakdown; silk films or microneedles guarantee that proteins stay active until they are released [133].

9.6.1.3 Biosensors and Medical Implants

Silk composites are also used in biodegradable implants and medical sensors. Silk fibroin substrates have been used to create flexible electrical sensors that combine conformability and biocompatibility, such as those that measure pH, temperature, or strain. Pure silk fibroin-based wearable piezoresistive sensors manufactured have been able to track human movement (such as bending a knuckle or straining a muscle) and then disintegrate safely after use [134, 135]. Similarly, silk has been used to create bioabsorbable implants including screws and bone fixation plates. According to many studies, silk plates and screws have great mechanical strength comparable to that of their metal counterparts. These implants helped in the development of bone initially, followed by disintegration into amino acids, preventing chronic inflammation and the need for second surgeries for removal [136, 137].

Innovative methods integrate silk with current implant technology. In this case, Polytech Health and AMSilk, a biotech company, collaborated to coat silicone implants (breast and soft tissue implants) with a recombinant spider-silk https://www.amsilk.com/polytech-and-amsilk-announce-start-of-joint-international-clinical-study/. The very tiny layer of silk protein makes a natural interface that is considered safe for the body. Compared to uncoated silicone, silk-coated SILKline® implants demonstrated better tolerability and a significantly reduced rate of fibroin capsule formation in preclinical investigations [138].

9.6.2 Textile and Structural Materials

The use of silk-reinforced polymer composites is expanding into high-performance textiles, automotive, and aerospace industries. Silk fiber has the ability to improve thermal and acoustic insulation and lighten composite products in the automotive and aerospace industries. In comparison to traditional laminates, experimental panels that combine silk with carbon or glass fibers have demonstrated strong tensile and flexural strength at a lower weight, which makes them appealing for use in many structural applications [139]. The inherent thermal stability of silk also makes textiles and composites more fire-resistant [140]. In a similar way, incorporating silk into synthetic fibers improves the comfort and longevity of sportswear; it keeps the wearer warm in winter (away from moisture) and cool and airy in summer. Silk-polymer layers give protective fabrics, toughness (military or industrial gear), and flame retardance. In particular, silk amino acid composition naturally gives it higher thermal stability and char formation than cotton, which improves heat protection in workwear or uniforms. All things considered, silk composites offer clothing that is comfortable, light, and robust in harsh environments [141].

9.6.3 Electronics and Optoelectronics

The usage of silk fibroin in flexible electronics and displays has increased due to its transparency and flexibility. Silk sheets can be used as biocompatible wearable device substrates, and it is perfect for foldable or flexible displays due to its high optical transmittance and mechanical durability [142]. The silk transparency provides researchers to create silk optical components (lenses or waveguides) and demonstrates silk-based e-link displays and light-emitting devices. Additionally, an electrode made of silk film applied to the skin can conduct electronic ink. It is significant that silk may be transformed using solution or printing procedures into a variety of shapes (fibers, films, sponges, and hydrogels) that allow for integration with conductive inks or nanoparticles [143]. Furthermore, silk-based devices, such as sensors and biodegradable batteries, can function and then innocuously dissolve. This transient electronics strategy is being investigated for environmental sensors and medicinal implants.

9.6.4 Environmental Applications

Silk-polymer composites help to purify water and make plastics more sustainable. In biodegradable packaging, silk is used to replace or reinforce traditional films. Silk fibroin can create flexible, robust films that are good for food packaging due to low permeability to water vapor and oxygen [144, 145]. According to recent assessments, Silk fibroin is a promising biopolymer for packaging because of its superior barrier qualities and biodegradability [146]. Silk-cellulose or silk-chitosan composites are used as crop wraps and mulching films that biodegrade after a period of time, preventing plastic residue in the soil.

Silk works well as an absorbent in water filtration and purification because of its large surface area and functional groups. Silk and cellulose nanofibril membranes that eliminate a variety of pollutants from water, such as heavy metals and PFAS (polyfluoroalkyl substances that are extremely persistent in the environment). Compared to synthetic resins, the material is more eco-friendly due to its natural origin and low toxicity byproducts [147]. All above reasons considered, absorbents and silk-polymer membranes are becoming eco-friendly filtration medium for portable filters and water treatment facilities [148].

9.6.5 Cosmetics

Silk-polymers improve texture and skin compatibility in cosmetics and personal care products. Silk fibroin is utilized to deliver active ingredients and enhance moisture retention in creams, serums, and masks. It provides sustained hydration by binding water with the support of its amino acids. The anti-inflammatory and antioxidant qualities of silk offer good health for skin. Review on wound healing shows that silk fibroin speeds up tissue repair and lowers inflammation, which may have comparable calming effects in skincare [149]. Additionally, silk microparticles act as transporters for peptides or vitamins, protecting the active ingredients and releasing them gradually onto the skin to increase efficacy. Cosmetics that feel light and non-greasy and have better ingredient stability provide good health for skin. Silk-polymer composites are therefore becoming more popular as clean ingredients in lotions, cosmetics, and hair care products for consumers looking for environmentally responsible formulas [150].

9.6.6 Art and Conservation

In modern times, cultural heritage is preserved and restored using materials derived from silk. On the other side, petroleum-derived polymers, such as PVC or acrylic glues, are frequently used in traditional conservation. Over a period of time, these polymers can discolor, break, or cause chemical harm. On the other hand, silk fibroin provides a colorless, reversible, and mechanically robust protein-based consolidate. For instance, silk adhesive has been used to re-bond fibers in weakened silk fabrics (such as antique tapestries or clothing) without changing their appearance [151]. Silk helps artifacts age (anything over 50 years old) more evenly because it is stable over time and compatible with natural fibers (linen, cotton, and wool). In conclusion, silk-polymer systems are making it possible to apply gentle, more resilient conservation treatments, protecting fabrics and artwork for upcoming generations without leaving harmful residues behind.

9.7 Challenges and Limitations of Silk-Polymer Composites

Silk-polymer composites still face practical challenges in spite of these advancements, such as

- Variation due to silk sources, such as diverse silk species leading to batch-to-batch variations.
- Complicated processes of extraction and purification (degumming, solvent exchange, sterilization).
- Scale up process needs to be optimized.
- Cost of production continues to be major obstacles.

At present medical-grade silk fibroin is more expensive than many petrochemical polymers due to the complicated processes of extraction and purification. Justified by industry report stating the pilot facility could only produce 30,000 doses of silk fibroin solution annually, emphasizing significant investment requirement to scale up and standardization of production process. Variability in silk sources, such as diverse silk species and batch-to-batch variations, makes regulatory approval and uniformity more difficult Medical Channel [152].

Additionally, the performance of silk composites is limited. Silk-polymers are susceptible to humidity due to their tendency to absorb moisture from the environment, leading to plasticization, swelling, and diminished mechanical strength [153]. Unmodified silk films might not last long outside or in high humidity without protective coatings. Similarly, regulating the rate of deterioration is challenging such as a biomedical implant needs to be strong for a few weeks or months before dissolving, and adjusting the breakdown of silk (vis crystallinity or crosslinking) involves careful processing [154, 155]. In order to achieve load-bearing requirements, hybridization blending with strong fibers or resins is often required because silk strength is sometimes inferior to premium synthetics [156]. Safety and regulatory testing take time and expense, especially when it comes to food and medicine. Current research in nanocomposite additives, advanced production, and bioengineering is making silk composite performance better, but these problems need to be addressed before it can be widely accepted.

9.8 Conclusion and Future Perspectives

Silk-polymer composites extend the boundaries of innovative materials by combining synthetic and natural materials. The quick development of fabrication technology is a significant trend. Researcher have demonstrated the print of silk fibre structures into intricate lattices, leading to advancement of 3D printing silk-based bioinks used to create layer-by-layer tissue scaffolds or implants with unique shapes. Future developments include 4D printing, which involves creating objects that alter their shape or characteristics over time. Silk matrices may be designed to expand or contract in response to external stimuli, such as heat or pH, allowing for the creation of smart textiles or self-adjusting implants. Similarly, fine silk sheets and meshes (with pore diameters ranging from nanometres to microns) are still produced via electrospinning and nanopatterning for use in tissue engineering, sensors, and filtration. Silk composites are anticipated to have a significant impact on coatings and membranes in nanotechnology. Nanofibers and nanoparticles made from silk can strengthen packaging films or provide antibacterial properties. Nanostructured silk filters could be scaled up for industrial usage in water treatment. Silk nanocomposites doped with magnetic or conductive nanoparticles hold promise for wearable power sources and flexible circuits in electronics. These developments could lead to greener electronics and packaging, including gadgets that is easily biodegradable. Hence, silk enhanced polymers are sustainable biocompatible material with wide range of applications from textiles, bioremediation, electronics and medical sector.

AI Usage Declaration

AI tools were used solely for language editing, grammar correction, and improving clarity and readability. The authors take full responsibility for the originality, accuracy, integrity, and scientific content of this work, including all data interpretation and conclusions.

References

1. Noorunnisa Khanam P, Al-Maadeed MA, Naseema Khanam P (2015) Silk as a reinforcement in polymer matrix composites. In: Advances in silk science and technology. Elsevier, pp 143–170. https://doi.org/10.1016/B978-1-78242-311-9.00008-2
2. Altman GH, Diaz F, Jakuba C, Calabro T, Horan RL, Chen J, Lu H, Richmond J, Kaplan DL (2003) Silk-based biomaterials. Biomaterials 24(3):401–416. https://doi.org/10.1016/S0142-9612(02)00353-8
3. Phiri R, Mavinkere Rangappa S, Siengchin S, Oladijo OP, Dhakal HN (2023) Development of sustainable biopolymer-based composites for lightweight applications from agricultural waste biomass: a review. Adv Ind Eng Polym Res 6(4):436–450. https://doi.org/10.1016/j.aiepr.2023.04.004
4. Liu Y, Ma Y, Xue C, Li X (2025) Effects of silk fibroin incorporation on the mechanical properties and osteogenic activity of bone repair materials. Med Novel Technol Dev 28:100399. https://doi.org/10.1016/j.medntd.2025.100399
5. Azimi B, Maleki H, Zavagna L, De la Ossa JG, Linari S, Lazzeri A, Danti S (2020) Bio-based electrospun fibers for wound healing. J Funct Biomater 11(3):67. https://doi.org/10.3390/jfb11030067
6. Low JT, Yusoff NISM, Othman N, Wong T, Wahit MU (2022) Silk fibroin-based films in food packaging applications: a review. Comprehens Rev Food Sci Food Saf 21(3):2253–2273. https://doi.org/10.1111/1541-4337.12939
7. Cao M, Zeng B, Zheng Y, Guo S (2023) Biocompatible shape-memory poly(propylene carbonate)/silk fibroin blend with body temperature responsiveness. RSC Adv 13(19):13120–13127. https://doi.org/10.1039/D3RA00670K
8. Vyas C, Zhang J, Øvrebø Ø, Huang B, Roberts I, Setty M, Allardyce B, Haugen H, Rajkhowa R, Bartolo P (2021) 3D printing of silk microparticle reinforced polycaprolactone scaffolds for tissue engineering applications. Mater Sci Eng C 118:111433. https://doi.org/10.1016/j.msec.2020.111433
9. Choi JH, Kim DK, Song JE, Oliveira JM, Reis RL, Khang G (2018) Silk fibroin-based scaffold for bone tissue engineering, pp 371–387. https://doi.org/10.1007/978-981-13-0947-2_20
10. Wongpinyochit T, Uhlmann P, Urquhart AJ, Seib FP (2015) PEGylated silk nanoparticles for anticancer drug delivery. Biomacromolecules 16(11):3712–3722. https://doi.org/10.1021/acs.biomac.5b01003
11. Hardy JG, Scheibel TR (2010) Composite materials based on silk proteins. Prog Polym Sci 35(9):1093–1115. https://doi.org/10.1016/j.progpolymsci.2010.04.005
12. Kiseleva AP, Krivoshapkin PV, Krivoshapkina EF (2020) Recent advances in development of functional spider silk-based hybrid materials. Front Chem 8:554. https://doi.org/10.3389/fchem.2020.00554
13. Sen K, Babu KM (2004) Studies on Indian silk. I. Macrocharacterization and analysis of amino acid composition. J Appl Polym Sci 92(2):1080–1097. https://doi.org/10.1002/app.13609
14. Ahsan F, Ansari T, Usmani S, Bagga P (2018) An insight on silk protein sericin: from processing to biomedical application. Drug Res 68(06):317–327. https://doi.org/10.1055/s-0043-121464

15. Zhang W, Fan Y (2021) Structure of animal silks. In: Fibrous Proteins. Methods in Molecular Biology, vol 2347. Humana, New York, NY, pp 3–15. https://doi.org/10.1007/978-1-0716-1574-4_1
16. Seo SJ, Das G, Shin HS, Patra JK (2023) Silk sericin protein materials: characteristics and applications in food-sector industries. Int J Mol Sci 24(5):4951. https://doi.org/10.3390/ijms24054951
17. Qiu W, Patil A, Hu F, Liu XY (2019) Hierarchical structure of silk materials versus mechanical performance and mesoscopic engineering principles. Small 15(51). https://doi.org/10.1002/smll.201903948
18. Tu H, Yu R, Lin Z, Zhang L, Lin N, Yu WD, Liu XY (2016) Programing performance of wool keratin and silk fibroin composite materials by mesoscopic molecular network reconstruction. Adv Func Mater 26(48):9032–9043. https://doi.org/10.1002/adfm.201603403
19. Koh L-D, Cheng Y, Teng C-P, Khin Y-W, Loh X-J, Tee S-Y, Low M, Ye E, Yu H-D, Zhang Y-W, Han M-Y (2015) Structures, mechanical properties and applications of silk fibroin materials. Prog Polym Sci 46:86–110. https://doi.org/10.1016/j.progpolymsci.2015.02.001
20. Du N, Yang Z, Liu XY, Li Y, Xu HY (2011) Structural origin of the strain-hardening of spider silk. Adv Func Mater 21(4):772–778. https://doi.org/10.1002/adfm.201001397
21. Liu R, Deng Q, Yang Z, Yang D, Han M, Liu XY (2016) "Nano-Fishnet" structure making silk fibers tougher. Adv Func Mater 26(30):5534–5541. https://doi.org/10.1002/adfm.201600813
22. Xu G, Gong L, Yang Z, Liu XY (2013) What makes spider silk fibers so strong? From molecular-crystallite network to hierarchical network structures. Soft Matter 10(13):2116–2123. https://doi.org/10.1039/C3SM52845F
23. Walter A (2024) The function of web decorations in orb web spiders. Front Arachnid Sci 3:1384128. https://doi.org/10.3389/frchs.2024.1384128
24. Hakimi O, Knight DP, Vollrath F, Vadgama P (2007) Spider and mulberry silkworm silks as compatible biomaterials. Compos B Eng 38(3):324–337. https://doi.org/10.1016/j.compositesb.2006.06.012
25. Brough HDA, Cheneler D, Hardy JG (2024) Progress in multiscale modeling of silk materials. Biomacromolecules 25(11):6987–7014. https://doi.org/10.1021/acs.biomac.4c01122
26. Gosline JM, Guerette PA, Ortlepp CS, Savage KN (1999) The mechanical design of spider silks: from fibroin sequence to mechanical function. J Exp Biol 202(23):3295–3303. https://doi.org/10.1242/jeb.202.23.3295
27. Shao Z, Young RJ, Vollrath F (1999) The effect of solvents on spider silk studied by mechanical testing and single-fibre Raman spectroscopy. Int J Biol Macromol 24(2–3):295–300. https://doi.org/10.1016/S0141 8130(98)00093 2
28. Li M, Ogiso M, Minoura N (2003) Enzymatic degradation behavior of porous silk fibroin sheets. Biomaterials 24(2):357–365. https://doi.org/10.1016/S0142-9612(02)00326-5
29. Asakura T (2021) Structure of Silk I (Bombyx mori silk fibroin before spinning)—Type II β-turn, not α-Helix-. Molecules 26(12):3706. https://doi.org/10.3390/molecules26123706
30. Zhang Y-Q (2002) Applications of natural silk protein sericin in biomaterials. Biotechnol Adv 20(2):91–100. https://doi.org/10.1016/S0734-9750(02)00003-4
31. Mishra N, Hazarika NC, Narain K, Mahanta J (2003) Nutritive value of non-mulberry and mulberry silkworm pupae and consumption pattern in Assam, India. Nutr Res 23(10):1303–1311. https://doi.org/10.1016/S0271-5317(03)00132-5
32. Bitar L, Isella B, Bertella F, Bettker Vasconcelos C, Harings J, Kopp A, van der Meer Y, Vaughan TJ, Bortesi L (2024) Sustainable Bombyx mori's silk fibroin for biomedical applications as a molecular biotechnology challenge: a review. Int J Biol Macromol 264:130374. https://doi.org/10.1016/j.ijbiomac.2024.130374
33. Cheung H-Y, Lau K-T, Tao X-M, Hui D (2008) A potential material for tissue engineering: silkworm silk/PLA biocomposite. Compos B Eng 39(6):1026–1033. https://doi.org/10.1016/j.compositesb.2007.11.009
34. Lee S, Cho D, Park W, Lee S, Han S, Drzal L (2005) Novel silk/poly(butylene succinate) biocomposites: the effect of short fibre content on their mechanical and thermal properties. Compos Sci Technol 65(3–4):647–657. https://doi.org/10.1016/j.compscitech.2004.09.023

35. Shubhra QTH, Alam AKMM, Beg MDH (2011) Mechanical and degradation characteristics of natural silk fiber reinforced gelatin composites. Mater Lett 65(2):333–336. https://doi.org/10.1016/j.matlet.2010.09.059
36. Wang G, Hu X, Lin W, Dong C, Wu H (2011) Electrospun PLGA–silk fibroin–collagen nanofibrous scaffolds for nerve tissue engineering. In Vitro Cell Dev Biol Anim 47(3):234–240. https://doi.org/10.1007/s11626-010-9381-4
37. Silva SS, Kundu B, Lu S, Reis RL, Kundu SC (2019) Chinese Oak Tasar Silkworm *Antheraea pernyi* Silk Proteins: current strategies and future perspectives for biomedical applications. Macromol Biosci 19(3). https://doi.org/10.1002/mabi.201800252
38. Bissoyi A, Kumar Singh A, Kumar Pattanayak S, Bit A, Kumar Sinha S, Patel A, Jain V, Kumar Patra P (2017) Understanding the molecular mechanism of improved proliferation and osteogenic potential of human mesenchymal stem cells grown on a polyelectrolyte complex derived from non-mulberry silk fibroin and chitosan. Biomed Mater 13(1):015011. https://doi.org/10.1088/1748-605X/aa890c
39. Xie H, Gu Z, Li C, Franco C, Wang J, Li L, Meredith N, Ye Q, Wan C (2016) A novel bioceramic scaffold integrating silk fibroin in calcium polyphosphate for bone tissue-engineering. Ceram Int 42(2):2386–2392. https://doi.org/10.1016/j.ceramint.2015.10.036
40. Li X, Zhang Q, Ye D, Zhang J, Guo Y, You R, Yan S, Li M, Qu J (2017) Fabrication and characterization of electrospun PCL/*Antheraea pernyi* silk fibroin nanofibrous scaffolds. Polym Eng Sci 57(2):206–213. https://doi.org/10.1002/pen.24402
41. Xiang P, Wu K-C, Zhu Y, Xiang L, Li C, Chen D-L, Chen F, Xu G, Wang A, Li M, Jin Z-B (2014) A novel Bruch's membrane-mimetic electrospun substrate scaffold for human retinal pigment epithelium cells. Biomaterials 35(37):9777–9788. https://doi.org/10.1016/j.biomaterials.2014.08.040
42. Ma C, Lv L, Liu Y, Yu Y, You R, Yang J, Li M (2014) *Antheraea pernyi* silk fibroin for targeted gene delivery of VEGF165-Ang-1 with PEI. Biomed Mater 9(3):035015. https://doi.org/10.1088/1748-6041/9/3/035015
43. Singh YP, Bhardwaj N, Mandal BB (2016) Potential of agarose/silk fibroin blended hydrogel for in vitro cartilage tissue engineering. ACS Appl Mater Interfaces 8(33):21236–21249. https://doi.org/10.1021/acsami.6b08285
44. Gogoi D, Choudhury AJ, Chutia J, Pal AR, Khan M, Choudhury M, Pathak P, Das G, Patil DS (2014) Development of advanced antimicrobial and sterilized plasma polypropylene grafted muga (*Antheraea assama*) silk as suture biomaterial. Biopolymers 101(4):355–365. https://doi.org/10.1002/bip.22369
45. Choudhury AJ, Gogoi D, Kandimalla R, Kalita S, Chaudhari YB, Khan MR, Kotoky J, Chutia J (2016) Penicillin impregnation on oxygen plasma surface functionalized chitosan/Antheraea assama silk fibroin: studies of antibacterial activity and antithrombogenic property. Mater Sci Eng C 60:475–484. https://doi.org/10.1016/j.msec.2015.11.070
46. Zhou B, Wang H (2020) Structure and functions of cocoons constructed by Eri silkworm. Polymers 12(11):2701. https://doi.org/10.3390/polym12112701
47. Shanmugam K, Sundaramoorthy S (2015) Development and characterization of an electrospun mat from Eri silk fibroin and PLA blends for wound dressing application. RSC Adv 5(40):31352–31364. https://doi.org/10.1039/C4RA15268A
48. Srisuwan Y, Kotseang N, Namtaku K, Simchuer W, Butiman C, Srihanam P (2012) Preparation of Eri silk fibroin and gelatin blend film loaded chlorhexidine using as model for hydrophilic drug release. Nat Sci 04(07):454–460. https://doi.org/10.4236/ns.2012.47061
49. Huang L, Shi J, Zhou W, Zhang Q (2023) Advances in preparation and properties of regenerated silk fibroin. Int J Mol Sci 24(17):13153. https://doi.org/10.3390/ijms241713153
50. Kumar R, Joseph E, Chaudhari A, Nisal A, Sharma KP (2022) Viscoelastic liquid from regenerated silk fibroin in the silk I conformation: a writeable and shapeable material. ACS Appl Polym Mater 4(7):4699–4708. https://doi.org/10.1021/acsapm.2c00290
51. Huang W, Rollett A, Kaplan DL (2015) Silk-elastin-like protein biomaterials for the controlled delivery of therapeutics. Expert Opin Drug Deliv 12(5):779–791. https://doi.org/10.1517/17425247.2015.989830

52. Dinjaski N, Kaplan DL (2016) Recombinant protein blends: silk beyond natural design. Curr Opin Biotechnol 39:1–7. https://doi.org/10.1016/j.copbio.2015.11.002
53. An B, DesRochers TM, Qin G, Xia X, Thiagarajan G, Brodsky B, Kaplan DL (2013) The influence of specific binding of collagen–silk chimeras to silk biomaterials on hMSC behavior. Biomaterials 34(2):402–412. https://doi.org/10.1016/j.biomaterials.2012.09.085
54. Ranakoti L, Gangil B, Rajesh PK, Singh T, Sharma S, Li C, Ilyas RA, Mahmoud O (2022) Effect of surface treatment and fiber loading on the physical, mechanical, sliding wear, and morphological characteristics of tasar silk fiber waste-epoxy composites for multifaceted biomedical and engineering applications: fabrication and characterizations. J Mater Res Technol 19:2863–2876. https://doi.org/10.1016/j.jmrt.2022.06.024
55. Reddy MSB, Ponnamma D, Choudhary R, Sadasivuni KK (2021) A comparative review of natural and synthetic biopolymer composite scaffolds. Polymers 13(7):1105. https://doi.org/10.3390/polym13071105
56. Jha S, Akula B, Enyioma H, Novak M, Amin V, Liang H (2024) Biodegradable biobased polymers: a review of the state of the art, challenges, and future directions. Polymers 16(16):2262. https://doi.org/10.3390/polym16162262
57. Perera KY, Jaiswal AK, Jaiswal S (2023) Biopolymer-based sustainable food packaging materials: challenges, solutions, and applications. Foods 12(12):2422. https://doi.org/10.3390/foods12122422
58. Elsabahy M, Heo GS, Lim S-M, Sun G, Wooley KL (2015) Polymeric nanostructures for imaging and therapy. Chem Rev 115(19):10967–11011. https://doi.org/10.1021/acs.chemrev.5b00135
59. Unterweger H, Dézsi L, Matuszak J, Janko C, Poettler M, Jordan J, Bäuerle T, Szebeni J, Fey T, Boccaccini A, Alexiou C, Cicha I (2018) Dextran-coated superparamagnetic iron oxide nanoparticles for magnetic resonance imaging: evaluation of size-dependent imaging properties, storage stability and safety. Int J Nanomed 13:1899–1915. https://doi.org/10.2147/IJN.S156528
60. Satchanska G, Davidova S, Petrov PD (2024) Natural and synthetic polymers for biomedical and environmental applications. Polymers 16(8):1159. https://doi.org/10.3390/polym16081159
61. Desai N, Rana D, Salave S, Gupta R, Patel P, Karunakaran B, Sharma A, Giri J, Benival D, Kommineni N (2023) Chitosan: a potential biopolymer in drug delivery and biomedical applications. Pharmaceutics 15(4):1313. https://doi.org/10.3390/pharmaceutics15041313
62. Luangbudnark W, Viyoch J, Laupattarakasem W, Surakunprapha P, Laupattarakasem P (2012) Properties and biocompatibility of chitosan and silk fibroin blend films for application in skin tissue engineering. Sci World J 2012:1–10. https://doi.org/10.1100/2012/697201
63. Viezuina D-M, Musa I, Aldea M, Matache I-M, Rotaru Zavaleanu A-D, Gresita A, Veronica S, Mitran SI (2025) Gelatin-based hydrogels for peripheral nerve regeneration: a multifunctional vehicle for cellular, molecular, and pharmacological therapy. Gels 11(7):490. https://doi.org/10.3390/gels11070490
64. Hadisi Z, Nourmohammadi J, Mohammadi J (2015) Composite of porous starch-silk fibroin nanofiber-calcium phosphate for bone regeneration. Ceram Int 41(9):10745–10754. https://doi.org/10.1016/j.ceramint.2015.05.010
65. Najdanović J, Rajković J, Najman S (2018) Bioactive biomaterials: potential for application in bone regenerative medicine. In: Biomaterials in clinical practice. Springer International Publishing, pp 333–360. https://doi.org/10.1007/978-3-319-68025-5_12
66. Farzi G, Gheysipour M (2023) Encapsulation with polymers. In: Principles of biomaterials encapsulation, vol 2. Elsevier, pp 3–38. https://doi.org/10.1016/B978-0-12-824345-9.00010-6
67. Taşdemır M, Koçak D, Usta İ, Akalin M, Merdan N (2007) Properties of polypropylene composite produced with silk and cotton fiber waste as reinforcement. Int J Polym Mater 56(12):1155–1165. https://doi.org/10.1080/00914030701323752
68. Zainab I, Naseem Z, Batool SR, Waqas M, Nazir A, Nazeer MA (2025) Polyurethane/silk fibroin-based electrospun membranes for wound healing and skin substitute applications. Beilstein J Nanotechnol 16:591–612. https://doi.org/10.3762/bjnano.16.46

69. Samir A, Ashour FH, Hakim AAA, Bassyouni M (2022) Recent advances in biodegradable polymers for sustainable applications. Npj Mater Degrad 6(1):68. https://doi.org/10.1038/s41529-022-00277-7
70. Balla E, Daniilidis V, Karlioti G, Kalamas T, Stefanidou M, Bikiaris ND, Vlachopoulos A, Koumentakou I, Bikiaris DN (2021) Poly(lactic acid): a versatile biobased polymer for the future with multifunctional properties—from monomer synthesis, polymerization techniques and molecular weight increase to PLA applications. Polymers 13(11):1822. https://doi.org/10.3390/polym13111822
71. Abpeikar Z, Moradi L, Javdani M, Kargozar S, Soleimannejad M, Hasanzadeh E, Mirzaei S. A., & Asadpour, S. (2021). Characterization of macroporous polycaprolactone/silk fibroin/gelatin/ascorbic acid composite scaffolds and *in vivo* results in a rabbit model for meniscus cartilage repair. CARTILAGE 13(2_suppl):1583S–1601S. https://doi.org/10.1177/19476035211035418
72. Rajasekaran R, Dutta A, Ray PG, Seesala VS, Ojha AK, Dogra N, Roy S, Banerjee M, Dhara S (2022) High fibroin-loaded silk-PCL electrospun fiber with core–shell morphology promotes epithelialization with accelerated wound healing. J Mater Chem B 10(46):9622–9638. https://doi.org/10.1039/D2TB01890J
73. Pryadko A, Surmeneva MA, Surmenev RA (2021) Review of hybrid materials based on polyhydroxyalkanoates for tissue engineering applications. Polymers 13(11):1738. https://doi.org/10.3390/polym13111738
74. Wei S, Wang Y, Sun Y, Gong L, Dai X, Meng H, Xu W, Ma J, Hu Q, Ma X, Peng J, Gu X (2023) Biodegradable silk fibroin scaffold doped with mineralized collagen induces bone regeneration in rat cranial defects. Int J Biol Macromol 235:123861. https://doi.org/10.1016/j.ijbiomac.2023.123861
75. Borbolla-Jiménez FV, Peña-Corona SI, Farah SJ, Jiménez-Valdés MT, Pineda-Pérez E, Romero-Montero A, Del Prado-Audelo ML, Bernal-Chávez SA, Magaña JJ, Leyva-Gómez G (2023) Films for wound healing fabricated using a solvent casting technique. Pharmaceutics 15(7):1914. https://doi.org/10.3390/pharmaceutics15071914
76. Cho SY, Lee ME, Choi Y, Jin H-J (2014) Cellulose nanofiber-reinforced silk fibroin composite film with high transparency. Fibers Polym 15(2):215–219. https://doi.org/10.1007/s12221-014-0215-y
77. Jiang S, Chen Y, Duan G, Mei C, Greiner A, Agarwal S (2018) Electrospun nanofiber reinforced composites: a review. Polym Chem 9(20):2685–2720. https://doi.org/10.1039/C8PY00378E
78. Wen D-L, Sun D-H, Huang P, Huang W, Su M, Wang Y, Han M-D, Kim B, Brugger J, Zhang H-X, Zhang X-S (2021) Recent progress in silk fibroin-based flexible electronics. Microsyst Nanoeng 7(1):35. https://doi.org/10.1038/s41378-021-00261-2
79. Xue J, Wu T, Dai Y, Xia Y (2019) Electrospinning and electrospun nanofibers: methods, materials, and applications. Chem Rev 119(8):5298–5415. https://doi.org/10.1021/acs.chemrev.8b00593
80. Luo CJ, Stoyanov SD, Stride E, Pelan E, Edirisinghe M (2012) Electrospinning versus fibre production methods: from specifics to technological convergence. Chem Soc Rev 41(13):4708. https://doi.org/10.1039/c2cs35083a
81. Li W-J, Mauck RL, Cooper JA, Yuan X, Tuan RS (2007) Engineering controllable anisotropy in electrospun biodegradable nanofibrous scaffolds for musculoskeletal tissue engineering. J Biomech 40(8):1686–1693. https://doi.org/10.1016/j.jbiomech.2006.09.004
82. Ma L, Dong W, Lai E, Wang J (2024) Silk fibroin-based scaffolds for tissue engineering. Front Bioeng Biotechnol 12:1381838. https://doi.org/10.3389/fbioe.2024.1381838
83. Van Vuure AW, Mosleh Y, Vanderbeke J, Verpoest I (2023) Highly impact-resistant silk fiber thermoplastic composites. Adv Eng Mater 25(20). https://doi.org/10.1002/adem.202300080
84. Guo C, Li C, Vu HV, Hanna P, Lechtig A, Qiu Y, Mu X, Ling S, Nazarian A, Lin SJ, Kaplan DL (2020) Thermoplastic moulding of regenerated silk. Nat Mater 19(1):102–108. https://doi.org/10.1038/s41563-019-0560-8

85. Tan L, Zhong J, Guo W, Gao Q, Li J, He Y, Huang J, Xu W, Xiao S, Yu H, Xu B (2024) A novel melt extrusion method for efficient and large-scale in-situ exfoliation of boron nitride to prepare high performance thermal conductive polymer composite. Polymer 315:127809. https://doi.org/10.1016/j.polymer.2024.127809
86. Guo F, Aryana S, Han Y, Jiao Y (2018) A review of the synthesis and applications of polymer-nanoclay composites. Appl Sci 8(9):1696. https://doi.org/10.3390/app8091696
87. Gao G, Dong H, Wu R, Wang J, Ao Y, Xu Y (2022) 3D-printed regenerative polycaprolactone/silk fibroin osteogenic and chondrogenic implant for treatment of hip dysplasia. Biochem Biophys Res Commun 636:96–104. https://doi.org/10.1016/j.bbrc.2022.10.046
88. Yang Z, Huang L, Song X, Wang M, Han X, Guan W, Luo C (2025) Properties of 3D-printed continuous silk fiber-reinforced poly(caprolactone). RSC Adv 15(19):14643–14654. https://doi.org/10.1039/D5RA01302J
89. Li Q, Xu S, Feng Q, Dai Q, Yao L, Zhang Y, Gao H, Dong H, Chen D, Cao X (2021) 3D printed silk-gelatin hydrogel scaffold with different porous structure and cell seeding strategy for cartilage regeneration. Bioact Mater 6(10):3396–3410. https://doi.org/10.1016/j.bioactmat.2021.03.013
90. Kristiawan RB, Imaduddin F, Ariawan D, Ubaidillah, Arifin Z (2021) A review on the fused deposition modeling (FDM) 3D printing: filament processing, materials, and printing parameters. Open Eng 11(1):639–649. https://doi.org/10.1515/eng-2021-0063
91. Mwema FM, Akinlabi ET (2020) Basics of fused deposition modelling (FDM), pp 1–15. https://doi.org/10.1007/978-3-030-48259-6_1
92. Chekkaramkodi D, Jacob L, C MS, Umer R, Butt H (2024) Review of vat photopolymerization 3D printing of photonic devices. Addit Manuf 86:104189. https://doi.org/10.1016/j.addma.2024.104189
93. Bukhari SMZS, Bukhari MA, Hossain M (2025) Vat photopolymerization of smart polymers for biomedical applications. ACS Appl Polym Mater 7(18):12117–12144. https://doi.org/10.1021/acsapm.5c01487
94. Awad A, Fina F, Goyanes A, Gaisford S, Basit AW (2021) Advances in powder bed fusion 3D printing in drug delivery and healthcare. Adv Drug Deliv Rev 174:406–424. https://doi.org/10.1016/j.addr.2021.04.025
95. Elkaseer A, Chen KJ, Janhsen JC, Refle O, Hagenmeyer V, Scholz SG (2022) Material jetting for advanced applications: a state-of-the-art review, gaps and future directions. Addit Manuf 60:103270. https://doi.org/10.1016/j.addma.2022.103270
96. Tyagi S, Yadav A, Deshmukh S (2022) Review on mechanical characterization of 3D printed parts created using material jetting process. Mater Today Proc 51:1012–1016. https://doi.org/10.1016/j.matpr.2021.07.073
97. Wei L, Wu S, Kuss M, Jiang X, Sun R, Reid P, Qin X, Duan B (2019) 3D printing of silk fibroin-based hybrid scaffold treated with platelet rich plasma for bone tissue engineering. Bioactive Mater 4:256–260. https://doi.org/10.1016/j.bioactmat.2019.09.001
98. Kolevar MP, Koshar A, Hirsch J, Choe RH, Wu J, Rocca MS, McLoughlin S, Venable-Croft A, Fisher JP, Packer JD (2024) Development of a patient specific cartilage graft using magnetic resonance imaging and 3D printing. J ISAKOS 9(4):519–525. https://doi.org/10.1016/j.jisako.2024.03.011
99. Agrawal K, Bhat AR (2025) Advances in 3D printing with eco-friendly materials: a sustainable approach to manufacturing. RSC Sustain 3(6):2582–2604. https://doi.org/10.1039/D4SU00718B
100. Silva EGS, Cardoso S, Bettencourt AF, Ribeiro IAC (2022) Latest trends in sustainable polymeric food packaging films. Foods 12(1):168. https://doi.org/10.3390/foods12010168
101. Siddiqui MAS, Rabbi MS, Ahmed RU, Billah MdM (2024) Biodegradable natural polymers and fibers for 3D printing: a holistic perspective on processing, characterization, and advanced applications. Clean Mater 14:100275. https://doi.org/10.1016/j.clema.2024.100275
102. Haider M, Leung V, Ferrari F, Crissman J, Powell J, Cappello J, Ghandehari H (2005) Molecular engineering of silk-elastinlike polymers for matrix-mediated gene delivery: biosynthesis and characterization. Mol Pharm 2(2):139–150. https://doi.org/10.1021/mp049906s

103. Lu Q, Hu X, Wang X, Kluge JA, Lu S, Cebe P, Kaplan DL (2010) Water-insoluble silk films with silk I structure. Acta Biomater 6(4):1380–1387. https://doi.org/10.1016/j.actbio.2009.10.041
104. Gu Y, Yu L, Mou J, Wu D, Zhou P, Xu M (2020) Mechanical properties and application analysis of spider silk bionic material. E-Polymers 20(1):443–457. https://doi.org/10.1515/epoly-2020-0049
105. Huang S, Fu Q, Yan L, Kasal B (2021) Characterization of interfacial properties between fibre and polymer matrix in composite materials—a critical review. J Mater Res Technol 13:1441–1484. https://doi.org/10.1016/j.jmrt.2021.05.076
106. Noishiki Y, Nishiyama Y, Wada M, Kuga S, Magoshi J (2002) Mechanical properties of silk fibroin–microcrystalline cellulose composite films. J Appl Polym Sci 86(13):3425–3429. https://doi.org/10.1002/app.11370
107. Yang Y, Chen X, Shao Z, Zhou P, Porter D, Knight DP, Vollrath F (2005) Toughness of spider silk at high and low temperatures. Adv Mater 17(1):84–88. https://doi.org/10.1002/adma.200400344
108. Dong Z, Lewis RV, Middaugh CR (1991) Molecular mechanism of spider silk elasticity. Arch Biochem Biophys 284(1):53–57. https://doi.org/10.1016/0003-9861(91)90262-H
109. Hamidi YK, Yalcinkaya MA, Guloglu GE, Pishvar M, Amirkhosravi M, Altan MC (2018) Silk as a natural reinforcement: processing and properties of silk/epoxy composite laminates. Materials 11(11):2135. https://doi.org/10.3390/ma11112135
110. Darshan GH, Kong D, Gautrot J, Vootla S (2017) Physico-chemical characterization of Antheraea mylitta silk mats for wound healing applications. Sci Rep 7(1):10344. https://doi.org/10.1038/s41598-017-10531-7
111. Temesgen AG, Turşucular ÖF, Turşucular ED (2024) A review on silk fibers and their medical applications. Recent Prog Mater 06(01):1–9. https://doi.org/10.21926/rpm.2401008
112. Zhao H-P, Feng X-Q, Shi H-J (2007) Variability in mechanical properties of Bombyx mori silk. Mater Sci Eng C 27(4):675–683. https://doi.org/10.1016/j.msec.2006.06.031
113. Shubhra QTH, Alam AKMM, Khan MA, Saha M, Saha D, Khan JA, Quaiyyum MA (2010) The preparation and characterization of silk/gelatin biocomposites. Polym-Plast Technol Eng 49(10):983–990. https://doi.org/10.1080/03602559.2010.482074
114. Srivastava A, Bhati P, Singh S, Agrawal M, Kumari N, Vashisth P, Chauhan P, Bhatnagar N (2024) A review on polylactic acid-based blends/composites and the role of compatibilizers in biomedical engineering applications. Polym Eng Sci 64(3):1003–1044. https://doi.org/10.1002/pen.26626
115. Wang Q, Zhang Y, Li B, Chen L (2017) Controlled dual delivery of low doses of BMP-2 and VEGF in a silk fibroin–nanohydroxyapatite scaffold for vascularized bone regeneration. J Mater Chem B 5(33):6963–6972. https://doi.org/10.1039/C7TB00949F
116. Omenetto FG, Kaplan DL (2010) New opportunities for an ancient material. Science 329(5991):528–531. https://doi.org/10.1126/science.1188936
117. Seib FP, Maitz MF, Hu X, Werner C, Kaplan DL (2012) Impact of processing parameters on the haemocompatibility of Bombyx mori silk films. Biomaterials 33(4):1017–1023. https://doi.org/10.1016/j.biomaterials.2011.10.063
118. Callaway KA, Xue Y, Altimari V, Jiang G, Hu X (2018) Comparative investigation of thermal and structural behavior in renewably sourced composite films of even-even nylons (610 and 1010) with silk fibroin. Polymers 10(9):1029. https://doi.org/10.3390/polym10091029
119. Tuan HA, Hirai S, Tamada Y, Akioka S (2019) Preparation of silk resins by hot pressing Bombyx mori and Eri silk powders. Mater Sci Eng C 97:431–437. https://doi.org/10.1016/j.msec.2018.12.060
120. Barreira-Pinto R, Carneiro R, Miranda M, Guedes RM (2023) Polymer-matrix composites: characterising the impact of environmental factors on their lifetime. Materials 16(11):3913. https://doi.org/10.3390/ma16113913
121. McGill M, Holland GP, Kaplan DL (2019) Experimental methods for characterizing the secondary structure and thermal properties of silk proteins. Macromol Rapid Commun 40(1). https://doi.org/10.1002/marc.201800390

122. Cho SY, Yun YS, Lee S, Jang D, Park K-Y, Kim JK, Kim BH, Kang K, Kaplan DL, Jin H-J (2015) Carbonization of a stable β-sheet-rich silk protein into a pseudographitic pyroprotein. Nat Commun 6(1):7145. https://doi.org/10.1038/ncomms8145
123. Plaza GR, Pérez-Rigueiro J, Riekel C, Perea GB, Agulló-Rueda F, Burghammer M, Guinea GV, Elices M (2012) Relationship between microstructure and mechanical properties in spider silk fibers: identification of two regimes in the microstructural changes. Soft Matter 8(22):6015. https://doi.org/10.1039/c2sm25446h
124. Mukasheva F, Adilova L, Dyussenbinov A, Yernaimanova B, Abilev M, Akilbekova D (2024) Optimizing scaffold pore size for tissue engineering: insights across various tissue types. Front Bioeng Biotechnol 12:1444986. https://doi.org/10.3389/fbioe.2024.1444986
125. Yazawa K, Ishida K, Masunaga H, Hikima T, Numata K (2016) Influence of water content on the β-sheet formation, thermal stability, water removal, and mechanical properties of silk materials. Biomacromolecules 17(3):1057–1066. https://doi.org/10.1021/acs.biomac.5b01685
126. Kim D-K, Lee JM, Jeong JY, Park HJ, Lee OJ, Chao J, Kim SH, Park HS, Khang G, Park CH (2018) New fabrication method of silk fibroin plate and screw based on a centrifugal casting technique. J Tissue Eng Regen Med 12(11):2221–2229. https://doi.org/10.1002/term.2752
127. Li M, You J, Qin Q, Liu M, Yang Y, Jia K, Zhang Y, Zhou Y (2023) A comprehensive review on silk fibroin as a persuasive biomaterial for bone tissue engineering. Int J Mol Sci 24(3):2660. https://doi.org/10.3390/ijms24032660
128. Rahman M, Kabir M, Li K, Li Y, Chen S, Wu S (2024) Electrospun zeolitic imidazole framework-8 loaded silk fibroin/polycaprolactone nanofibrous scaffolds for biomedical application. J Mech Behav Biomed Mater 160:106769. https://doi.org/10.1016/j.jmbbm.2024.106769
129. Mazurek Ł, Szudzik M, Rybka M, Konop M (2022) Silk fibroin biomaterials and their beneficial role in skin wound healing. Biomolecules 12(12):1852. https://doi.org/10.3390/biom12121852
130. Kornstein A (2014) SERI surgical scaffold as an adjunct for circumferential abdominoplasty and lower body lift. Plast Reconstr Surg Glob Open 2(11):e244. https://doi.org/10.1097/GOX.0000000000000214
131. Zhao Z, Li Y, Xie M-B (2015) Silk fibroin-based nanoparticles for drug delivery. Int J Mol Sci 16(3):4880–4903. https://doi.org/10.3390/ijms16034880
132. Florczak A, Deptuch T, Kucharczyk K, Dams-Kozlowska H (2021) Systemic and local silk-based drug delivery systems for cancer therapy. Cancers 13(21):5389. https://doi.org/10.3390/cancers13215389
133. Yucel T, Lovett ML, Kaplan DL (2014) Silk-based biomaterials for sustained drug delivery. J Control Rel 190:381–397. https://doi.org/10.1016/j.jconrel.2014.05.059
134. Pang C, Li F, Hu X, Meng K, Pan H, Xiang Y (2024) Degradable silk fibroin based piezoresistive sensor for wearable biomonitoring. Discov Nano 19(1):55. https://doi.org/10.1186/s11671-024-04001-z
135. Xu M, Yadavalli VK (2019) Flexible biosensors for the impedimetric detection of protein targets using silk-conductive polymer biocomposites. ACS Sens 4(4):1040–1047. https://doi.org/10.1021/acssensors.9b00230
136. Cao Y, Wang B (2009) Biodegradation of silk biomaterials. Int J Mol Sci 10(4):1514–1524. https://doi.org/10.3390/ijms10041514
137. Perrone GS, Leisk GG, Lo TJ, Moreau JE, Haas DS, Papenburg BJ, Golden EB, Partlow BP, Fox SE, Ibrahim AMS, Lin SJ, Kaplan DL (2014) The use of silk-based devices for fracture fixation. Nat Commun 5(1):3385. https://doi.org/10.1038/ncomms4385
138. Holland C, Numata K, Rnjak-Kovacina J, Seib FP (2019) The biomedical use of silk: past, present, future. Adv Healthc Mater 8(1). https://doi.org/10.1002/adhm.201800465
139. Yang K, Guan J, Numata K, Wu C, Wu S, Shao Z, Ritchie RO (2019) Integrating tough Antheraea pernyi silk and strong carbon fibres for impact-critical structural composites. Nat Commun 10(1):3786. https://doi.org/10.1038/s41467-019-11520-2

140. Ji Y, Li Y, Chen G, Xing T (2017) Fire-resistant and highly electrically conductive silk fabrics fabricated with reduced graphene oxide via dry-coating. Mater Des 133:528–535. https://doi.org/10.1016/j.matdes.2017.08.006
141. Hasan MK, Kayumov J, Zhu G, Khatun M, Nur A, Ding X (2019) An experimental investigation to examine the wicking properties of silk fabrics. J Text Sci Technol 05(04):108–124. https://doi.org/10.4236/jtst.2019.54010
142. Chen Y, Duan L, Ma Y, Han Q, Li X, Li J, Wang A, Bai S, Yin J (2020) Preparation of transient electronic devices with silk fibroin film as a flexible substrate. Colloids Surf A 600:124896. https://doi.org/10.1016/j.colsurfa.2020.124896
143. Xu Z, He Z, Huang J, Qiu W, Cao LNY, Patil A, Yang B-R, Liu XY, Guo W (2023) Flexible, biocompatible, degradable silk fibroin based display. Chem Eng J 464:142477. https://doi.org/10.1016/j.cej.2023.142477
144. Marelli B, Brenckle MA, Kaplan DL, Omenetto FG (2016) Silk fibroin as edible coating for perishable food preservation. Sci Rep 6(1):25263. https://doi.org/10.1038/srep25263
145. Sudsandee S, Hu C-C, Liu Y-L, Worakhunpiset S, Loahaprapanon S, Hung W-S, Lee K-R, Lai J-Y (2019) Improving barrier performance of transparent polymeric film using silk nanofibril combine graphene oxide. J Taiwan Inst Chem Eng 95:332–340. https://doi.org/10.1016/j.jtice.2018.07.024
146. Zha L, Li K, Wang S, Zhou Q (2025) Rehydration of nanocellulose films in an aqueous silk fibroin solution for facile fabrication of strong composites. ACS Sustain Chem Eng 13(29):11348–11361. https://doi.org/10.1021/acssuschemeng.5c02888
147. Liu L, Zhang S, Huang J (2019) Progress in modification of silk fibroin fiber. Sci China Technol Sci 62(6):919–930. https://doi.org/10.1007/s11431-018-9508-3
148. Wang X, Meng Z, Wang X, Cai W, Liu K, Wang D (2023) Silk nanofibril-palygorskite composite membranes for efficient removal of anionic dyes. Nanomaterials 13(2):247. https://doi.org/10.3390/nano13020247
149. Chouhan D, Mandal BB (2020) Silk biomaterials in wound healing and skin regeneration therapeutics: from bench to bedside. Acta Biomater 103:24–51. https://doi.org/10.1016/j.actbio.2019.11.050
150. Qadir J, Islam T (2024) Potential of silk proteins in cosmetics. J Sci Agric 79–87. https://doi.org/10.25081/jsa.2024.v8.9070
151. Caruso MR, D'Agostino G, Milioto S, Cavallaro G, Lazzara G (2023) A review on biopolymer-based treatments for consolidation and surface protection of cultural heritage materials. J Mater Sci 58(32):12954–12975. https://doi.org/10.1007/s10853-023-08833-5
152. Kash S (2024) How Thai silk is shaping the future of medicine delivery. Medical Channel Asia. Available from: https://medicalchannelasia.com/how-thai-silk-is-shaping-the-future-of-medicine-delivery/
153. Lawrence BD, Wharram S, Kluge JA, Leisk GG, Omenetto FG, Rosenblatt MI, Kaplan DL (2010) Effect of hydration on silk film material properties. Macromol Biosci 10(4):393–403. https://doi.org/10.1002/mabi.200900294
154. Nguyen TP, Nguyen QV, Nguyen V-H, Le T-H, Huynh VQN, Vo D-VN, Trinh QT, Kim SY, Le QV (2019) Silk fibroin-based biomaterials for biomedical applications: a review. Polymers 11(12):1933. https://doi.org/10.3390/polym11121933
155. You R, Xu Y, Liu G, Liu Y, Li X, Li M (2014) Regulating the degradation rate of silk fibroin films through changing the genipin crosslinking degree. Polym Degrad Stab 109:226–232. https://doi.org/10.1016/j.polymdegradstab.2014.07.029
156. Eom J, Park S, Jin H-J, Kwak HW (2020) Multiscale hybridization of natural silk-nanocellulose fibrous composites with exceptional mechanical properties. Front Mater 7:98. https://doi.org/10.3389/fmats.2020.00098

Chapter 10
Technical Textiles and Silk: Flexible Electronics and Smart Wearable Textiles

Siam Sarker, Borhan Uddin Khan, Tawsif Sikder Adib, Shumya Akther, and Md Rashedul Islam

10.1 Introduction

Clothing serves as a dynamic interface between humans and the surrounding environment. Human beings usually choose clothes, considering different colors and styles, to express identity or respond to changes in the external environment [1]. However, the concept of clothing has undergone a significant evolution with time, introducing technical textile technology [2–4]. The development of flexible and wearable smart textiles has become the solution for numerous promising applications [5–7]. Since conventional wearable devices typically incorporate rigid capacitive and sensory components that limit user movement, textiles have become the preferred platform for smart wearables and bendable electronics. Textiles provide comfort, moisture transport, soft texture, and adapt to the human body. As a result, textiles are often conceptualized as a secondary skin of humans [8, 9]. The textiles transform into smart wearable devices after the incorporation of bendable electronics.

Silk, the natural filament, is derived from the silkworm and extensively used in the technical textile industry because of its great mechanical strength, lightweight, biodegradability, biocompatibility, conductivity, noninflammatory degradation, and participation in sustainability practices [10]. Silk shows a more complex and organized hierarchical structure compared to other cellulosic fibers. This structure ensures a durable structural framework and controlled biodegradation, which satisfies the

S. Sarker · T. S. Adib · S. Akther
Department of Apparel Engineering, Bangladesh University of Textiles, Dhaka, Bangladesh

B. U. Khan
Department of Textile Engineering, International Standard University, Dhaka, Bangladesh

Md R. Islam (✉)
Department of Wet Process Engineering, Bangladesh University of Textiles, Dhaka, Bangladesh
e-mail: rashedul.islam@wpe.butex.edu.bd

M. Shahid et al. (eds.), *Science and Technology of Silk*, Textile Science and Clothing Technology, https://doi.org/10.1007/978-3-032-26880-8_10

requirements of different application sectors [11]. Silk fibroin solution offers regenerated silk fibers, silk films, silk sponges, and silk hydrogels. These silk-based materials are widely used to produce wearable, flexible smart textiles. Silk is characterized by hygroscopic swelling and the piezoelectric effect, showing variance in water molecular absorption between the liquid and vapor phases [9]. The application area of silk-based technical textiles follows a wide range, starting from sensors, energy-harvesting materials, and shape-memory devices to biomedical science [12]. However, in-depth research into silk-based flexible electronics and smart wearable textiles can pave the way for advancement in economic aspects and scalable material processing techniques, ultimately fulfilling the demand for high-performance and sustainable technical textiles.

This chapter aims to provide insights into silk-based technical textiles and their influence on flexible electronics and smart wearable textiles. In addition, this chapter will present an overview of silk fiber by discussing morphological structure, physical and chemical properties, and the relationship between silk structure and its inherent properties. Moreover, it discusses the suitable properties of silk, which makes it the ideal choice for technical textiles and the principles of different preparation methods of silk-based materials from silk fibroin. The chapter provides a brief overview of the application areas of silk-based smart wearable textiles, which include bendable electronics such as actuators, sensors, energy applications, electronic skin, shape-memory devices, radiative cooling smart clothing, implantable flexible electronics, and medical applications. The chapter concludes by discussing the challenges associated with silk-based technical textiles and suggesting future research directions for enthusiastic scholars.

10.2 Overview of Silk Fiber

10.2.1 Morphological Structure of Silk

Silk is an animal fiber produced from silkworms through a distinctive spinning process, which ultimately results in a sheath-core structure [9]. The silk fibers usually have a triangular cross section or a flat elliptical cross section. Moreover, an even or irregular longitudinal appearance can be observed. These aspects depend on the type of fiber or type of cocoon [9, 13]. The hierarchical structure of silk fibroin consists of five levels, from the first level to the fifth level, named amino acid sequence, sheet and helix, crystalline, crystal network, and nanofibril network, consecutively [14]. Figure 10.1 shows the morphological structure of the silk fiber.

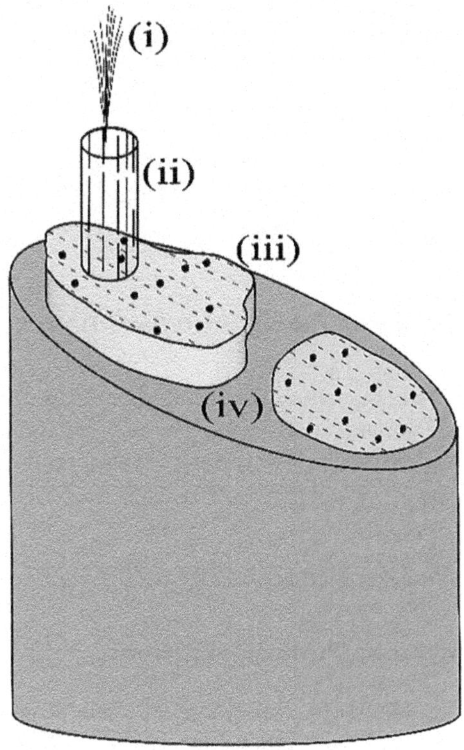

Fig. 10.1 Structure of raw silk filament: (i) Microfibrils, (ii) Fibril bundle, (iii) Fibroin brins, (iv) Sericin [13]

10.2.2 Chemical Structure of Silk

The two main components of raw silk fiber are fibroin and sericin. Fibroin and sericin are both composed of amino acids. The acid sequences of amino acids present in fibroin have an ordered pattern, while sericin has a random pattern [13]. Fibroin consists of both light- and heavy-chain polypeptides in a ratio of 1:1. The bonding is formed by disulfide bonds. Glycine, alanine, serine, and tyrosine participate in the formation of the heavy chain at a ratio of 8:5:2:1 [15–17]. Twelve hydrophobic repetitive domains present in the heavy chain result in the crystalline part, and 11 hydrophilic repetitive domains present in the heavy chain result in the non-crystalline part in the fibroin. The peptide chains are constructed through hydrogen bonds between carboxyl and amino groups following a definite order [18]. Two hexapeptides composed of glycine and other amino acids sequenced as (-G-A-G-A-G-Y-)n and (-G-A-G-A-G-S-)n are mainly responsible for almost 72% of all hydrophobic repeating domains called dipeptide unit [19].

Table 10.1 Composition of various types of silk cocoons [13]

Silk cocoon	Components			
	fibroin (%)	Sericin (%)	Wax (%)	Ash, minerals, and others (%)
Mulberry	66–72	25–32	0.3–0.4	0.7–0.8
Tasar	78–85	14–17	1–2	3–4
Muga	80–86	12–16	0.5–1	2–3
Eri	82–88	11–13	1.5–2.2	2–3

10.2.3 Chemical Composition of Silk

The composition of silk fiber varies depending on the sourcing position, environmental and atmospheric factors, processing methods, etc. [20]. Table 10.1 displays the composition of different types of silk cocoons.

10.2.4 Properties of Silk Fiber

10.2.4.1 Physical Properties

The physical properties of silk cocoons vary with types. The fiber density (expressed by g/cm^3) of Mulberry (bivoltine, multivoltine) silk is 1.34, Tasar and Oak Tasar silk is 1.31, and Muga and Eri silk is 1.30. The value of fiber fineness of Mulberry silk is 2–3 den, Muga silk 4–7 den, Eri silk 3–4 den, and Tasar silk 8–12 den. Mulberry silks are the finest among all the silk cocoons. A gradual decline in fiber fineness can be seen from the outer layer to the inner layer [13].

10.2.4.2 Chemical Properties

Twenty different amino acids are present in the fibroin and can be further classified into three types: acidic, basic, and neutral. As a result, silk fiber is compatible with any dyes. Amino acids contain both acidic and basic groups, forming a zwitterion. Therefore, amino acid demonstrates an amphoteric nature in solutions. The silk fibers show good resistive properties against acids and reducing agents, but poor resistance toward alkali and oxidizing agents. Concentrated acid in high temperatures and strong bleaches pose a danger to silk and can damage it severely [13].

The silk fibers exhibit good moisture regain quality. The hydrophilic groups present in fibroin and sericin are responsible for good water absorbance [21–23]. Degumming is the process in which sericin is removed from the raw silk strand. The removal of sericin can result in a slight decrease in water absorbency. The moisture regain varies based on the types of silk cocoons. For example, the moisture regain of Mulberry silk is 11%, Tasar silk is 10.5%, and Muga and Eri is 10%. Moisture regain value changes inside the silk fiber, decreasing from the exterior to the interior layers [13]. Water absorption causes swelling in silk fiber, and the highest water pick-up results in increased diameter (around 15%), volume (around 34%), and length (around 1%) [21]. Silk demonstrates controllable degradation. Moreover, dissolution properties are scaled down due to the presence of contents of α-sheet and ß-crystallites, along with higher orientation [14].

10.2.5 Nexus Between the Structure and Property of Silk Fiber

The fibrillar structure and orientation of silk are responsible for its highly tenacious, flexible nature, and impressive wrinkle recovery property. The bonding between fibrillar regions by microfibrils and polypeptide chains is not strong enough, causing the easy occurrence of fibrillation under abrasive force. This is the reason behind the mediocre abrasion-resistance behavior of silk. Silk is susceptible to degradation in boiling water. However, silk fibroin is quite resistant to enzymes due to its organized and compact structure, which is impermeable to large molecules of enzymes [13].

10.3 Suitable Properties of Silk for Technical Textile Applications

10.3.1 Mechanical Property

The hierarchical architecture of silk's sericin fibers is responsible for the exceptional mechanical performance of silk, which includes tensile strength, elastic modulus, toughness, and durability [24]. Eisaku Iizuka researched various mulberry and non-mulberry silk varieties and demonstrated that the tenacity, elongation, and elastic modulus of silk filaments are correlated with their linear density. The race of silkworms also influences the linear density or mean filament size. Both tenacity and elastic modulus show a negative linear correlation with linear density. When linear density increases, the silk fiber's tenacity and elastic modulus decrease. However, elongation shows a positive correlation, increasing with higher linear density [25].

Variations in tenacity and elongation at break can be observed across different silk varieties. For example, B. mori silk has an ultimate tensile strength (UTS) of 208.45 MPa, with elongation at a break of 19.55%, and Young's modulus of 6.10 GPa. Twisted B. mori silk has the lowest UTS of 165.27 MPa and an elongation at break of 20.57%. On the other hand, Tussah silk shows the highest UTS of 248.77 MPa and the highest percentage on the elongation at a break (33.48%), with a Young's modulus of 5.79 GPa [26]. The remarkable mechanical properties of silk lead to a diversity of applications across different fields, including smart technical textiles, engineering elements, and biomedical procedures [9, 12, 27].

10.3.2 Conductivity of Silk

Electrical conductivity significantly influences the functionality of wearable electronic systems. Though silk fibers lack inherent electrical conductivity, they can be modified to impart conductive properties. As a result, the resulting silk can be used to complete the fabrication of wearable textiles that contain silk-derived electronics [9].

Conductive substances like graphene [28], carbon nanotubes (CNTs) [29], metallic elements [30], and conductive polymers [31] are incorporated into the surface of silk fibers, resulting in silk-derived conductive fibers. Graphene, a two-dimensional nanomaterial composed of a single atomic carbon layer, shows exceptional mechanical strength, electrical, and thermal conductivity [32]. Graphene lacks surface functional groups and inherent chemical stability, limiting direct adhesion to silk fibers. However, graphene oxide (GO) shows enhanced adhesion to silk surfaces through mixtures of hydrogen and covalent bonds that strengthen the interfacial attachment [32]. The forces acting between the surfaces of silk and GO remain insufficient for some applications so it is necessary to utilize coupling and cross-linking agents to further improve the interfacial adhesion [33, 34]. Lu et al. [28] experimented with reduced graphene oxide (rGO)-coated fibers and fabrics, which were usable as performance aid matrices and wearable smart textiles with electrically conductive thread or fabric. The technique of incorporating GO, silver, and copper nanoparticles along with coupling agent 3-glycidyloxypropyl trimethoxysilane, followed by the reduction of GO to rGO with sodium sulfate, yields a conductive silk fabric with enhanced functionalities. This incorporation results in notable electrothermal properties, ultraviolet shielding, improved thermal stability, pronounced hydrophobicity, and reduced surface resistance, which offer promising applications in the development of electronic textiles [35]. Silk fabric coated with carbon nanotubes (CNTs) also displayed heat absorptivity, protection from water and solvent, and outstanding electrical conductivity [36]. Moreover, poly (3,4-ethylene dioxythiophene) (PEDOT) and polyelectrolyte poly (styrene sulfonate) (PSS) with electrical conductivity can be coated on silk fabric to get silk-derived conductive thread with resistance against abrasion and washing [31].

10.3.3 Optical Transparency of Silk

The inherent optical characteristics of silk, including luster and refractive properties, are directly linked to its fiber morphology and chemical composition. Mulberry, Muga, and Tasar silks have uniform cross-sectional areas and continuous filament structures, ensuring high luster through consistent light reflection. These optical properties can be further enhanced through chemical treatments. For example, weak acid treatments enhance luster by improving fibroin molecular chain alignment. On the other hand, alkali treatments reduce the luster by damaging the fibroin molecular chains and the hydrolysis process [13]. Moreover, silk fibroin can be nanopatterned by enabling the fabrication of diverse optical structures for biomedical, photonic, and nanophotonic applications. It is possible to introduce fluorescent materials to the surface patterning of silk, which opens opportunities for advanced silk-based bio-optical devices and sensing technologies in technical textiles [37].

10.3.4 Biodegradability of Silk

The biodegradation process involves the decomposition of polymers into small-molecule compounds. Biodegradability is a crucial parameter in the field of medical textiles. Silk offers superior biodegradation performance compared to other biomaterials due to its ability to control degradation rates and release acidic by-products [12, 38]. Wang et al. [39] conducted in vivo experiments to evaluate the degradability of silk fibroin. They used regenerated silk as a base material and fabricated three-dimensional porous scaffolds by aqueous processing technique or organic solvent (1,1,1,3,3,3-hexafluoro-2-propanol) method. Later, these scaffolds were embedded into rat subjects. According to their findings, the host animal could tolerate every scaffold type throughout the implantation period with minimal observed immune response. Moreover, scaffolds produced using the aqueous method showed complete degradation within 2–6 months, while scaffolds prepared with organic solvents took 1 year to complete degradation. Their experiment proved that the degradation property of silk scaffolds in vivo can be predicted and controlled for a variety of biomedical research. Therefore, the biodegradable properties of silk make it an ideal material for biomedical textiles, implants, and smart fibers [12].

10.4 Preparation of Silk-Based Materials for Technical Textiles

Silk fibroin is a promising biomaterial because of specific properties like biocompatibility, adjustable degradation rate, minimal toxicity, and desirable mechanical properties. Silk fibroin is extracted from silkworm cocoons after removing sericin and impurities. They are later converted into a solution using water, chemical solvents, and high-power equipment. Silk-based materials such as fiber, film, sponge, and hydrogel can be directly prepared from the fibroin solution [40] (Fig. 10.2).

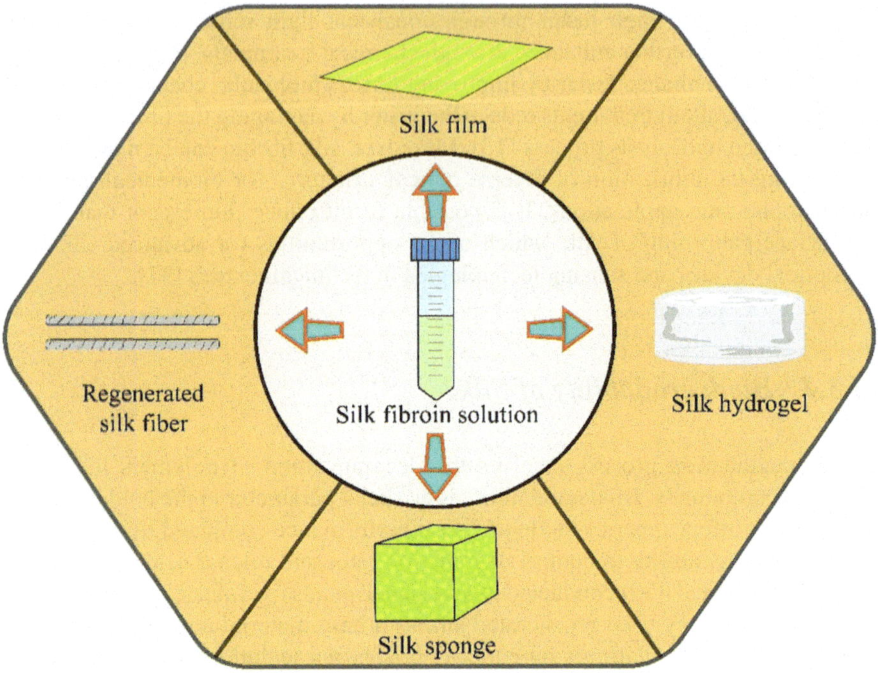

Fig. 10.2 Preparation of silk-based materials from silk fibroin solution

10.4.1 Preparation of Silk Fibroin Solution

To prepare the silk fibroin solution, silk cocoons are first shredded into finer pieces. Then, the shredded pieces are boiled in a pre-prepared sodium carbonate solution (0.02 M) for 30 to 45 min. The boiled silk fibers are washed at least four times in deionized (DI) water to remove the sericin material. After draining the excess water, the fibers are dried under normal conditions or inside a fume hood.

The fibers are then put in a beaker, and a 9.3 M lithium bromide (LiBr) solution is applied to them. The beaker is heat-treated inside an oven for 4 h at 60 °C. Later, the silk/lithium bromide solution is pushed inside a dialysis film or a dialysis cassette using a syringe and dialyzed in deionized water for 48 h. As a result, a primary silk fibroin solution is produced, free from lithium bromide ions. The use of a dialysis cassette with a large margin is essential as it can prevent breakage from water absorption. Moreover, the deionized water must be replaced five times throughout the process. Afterward, the primary solution is collected in a centrifuge tube from the dialysis cassette. Then, the solution is centrifuged at 9000 rpm for two 20 min rounds, which removes the contaminants. Finally, the pure silk fibroin solution is stored under 4 °C in a temperature-controlled area [41, 42]. The prepared fibroin solution serves as a precursor for most silk-based materials.

10.4.2 Regenerated Silk Fiber

Textiles woven from regenerated silk fiber show exceptional mechanical properties. These fibers have been extensively used in flexible electronic devices. Wet spinning and electrospinning are the two main methods for preparing regenerated silk fiber [43].

In the wet spinning process, the polymers are first melted in a solvent to prepare a dope solution. The solution is then dropped into a spin cell equipped with a spinneret. A nitrogen gas cylinder is connected to the spin cell. The cylinder generates the pressure and forces the solution out through the spinneret. The spinneret is placed inside a coagulation bath. The solution precipitates and solidifies into fibers as it comes out from the spinneret into the coagulation bath. These fibers are subsequently washed in a water bath to remove the contaminants and remaining solvents [44, 45].

Electrospinning utilizes a powerful electric field to transform silk fibroin solution into fibers. The entire method is completed in four steps. Firstly, a highly viscous polyethylene oxide is mixed with a silk fibroin solution to create a mixture of silk/PEO. This step gives the silk fibroin solution its desirable viscosity and spinning properties. Secondly, the syringe is loaded with the correct quantity of silk/PEO solution and pushed through the needle onto a collector, forming a silk/PEO fiber textile. The positive and the negative terminals of a high-voltage power supply must be connected to the needle and the collector. The solution discharge rate, supplied voltage, and the gap between the needle tip and collector also influence the shape and diameter of the formed textile. Thirdly, annealing with water vapor or alcohol is carried out as an after-treatment for the textile made of silk/PEO fibers formed on the collector. These annealing processes create higher β-sheet crystals in the silk fiber, which increases mechanical strength and reduces fiber breakdown. Finally, the post-treated material is kept inside a shaker filled with deionized water. It is shaken overnight to ensure the complete removal of polyethylene oxide (PEO) [46–50]. The two methods mentioned above produce aligned fibers with higher strength and flexibility, which make them ideal for technical textiles.

10.4.3 Silk Films

Silk films are becoming popular for developing wearable electronic systems that are integrated into technical textiles. These films are highly flexible, lightweight, and provide variable transmission. There are two types of silk films: smooth surface silk films and patterned surface silk films [45].

Smooth surface silk films are prepared in three steps. First, silk fibroin solution is spread on the surface of a plastic petri dish using spin coating. Next, the film is dried at 25 °C to stabilize the structure. Finally, the dried film crystallizes when kept in an alcohol solution or an environment filled with water vapor. The film developed in this way shows improved mechanical performance and degrades at a preferred rate.

Moreover, solution density impacts the thickness of the film. For example, a highly dense fibroin solution produces thicker films, but a lesser one produces thinner films [50].

Patterned silk films are more commonly used than smooth silk films. Generally, a patterned silk film is prepared using two methods. In the first method, the silk fibroin solution is spread over the surface of a patterned substrate. As a result, the pattern on the surface of the substrate is transferred to the film's surface. Finally, the pattern gets fixed upon drying [51, 52]. The second method involves surface coating of silk fibroin solution on a flat substrate. Then, a patterned layer is directly pressed on the coated solution to imprint the pattern. After drying, the pattern gets fixed on the surface, and the film is carefully removed from the mold [53]. Both smooth and patterned surface silk films are a basic version of silk-based materials. They are prepared without harmful chemicals or high-energy equipment, which makes them eco-friendly [40].

10.4.4 Silk Sponges

Silk sponges are highly cross-linked silk-based material with a broad specific surface area. Their structural network is filled with size-controllable pores [54, 55]. Different sponge preparation methods include freeze-drying [56], gas foaming [57, 58], and particulate leaching [59, 60] (i.e., salt leaching and polymer particulate leaching). However, particulate leaching is the most beneficial due to its simplicity, easier size control of pores, and need for less equipment.

The particulate leaching method involves the following steps. Polystyrene and NaCl are prepared with a specific diameter suitable for the desired pore size and placed inside a container. In the second step, a silk fibroin solution is gradually added inside the container, where it crystallizes at room temperature. The container must be rotated carefully to ensure uniform dispersion of particulates in the solution. Then, the fibroin solution and particulates mixture are dipped in specific solvents. The mixture is then gently stirred to remove the particulates. Changing the solvent five to six times is necessary for complete removal. When the silk sponge is purified and ready, it is dried and stored at room temperature. Before drying, the silk sponge must be rinsed in deionized water if an organic solvent is used for particulate removal. Finally, annealing is performed so the dried silk sponge breaks down at a controlled rate [43, 59–61]. This unique porous structure of the silk sponge makes it ideal for technical textiles utilized in the medical sector [50].

10.4.5 Silk Hydrogels

Hydrogels are water-rich polymers with a three-dimensional structural network that form through the physical or chemical cross-linking of molecules. They hold water, retain shape, and offer opportunities for functional improvements. As per cross-linking, silk-protein-based hydrogels are either physical or chemical gels [62–65].

The physical cross-linking method relies on non-covalent bonds like hydrogen bonds, hydrophobic interaction, electrostatic interaction, ionic interaction, and chain winding [64]. Techniques for physical cross-linking include self-assembly [66, 67], ultrasonic treatment [67, 68], shear action [69], electric field effect [70], temperature effect [71], and the effect of pH adjustment [72, 73].

In the self-assembly technique, the silk fibroin solution converts into β-sheet from a coiled state after the solution evaporates or concentrates at room temperature. Eventually, the molecules self-cluster into hydrogels. During ultrasonic treatment, intermolecular interactions improve through ultrasound features. As a result, the silk fibroin structure is modified into a hydrogel. In the shear action technique, the silk fibroin solution is placed in a mixer, which produces a rapid vortex. The continuous rotation of the vortex accumulates the solution into a hydrogel. The electric field effect forces protons in the solution to migrate toward the anode. Therefore, the pH value reduces below the isoelectric point, and the molecules acquire a negative charge. These electronegative molecules create a structure similar to a micelle. Finally, a hydrogel forms when physical interaction increases in this structure [65, 74, 75].

Temperature change affects hydrogel formation. As the temperature rises, molecules vibrate highly, leading to more frequent molecular collisions and clustering. Adjusting the pH value also helps in hydrogel formation. When the pH value of the solution is equal to the isoelectric point, charged particles face negligible repulsion. Therefore, particles in the fibroin solution gather and develop a gel structure [65].

The chemical cross-linking method forms three-dimensional network structures through covalent bonding. Popular chemical cross-linking methods include photopolymerization [76, 77], gamma irradiation [78], chemical cross-linking [79, 80], and enzyme cross-linking [81–84].

In photopolymerization, a photoinitiator is responsible for the polymeric reaction. The initiator absorbs visible or ultraviolet light. As a result, excited molecules detach from the initiator surface after absorbing the energy. These separated molecules produce free radicals, which trigger free radical polymeric reactions between unstable bonds. Therefore, cross-linking increases and forms hydrogel. Hydrogels created from both physical and chemical cross-linking provide improved durability and mechanical strength. However, these materials need further development because they are beneficial for technical textiles [65, 85, 86].

10.5 Application of Silk-Based Flexible Electronics and Smart Wearable Textiles

Silk-based technical textiles have gained significant research attention in the flexible electronics and smart wearable sectors due to the flexibility, mechanical strength, biocompatibility, and lightweight of silk. Therefore, silk-based flexible electronics and smart wearable textiles have demonstrated great potential in various application sectors, such as sensor, energy, environment, and medical.

10.5.1 Actuators

An actuator acts as a functional equipment that converts energy into motion. Actuators can be classified as electromechanical, pneumatic, and hydraulic. Humidity-responsive actuators swell or shrink with exposure to water, thermal-responsive actuators change shapes upon temperature variations, and electrically responsive actuators respond to electrical activations [87, 88].

The silk-based actuator is the next generation of bio-inspired flexible actuators. The silk properties of biocompatibility, flexibility, and environment sustainability make it more popular in wearable applications [89]. Manikandan et al. [90] developed an entirely integrated silk-based actuator from pure silk fibroin films (Fig. 10.3a). The actuator shows the humidity-sensitive, ultrafast actuation characteristic with a response time of 0.1 s and an actuation time of 0.3 s. These actuators were reported to have good performance, such as bending angle, durability (up to 1000 exposure cycles), and variation in stiffness in response to water vapor exposure. Their study also showcased potential application areas such as smart textiles in soft robotics and wearables. Ganesan et al. [91] reported multiple vapor-responsive bidirectional macroscale actuation behaviors of single-layer regenerated silk fibroin films (Fig. 10.3b). The silk fibroin films show antagonistic actuation characteristics by showing upward and downward bending upon exposure to water and ethanol, respectively. Moreover, the magnitude of actuation showed a positive correlation with increasing water vapor but a negative correlation with elevated ethanol vapor pressure. Therefore, the demand for silk-based actuators is becoming popular in the scientific community [12, 87].

10.5.2 Sensors

A sensor is a device that detects and responds to different stimuli such as touch, light, pressure, temperature, and gases. They work as the key component of modern technology and assist in performing everyday activities. Silk sensors are a highly promising technology in the advancement of smart clothing solutions in pressure,

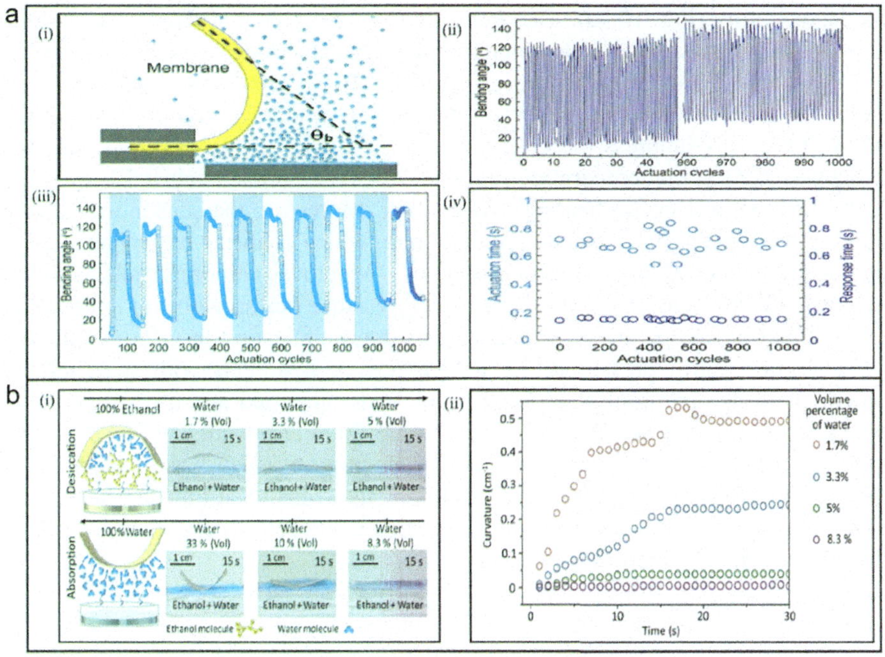

Fig. 10.3 **a** (i) Diagram depicting the film's bending angle (θb) in response to water vapor. (ii) Bending angle variations observed over repeated exposure cycles. (For display purposes, only cycles 1–40 and 960–1000 are shown.) The film experienced 5 s moisture treatments, regulated by an automated procedure. (iii) Actuation behavior shown for every 100th cycle performed. (iv) Response and actuation duration (time to peak bending) plotted across exposure cycles. These times remained consistent over 1000 cycles. Film dimensions: 1 cm × 1.5 cm × 10 μm [90].; **b** (i) Actuation characteristics of the silk fibroin film in different ethanol–water compositions. Water percentages are specified. The film exhibited an upward bend in pure ethanol. The bending radius lessened at 3.3% water (v/v), and actuation ceased near 8.3% water. At 10% water (v/v), upward bending resumed, signifying water vapor's greater effect. (ii) Temporal evolution of the film's curvature when subjected to vapor from ethanol–water mixtures containing 1.7, 3.3, 5, or 8.3% water (v/v) [91]

temperature, and humidity sensing [92]. The hydrophilicity of silk fibroin makes humidity sensors highly sensitive in detecting water vapor concentration. As a result, they could be a fine arrangement for climate control, monitoring the respiratory system and agricultural irrigation management. Conductive polymer materials like PEDOT:PSS are mostly used in thermal sensors, which make them available in wearable healthcare, adaptive dress, and temperature control systems. Pressure sensors work with the mechanical toughness of silk fibroin to detect mechanical force, making them ideal for different sensing applications such as gesture recognition, rehabilitation process monitoring, and sensory fabric systems [93].

Pradhan and Yadavalli [94] developed a fully organic, silk-protein-based temperature sensor with high flexibility, transparency, conformability, durability, and biodegradability (Fig. 10.4a). The architecture of the sensor comprised a photoactive silk fibroin substrate, upon which interdigitated electrodes were micropatterned via a photolithographic technique using PEDOT:PSS conductive ink, and a silk sericin and rGO-coated thermo-responsive coating, which worked as encapsulation within fibroin to suppress the effect of humidity. The sensor was more sensitive ($-0.99\%\ °C^{-1}$) in the temperature range of 20–50 °C, stable against 10–90% RH, and robust against cyclic heating and cooling. Moreover, the sensor was able to measure both the surface and skin temperature, which confirms its probable use in personalized medical care and disease diagnosis. Moreover, carbonized silk nanofiber and PEDOT:PSS sensors show excellent piezoresistive performance for temperature sensing and biocompatibility, which widens their application in wearable devices [95]. He et al. [96] presented silk fibroin-derived hydrogels with stretchability, biocompatibility, and multifunctionality for wearable strain and pressure sensing, which showed a wide sensing range (strain, 2%–600%: pressure, 0.5%–119.4 kPa), greater mechanical stability, and quick recovery characteristics during deformations (Fig. 10.4b). Researchers have reported numerous works regarding silk-based flexible sensors, which indicate fabrication techniques and prospects of these sensors in different wearable electronics [9, 12].

10.5.3 Energy Applications

Energy storage and harvesting devices are integrated with the design of wearable and portable electronics because they provide energy for extended usage without the need for an external power supply. Energy storage devices such as batteries and supercapacitors store energy to be utilized later, while energy-harvesting devices such as solar cells or piezoelectric devices harvest energy from the environment or body motions [97]. Silk can be used as a substrate for flexible energy storage devices due to its good mechanical flexibility, biocompatibility, and renewability. Silk can further add different benefits, such as tunability, multifunctionality, and versatility, which indicates silk's brilliance in hybrid energy systems [98].

For instance, [99] created a flexible thermoelectric power generator by coating. Bi_2Te_3 and Sb_2Te_3 nanostructures on silk fabric, which could be integrated with rechargeable batteries to create self-chargeable systems for wearable devices (Fig. 10.5a). The 12-thermocouple prototype efficiently converted heat to electricity in the temperature difference range of 5–35 K. The maximum voltage output was ~10 mV and the power output was ~15 nW. The generator also showed stable performance after 100 cycles of twisting and bending. Kim et al. [100] presented an eco-friendly silk fibroin bio-triboelectric generator (silk Bio-TEG) based on an electrospun nanofiber-networked film (Fig. 10.5b). They implemented a cost-effective electrospinning method to generate a higher surface-to-volume ratio and rough surfaces compared to smooth-cast silk film. This device showed a maximum of 1.86 $\mu C\ m^{-2}$ triboelectric surface charge density and 4.3 mW m^{-2} electrical power output at a resistive load of 5MΩ. Since this silk Bio-TEG was prepared with the help of a

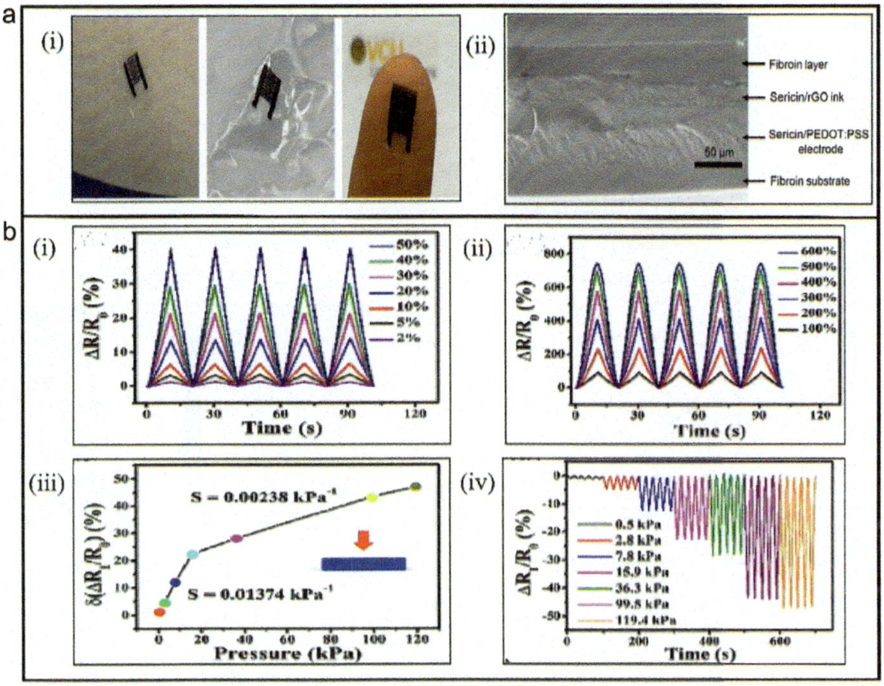

Fig. 10.4 **a** Illustrations of a silk-protein-based temperature sensor's flexibility (i) Showing skin and textile integration, (ii) A cross-sectional SEM image highlighting the sensor's layered construction, where covalent linkages contribute to enhanced durability and resistance to delamination [94], **b** Analysis of the PSGP hydrogel sensor's electrical response to mechanical stress (i, ii) The relationship between stretching and the sensor's relative resistance change, (iii) Depiction of the sensor's response to pressure, illustrating its sensitivity to differing magnitudes of force [96]

simple, eco-friendly, and sustainable biomaterial-based production process, it can be used in living organisms without any harm. Moreover, researchers have discussed numerous potential applications of silk composites with different hybrid materials for energy conversion, which indicates the popularity of silk technology for energy applications [10, 12, 98, 101].

10.5.4 Electronic Skin

Electronic skin (E-skin) or artificial skin is a flexible, stretchy device that simulates skin and can respond to and detect stimuli such as pressure, temperature, strain, and humidity. The E-skin contains sensors and occasionally actuators, allowing it to interact with the environment like the human sense system. Therefore, E-skin is a major field for wearable electronics, prosthetics, robotics, and human–machine interface [102]. Unlike conventional E-skins, multifunctional E-skins are advanced

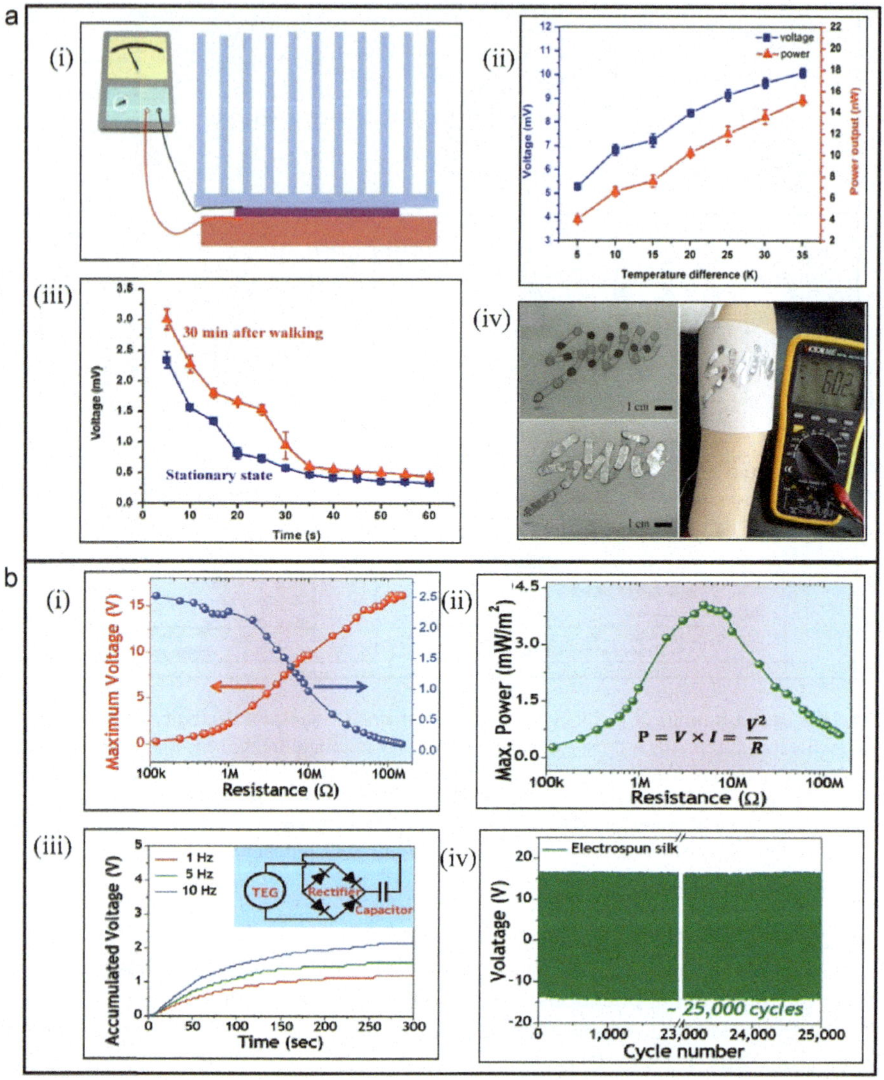

Fig. 10.5 **a** (i) Illustration of the custom-built experimental setup for measurement. (ii) Electrical output behavior of the silk fabric thermoelectric energy source. (iii) Recorded voltage from the silk-based thermoelectric generator on a human subject, comparing the starting voltage (blue) and the voltage after a 30-min walk (red) at a temperature of 20 °C. (iv) The voltage generated by a custom-engineered silk thermoelectric generator affixed to the human arm, measured at 20 °C [99], **b** Analysis of the silk Bio-TEG's operational behavior: Investigating how (i) Voltage, (ii) Power when driven by a pulsed signal, (iii) Voltage accumulation on a 10 μF capacitor driven by the silk Bio-TEG, demonstrating its response to varying pulse frequencies, (iv) Analysis of the silk Bio-TEG's operational longevity and resilience, including the entire signal waveform [100]

alternatives that can detect more than one type of stimulus and show more accurate feedback. Despite being an interesting idea, multifunctional E-skin fabrication faces different challenges, such as sensor interaction, complex structure designs, and cost-effective and scalable processing strategies [103]. Silk-based textiles are becoming popular as appropriate intermediates for E-skin fabrication. Since silk possesses properties like biocompatibility, higher mechanical properties, and high optical transmittance, they are becoming popular as an intermediate for E-skin. Silk-based materials like fiber, films, and foams can easily be integrated with multifunctional sensors. These attributes of silk make it a viable and sustainable choice for E-skin technology [104].

Wang et al. [60] developed an E-skin, which was a combo temperature–pressure sensor driven from silk-nanofiber-derived carbon fiber membranes (SilkCFM) (Fig. 10.6a). The silk-based temperature sensor showed a sensitivity of approximately $0.81\%\ °C^{-1}$ and a negligible response to external pressure. The silk-based strain sensor displayed a broad sensing range exceeding 50% strain and high stability and durability. Moreover, the strain sensor's performance remained unaffected by ambient temperature variations. This combined E-skin can successfully monitor physiological activities (respiration and tactile pressure), which highlights its potential for applications in human–machine interfaces and human activity monitoring. Huang et al. [105] developed a heat-proof and flexible silk fibroin composite membrane doped with polyurethane for the surface of electronic skins (Fig. 10.6b). These SFCMs stretched over 200%, endured temperatures above 160 °C, and had high transmittance (>90%). Silver (Ag) and platinum (Pt) nanofiber networks were used for heating and temperature sensing. Moreover, no immune response was observed when this E-skin was worn on the human arm for 20 days.

Researchers have gathered numerous examples, showcasing the success of silk-based flexible E-skin technology. They also address challenges and future direction considering the sustainability practices of flexible wearable electronics [9, 10, 43].

10.5.5 Shape-Memory Devices

Shape-memory devices are materials or architectures that can revert to their original shape under certain stimuli, such as temperature, light, electricity, or moisture. They are based on shape-memory alloys, shape-memory polymers, and other responsive biomaterials [106]. The development of silk-based textiles as shape-memory material components can be done mostly through the modification of their structure or by incorporating functional components [107]. Conductive-type elements like graphene and carbon nanotubes can be fabricated with silk to produce shape-memory textiles that will respond to electrical or infrared stimulation changes. Infusing hydrogel on silk fibers will make these fibers swell/collapse with humidity variations, allowing for smart clothing and medical applications [108]. Incorporating shape-memory characteristics into silk-based textiles can be adopted in the electronics and biomedical sectors.

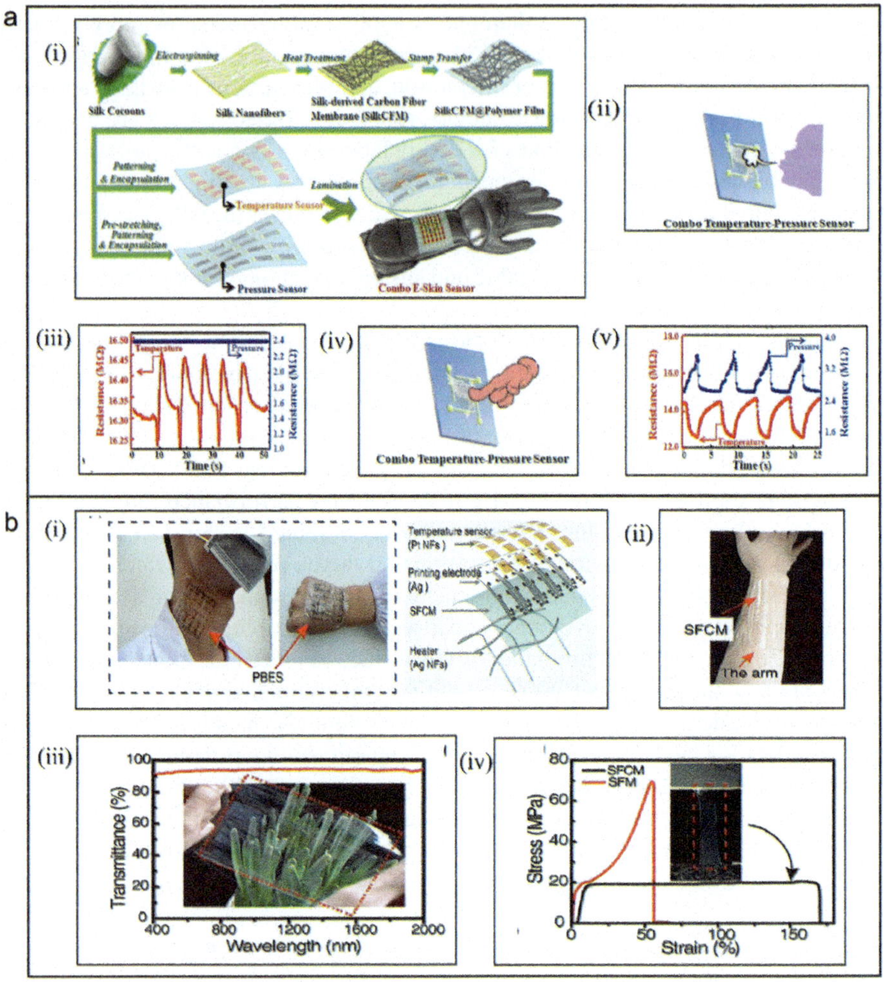

Fig. 10.6 **a** (i) Diagram of the manufacturing steps for the silk-based temperature and pressure sensitive electronic skin, (ii) Diagram and simultaneous sensing behavior of the integrated electronic skin sensor under different stimuli. (ii–v) Measurements during exhalation (ii, iii) and finger pressure application (iv, v). The red and blue lines in (iii) and (v) depict the electrical signals from the temperature and strain sensors, respectively [60]; **b** Manufacturing of the PBES device based on SFCM (i) Pictures illustrating the PBES in close contact with a human neck and hand. (ii) Photograph showing SFCM's flexible fit on a human arm (approximately 23 μm thick). (iii) Optical transmission assessment and a visual representation of SFCM. (iv) Tensile stress–strain curves comparing SFCM and pure SFM [105]

A recent study on silk fibroin fiber-based shape-memory membranes explains the huge opportunities of silk as a shape-memory device, considering multiple application areas. [107] developed a silk fibroin nanofiber-based membrane that has Janus wettability, self-cleaning, waterproof, and antibacterial features (Fig. 10.7). This material exhibited a self-cleaning physical barrier (an approximate 98.1% filtration efficiency for particulate matter around 2.5 μm) and a biological barrier (sterilizing and inhibiting bacterial proliferation, exceeding 95%). Therefore, their work suggests a lightweight and skin-friendly material with air-filtration and antibacterial capabilities that can be used as wound care products, medical-grade masks, protective clothing, and advanced filtration devices. There is also another example where regenerated silk fibers incorporate water-responsive shape-memory behavior and show excellent mechanical properties [109]. Moreover, researchers have explored numerous possibilities of the characteristics of silk-based shape-memory devices and emphasized the opportunity of silk in the shape-memory technology, particularly for biomedical applications by considering the biocompatible and biodegradable background of silk [108, 110].

10.5.6 Radiative Cooling Smart Clothing

Radiative cooling textiles are an advanced type of clothing that enables the passive regulation of body temperature through efficient heat dissipation by infrared radiation. Unlike conventional fabrics, which are primarily used to insulate the body, radiative cooling textiles employ mid-infrared (MIR) emissivity to dissipate body heat directly to the environment and reflect solar radiation to minimize thermal absorption. As a result, radiative cooling textiles can be used for personal thermal management systems in high-temperature environments and outdoor applications without the need for any electrical power consumption [111]. Silk-based clothing possesses intrinsic thermal and photonic properties. Silk promotes natural high and mid-infrared emissivity, which ensures successful thermal dissipation. Moreover, the transparency and lightness of silk ensure its compatibility for wear. However, the functionality of silk-based radiative cooling smart textiles can also be enhanced by the addition of different nanoparticles, such as boron nitride, silica, or aluminum oxide, into the silk matrix [112]. The incorporation of conductive components such as graphene or silver nanowires into silk-based textiles also enables the development of thermo-responsive fabrics that can actively sense and react to temperature changes [113]. Therefore, silk-based radiative textiles can be adopted as personalized thermal management systems, ensuring thermal comfort and eco-friendly solutions.

Zhang et al. [97] explored the use of electro-flocked short silk fibers (SSFs) for passive radiative cooling (Fig. 10.8a). They demonstrated that these SSFs were able to enhance both reflectance in the visible to near-infrared range and emittance in the mid-infrared range. This experiment showed improvement in cooling performance by reducing the temperature of the substrate by 7.6 °C (applied on PDMS film) and 5.6 °C (applied on cotton fabric) compared to the untreated samples. This approach

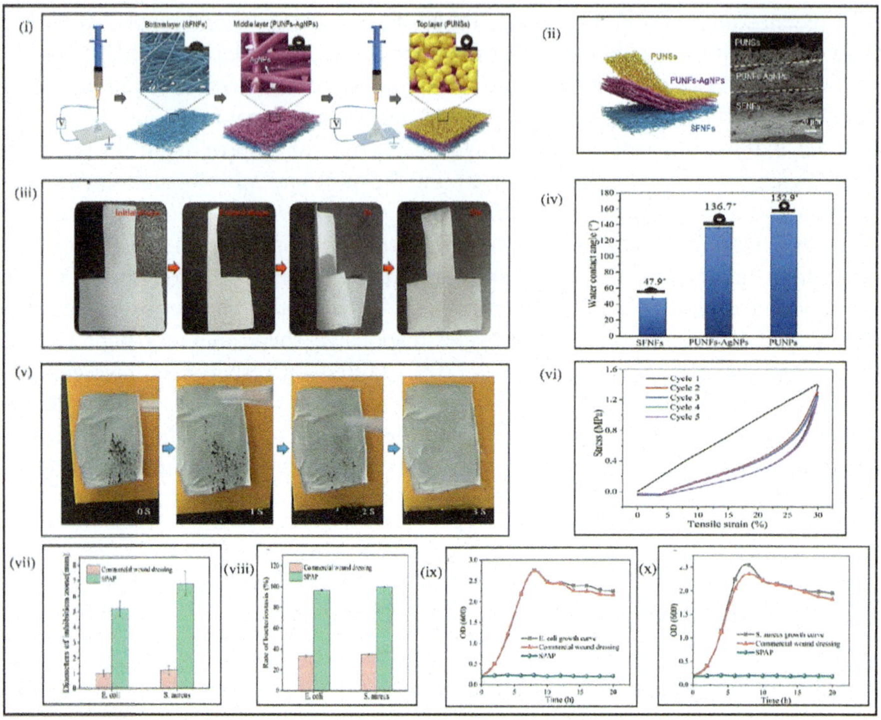

Fig. 10.7 Shape-memory-enabled Janus SPAP membrane. (i) Illustration of the fabrication process of an SPAP membrane through electrospinning and electrospraying techniques. (ii) Cross-sectional image of the SPAP membrane, highlighting its layered structure. (iii) Demonstrate the SPAP membrane's thermal shape-memory property. (iv) Evaluation of the SPAP membrane's self-cleaning properties, stretchability, and mechanical stability. Water contact angle (WCA) measurements for SFNFs, PUNFs-AgNPs, and PUNSs layers are presented. Error bars indicate two standard deviations (SD). (v) Visual demonstration of the SPAP membrane's self-cleaning capability through photographs. (vi) Assessment of the SPAP membrane's elasticity under cyclic stretching. (vii) Analysis of the SPAP membrane's antibacterial properties. Comparative analysis of antimicrobial ring diameters between the SPAP membrane and a traditional wound dressing. (viii) Graphical analysis of the bacterial inhibition rates for the SPAP membrane and a typical wound dressing. (ix, x) Comparative growth curves of (ix) E. coli and (x) S. aureus in the presence of the commercial wound dressing and the SPAP membrane [107]

of electro-flocked SSFs offers a robust and scalable solution for creating efficient silk-based radiative cooling materials. [114] investigated passive personal thermal management and developed nano-processed silk with enhanced radiative cooling properties (Fig. 10.8b). They used a molecular bonding design and a scalable dip-coating method and were able to reduce skin temperature by 8 °C compared to natural silk. Their findings show a sustainable and energy-free approach to thermoregulatory materials without compromising wearability.

Researchers around the world are still researching to find out the best strategies to incorporate silk as a radiative cooling material. There are numerous works

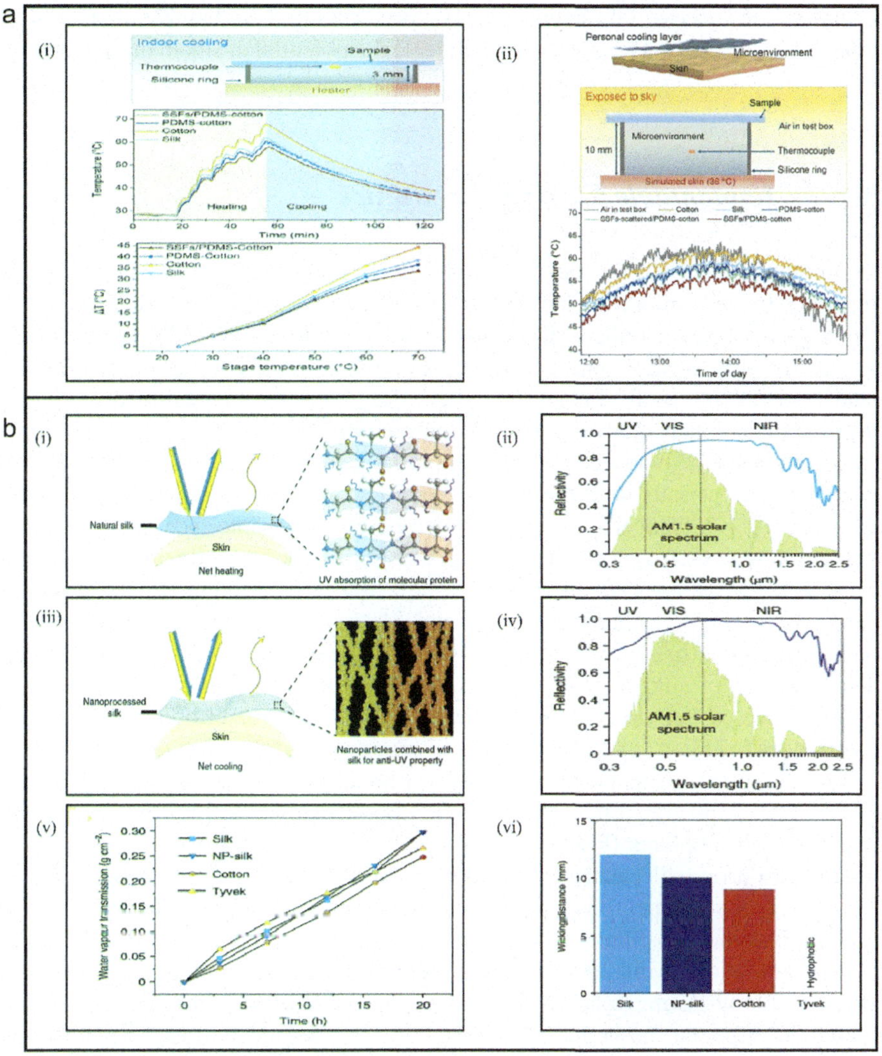

Fig. 10.8 **a** Applicability of SSFs/PDMS as a fabric coating (i) Temperature variations at the bottom surface of the sample and variation between initial and instant temperatures of samples, examined by an indoor cooling test where the sample lies on a stage heated from 70 °C room temperature. (ii) Assessment of outdoor cooling performance under clear summer skies, with samples positioned 10 mm above a 36 °C simulated skin [116]; **b** (i) Diagrammatic representation of the net thermal performance of nature silk due to the elevated absorption in the UV range. (ii) The reflectance spectrum of natural silk, characterized within 0.3–2.5 μm wavelength range. (iii) Diagrammatic representation of the net cooling performance of nano-processed silk due to enhanced UV light reflection. (iv) The reflectance spectrum of nano-processed silk, characterized within 0.3–2.5 μm [114]

regarding the success, fabrication process, performances, and applications of silk-based radiative cooling textiles, which indicate the potential of silk in this technology [9, 115].

10.5.7 Implantable Flexible Electronics

An implantable electronic device is any electronic subsystem or module to be implanted in the human body for any number of reasons, ranging from data recording and acquisition to the enhancement of physiological function [117]. These devices are also used in medical and non-medical applications like cybersecurity, human enhancement, and biohacking [118, 119]. They have the capability of capturing physiological signals, allowing inter-operation with extraneous systems, or amplification of sensory experience. The implantation of these devices enables integration over extended periods with the human body and thus has been utilized in wearable sensors, bio-feedback systems, and communication interfaces with seamless interaction with biological systems [120]. Currently, silk-based flexible implantable electronic devices are becoming popular. Silk fibroin is a tough structural protein and possesses high mechanical strength and minimal immunogenicity. On the other hand, silk sericin has proved to be a material of great biological value. All these characteristics of silk make it an ideal material for implantable devices, particularly for devices that would be self-powered by body motion or other physiological occurrences [121].

Tang-Schomer et al. [122] established a novel strategy of neural engineering by fabricating integrated neuron–electrode interfaces on silk films that were organized as brain implants (Fig. 10.9a). The silk film had compartmentalized neural circuits and arrays of electrodes. Electrical stimulation of the electrodes properly activated proximate neurons, causing calcium responses. Upon placement within mice brains, the silk films created chronic contact, maintained brain cells under low inflammation levels, and remained viable for weeks. This integration was an example of long-lasting, focused brain–machine interfaces with excellent control and interaction between the brain and external devices. Moreover, [123] developed an implantable bio-piezoelectric T-silk nanoribbons (T-SNR) film that displayed high-energy output, quick response, mechanical flexibility, biodegradability, and biocompatibility (Fig. 10.9b). They worked with the meso- and nano-structure of silk materials, which influenced the final piezoelectric performance. They conducted an in vivo assessment on adult rats to understand the potential of biological systems. Their findings showed that T-SNR films were implantable into a wide range of human tissues as a self-powered device to ensure proper healthcare by creating human–machine interfaces.

Researchers have analyzed numerous silk biomaterial-based implantable electronic devices. Their works highlight various fabrication methods, such as self-assembly, cross-linking, and soft lithography, to process silk into functional devices. Silk-based hydrogels and films are also utilized as active components (electrodes,

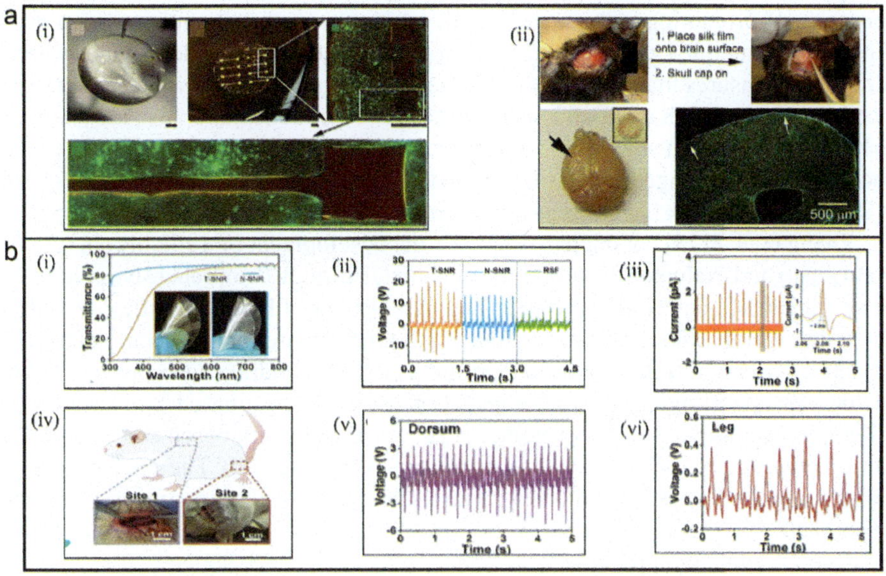

Fig. 10.9 **a** (i) Integration of film-based neuro-electrode interface. (ii) Placing the patterned silk film on the brain surface of a rat [122]; **b** (i) UV–visible transmission curves of T-SNR and N-SNR films. Images represent transparent and flexible T-SNR and N-SNR films. (ii) T-SNR, N-SNR, and RSF films' resultant voltage response under ~5 Hz hand tapping motion, illustrating their piezoelectric response. (iii) Current generated by T-SNR films, with the inset displaying their recorded response time. (iv) Illustration of encapsulated biocompatible piezoelectric T-SNR films for use as in vivo monitoring tools, self-powered energy-harvesting devices, and sensing technologies. Schematic and photographic views showing the packaged T-SNR film implants located in specific areas of SD rats. (v) Minimal finger taps generating piezoelectric voltage in T-SNR films placed subcutaneously in SD rats. (vi) Piezoelectric voltage response resulting from the gentle stretching of T-SNR films placed in the thigh areas of SD rats [123]

power supplies) and passive components (scaffolds, substrates) in implantable devices. Therefore, silk shows great promise for advancing flexible implantable technology [12, 27, 121].

10.5.8 Medical Applications

Silk-based medical applications rely on the distinct features of silk fibroin, which is a naturally occurring protein from silkworms. The biodegradable, biocompatible, and mechanical properties of silk make it suitable for wound healing, scaffolds, fracture fixation, and drug delivery systems [124]. Silk fibroin consists of a β-sheet domain structure, which gives rigidity and stability. These properties are prone to alteration by process techniques like degumming, spinning, and electrospinning for managing the mechanical, chemical, and biological characteristics of the material [16]. The property of promoting cell attachment, development, and tissue regeneration gives silk-based material a great advantage in the medical field. The biodegradability property of silk ensures that it will naturally break down in the body without leaving any toxic residues, which makes it a safe alternative to synthetic materials [125]. Studies have also shown that silk-based materials can engage harmoniously with the human body, ensure prolonged performance, and eliminate the danger of rejection [126, 127]. Therefore, silk has become an important candidate for medical technology.

Zhang et al. [128] conducted a study on the development of a silk fibroin film for skin repair and evaluated its translational potential through comprehensive preclinical and clinical studies (Fig. 10.10a). They assessed the film's properties, including biocompatibility, bacterial barrier function, and moisture management, and tested its effectiveness in rabbit and porcine models. Their findings also provide strong evidence supporting the clinical application of silk fibroin-based biomaterials for skin regeneration and wound healing. [129] presented a silk-based screw model as an effective device for bone-remodeling processes (Fig. 10.10b). The silk screws fixed bone for up to 8 weeks and showed comparable mechanical properties to traditional resorbable systems. Silk screws were also self-tapping, eliminating the need for a two-step implantation process.

Researchers are exploring more diverse applications of silk-based implementations in the medical sector, working with silk fibers, films, hydrogels, and scaffold materials. There have always been limitations, and numerous studies have drawn valuable conclusions, considering the excellent possibilities of silk-based materials in medical technology [12, 126, 130].

10.6 Future Perspectives and Challenges

10.6.1 Economic and Scalability Constraints

Silk-based wearable textile manufacturing faces some limitations, including high cost, market competition, relatively high price compared to synthetic fibers, etc. When the harvesting of silk fiber is already costly, additional cost is observed with the integration of sensors, actuators, and several other electronic components into the clothing. Since the initial investment is high, the development of this sector is getting slow [131]. The production of sustainable products also incurs additional costs. Polymer conductors or metal electrodes are preferred for their high conductivity, but their biodegradability and biocompatibility properties are not up to the mark and raise concerns. Therefore, the search for sustainable electronic components is continued, and the process is rather costly [132]. The waste management of wearable production is also responsible for increased costs. Even though silk fiber is biodegradable and can be repurposed, the other components used in the production of smart wearable textiles are not always easily disposable, and the separation of these materials from silk is quite complex [9].

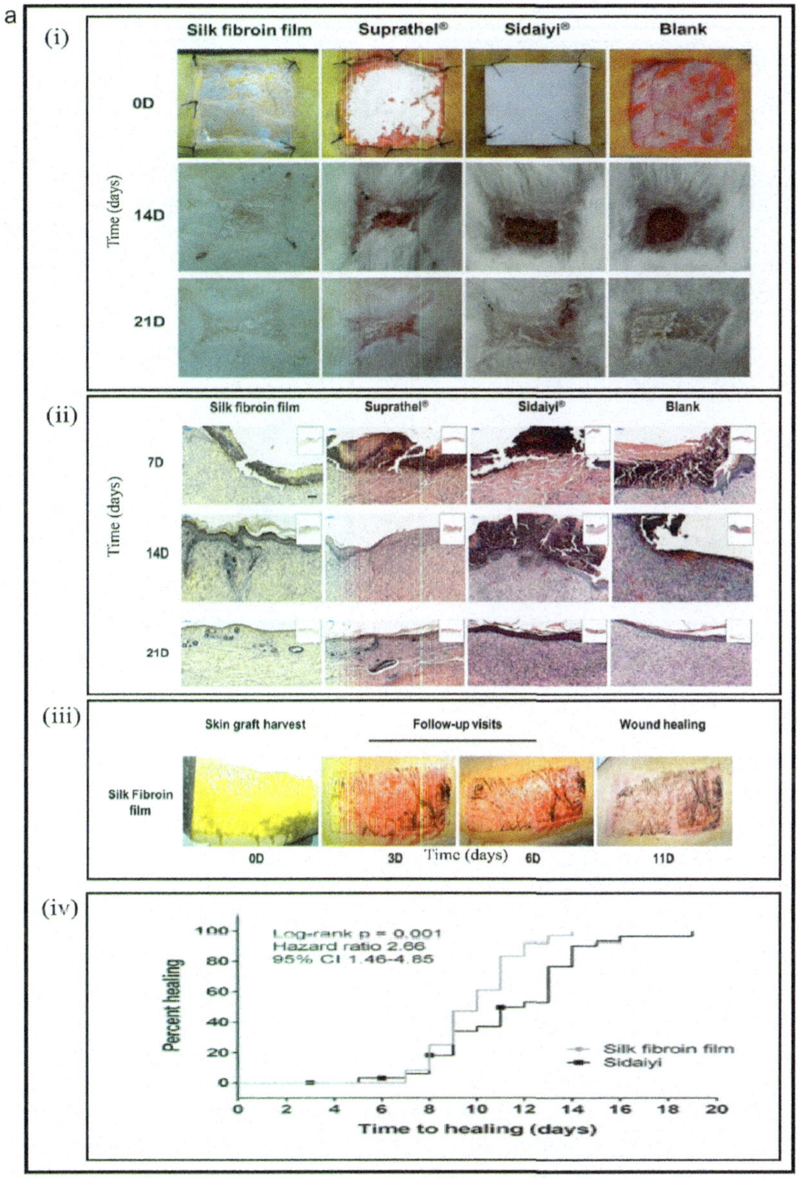

Fig. 10.10 **a** (i) Experimental setup with a rabbit model and macroscopic evaluation. General observation of wound healing morphology across days 0–21. (ii) Histological evaluation of the rabbit model using HE staining. Images with scale bars of 100 μm. The inset displays the complete wound area, with the boxed regions magnified tenfold. (iii) Clinical trial design and follow-up assessment. Representative case from the silk fibroin film group. Macroscopic images were obtained immediately after skin graft removal (day 0), at the first follow-up (day 3), second follow-up (day 6), and at full wound closure (day 11). (iv) Comparison of cumulative wound healing rates between treatment groups using Kaplan–Meier survival curves [128], **b** Silk-based devices for fracture fixation (i) An SEM image of a silk-based fibroin screw (scale bar 1 mm). (ii) Silk-based fibroin screw inserted into a rat femur at 4 weeks post-surgery. (iii and iv) Cross sections of the silk fibroin screw inserted into a rat femur at 4 weeks post-surgery [129]

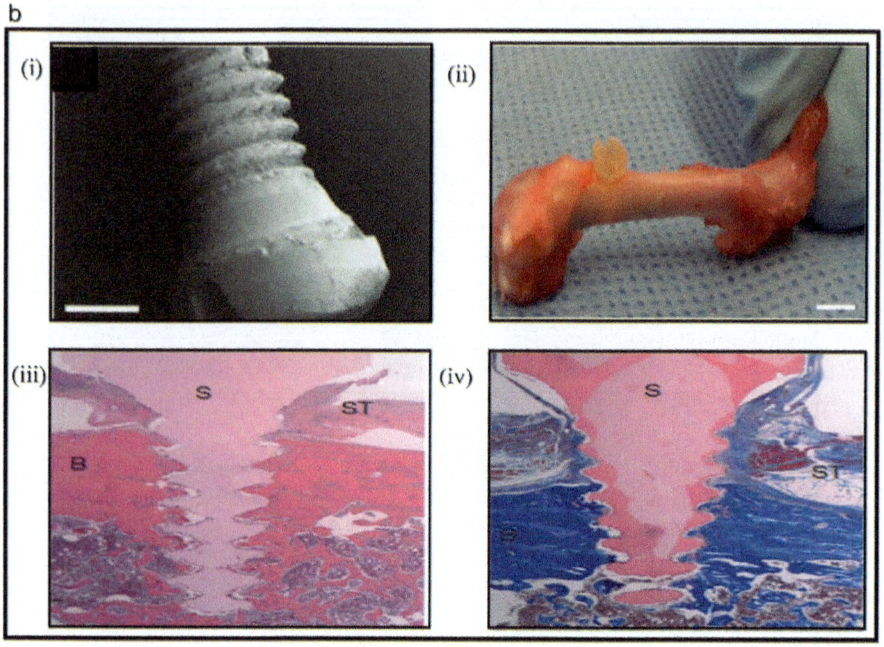

Fig. 10.10 (continued)

The large-scale production of silk-based wearable textiles is hindered due to several causes. The silk industry was overlooked for a long time, and it has only grown based on silk fibroin. The use of sericin protein is not widespread due to not having enough reports about its environmental effect and degradation during production [132]. Another reason for the scalability issue is that the harvesting process is quite intricate. For example, harvesting synthetic polymers and other natural fibers is more efficient than harvesting spider silk. Durability is an important aspect of wearable textiles as they undergo serious bending, stretching, and twisting in their useful lifetime. So, the washing and mechanical properties of silk-based wearable textiles need to be good enough to last for a long time, which is a huge challenge as textile-based sensors are susceptible to dust, sweat, heat cycling, and washing [133–138]. Furthermore, the competency and operation of hybrid devices are affected because pre- or post-treatment is compulsory for the conductivity of silk. Conductive fillers that are incorporated into the devices possess mechanical properties quite different from those of silk. The inconsistency of those properties causes the deterioration of the conductive layer or the detachment of silk [139].

The environmental and health effects also need to be considered. The large-scale production of wearable textiles becomes difficult due to accidental electric shock, the absence of an emergency shut-off system during malfunctioning, the probable toxicity of metals and nanoparticles during lifetime, and a great deal of waste generation after the end of use [8, 140, 141]. The short lifetime of these wearable textiles also causes a great deal of waste formation [9]. The short life span of silk-based wearable textiles is caused by the high degradation properties of silk, which hinder the mass production of silk-based wearable textiles [142].

10.6.2 Consumer Adoption and Awareness

Several factors need to be considered while keeping consumer satisfaction in mind. For example, washability, cleaning facility, strength and stability, dyeability, easily understandable and usable, easily wearable, comfortability, physical aesthetic, durability, multifunctionality, mobility, and user empowerment [131, 143, 144]. Consumers have several concerns when it comes to smart textiles, such as the adverse effect of nanocomponents used in the textiles on human health and the environment; the leak of electrolytes used for conversion and storage of energy, which can pose a serious threat to the wearer; the sweat can destroy the protective layer of electronic components during long time wearing and can cause electrical leakage; burning of skin can occur due to heat incurring during operation of smart textile, which is quite dangerous, especially for babies; fiber-shaped electronic materials containing combustible and harmful organic electrolytes have the possibility of ignition and eruption under deformation and lastly dyeing [145, 146]. Another big concern is the leakage of personal data, as the operation and application of smart textiles depend heavily on data collection and processing. As a result, it becomes important to ensure one hundred percent data security to gain consumer trust and prevent any possibility of future social problems [145–148]. Otherwise, the future of silk-based smart textiles will face a serious threat.

10.6.3 Regulatory and Ethical Considerations

The operation of smart textiles relies on data accumulation, precise analysis, and proper use of analyzed results. The whole process needs to be accurate with zero error tolerance because the major application sector of smart textiles is still the biomedical sector. Any error can cause severe health risks and can even cost a life. However, the current situation of smart textiles is still lacking in terms of accuracy, consistency, and reproducibility. Direct and strong skin contact is not always easy to obtain, which costs accuracy. Additionally, other factors like rain and sweating can hinder the function of smart textiles [141, 149, 150]. Furthermore, the insufficiency of regulatory structures and standardized test methods may potentially threaten health [141]. Therefore, further research, exploration, and thorough testing are required to ensure precise operation. These initiatives are crucial while being costly and labor-intensive [147]. The environmental impact also needs to be considered with great importance, such as waste generation, power consumption, and electronic waste.

Despite the above-mentioned constraints, factors like the demand for green technology, silk's strength, and its versatility create hope for the flourishing future and further development of this sector. The connectivity between the silk fiber structure and properties is fascinating, and it always motivates scientists to continue their efforts toward the exploration and development of silk-based sectors. Therefore, to overcome the challenges faced in the fabrication and application of silk-based smart textiles, more research on the cultivation and integration process, and on incorporating material, data gathering, development of an efficient waste disposal process, and appropriate test facilities are essential.

10.7 Conclusion

The incorporation of silk-derived materials into functional textiles and bendable electronics presents a pivotal role in the aspect of technical textiles. The property of silk fiber facilitates diverse applications, ranging from regular sensors to biomedical technology. Materials prepared from silk fibroin solution (regenerated silk fiber, silk sponges, silk films, silk hydrogels) have specific attributes compatible with the special requirements of wearable smart textiles. However, silk-based technical textiles face both economic and scalability issues. The initial investment of this section has become a major constraint. The large-scale production of these technical textiles also proved less efficient, and the deterioration of suitable properties in the end product is also challenging. Moreover, customer satisfaction also needs to be maintained at different levels during the life cycle of these technical textiles. Since the demand for wearable smart textiles is growing, silk-based technical textiles present significant opportunities for innovation and development in the textile industry, making them a valuable asset for future textile technologies. Therefore, more intensive study is required to functionalize more smart properties into silk-based technical textiles, and more research is mandatory to improve the current limitations to ensure their long-term performance.

Declaration of AI Usage The authors declare that generative AI and AI-assisted technologies were used solely to enhance the readability and language of the text. The authors have thoroughly reviewed and edited the content as necessary and take full responsibility for the final version of the publication.

References

1. Chen M, Ma Y, Song J, Lai C-F, Hu B (2016) Smart clothing: connecting human with clouds and big data for sustainable health monitoring. Mob Netw Appl 21:825–845
2. Gaffar MA, Mahin M, Islam MS, Raha MR, Islam MR (2025) Conductive and smart textiles: innovations and environmental impact. In: Advancements in Textile Finishing. Springer, Singapore, pp 305–340.

3. Khan BU, Tanha JA, Amin J, Fahim TH, Islam MR (2025) Thermal regulation in textiles: phase change materials and sustainability. In: Sustainable Finishing Techniques in Textiles. Springer, Singapore, pp 179–213
4. Syduzzaman M, Hassan A, Anik HR, Akter M, Islam MR (2023) Nanotechnology for high-performance textiles: a promising frontier for innovation. ChemNanoMat 9(9):e202300205
5. Ghahremani Honarvar M, Latifi M (2017) Overview of wearable electronics and smart textiles. J TextE Inst 108(4):631–652
6. Islam MR, Afroj S, Novoselov KS, Karim N (2022) Smart electronic textile-based wearable supercapacitors. Advanced Science 9(31):2203856
7. Islam MR, Afroj S, Yin J, Novoselov KS, Chen J, Karim N (2024) Advances in printed electronic textiles. Adv Sci 11(6):2304140
8. Dulal M, Modha HRM, Liu J, Islam MR, Carr C, Hasan T, Thorn RMS, Afroj S, Karim N (2025) Sustainable, wearable, and eco-friendly electronic textiles. Energy Environ Mater 8(3):e12854
9. Xing T, He A, Huang Z, Luo Y, Zhang Y, Wang M, Shi Z, Ke G, Bai J, Zhao S (2023) Silk-based flexible electronics and smart wearable textiles: progress and beyond. Chem Eng J 474:145534
10. Wang C, Xia K, Zhang Y, Kaplan DL (2019) Silk-based advanced materials for soft electronics. Acc Chem Res 52(10):2916–2927
11. Sahoo JK, Hasturk O, Falcucci T, Kaplan DL (2023) Silk chemistry and biomedical material designs. Nat Rev Chem 7(5):302–318
12. Yang X-C, Wang X-X, Wang C-Y, Zheng H-L, Yin M, Chen K-Z, Qiao S-L (2024) Silk-based intelligent fibers and textiles: structures, properties, and applications. Chem Commun 60(61):7801–7823
13. Padaki NV, Das B, Basu A (2015) Advances in understanding the properties of silk. In: Advances in Silk Science and Technology. Woodhead Publishing, Cambridge, pp 3–16
14. Qiu W, Patil A, Hu F, Liu XY (2019) Hierarchical structure of silk materials versus mechanical performance and mesoscopic engineering principles. Small 15(51):1903948
15. Huang W, Ling S, Li C, Omenetto FG, Kaplan DL (2018) Silkworm silk-based materials and devices generated using bio-nanotechnology. Chem Soc Rev 47(17):6486–6504
16. Koh L-D, Cheng Y, Teng C-P, Khin Y-W, Loh X-J, Tee S-Y, Low M, Ye E, Yu H-D, Zhang Y-W (2015) Structures, mechanical properties and applications of silk fibroin materials. Prog Polym Sci 46:86–110
17. Zhou CZ, Confalonieri F, Jacquet M, Perasso R, Li ZG, Janin J (2001) Silk fibroin: structural implications of a remarkable amino acid sequence. Proteins: Struct, Funct, Bioinform 44(2), 119–122
18. Nguyen AT, Huang QL, Yang Z, Lin N, Xu G, Liu XY (2015) Crystal networks in silk fibrous materials: from hierarchical structure to ultra performance. Small 11(9–10):1039–1054
19. Ha S-W, Gracz HS, Tonelli AE, Hudson SM (2005) Structural study of irregular amino acid sequences in the heavy chain of Bombyx m ori silk fibroin. Biomacromol 6(5):2563–2569
20. Harapanahalli M, Raaja N (2016) Eri: the silk of the century. Int J Eng Technol Sci Res 3(5):155–160
21. Basu A (2015) Advances in silk science and technology. Woodhead Publishing
22. Robson RM (1999) Microvoids in Bombyx mori silk. An electron microscope study. Int J Biol Macromol 24(2–3):145–150
23. Zhu Z, Gong D, Liu L, Wang Y (2014) Microstructure elucidation of historic silk (Bombyx mori) by nuclear magnetic resonance. Anal Bioanal Chem 406:2709–2718
24. Porter D, Vollrath F (2009) Silk as a biomimetic ideal for structural polymers. Adv Mater 21(4):487–492
25. IIZUKA E (1996) Size dependency of the physical properties of Bombyx silk. J Sericultural Sci Jpn 65(2), 102–108
26. Cheung H-Y, Lau K-T, Ho M-P, Mosallam A (2009) Study on the mechanical properties of different silkworm silk fibers. J Compos Mater 43(22):2521–2531

27. Li G, Li Y, Chen G, He J, Han Y, Wang X, Kaplan DL (2015) Silk-based biomaterials in biomedical textiles and fiber-based implants. Adv Healthcare Mater 4(8):1134–1151
28. Lu Z, Mao C, Zhang H (2015) Highly conductive graphene-coated silk fabricated via a repeated coating-reduction approach. J Mater Chem C 3(17):4265–4268
29. Ye C, Ren J, Wang Y, Zhang W, Qian C, Han J, Zhang C, Jin K, Buehler MJ, Kaplan DL (2019) Design and fabrication of silk templated electronic yarns and applications in multifunctional textiles. Matter 1(5):1411–1425
30. Chiu W-T, Chen C-Y, Chang T-FM, Tahara Y, Hashimoto T, Kurosu H, Sone M (2018) Platinum coating on silk by a supercritical CO2 promoted metallization technique for applications of wearable devices. Surf Coat Technol 350:1028–1035
31. Ryan JD, Mengistie DA, Gabrielsson R, Lund A, Müller C (2017) Machine-washable PEDOT: PSS dyed silk yarns for electronic textiles. ACS Appl Mater Interfaces 9(10):9045–9050
32. Georgakilas V, Otyepka M, Bourlinos AB, Chandra V, Kim N, Kemp KC, Hobza P, Zboril R, Kim KS (2012) Functionalization of graphene: covalent and non-covalent approaches, derivatives and applications. Chem Rev 112(11):6156–6214
33. Cao J, Wang C (2018) Highly conductive and flexible silk fabric via electrostatic self assemble between reduced graphene oxide and polyaniline. Org Electron 55:26–34
34. Zulan L, Zhi L, Lan C, Sihao C, Dayang W, Fangyin D (2019) Reduced graphene oxide coated silk fabrics with conductive property for wearable electronic textiles application. Adv Electron Mater 5(4):1800648
35. Bhattacharjee S, Macintyre CR, Bahl P, Kumar U, Wen X, Aguey-Zinsou KF, Chughtai AA, Joshi R (2020) Reduced graphene oxide and nanoparticles incorporated durable electroconductive silk fabrics. Adv Mater Interfaces 7(20):2000814
36. Zhou J, Zhao Z, Hu R, Yang J, Xiao H, Liu Y, Lu M (2020) Multi-walled carbon nanotubes functionalized silk fabrics for mechanical sensors and heating materials. Mater Des 191:108636
37. Babu KM (2020) Silk fibres–structure, properties and applications. In: Handbook of Natural Fibres. Elsevier, pp 385–416
38. Jin HJ, Park J, Karageorgiou V, Kim UJ, Valluzzi R, Cebe P, Kaplan DL (2005) Water-stable silk films with reduced β-sheet content. Adv Func Mater 15(8):1241–1247
39. Wang Y, Rudym DD, Walsh A, Abrahamsen L, Kim H-J, Kim HS, Kirker-Head C, Kaplan DL (2008) In vivo degradation of three-dimensional silk fibroin scaffolds. Biomaterials 29(24–25):3415–3428
40. DeBari MK, King CI III, Altgold TA, Abbott RD (2021) Silk fibroin as a green material. ACS Biomater Sci Eng 7(8):3530–3544
41. Malinowski C, He F, Zhao Y, Chang I, Hatchett DW, Zhai S, Zhao H (2019) Nanopatterned silk fibroin films with high transparency and high haze for optical applications. RSC Adv 9(70):40792–40799
42. Wen D-L, Liu X, Deng H-T, Sun D-H, Qian H-Y, Brugger J, Zhang X-S (2019) Printed silk-fibroin-based triboelectric nanogenerators for multi-functional wearable sensing. Nano Energy 66:104123
43. Wen D-L, Sun D-H, Huang P, Huang W, Su M, Wang Y, Han M-D, Kim B, Brugger J, Zhang H-X (2021) Recent progress in silk fibroin-based flexible electronics. Microsyst Nanoeng 7(1):35
44. Ha S-W, Tonelli AE, Hudson SM (2005) Structural studies of bombyx m ori silk fibroin during regeneration from solutions and wet fiber spinning. Biomacromol 6(3):1722–1731
45. Srivastava CM, Purwar* R (2014) Recent developments in regenerated silk fibroin fibers. Int J Res 2(1)
46. Cao Y, Wang B (2009) Biodegradation of silk biomaterials. Int J Mol Sci 10(4):1514–1524
47. Carrasco-Torres G, Valdés-Madrigal MA, Vásquez-Garzón VR, Baltiérrez-Hoyos R, De la Cruz-Burelo E, Román-Doval R, Valencia-Lazcano AA (2019) Effect of silk fibroin on cell viability in electrospun scaffolds of polyethylene oxide. Polymers 11(3):451
48. Jin H-J, Fridrikh SV, Rutledge GC, Kaplan DL (2002) Electrospinning Bombyx mori silk with poly (ethylene oxide). Biomacromol 3(6):1233–1239

49. Keten S, Xu Z, Ihle B, Buehler MJ (2010) Nanoconfinement controls stiffness, strength and mechanical toughness of β-sheet crystals in silk. Nat Mater 9(4):359–367
50. Rockwood DN, Preda RC, Yücel T, Wang X, Lovett ML, Kaplan DL (2011) Materials fabrication from Bombyx mori silk fibroin. Nat Protoc 6(10):1612–1631
51. Mehrotra S, Nandi SK, Mandal BB (2017) Stacked silk-cell monolayers as a biomimetic three dimensional construct for cardiac tissue reconstruction. J Mater Chem B 5(31):6325–6338
52. Omenetto FG, Kaplan DL (2008) A new route for silk. Nat Photonics 2(11):641–643
53. Ding G, Jin Q, Chen Q, Hu Z, Liu J (2015) The fabrication of ordered bulk heterojunction solar cell by nanoimprinting lithography method using patterned silk fibroin mold at room temperature. Nanoscale Res Lett 10:1–10
54. Lee KY, Chun J, Lee J-H, Kim KN, Kang N-R, Kim J-Y, Kim MH, Shin K-S, Gupta MK, Baik JM (2014) Hydrophobic sponge structure-based triboelectric nanogenerator. Adv Mater 26(29):5037–5042
55. Yin X, Vinod T, Jelinek R (2015) A flexible high-sensitivity piezoresistive sensor comprising a Au nanoribbon-coated polymer sponge. J Mater Chem C 3(35):9247–9252
56. Gao K, Guo Y, Niu Q, Han L, Zhang L, Zhang Y, Wang L (2018) Cellulose nanofibers/silk fibroin nanohybrid sponges with highly ordered and multi-scale hierarchical honeycomb structure. Cellulose 25:429–437
57. Maniglio D, Bonani W, Migliaresi C, Motta A (2018) Silk fibroin porous scaffolds by N2O foaming. J Biomater Sci Polym Ed 29(5):491–506
58. Nazarov R, Jin H-J, Kaplan DL (2004) Porous 3-D scaffolds from regenerated silk fibroin. Biomacromol 5(3):718–726
59. Park HJ, Lee OJ, Lee MC, Moon BM, Ju HW, Lee JM, Kim JH, Kim DW, Park CH (2015) Fabrication of 3D porous silk scaffolds by particulate (salt/sucrose) leaching for bone tissue reconstruction. Int J Biol Macromol 78:215–223
60. Wang C, Xia K, Zhang M, Jian M, Zhang Y (2017) An all-silk-derived dual-mode e-skin for simultaneous temperature–pressure detection. ACS Appl Mater Interfaces 9(45):39484–39492
61. Ribeiro VP, da Silva Morais A, Maia FR, Canadas RF, Costa JB, Oliveira AL, Oliveira JM, Reis RL (2018) Combinatory approach for developing silk fibroin scaffolds for cartilage regeneration. Acta Biomater 72:167–181
62. Gholipourmalekabadi M, Sapru S, Samadikuchaksaraei A, Reis RL, Kaplan DL, Kundu SC (2020) Silk fibroin for skin injury repair: where do things stand? Adv Drug Deliv Rev 153:28–53
63. Han L, Lu X, Wang M, Gan D, Deng W, Wang K, Fang L, Liu K, Chan CW, Tang Y (2017) A mussel-inspired conductive, self-adhesive, and self-healable tough hydrogel as cell stimulators and implantable bioelectronics. Small 13(2):1601916
64. Wang HY, Zhang YQ (2015) Processing silk hydrogel and its applications in biomedical materials. Biotechnol Prog 31(3):630–640
65. Zheng H, Zuo B (2021) Functional silk fibroin hydrogels: preparation, properties and applications. J Mater Chem B 9(5):1238–1258
66. Chouhan D (2018) Lohe Tu; Samudrala PK; Mandal, BB In situ forming injectable silk fibroin hydrogel promotes skin regeneration in full thickness burn wounds. Adv Healthc Mater 7(1801092.10):1002
67. Ming J, Zuo B (2014) Crystal growth of calcium carbonate in silk fibroin/sodium alginate hydrogel. J Cryst Growth 386:154–161
68. Vu T, Xue Y, Vuong T, Erbe M, Bennet C, Palazzo B, Popielski L, Rodriguez N, Hu X (2016) Comparative study of ultrasonication-induced and naturally self-assembled silk fibroin-wool keratin hydrogel biomaterials. Int J Mol Sci 17(9):1497
69. Chen D, Yin Z, Wu F, Fu H, Kundu SC, Lu S (2017) Orientational behaviors of silk fibroin hydrogels. J Appl Polym Sci 134(32):45050
70. Wang L, Song D, Zhang X, Ding Z, Kong X, Lu Q, Kaplan DL (2018) Silk–graphene hybrid hydrogels with multiple cues to induce nerve cell behavior. ACS Biomater Sci Eng 5(2):613–622

71. Kim U-J, Park J, Li C, Jin H-J, Valluzzi R, Kaplan DL (2004) Structure and properties of silk hydrogels. Biomacromol 5(3):786–792
72. Floren ML, Spilimbergo S, Motta A, Migliaresi C (2012) Carbon dioxide induced silk protein gelation for biomedical applications. Biomacromol 13(7):2060–2072
73. Nagarkar S, Patil A, Lele A, Bhat S, Bellare J, Mashelkar R (2009) Some mechanistic insights into the gelation of regenerated silk fibroin sol. Ind Eng Chem Res 48(17):8014–8023
74. Aigner TB, DeSimone E, Scheibel T (2018) Biomedical applications of recombinant silk-based materials. Adv Mater 30(19):1704636
75. Lu Q, Huang Y, Li M, Zuo B, Lu S, Wang J, Zhu H, Kaplan DL (2011) Silk fibroin electrogelation mechanisms. Acta Biomater 7(6):2394–2400
76. Dorishetty P, Balu R, Sreekumar A, de Campo L, Mata JP, Choudhury NR, Dutta NK (2019) Robust and tunable hybrid hydrogels from photo-cross-linked soy protein isolate and regenerated silk fibroin. ACS Sustain Chem Eng 7(10):9257–9271
77. Zhou Y, Liang K, Zhao S, Zhang C, Li J, Yang H, Liu X, Yin X, Chen D, Xu W (2018) Photopolymerized maleilated chitosan/methacrylated silk fibroin micro/nanocomposite hydrogels as potential scaffolds for cartilage tissue engineering. Int J Biol Macromol 108:383–390
78. Kim MH, Kim BS, Lee J, Cho D, Kwon OH, Park WH (2017) Silk fibroin/hydroxyapatite composite hydrogel induced by gamma-ray irradiation for bone tissue engineering. Biomater Res 21(1):12
79. Lin Y, Xia X, Shang K, Elia R, Huang W, Cebe P, Leisk G, Omenetto F, Kaplan DL (2013) Tuning chemical and physical cross-links in silk electrogels for morphological analysis and mechanical reinforcement. Biomacromol 14(8):2629–2635
80. Zhou H, Wang Z, Cao H, Hu H, Luo Z, Yang X, Cui M, Zhou L (2019) Genipin-crosslinked polyvinyl alcohol/silk fibroin/nano-hydroxyapatite hydrogel for fabrication of artificial cornea scaffolds—a novel approach to corneal tissue engineering. J Biomater Sci Polym Ed 30(17):1604–1619
81. Chirila TV, Suzuki S, Papolla C (2017) A comparative investigation of Bombyx mori silk fibroin hydrogels generated by chemical and enzymatic cross-linking. Biotechnol Appl Biochem 64(6):771–781
82. Hasturk O, Jordan KE, Choi J, Kaplan DL (2020) Enzymatically crosslinked silk and silk-gelatin hydrogels with tunable gelation kinetics, mechanical properties and bioactivity for cell culture and encapsulation. Biomaterials 232:119720
83. Hennink WE, van Nostrum CF (2002) Novel crosslinking methods to design hydrogels. Adv Drug Deliv Rev 54(1):13–36
84. Raia NR, Partlow BP, McGill M, Kimmerling EP, Ghezzi CE, Kaplan DL (2017) Enzymatically crosslinked silk-hyaluronic acid hydrogels. Biomaterials 131:58–67
85. Ahmed EM (2015) Hydrogel: preparation, characterization, and applications: a review. J Adv Res 6(2):105–121
86. Wu N, Yu H, Sun M, Li Z, Zhao F, Ao Y, Chen H (2019) Investigation on the structure and mechanical properties of highly tunable elastomeric silk fibroin hydrogels cross-linked by γ-ray radiation. ACS Appl Bio Mater 3(1):721–734
87. Enyan M, Bing Z, Amu-Darko JNO, Issaka E, Otoo SL, Agyemang MF (2025) Advances in smart materials soft actuators on mechanisms, fabrication, materials, and multifaceted applications: a review. J Thermoplast Compos Mater 38(1):302–370
88. Jayathilaka WADM, Qi K, Qin Y, Chinnappan A, Serrano-García W, Baskar C, Wang H, He J, Cui S, Thomas SW (2019) Significance of nanomaterials in wearables: a review on wearable actuators and sensors. Adv Mater 31(7):1805921
89. Shen X, Shi H, Wei H, Wu B, Xia Q, Yeo J, Huang W (2022) Engineering natural and recombinant silks for sustainable biodevices. Front Chem 10:881028
90. Manikandan G, Murali A, Kumar R, Satapathy DK (2021) Rapid moisture-responsive silk fibroin actuators. ACS Appl Mater Interfaces 13(7):8880–8888
91. Ganesan M, Kumar R, Satapathy DK (2022) Bidirectional actuation of silk fibroin films: role of water and alcohol vapors. Langmuir 38(19):6066–6075

92. Javaid M, Haleem A, Rab S, Singh RP, Suman R (2021) Sensors for daily life: a review. SensS Int 2:100121
93. Pragya A, Chatterjee K, Ghosh TK (2023) Sensors and actuators for textiles: from materials to applications. In: Smart and Functional Textiles. De Gruyter, Berlin, Boston, pp 469–532
94. Pradhan S, Yadavalli VK (2020) Photolithographically printed flexible silk/PEDOT: PSS temperature sensors. ACS Appl Electron Mater 3(1):21–29
95. Yang Z, Huang T, Cao P, Cui Y, Nie J, Chen T, Yang H, Wang F, Sun L (2022) Carbonized silk nanofibers in biodegradable, flexible temperature sensors for extracellular environments. ACS Appl Mater Interfaces 14(16):18110–18119
96. He F, You X, Gong H, Yang Y, Bai T, Wang W, Guo W, Liu X, Ye M (2020) Stretchable, biocompatible, and multifunctional silk fibroin-based hydrogels toward wearable strain/pressure sensors and triboelectric nanogenerators. ACS Appl Mater Interfaces 12(5):6442–6450
97. Zhang Q, Soham D, Liang Z, Wan J (2025) Advances in wearable energy storage and harvesting systems. Med-X 3(1):3
98. Strassburg S, Zainuddin S, Scheibel T (2021) The power of silk technology for energy applications. Adv Energy Mater 11(43)
99. Lu Z, Zhang H, Mao C, Li CM (2016) Silk fabric-based wearable thermoelectric generator for energy harvesting from the human body. Appl Energy 164:57–63
100. Kim HJ, Kim JH, Jun KW, Kim JH, Seung WC, Kwon OH, Park JY, Kim SW, Oh IK (2016) Silk nanofiber-networked bio-triboelectric generator: silk Bio-TEG. Adv Energy Mater 6(8)
101. Islam MR, Afroj S, Novoselov KS, Karim N (2024) Inkjet-printed 2D heterostructures for smart textile micro-supercapacitors. Adv Func Mater 34(52):2410666
102. Lee S, Ozlu B, Eom T, Martin DC, Shim BS (2020) Electrically conducting polymers for bio-interfacing electronics: from neural and cardiac interfaces to bone and artificial tissue biomaterials. Biosens Bioelectron 170:112620
103. Li G, Liu S, Mao Q, Zhu R (2022) Multifunctional electronic skins enable robots to safely and dexterously interact with human. Adv Sci 9(11):2104969
104. Zheng Y, Wang D, Zhao L, Wang X, Han W, Wang L (2022) Chemically modified silk fibroin hydrogel for environment-stable electronic skin. Sens Actuators Rep 4:100089
105. Huang J, Xu Z, Qiu W, Chen F, Meng Z, Hou C, Guo W, Liu XY (2020) Stretchable and heat-resistant protein-based electronic skin for human thermoregulation. Adv Func Mater 30(13):1910547
106. Huang W, Ding Z, Wang C, Wei J, Zhao Y, Purnawali H (2010) Shape memory materials. Materials today 13(7–8):54–61
107. Zhang Y, Zhou J, Deng H, Fang Y, Qiao N, Ren M, Zhang Y, Zhang D, Lin H, Chen Y (2023) Silk fibroin fibers-based shape memory membrane with Janus wettability for multitiered wearable protection. J Mater Res 38(3):633–643
108. Oral CB, Su E, Okay O (2024) Silk fibroin-based multiple-shape-memory organohydrogels. ACS Appl Mater Interfaces 16(41):56146–56158
109. Xu X, Wang Z, Su Y, Zhang K, Li M, Zhang Q, Zhang S, Zhao Y, Ke Q, Hu H (2024) Biomimetic design of hydration-responsive silk fibers and their role in actuators and self-modulated textiles. Adv Func Mater 34(33):2401732
110. Jaya Prakash N, Sarkar SS, Kandasubramanian B (2023) Emerging strategies in stimuli-responsive silk architectures. Macromol Biosci 23(6):2200573
111. Hasan KF, Bai S, Chen S, Lin K, Ahmed T, Chen J, Pan A, Zhu Y, Lin CSK, Tso CY (2024) Nanotechnology-empowered radiative cooling and warming textiles. Cell Rep Phys Sci 5(9):102108
112. Li X, Cui R, Liu W, Sun L, Yu B, Fan Y, Feng Q, Cui F, Watari F (2013) The use of nanoscaled fibers or tubes to improve biocompatibility and bioactivity of biomedical materials. J Nanomater 2013(1):728130
113. Ma R, Xu C, Li D, Yang J, Huang J, Guo Z (2024) Fabricated advanced textile for personal thermal management, intelligent health monitoring and energy harvesting. Adv Colloid Interface Sci 332:103252

114. Zhu B, Li W, Zhang Q, Li D, Liu X, Wang Y, Xu N, Wu Z, Li J, Li X (2021) Subambient daytime radiative cooling textile based on nanoprocessed silk. Nat Nanotechnol 16(12):1342–1348
115. Jiang S, Zhang K, Wang C-F, Li Q, Zhu L, Chen S (2024) Recent advancements in radiative cooling textiles for personal thermal management. J Mater Chem A 12:14866–14884
116. Zhang H, Xu Z, Wei Z, Zhang T, Wang X, Zhao Y (2025) Electrostatic flocking of hierarchically micro/nanostructured natural silk fibers for efficient passive radiative cooling. ACS Appl Mater & Interfaces 17(11):17580–17589
117. Strydis C (2012) Implantable microelectronic devices. Master's thesis, Delft University of Technology, Delft, Netherlands
118. Bradley-Munn SR (2017) Body-Modifying Technologies and the Individual's Right of Moral Autonomy University of Wollongong].
119. Klich J (2024) Human enhancement technologies and healthcare policy. Taylor & Francis
120. Li X, Huang X, Yang L, Jung S, Wang J, Zhao H (2025) Implantable physical sensors for in vivo organ monitoring. Med-X 3(1):1
121. Lv Q, Li Q, Cao P, Wei C, Li Y, Wang Z, Wang L (2024) Designing silk biomaterials toward better future healthcare: the development and application of silk-based implantable electronic devices in clinical diagnosis and therapy. Adv Mater 37:2411946
122. Tang-Schomer MD, Hu X, Hronik-Tupaj M, Tien LW, Whalen MJ, Omenetto FG, Kaplan DL (2014) Film-based implants for supporting neuron–electrode integrated interfaces for the brain. Adv Func Mater 24(13):1938–1948
123. Niu Q, Chen J, Fan S, Yao X, Gu Y, Hsiao BS, Wei H, Zhang Y (2024) Silk nanoribbon films with enriched silk II structure and enhanced piezoelectricity for self-powered implantable and wearable devices. Nano Today 56:102228
124. Lujerdean C, Baci G-M, Cucu A-A, Dezmirean DS (2022) The contribution of silk fibroin in biomedical engineering. Insects 13(3):286
125. Kasoju N, Bora U (2012) Silk fibroin in tissue engineering. Adv Healthcare Mater 1(4):393–412
126. Leal-Egaña A, Scheibel T (2010) Silk-based materials for biomedical applications. Biotechnol Appl Biochem 55(3):155–167
127. Nguyen TP, Nguyen QV, Nguyen V-H, Le T-H, Huynh VQN, Vo D-VN, Trinh QT, Kim SY, Le QV (2019) Silk fibroin-based biomaterials for biomedical applications: a review. Polymers 11(12):1933
128. Zhang W, Chen L, Chen J, Wang L, Gui X, Ran J, Xu G, Zhao H, Zeng M, Ji J (2017) Silk fibroin biomaterial shows safe and effective wound healing in animal models and a randomized controlled clinical trial. Adv Healthcare Mater 6(10):1700121
129. Perrone GS, Leisk GG, Lo TJ, Moreau JE, Haas DS, Papenburg BJ, Golden EB, Partlow BP, Fox SE, Ibrahim AM (2014) The use of silk-based devices for fracture fixation. Nat Commun 5(1):3385
130. Holland C, Numata K, Rnjak-Kovacina J, Seib FP (2019) The biomedical use of silk: past, present, future. Adv Healthcare Mater 8(1):1800465
131. Dias T (2015) Electronic textiles: Smart fabrics and wearable technology. Woodhead Publishing
132. Song Y, Hu C, Wang Z, Wang L (2024) Silk-based wearable devices for health monitoring and medical treatment. IScience 27(5):109686
133. Islam GN, Ali A, Collie S (2020) Textile sensors for wearable applications: a comprehensive review. Cellulose 27:6103–6131
134. Ismar E, Kurşun Bahadir S, Kalaoglu F, Koncar V (2020) Futuristic clothes: electronic textiles and wearable technologies. Global Chall 4(7):1900092
135. Merhi Y, Mikkelsen PH, Suetta C, Nygaard JV, Agarwala S (2020) Mechanical performance of electronically functional smart textiles. Trans Addit Manuf Meets Med 2(1):012
136. Park K-J, Gong M-S (2017) A water durable resistive humidity sensor based on rigid sulfonated polybenzimidazole and their properties. Sens Actuators, B Chem 246:53–60
137. Shahariar H, Soewardiman H, Muchler CA, Adams JJ, Jur JS (2018) Porous textile antenna designs for improved wearability. Smart Mater Struct 27(4):045008

138. Zhuang Z, Li Y, Qi D, Zhao C, Na H (2017) Novel polymeric humidity sensors based on sulfonated poly (ether ether ketone) s: Influence of sulfonation degree on sensing properties. Sens Actuators, B Chem 242:801–809
139. Hou J, Cao C, Idrees F, Ma X (2015) Hierarchical porous nitrogen-doped carbon nanosheets derived from silk for ultrahigh-capacity battery anodes and supercapacitors. ACS Nano 9(3):2556–2564
140. Ahmed DM, Gaffar MA, Mahin M, Khan BU, Rayyaan R, Mobin A (2025) Advances in bio-based regenerated and synthetic fibres and their sustainability issues. In: SDG 12 and Global Fashion Textiles Production. Springer, pp 65–100
141. Dolez PI, Decaens J, Buns T, Lachapelle D, Vermeersch O (2020) Applications of smart textiles in occupational health and safety. IOP Conf Ser Mater Sci Eng 827:012014
142. Lee S, Kim SH, Jo Y-Y, Ju W-T, Kim H-B, Kweon H (2021) Effects of ultraviolet light irradiation on silk fibroin films prepared under different conditions. Biomolecules 11(1):70
143. Jung S, Lauterbach C, Strasser M, Weber W (2003) Enabling technologies for disappearing electronics in smart textiles. In: 2003 IEEE International Solid-State Circuits Conference (ISSCC), Digest of Technical Papers, pp 386–387
144. Kan C-W, Lam Y-L (2021) Future trend in wearable electronics in the textile industry. Appl Sci 11(9):3914
145. Shi J, Liu S, Zhang L, Yang B, Shu L, Yang Y, Ren M, Wang Y, Chen J, Chen W (2020) Smart textile-integrated microelectronic systems for wearable applications. Adv Mater 32(5):1901958
146. Wang L, Fu X, He J, Shi X, Chen T, Chen P, Wang B, Peng H (2020) Application challenges in fiber and textile electronics. Adv Mater 32(5):1901971
147. Loncar-Turukalo T, Zdravevski E, da Silva JM, Chouvarda I, Trajkovik V (2019) Literature on wearable technology for connected health: scoping review of research trends, advances, and barriers. J Med Internet Res 21(9):e14017
148. Xu K, Lu Y, Takei K (2019) Multifunctional skin-inspired flexible sensor systems for wearable electronics. Adv Mater Technol 4(3):1800628
149. Firšt Rogale S, Rogale D, Nikolić G (2018) Intelligent clothing: first and second generation clothing with adaptive thermal insulation properties. Text Res J 88(19):2214–2233
150. Jiang Y, Pan K, Leng T, Hu Z (2019) Smart textile integrated wireless powered near field communication body temperature and sweat sensing system. IEEE J Electromagn, RF MicrowS Med Biol 4(3):164–170
151. Wang Y, Aurelio D, Li W, Tseng P, Zheng Z, Li M, Kaplan DL, Liscidini M, Omenetto FG (2017) Modulation of multiscale 3D lattices through conformational control: painting silk inverse opals with water and light. Adv Mater 29(38):1702769

Chapter 11
Silk for Flexible Electronics and Multifunctional Energy Harvesting Applications

M. Amarnath, Afreen Begum, and Shital Palaskar

11.1 Introduction

The development of electronic textiles (e-textiles) has accelerated over the past few years, despite the idea being decades old. In the past, textiles were incorporated with or assembled to feature tiny electronic devices in their surface structure; however, total integration remains a significant challenge [1]. And so, it is not easy to reconcile high device functionality with maintaining qualified textile properties, such as toughness, flexibility, and comfort, particularly when considering large electronics and wired connections over it. One new strategy is to introduce the textiles themselves as an active device, originating from the development of nanotechnology, which transforms fibers into woven, knitted, or non-woven states with in-built electronic functionalities [2]. This enables e-textiles to contribute to a range of tasks, including energy harvesting, conversion, sensing, and information transmission. Energy-harvesting e-textiles have been revolutionizing modern wearable electronics due to their mechanical flexibility, durability, and vast surface area. Particularly, the use of textile materials has become more user-friendly across various fields, ranging from everyday portable energy essentials to continuous human health monitoring. Now, a question arises: How can this energy be harnessed? A simple answer is that for low-power devices, only a very small amount of energy is required for their action, which can be easily generated from the ambient environment by using this textile substrate when functionalized with novel conducting materials [3–5]. We can harness and convert energy from various readily available sources, such as light [6], heat [7], and mechanical motion [8]. However, if this energy was not utilized, it would be automatically dissipated into the atmosphere. This brings forth the concept of waste-to-use, where simple modifications can lead to significant changes, helping to reduce the heavy consumption of non-renewable energy sources. Now, in the twenty-first

M. Amarnath · A. Begum · S. Palaskar (✉)
Plasma Research Lab, The Bombay Textile Research Association, Mumbai, India
e-mail: pmebtra@btraindia.com

M. Shahid et al. (eds.), *Science and Technology of Silk*, Textile Science and Clothing Technology, https://doi.org/10.1007/978-3-032-26880-8_11

century, e-textiles have undergone significant advancements through breakthroughs in inventions and preparation methods of advanced 2D and 3D material developments using nanotechnology [9]. Miniaturization of this device system and integration in applications make them more affordable and user-friendly. The pursuit of ambient energy conversion has led to the emergence of diverse energy-harvesting technologies. Most common energy-harvesting principles were based on thermoelectric [10], piezoelectric [11], triboelectric [12], and photovoltaic mechanisms [13]. Substantial progress in materials science and device engineering has significantly enhanced the efficiency and functionality of these technologies. Notably, the invention of the piezoelectric nanogenerators (PENGs) in 2006 and the triboelectric nanogenerators (TENGs) in 2012 marked a transformative era in mechanical energy harvesting, which can be directly done by a user who is wearing the e-textiles. Concurrently, advancements in nanoengineered materials with controlled architectures have optimized the performance of solar cells (SCs) and thermoelectric devices, further expanding the potential of wearable energy-harvesting systems. Excitingly, ongoing research is exploring groundbreaking technologies to harness energy in novel forms, including wearable and textile-integrated biofuel cells. These innovations promise to revolutionize the field of e-textiles by enabling self-sustaining, multifunctional smart textiles for a wide range of applications [14].

A wide range of textile materials, both natural (cotton, linen, silk, wool, hemp, leather, cashmere, bamboo, sisal, and jute) and synthetic (nylon, polyester, acetate, acrylic, polar fleece, rayon, and spandex), are available in the market. Some of them are being used as active substrates for e-textiles. Of these, silk is a natural protein fiber and the sole natural fiber to occur in filament form [15]. Silk originated in ancient China and later extended to other parts of the world. Mention of silk evokes an image of luxury due to its tender nature and quality. Its inherent characteristics, like comfort and protection from heat, have made silk a sought-after fiber in upmarket fashion clothing. Additionally, its natural piezoelectricity and elasticity also make silk a sought-after material to be utilized for energy harvesting applications in wearable electronics [16]. If we analyse the structure of silk, it is made of two proteins: fibroin and sericin. Fibroin contains the core filaments, while sericin is a natural adhesive, binding the filaments together. Both are composed of amino acids, but it is the unique arrangement of fibroin that imparts silk with its marvelous crystalline property. Silk fiber is made up of amino acids containing acidic (COO^-) and basic (NH_3^+) reactive groups in its molecular structure [17]. While silk is resistant to acids, it is somewhat sensitive toward alkalis, and hence, special precautions need to be taken while handling or functionalizing it. For more details about the basics of silk, the authors can refer to [18]. For energy-harvesting silk textiles, the incorporation of advanced nanomaterials has become a fashionable trend.

Electrically and thermally conductive materials as coating and impregnations over silk fabric are essential for fabricating energy harnessing components. In the realm of flexible energy harvesting silk textiles, reducing the dimensions of conventional electronic components is the primary approach. However, integrating flexible conductive materials offers a more sustainable solution for the long-term adaptability of silk textiles. Conductivity can be introduced at various stages from raw materials, yarns, or fabrics allowing through layered conductive fabrics. The selection of an appropriate method for imparting conductivity depends on several factors, including the available equipment, desired conductivity level, percolation threshold, and the mechanical properties required for fabric flexibility and durability [19]. The application of silk fibers and fabrics with conductive nanomaterials has drawn considerable attention as a viable means of producing next-generation energy-harvesting textiles (EHTs) and electronic textile sensors (ETS) [20]. The process enhances energy conversion efficiency by leveraging the unique properties of the materials. EHTs are generally fabricated by the blending of conducting polymers and inorganic nanomaterials through solution-based coatings or atomic layer deposition. However, structural stability and mechanical durability are major issues, especially for applications requiring extended use. One of the main limitations of EHTs made via coating and deposition techniques is that they are prone to wear and tear, especially during repeated washing. Besides, other wearable energy and sensing technologies like fiber-based energy storage and flexible sensors are unable to provide durability in addition to function. Most wearable sensors and most fiber-shaped electrodes provide only a few cycles of deformation, which is far from the requirement for day-to-day textiles. Ideally, smart textiles must be able to maintain function even after millions of bending and friction cycles and long-term reliability. For overcoming this issue, nature has some really profound design principles to extend longevity as well as function in materials. Silk, for instance, possesses great mechanical strength, toughness, and resilience and is therefore a highly promising candidate for wearable electronics. Inspiration from hierarchical structures in nature can provide a template to synthesize materials in which internal constitution is synergistic with external performance, paving the way for creative solutions to address practical requirements. This chapter will attempt to summarize the basic principles of choosing and functionalizing suitable materials for energy-harvesting technologies and their application in sensing technologies, with special focus on silk-based materials. As a natural biopolymer, silk exhibits unique merits, such as flexibility, biocompatibility, and tunability of mechanical and optical properties, making it a prime candidate to be utilized in sensing and energy-harvesting technologies. We will discuss its use in piezoelectric, thermoelectric, triboelectric energy harvesters, how to modify it to optimize its performance, and how to integrate it into electronics. In addition, we will discuss its potential for gas and biochemical sensing with a focus on recent advancements, challenges, and potential future applications of silk-based multifunctional materials. A model for silk triboelectric energy harvesting textile is shown in (Fig. 11.1).

11.2 Silk as a Functional Textile

Eco-friendly energy solutions are gaining significant attention as a means to combat global warming. Among these, renewable energy harvesting and conversion using natural materials, such as silk, have emerged as a promising and innovative choice. This approach not only aligns with sustainability goals but also offers a unique and

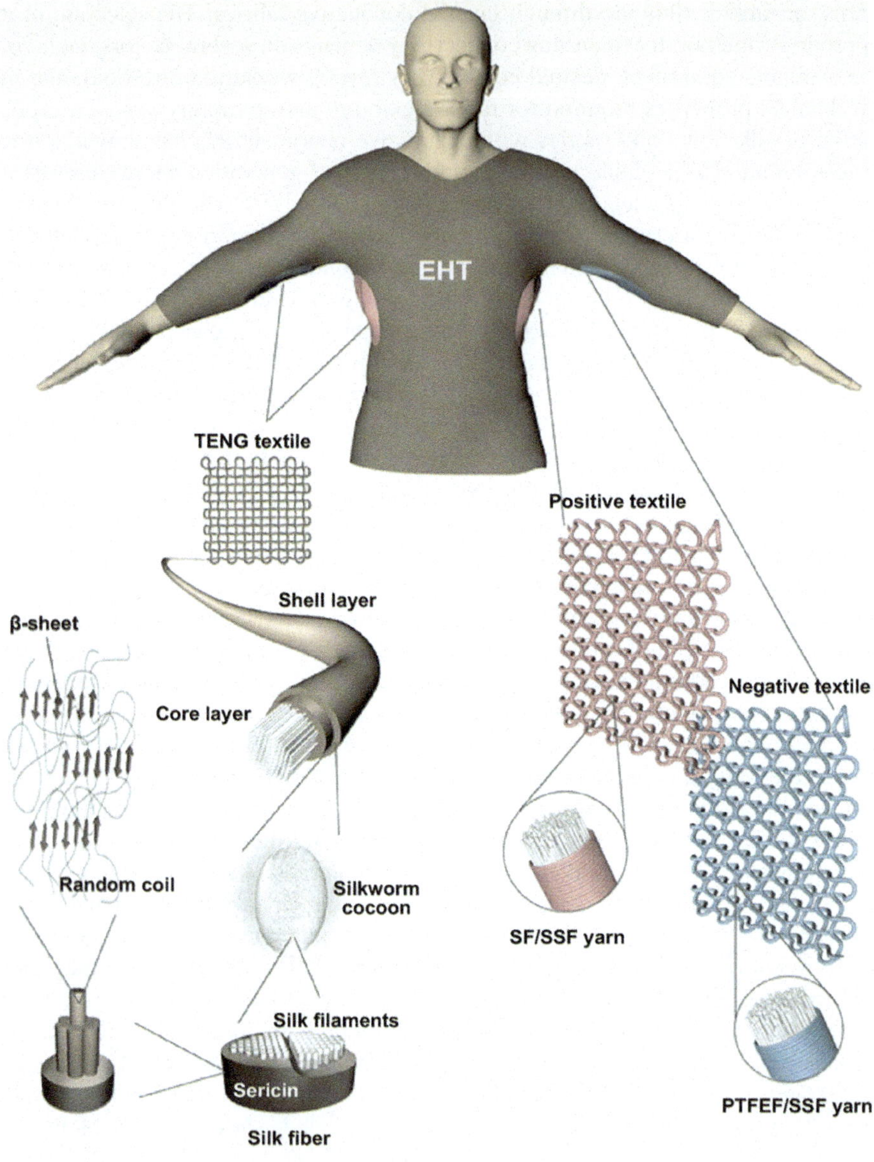

Fig. 11.1 Model silk-based triboelectric energy-harvesting textile with hierarchical design. Reproduced with permission obtained from [21]. Under the terms of the Creative Commons cc by license, Springer Nature© 2019

environmentally friendly alternative to traditional energy sources. Silk can be utilized in various forms, such as fibers, fabrics, and thin films, depending on the intended application. For wearable technologies, its high inherent flexibility makes it an ideal material, enabling seamless integration into everyday use. A well-suited e-textile should possess key properties such as stretchability, breathability, washability, and water resistance. Silk, as a material, has the potential to meet these requirements, making it an excellent candidate for advanced e-textile applications. Here, we explore how silk can be utilized as an efficient and versatile material for e-textiles.

11.2.1 Mechanism behind the Energy Harvesters

The working mechanisms of the four major energy harvesters are intrinsically linked to the individual materials that are functionalized or coated onto fabrics and silk fibers [22]. The inherent mechanisms and operational modes of the energy-harvesting devices with silk fabrics as the platform are as follows. Triboelectric energy harvesters operate based on the triboelectric effect that induces a triboelectric charge by friction between two dissimilar materials; they are defined by the lightweight and the ability for energy harvesting from body movement [23]. Piezoelectric energy harvesters operate based on piezoelectric materials in generating electricity with mechanical stress, such as deformation or pressure, and are thus especially effective for energy harvesting from kinetic motion or vibrational processes [24]. Thermoelectric energy harvesters operate based on the use of temperature differentials to harvest electrical energy through the Seebeck effect, thereby achieving power harvesting from the temperature differences between the body and the environment [25]. Photovoltaic energy harvesters integrate solar cells into fabric to harness solar light and convert it into electricity, thereby providing a sustainable source of power, particularly for outdoor use. These technologies all have specific merits and demerits; collectively, they form a compelling development in wearable technology and clean power solutions [26]. As shown in Table 11.1, we can identify the advantages and challenges associated with each type and its applications.

11.2.2 Materials Actively Involved

Now, each of these four methods will be executed using appropriate materials, each of which possesses its own distinct properties. It is important to note that not all materials can be used for all the methods mentioned above. The selection of materials depends on their compatibility with the specific requirements of each method, such as durability, flexibility, conductivity, or other relevant characteristics. Therefore, careful consideration must be given to the properties of the materials to ensure optimal performance and effectiveness in their respective applications. In the triboelectric effect, a wide range of materials can be used to induce this effect, and they are

Table 11.1 Advantages and challenges of the four types of energy harvesters

Type of energy generators	Primary energy source	Advantages	Challenges
Triboelectric	Mechanical motion	High output from minimum movement	Durability and environmental sensitivity
Piezoelectric	Mechanical stress	Highly efficient in vibration movement	Limited flexibility, temperature sensitivity
Thermoelectric	Temperature gradient	High energy conversion from body temperature	Low efficiency, requires temperature difference
Photovoltaic	Sunlight	High output in presence of hot daylight	Low efficiency in low light, durability

usually listed in a triboelectric series in order of the degree to which they tend to gain or lose electrons. On one end of the series, materials such as glass, nylon, and wool tend to be inclined to lose electrons and become positively charged when rubbed [27]. Halfway down, materials such as silicon, polyester, and paper have a mid-tendency to gain or lose electrons depending on what they are rubbed against. On the other end of the series, materials such as Teflon, silicone rubber, and polyvinyl chloride tend to gain electrons and become negatively charged. Other common triboelectric materials include metals (aluminum and copper), natural fibers (cotton and silk), and synthetic polymers (polyethylene and polypropylene). The materials chosen are dependent on what they will be used for because the efficiency of the triboelectric effect is dependent on parameters such as surface roughness, environmental conditions, and position in the triboelectric series. By selecting and combining materials from opposite ends of the series, maximizing the creation of triboelectric charges is possible for the application in devices such as energy harvesting, sensors, and self-powered devices. Piezoelectric materials find universal application in sensors, actuators, energy-harvesting components, and other electronic devices. Well-known piezoelectric materials are natural crystals such as quartz and Rochelle salt, which are the first materials found with piezoelectric properties [28]. Synthetic ceramics such as lead zirconate titanate are also widely used because they possess a high piezoelectric effect and mechanical stability. Other ceramics such as barium titanate and zinc oxide are also well known for their piezoelectric effects. Some polymers such as polyvinylidene fluoride are also preferred due to their flexibility and ease of handling, therefore suitable for wearable and flexible electronics wear. Special advanced materials such as aluminum nitride and lithium niobate are utilized in high-temperature and high-frequency applications based on their excellent thermal and mechanical stability. Other biocompatible materials such as gallium phosphate and certain composites are also studied for potential medical devices. In thermoelectric generators, thermoelectric materials will be seamlessly woven over the silk fabric. Bismuth telluride and other derivatives were extensively used owing to its high efficacy at room temperature. These materials can be functionalized as thin films or nanowires on silk fibers. Moreover, conducting polymer, PEDOT:PSS [29], is incorporated for its remarkable flexibility and straightforward processing capabilities, enabling it to be printed or coated onto silk. In photovoltaic textiles, various organic materials such as (P3HT:PCBM and PTB7:PCBM) [30] are commonly used, and non-fullerene acceptors like ITIC and Y6 [31] further improve efficiency and stability. Perovskite solar cells are made from materials such as methylammonium lead iodide ($MAPbI_3$) and formamidinium lead iodide ($FAPbI_3$). These materials can be processed into thin films that adhere well to silk, allowing for high-performance energy harvesting. With these aspects, we will see in detail how these materials were actively incorporated into silk materials.

11.3 Harnessing Energy with Silk Textiles

11.3.1 Silk Fibroin as a High-End Energy Harvester

Fibroin was collected from silkworm cocoons and appears to be a semicrystalline material with ordered lamellae held together by an amorphous matrix structure. Its mechanical properties, charge transport capabilities, and strong electron loss characteristics make it a good biodegradable material for constructing robust energy harvesters [32]. Figure 11.2 shows a mode scheme for the silk fibroin preparation. It involves four main steps. First, silk is extracted from Bombyx mori cocoons by cutting them into small pieces and boiling them in a sodium carbonate solution for about 30–45 minutes to remove a protein called sericin. After boiling, the silk fibers are thoroughly rinsed with clean water and dried. Next, the dried fibers are mixed with a lithium bromide (LiBr) solution and heated at 60 °C for 4 hours until they completely dissolve. To remove any remaining LiBr, the solution is placed in a special dialysis bag and soaked in clean water for 48 hours, with the water being changed five times. Finally, the solution is purified by spinning it at high speed in a centrifuge to remove any leftover impurities, and additional filtering can be done if needed. To keep the SF solution stable and prevent it from thickening, it is stored at temperatures below 4 °C. After preparations, the application of these silk fibroin was not limited across wearable technology, health care, and energy harvesting. In biomedical and wearable devices, it is used to develop electronic skin, electronic tattoos, wireless therapy devices, and drug delivery systems [33]. Its excellent sensitivity and biocompatibility make it ideal for sensors, including water molecule sensors, temperature sensors, temperature-pressure hybrid sensors, and mechanical sensors. In the field of energy harvesting and electronics, silk fibroin plays a key role in the development of film-based and fiber-based thermoelectric generators (TEGs), fiber-based PEG devices, and transient electronic devices that dissolve after use. Beyond these applications, it is also utilized in advanced technologies like LED lenses, air filters, and bio memristors. The combination of biocompatibility, biodegradability, transparency, and strength makes silk fibroin a promising material for next-generation innovations in health care, environmental monitoring, and electronic systems.

In the triboelectrification process, silk fibroin tends to lose electrons and transfer them to the other material, which has high electronegativity. This charge difference induces charge polarization and creates an electric field, which is then collected through the external circuit. Interestingly, silk fibroin exhibits dimorphism in its crystallinity. It consists of both monoclinic and orthorhombic crystal systems, which lack a center of symmetry, giving rise to a piezoelectric effect under mechanical stress [34]. Apart from these intrinsic properties, it can be regenerated into various

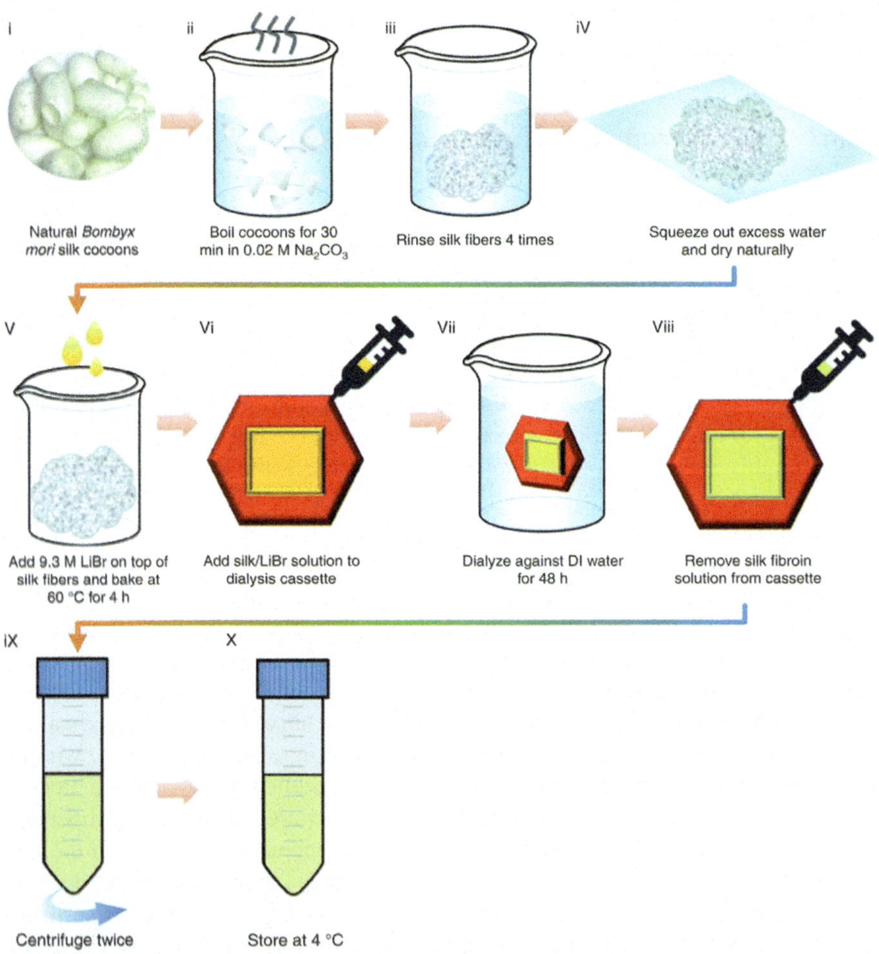

Fig. 11.2 Step-by-step schematic representation of a flow chart for the preparation of silk fibroin solution prepared from natural Bombyx mori cocoons. Reproduced with permission obtained from [33]. Under the terms of the Creative Commons cc by license, Springer Nature© 2021

forms by dissolving into a solution state. It will create rich unfolded molecules over it, providing convenient mesoscopic structure modification and functionalization.

Here, we will explore the possibilities of how this silk fibroin is utilized. The very first work on triboelectrification using these silk fibroins was done by Kim et al. in 2016 [35]. They made the way for others to actively work on this domain. A pioneering study by Zhelin Li et al. introduces the possibility of silk fibroin as an ideal material for medical electronic devices (Fig. 11.3). The researchers detail its application in therapeutic methods for wound healing, including skin wounds, infections, and burns. Inherent to the body is a bioelectrical phenomenon known as transepithelial electric potential that is controlled by the ionic migration of Na^+/K^+ ATPase and acts as a directional force for cellular migration for tissue regeneration. This physiological effect can be augmented by the incorporation of electrical stimulation, which causes ion migration initiated by wearable triboelectric nanogenerators that convert body movement into an electrical signal to promote sustainable wound healing. This device was capable of generating a maximum voltage of 166.4 V under a force of 25 N at a frequency of 1 Hz. Scratch assays of vascular endothelial cells, fibroblasts, Schwann cells were taken as the models, and they evaluated the electrical stimulation which promotes migration and proliferation for wound healing. These findings show the prospective practical application of biodegradable silk fibroin [36].

One of the biggest challenges with piezoelectric textiles is maintaining strong output performance while keeping them flexible, durable, and biocompatible. Their brittleness and stiffness often make them less than ideal for wearable electronics.

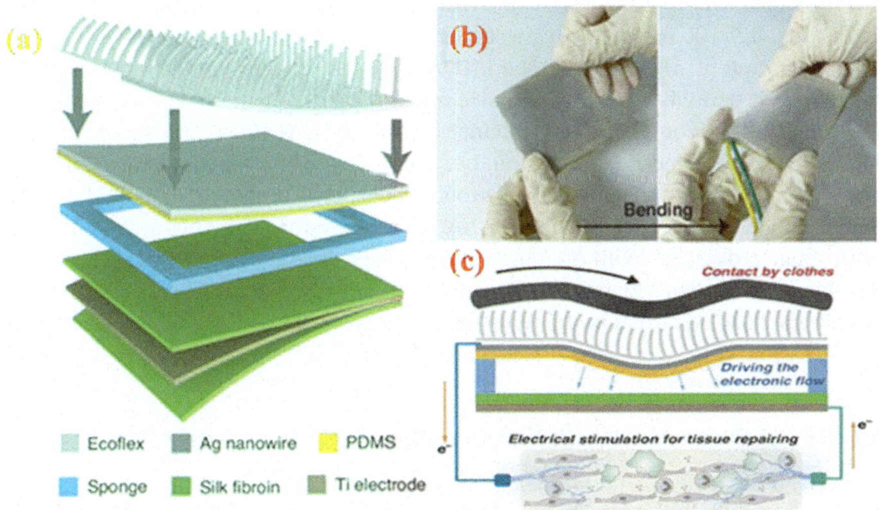

Fig. 11.3 Device structure and working mechanism of the silk fibroin-based triboelectric generator. a) Assembly using silk fibroin. b) Photo of a flexible and biocompatibility. c) Assembling and working process over skin. Reused with permission obtained from [36]. Under the terms of the creative commons cc by license, Springer Nature© 2024

Researchers Siqi Wang et al. have developed flexible core-shell nanofibers made of silk fibroin and PVDF using a special spinning technique. By combining these materials, they boosted the piezoelectric effect where mechanical movement is converted into electricity without needing extra fillers. The fibers with 14% SF generated an output voltage of 16.5 V and maintained their performance over time, even after repeated bending. These nanofibers can store energy in capacitors and work as self-powered sensors. They were successfully used to track human joint movements and even control a smart car through simple gestures, showing great promise for wearable electronics and energy-harvesting devices [37].

Silk fibroin has vast potential for enhancing the way supercapacitors store charge. Supercapacitors charge fast, have an incredibly long lifespan, and are great energy storage devices. Scientists enhanced their energy-storing ability by employing silk fibroin in electrodes as thin films. With the addition of potassium hydroxide to the films, increasing the surface area so that it can store more energy, this was attained. A breakthrough was realized when silk fibroin was mixed with graphene to produce a device that retained 96.3% capacity after 10,000 charging cycles. This indicates the future of silk fibroin supercapacitors in sustainable energy storage. Another new concept in harvesting osmotic energy is tapping clean energy from the natural power difference in seawater and river water. The challenge here is that most of the materials employed are brittle and hard, making it not easy to employ in real applications. Overcoming this, Jianjun Chen and his team created a very thin membrane composed of silk fibroin. The membrane is only 10 nm thick per layer and is an ion-permeable membrane whose resistance is negligible. They attained one of the highest energy yields ever recorded at 21.66 W m^{-2}. Optimizing the membrane thickness to approximately 100 nm gives the best energy efficiency and lifespan [38].

Application of silk fibroin-based materials in new energy harvesting and storage technologies demonstrates the extensive use of bio-based nanomaterials in advanced electronics. This composite can enhance piezoelectric textiles, and supercapacitors facilitate efficient energy conversion, providing flexibility, toughness, and functionality. The capability for enhanced electrochemical performance, mechanical strength, and ion selectivity is a significant step toward the development of eco-friendly, high-performance devices. With ongoing research, silk fibroin plays a significant role in addressing global energy issues, enhancing wearable electronics, and facilitating the development of self-sustaining, renewable energy systems.

11.3.2 Advances in Silk Triboelectric Textile Engineering

Skin electronics will provide an ideal platform for self-powered sensors. In this emerging field, stretchable triboelectric nanogenerators with low stiffness, water resistance, and breathability are a major concern that needs to be addressed. To obtain it, a prominent work done by Yi Li et al. explains an electrospinning strategy for preparing super-elastic hydrophobic fibers. These fibers were compatible with different deformable conductors to make a versatile, permeable, and stretchable membrane-based triboelectric nanogenerator. The idea originated from the authors' examination of the structure of rough-wrought silk craft made from boiled silkworm cocoons. With a synchronous electrospinning and electro-spraying process, a flexible styrene-isoprene-styrene block copolymer was compounded with fluorinated silica nanoparticles, thereby resulting in in situ welding at fiber joints and embedding of nanoparticles. The obtained triboelectric generators showed higher mechanical properties, including an elongation of over 3600%, a strain recovery of 2000%, a contact angle of 154.2°, and a permeability of over 90% compared with cotton. To explore its functional applications, it was blended with conductive materials, such as sputtered gold, vacuum-filtered liquid metal-silver nanowires, and printed conductive inks, to create flexible and breathable electrodes. The electrodes generated 40 V and 36 $\mu W\ cm^{-2}$ under mechanical stimulation by tapping, stretching, and bending. This device was used to power a 60-LED array and exhibited applications in self-powered sensing for material identification and hand posture detection. To make triboelectric nanogenerators more wearable electronics-friendly, we need to improve their fabrication process, performance, and integration with other functions [39]. Dan-Liang Wen et al. fabricated a printed silk fibroin-based TENG that is tailored for various wearable sensing types. The screen-printing technique was used to print graphite-based conductive patterns on a soft polydimethylsiloxane substrate. To improve the printability of the electrodes on the soft substrate, a micro-structuring pre-treatment was given. Thanks to this, it showed an outstanding power density of 412 μWcm^{-2}, proving it to be suitable for energy harvesting. Besides its electrical performance, using silk fibroin—a biocompatible polymer that absorbs liquid water well—equipped the device with high-end environmental sensing capabilities. In contrast to common humidity sensors that sense liquid and gas water, it was capable of distinguishing between the two, allowing for improved environmental monitoring. Additionally, the device was capable of monitoring breathing and sensing joint movement on the same platform, proving it to be a wearable sensing system-friendly. Most researchers primarily focus on enhancing the output performance, although substantial progress is needed to ensure user thermal comfort when using these triboelectric nanogenerators [40]. The work done by Mengqi Wang et al. introduces HT-SF, a hybrid material consisting of TiO_2-coated silk protein films on one half-side. The device is fabricated using solution-blow-spinning to produce a silk fibroin substrate and depositing TiO_2 using chemical vapor deposition. The robust silk fibroin-TiO_2 interaction provides a stable triboelectric output (~40 V), even at high humidity, making it an efficient human motion sensor. Additionally, in comparison to conventional UV-blocking textiles, it boasts superior cooling capability, with a temperature decrease of up to 15 °C, thus ensuring thermal comfort for wearable TENG devices [41].

Sericin, which is obtained from silk, acts as a natural binder in a composite. It has high amorphous-domain content and can facilitate enhanced triboelectrification with amorphous-crystalline interaction under pressure. In contrast, the conventional fabrication of silk-TENG typically involves the removal of sericin to avoid an immunological response in biomedical applications, which requires complicated processing operations such as degumming, rehydration, and dialysis. In a work by Qian wang et al. a structure is proposed that maintains the structural stability of silk while maximizing the performance output, as shown in Fig. 11.4a. They have fabricated a robust silkworm cocoon layer with polydimethylsiloxane to efficiently harvest mechanical energy from human movement. The device operates in a contact-separation mode, exhibiting an open-circuit voltage of 126 V, a short-circuit current of 3 μA, and a power density of 216 mW/m^2. Additionally, the integrated self-charging TENG can power small electronic devices and monitor human movement (Fig. 11.4b, c). This work illustrates the capability of silk, which maintains its structural stability as an efficient dielectric material, paving the way for enhanced TENG functionality in real-world applications [42].

Beyond the direct usage of silks, conductive hydrogels with silk offer mechanical flexibility, which is promising for various applications. However, they often lose elasticity and conductivity at extremely low temperatures, limiting their use in harsh environments such as Arctic exploration. A double-network hydrogel, as described by Xiaotian Wang et al. works by incorporating silk fibroin and PEDOT: PSS into a lithium bromide solution, achieving exceptional anti-freezing properties (freezing point of −108 °C) and high conductivity (16.33 S m^{-1}). This hydrogel maintains stable performance under compression (0.5–60%) and tension (1.0–100%), with a fast response time of ~70 ms. Notably, it retains conductivity and mechanical robustness even at −80 °C which demonstrates applications for flexible sensors in extreme cold environments [43].

These outcomes demonstrate the capabilities in fabricating triboelectric energy generation systems, particularly in applications such as energy harvesting and self-powered sensing systems.

11.3.3 Advances in Silk Piezoelectric Textile Engineering

Trends in silk-based piezoelectric energy harvesting and sensor development will focus on improving materials, integrating them with smart electronics, and exploring other applications in the medical field. To improve the piezoelectric performance of silk fibroin sensors, scientists are exploring new methods of incorporating molecules, reinforcing with nanocomposites, and orienting the β-sheet structure. The immersion of biocompatible piezoelectric nanoparticles, such as $BaTiO_3$ and ZnO, into silk fibroin can enhance energy conversion while maintaining flexibility and biodegradability [44]. New processing methods, such as controlled fiber alignment, electrospinning, and 3D printing of silk structures, are also being developed to enhance its strength and sensitivity. To be practically applicable, silk-based piezoelectric sensors must be compatible with wireless communications and electronic technologies. Scientists are creating stretchable and flexible electronic circuits that can be directly interfaced with silk fibroin sensors. Advanced signal processing techniques, such as AI-driven algorithms, can be utilized to process sensor signals in real time. This makes the devices suitable for personal health monitoring and remote diagnosis. These technologies address the increasing demand for smart healthcare solutions that

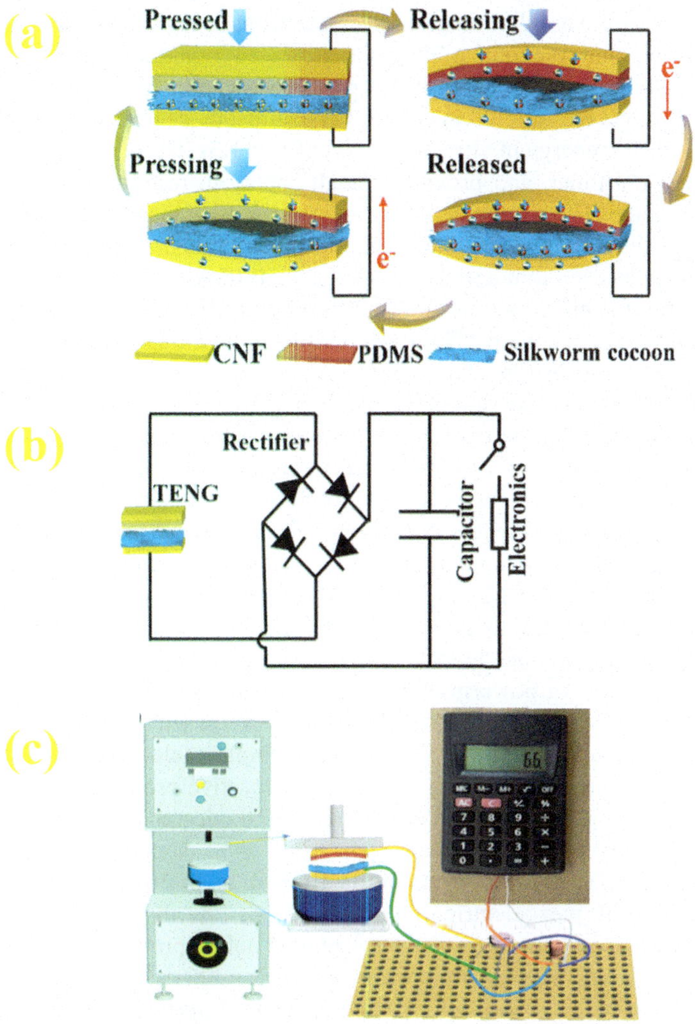

Fig. 11.4 TENG and the working principle of using silk and its electronic application. Reprinted with permission from [42] Copyright © 2023 American Chemical Society

enable continuous monitoring without the need for invasive interventions. Some of the new applications include wearable health monitoring systems that monitor vital signs in real time, muscle activity, and joint movement; implantable bioelectronics, using silk sensors integrated into degradable medical implants to monitor the healing process after surgery and induce tissue regrowth; soft robots, in which silk sensors are used as sensing and feedback components; and drug delivery systems, in which piezoelectric sensor-enabled intelligent dressings for wounds allow for controlled drug delivery based on body signals [45].

Existing cardiovascular implantable electronic devices need battery-free, biodegradable, and biocompatible alternatives to minimize undesired in vivo responses. Qiying Lv et al. prepared a piezoelectric film from silk sericin with a high d33 of 12 pC N^{-1} allows an energy-harvesting device to transform mechanical stress into electric energy in vitro and in vivo (Fig. 11.5). The device is able to power LEDs, operate a timer, and provide enough energy to restore heart function in a preclinical model. The biocompatibility and flexibility of piezoelectric materials are crucial in facilitating effective interaction between implantable power sources and tissues. Piezoelectric biomaterials such as glycine-PVA thin films and silk-derived materials offer tunable properties through structural adjustments. Experiments have shown that heterogeneous β-glycine-PVA films have a d33 value of 5.3 pC N^{-1}, whereas γ-glycine/PVA films have a d33 value of 10.4 pC N^{-1}. Silk fibroin, whose crystalline β-sheet structures are stabilized through hydrogen bonding, exhibits native piezoelectricity that can be enhanced by molecular interactions. In this study, silk sericin was functionalized with aminated reduced graphene oxide to enhance its crystallinity and charge storage capacity, forming functionalized silk sericin for use in an energy-generating device. A piezoelectric film was fabricated by combining F-SS, polyvinyl alcohol, and ZnO nanorods, achieving a high longitudinal piezoelectric coefficient (d33 = 12.1 pC N^{-1}), surpassing most bio-organic films. The EG device generated 850 mV and 153 nA under mechanical pressure, while subcutaneous implantation produced stable outputs (36 mV and 16 mV) from respiratory movements, capable of powering LEDs and restoring cardiac function in rats. Due to its biocompatibility and biodegradability, the F-SS-based piezoelectric film holds significant potential for next-generation cardiac implantable energy devices [46].

Changwan Sohn et al. investigated the electroactive properties of silk fibroin, focusing on hydrogen bonding to create efficient, biodegradable devices to capture mechanical energy. Scientists electrospun silk fibroin membranes and immersion in ethanol (EtOH) triggered a transition from α to β structures, resulting in a tenfold increase in piezoelectric output, producing 7 V and 150 nA, compared to untreated membranes. In contrast to conventional poling protocols, electrical poling destroyed the piezo response by breaking hydrogen bonds, illustrating the importance of solvent-induced polarization in performance improvement. The electrospun silk piezoelectric generator showed promise as a motion sensor, detecting motion through joints and powering LEDs [47].

Sericin degumming is an effective strategy for using silk as a piezoelectric generator. A study by Min-Soo Kim et al. investigates the development of piezoelectric nanogenerators from woven silk fabric and analyzes the influence of sericin removal on their piezoelectric performance. Silk contains 25–30% sericin and 70–75% fibroin by weight, and sericin is removed using acid, alkali, or high-temperature (120–140 °C) and high-pressure (0.1–0.2 MPa) treatments. Fibroin exists in two forms, Silk I and Silk II, and the β-sheet-rich Silk II structure can generate electricity with an output voltage of approximately 2.5 V and an output current of approximately 200 nA under mechanical stress. Conventional methods employed to develop these devices, such as electrospinning and photolithography, are complex and costly, making them difficult to scale up. This study compares the variation in output characteristics and

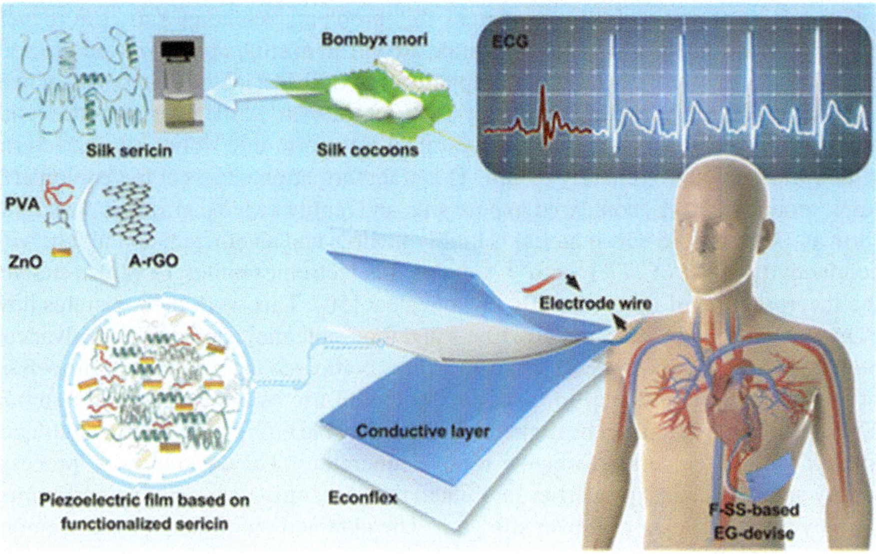

Fig. 11.5 Silk fibroin-based piezoelectric generator for an implantable device. Reprinted with permission from [46]. Copyright© 2025 Wiley-VCH GmbH

structure of silk with the degumming process and how it improves PENG performance [48].

11.3.4 Advances in Silk Thermoelectric Textile Engineering

Thermoelectric materials can be integrated into silk textiles through coating, printing, stitching, embroidery with conductive yarns, or knitting and weaving with conducting yarns. Thermoelectric materials should be light, flexible, and non-toxic to provide comfort and safety. Inorganic semiconductors, although possessing high thermoelectric performance, are made up of rare and toxic elements and are treated with high energy. Carbon nanomaterials such as carbon nanotubes and graphene also pose toxicity issues. Direct electricity generation from the human body using the thermoelectric effect has attracted significant interest for wearable textile electronics. Direct deposition of the thermoelectric materials on both sides of the silk fabrics was an easy way to produce an efficient energy generation process. A prototype consisting of multiple arrays of thermocouples was prepared by Zhisong Lu et al. with nanostructured Bi_2Te_3 and Sb_2Te_3 deposited over silk. This device provides an output voltage and power of 10 mV and 15nW with high durability. Through rapid research attempts, p-type thermoelectric devices consisting of silk threads/polyaniline/amino multiwalled carbon nanotubes were prepared using a dyeing process to enhance their high performance. Failure will be obtained while using only one type of material, either p-type or n-type [49]. To address this problem, Feichong Yao et al. developed the above-mentioned material composite via an in situ polymerization method. This material assembly exhibits an output voltage of 0.749 mV and an output power of 0.326 nW, with a power density of 1087.532 nW/m^2. Silk fibers blended with polydopamine and subjected to heat treatment transform into ribbon-like structures with a width of approximately 65 nm. This transformation occurs at lower temperatures, around 1000 °C, compared to pure silk, and highly ordered graphitic structures form at 1300 °C. The new material is highly flexible and an efficient conductor, with conductivity values of 277.3 S·cm^{-1}, suitable for electronic textiles. PDA also enables the incorporation of Fe_2O_3 and Bi_2Te_3 powders [50]. This work demonstrates how PDA enables the separation of silk fibers and their potential application in advanced purposes [51]. In a recent research, researchers compared the mechanical properties of poly(benzodifurandione) to clarify its potential use as a yarn coating material. They discovered that it can be transformed into flexible bulk films with some stretchability, even at temperatures below room temperature. The advantageous processability and mechanical properties facilitated the preparation of an n-type conducting yarn by coating a layer of it over silk yarn. The obtained multifilament yarn demonstrated electrical conductivity comparable to that of PEDOT:PSS-coated yarn and exhibited stability for at least 14 months under ambient conditions, without the need for encapsulation. Additionally, the yarn coated with it remained stable for at least seven machine wash cycles and could be successfully utilized for sewing buttons and for the fabrication of textile thermoelectric devices via embroidery, indicating its applicability in electronic textiles [52].

New advances in conductive coatings, nanostructured modifications, and hybrid composites have significantly improved the mechanical and electrical performance of silk fabric. Such advances enabled the development of washable, flexible, and strong thermoelectric fabrics with high potential for energy conversion efficiency. Challenges in performance improvement, scalability, and compatibility with existing textile manufacturing processes remain, however despite such encouraging advances. Future research should focus on enhancing the stability, efficiency, and sustainability of silk-based thermoelectric fabrics to make them compatible with mass adoption in smart fabrics and wearable electronic devices.

11.3.5 Advances in Silk Photovoltaic Textile Engineering

Direct electricity generation from sunlight will be an indispensable option for on-the-go charging of future devices. In this realm, the use of solar cells on clothing would be a suitable match. For solar energy conversion, most research efforts are focused on synthetic polymers; silk is a promising tool owing to its greenness and biocompatibility. But it is not studied in well detail. A silk-based perovskite solar cell design was developed by Ma et al. They combine silver nanowires with silk fibroin and create various structures capable of achieving maximum solar absorbance and conversion. The interesting part lies in the continued resistance of the structure, still

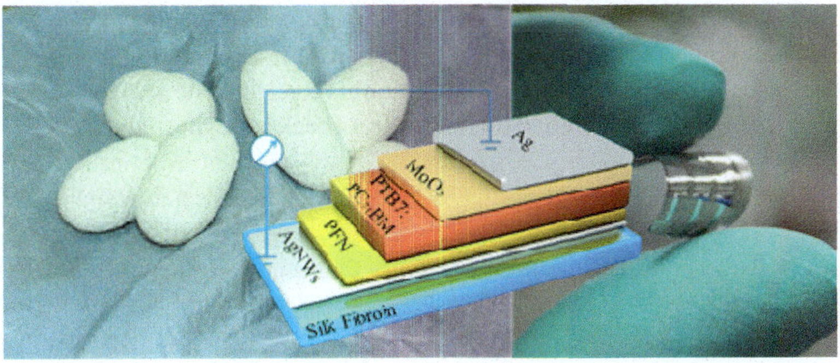

Fig. 11.6 Silk fibroin as a flexible solar cell substrate. Reprinted with permission from [54]

under severe deformation, after bending and stretching, which demonstrates the good durability of silk with silver. These combined sheets were employed in photovoltaic cells, utilizing layers of PEDOT: PSS, $CH_3NH_3PbI_3$, phenyl-C61-butyric acid methyl ester. The final output of the photovoltaic cell obtained a respectable power conversion efficiency of 10.4% [53]. Similarly, Yuqiang Liu et al. prepared silk fibroin integrated with a mesh of silver nanowires to build up a flexible organic solar cell with a maximum power conversion efficiency of up to 6.62%. This device exhibits a conductivity of ~approximately 11.0 Ω sq.$^{-1}$ and a transmittance of ~approximately 80% in the visible light range [54]. The device setup was shown in Fig. 11.6.

11.4 Future Outlook and Conclusion

Silk-based energy harvesting and sensors have arisen as a hopeful solution for flexible wearable electronics. The unique and inherent properties make them an ideal solution platform self-powered device. To unlock its full potential, a primary understanding of the functionalized surface of silk is necessary. One of the primary goals of energy harvesters is to enhance both their power output and conversion efficiency. The integration of silk with advanced materials can further improve the high electrical conductivity and sensing performance. Their natural biocompatibility makes them particularly suitable for wearable and implantable biomedical applications, such as smart textiles that can monitor body movements, temperature, or physiological signals. Additionally, implantable silk-based sensors could be developed for real-time tracking of biological markers, revolutionizing personalized medicine. The advent of the Internet of Things has opened the door to embedding silk-based devices into smart textiles and wireless networks, which can harvest energy from human motion and convey information for remote health monitoring and environmental observation. Collectively, silk-based energy harvesting systems and sensors hold promise as a key to realizing sustainability in autonomous electronic devices. Silk-derived technologies will continue to play a central role in the innovation of smart wearables, healthcare technologies, and sustainable electronic systems, connecting natural processes with innovative technological breakthroughs.

Acknowledgements The authors express their heartfelt thanks to Dr. T. V. Sreekumar, Director, BTRA, for his invaluable guidance and support.

AI Usage Declaration AI tools were utilized solely for the purpose of improving the language and readability of this manuscript. The authors have thoroughly reviewed and verified the content, taking full responsibility for the final content of the chapter.

References

1. Hughes-Riley T, Dias T, Cork C (2018) A historical review of the development of electronic textiles. Fibers 6:34
2. Roy Choudhury AK (2013) Green chemistry and the textile industry. Text Prog 45:3–143
3. Ramasubramanian B, Sundarrajan S, Rao RP, Reddy MV, Chellappan V, Ramakrishna S (2022) Novel low-carbon energy solutions for powering emerging wearables, smart textiles, and medical devices, energy. Environ Sci 15:4928–4981
4. Gao M, Wang P, Jiang L, Wang B, Yao Y, Liu S, Chu D, Cheng W, Lu Y (2021) Power generation for wearable systems, energy. Environ Sci 14:2114–2157
5. Ali I, Dulal M, Karim N, Afroj S (2024) 2D material-based wearable energy harvesting textiles: a review. Small Struct 5:2300282
6. Satharasinghe A, Hughes-Riley T, Dias T (2020) A review of solar energy harvesting electronic textiles. Sensors 20:5938
7. Chen G, Li Y, Bick M, Chen J (2020) Smart textiles for electricity generation. Chem Rev 120:3668–3720
8. Wang Q, Liu X, Han J, Xiao Y, Tan D, Yang Y, Zhang J, Xu B (2025) High-performance naturally crosslinked silk-based triboelectric nanogenerators for multimodal sensing and energy harvesting. Nano Energy 135:110620
9. Mondal K (2018) Recent advances in soft E-textiles. Inventions 3:23
10. Sootsman JR, Chung DY, Kanatzidis MG (2009) New and old concepts in thermoelectric materials. Angew Chem Int Ed 48:8616–8639
11. Uchino K (2017) Advanced piezoelectric materials: science and technology. Woodhead Publishing
12. Cheng T, Shao J, Wang ZL (2023) Triboelectric nanogenerators. Nature Reviews Methods Primers 3:39
13. Amarnath M, Mohite S, Palaskar S (2025) Energy harvesting/storage textiles: integrating sustainability into wearable technology. BTRA Scan 54:16–27
14. Costa RS, Pires AL, Pereira AM, Pereira CR (2023) Multifunctional energy harvesting and storage textile technology based on thermionic effect. J Power Sources 587:233712
15. Wang C, Xia K, Zhang Y, Kaplan DL (2019) Silk-based advanced materials for soft electronics. Acc Chem Res 52:2916–2927
16. Yucel T, Cebe P, Kaplan DL (2011) Structural origins of silk piezoelectricity. Adv Funct Mater 21:779–785
17. Saikia S, Saikia M (2022) Physical and chemical properties of Indian silk fibers. The Pharma Innovation Journal 11:732–738
18. Padaki NV, Das B, Basu A (2015) Advances in understanding the properties of silk. In: Advances in Silk Science and Technology. Woodhead Publishing, Cambridge, UK, pp 3–16

19. Gregory RV, Kimbrell WC, Kuhn HH (1989) Conductive textiles. Synth Met 28:823–835
20. Chai Z, Zhang N, Sun P, Huang Y, Zhao C, Fan HJ, Fan X, Mai W (2016) Tailorable and wearable textile devices for solar energy harvesting and simultaneous storage. ACS Nano 10:9201–9207
21. Ye C, Dong S, Ren J, Ling S (2020) Ultrastable and high-performance silk energy harvesting textiles. Nanomicro Lett 12:1–15
22. Huang L, Lin S, Xu Z, Zhou H, Duan J, Hu B, Zhou J (2020) Fiber-based energy conversion devices for human-body energy harvesting. Adv Mater 32:1902034
23. Paosangthong W, Torah R, Beeby S (2019) Recent progress on textile-based triboelectric nanogenerators. Nano Energy 55:401–423
24. Mokhtari F, Spinks GM, Fay C, Cheng Z, Raad R, Xi J, Foroughi J (2020) Wearable electronic textiles from nanostructured piezoelectric fibers. Adv Mater Technol 5:1900900
25. Wang L, Zhang K (2020) Textile-based thermoelectric generators and their applications. Energy Environ Mater 3:67–79
26. Mather RR, Wilson JIB (2017) Fabrication of photovoltaic textiles. Coatings 7:63
27. Laudermilch RB (2019) Molecular structures and chemical forces in textiles. In: Big Molecules, Big Problems: Curriculum Units by Fellows of the Yale-New Haven Teachers Institute, Volume IV. Yale-New Haven Teachers Institute, Yale University, New Haven, CT
28. Sekhar MC, Veena E, Kumar NS, Naidu KCB, Mallikarjuna A, Basha DB (2023) A review on piezoelectric materials and their applications. Cryst Res Technol 58:2200130
29. Tseghai GB, Mengistie DA, Malengier B, Fante KA, Van Langenhove L (2020) PEDOT: PSS-based conductive textiles and their applications. Sensors 20:1881
30. Ciammaruchi L, Brunetti F, Visoly-Fisher I (2016) Solvent effects on the morphology and stability of PTB7: PCBM based solar cells. Sol Energy 137:490–499
31. Kupgan G, Chen X-K, Brédas J-L (2021) Molecular packing of non-fullerene acceptors for organic solar cells: distinctive local morphology in Y6 vs. ITIC derivatives, Mater Today Adv 11:100154
32. Koh L-D, Cheng Y, Teng C-P, Khin Y-W, Loh X-J, Tee S-Y, Low M, Ye E, Yu H-D, Zhang Y-W (2015) Structures, mechanical properties and applications of silk fibroin materials. Prog Polym Sci 46:86–110
33. Wen D-L, Sun D-H, Huang P, Huang W, Su M, Wang Y, Han M-D, Kim B, Brugger J, Zhang H-X (2021) Recent progress in silk fibroin-based flexible electronics. Microsyst Nanoeng 7:35
34. Sencadas V, Garvey C, Mudie S, Kirkensgaard JJK, Gouadec G, Hauser S (2019) Electroactive properties of electrospun silk fibroin for energy harvesting applications. Nano Energy 66:104106
35. Zhang X-S, Brugger J, Kim B (2016) A silk-fibroin-based transparent triboelectric generator suitable for autonomous sensor network. Nano Energy 20:37–47
36. Li Z, Xu S, Xu Z, Shu S, Liu G, Zhou J, Lin D, Tang W (2024) Enhancing cellular behavior in repaired tissue via silk fibroin-integrated triboelectric nanogenerators. Microsyst Nanoeng 10:68
37. Wang S, Shi K, Chai B, Qiao S, Huang Z, Jiang P, Huang X (2022) Core-shell structured silk fibroin/PVDF piezoelectric nanofibers for energy harvesting and self-powered sensing, Nano. Mater Sci 4:126–132
38. Chen J, Xin W, Kong X-Y, Qian Y, Zhao X, Chen W, Sun Y, Wu Y, Jiang L, Wen L (2019) Ultrathin and robust silk fibroin membrane for high-performance osmotic energy conversion. ACS Energy Lett 5:742–748
39. Li Y, Xiao S, Zhang X, Jia P, Tian S, Pan C, Zeng F, Chen D, Chen Y, Tang J (2022) Silk inspired in-situ interlocked superelastic microfibers for permeable stretchable triboelectric nanogenerator. Nano Energy 98:107347
40. Wen D-L, Liu X, Deng H-T, Sun D-H, Qian H-Y, Brugger J, Zhang X-S (2019) Printed silk-fibroin-based triboelectric nanogenerators for multi-functional wearable sensing. Nano Energy 66:104123
41. Wang M, Lu S, Liu Y, Liu Y (2025) Half-side deposited TiO2 on silk fibroin films: for thermal comfort wearable triboelectric nanogenerators. Chem Eng J 505:159120

42. Wang Q, Xu B, Huang J, Tan D (2023) Natural silkworm cocoon-based hierarchically architected composite triboelectric nanogenerators for biomechanical energy harvesting. ACS Appl Mater Interfaces 15:9182–9192
43. Wang X, Zhou Y, Li X, Zou M, Zhang Q, Xu W, Feng Y, Zhang Y, You R (2025) Silk fibroin-based Antifreezing and highly conductive hydrogel for sensing at ultralow temperature. ACS Sensors 10(3), 2297–2308
44. Ali M, Bathaei MJ, Istif E, Karimi SNH, Beker L (2023) Biodegradable piezoelectric polymers: recent advancements in materials and applications. Adv Healthc Mater 12:2300318
45. Shajari S, Kuruvinashetti K, Komeili A, Sundararaj U (2023) The emergence of AI-based wearable sensors for digital health technology: a review. Sensors 23:9498
46. Lv Q, Chen S, Luo D, Liu H, Song Y, Liu M, Xiao F, Wang Z, Wang L (2025) An implantable and degradable silk Sericin protein film energy harvester for next-generation cardiovascular electronic devices. Adv Mater 2413610
47. Sohn C, Kim H, Han J, Lee K-T, Šutka A, Jeong CK (2022) Generating electricity from molecular bonding-correlated piezoresponse of biodegradable silk nanofibers. Nano Energy 103:107844
48. Kim M-S, Commerell W, Roh J-W, Park S-S (2023) Degumming effects of silk fabrics as piezoelectrics for nanogenerators. Mater Sci Eng B 298:116898
49. Lu Z, Zhang H, Mao C, Li CM (2016) Silk fabric-based wearable thermoelectric generator for energy harvesting from the human body. Appl Energy 164:57–63
50. Yao F, Xie W, Zhao X, Gu H (2025) A p-type thermoelectric fabric power supply device. Adv Compos Hybrid Mater 8:1–12
51. Jang H-S, Kim S, Park I, Jung WT, Seo JH, Kwon J-H, Hong WG, Mrówczyński R, Lee H, Choi SB (2024) Long-range ordered graphitic structure in silk fibers delaminated using dopamine and thermal treatment for super-flexible electronic textiles: possible applications for magnetic and thermoelectric textiles. Adv Compos Hybrid Mater 7:37
52. Craighero M, Li Q, Zeng Z, Choi C, Kim Y, Yoon H, Liu T, Sowinski P, Haraguchi S, Hwang B (2024) Poly (benzodifurandione) coated silk yarn for thermoelectric textiles. Adv Sci 11:2406770
53. Ma P, Lou Y, Cong S, Lu Z, Zhu K, Zhao J, Zou G (2020) Malleability and pliability of silk-derived electrodes for efficient deformable perovskite solar cells. Adv Energy Mater 10:1903357
54. Liu Y, Qi N, Song T, Jia M, Xia Z, Yuan Z, Yuan W, Zhang K-Q, Sun B (2014) Highly flexible and lightweight organic solar cells on biocompatible silk fibroin. ACS Appl Mater Interfaces 6:20670–20675

Chapter 12
Silk-Based Materials in Energy Harvesting and Sensing Technologies

Md. Kamrul Hassan Chowdhury, Siam Sarker, Mila Akter, Md. Mubin Atik Soscho, Shah Md. Maruf Hasan, and Mahmuda Akter

12.1 Introduction

Silk fibres are naturally strong and protein-based filament fibres that can be further improved through genetic engineering. Their renewable nature, high tensile strength and biocompatibility make them exceptional candidates for flexible electronics, energy storage devices and hybrid solar technologies, but there has been little exploration in this field [1]. Hydrogel synthesis and electrospinning are methods those are capable of allowing the functionalization of silk in a wide range of applications such as sensors, photovoltaic cells, batteries, supercapacitors and water splitting systems [2]. As sustainable materials in energy harvesting and sensing technologies, materials based on silk are gaining academic popularity. Being a protein biopolymer, silk has high strength, biocompatibility and renewability thus making it amenable to next-generation electronic and energy systems [3]. The material is capable of being processed into films, gels, foams, capsules and nanofibres, hence allowing it to be integrated into devices like sensors, supercapacitors and energy harvesting elements [4, 5]. Such versatility makes silk a potential candidate for advanced and environmentally friendly technological applications. The possibility of using silk hybrids in flexible electronics has already been shown by early research, with the initial contribution of Kang et al. [6] who made one of the earliest electrospun silk carbon nanotube membranes with an observable electrical conductivity [6]. Ever since, flexible electronics based on silk are now extended to systems that use ZnO nanorods, CdS nanostructures and Ag nanowires. Each study provided a distinct technique for using silk in flexible electronics [7–9]. These studies collectively show that silk can be engineered into photodetectors and conductive substrates that retain stable

Md. K. H. Chowdhury · S. Sarker · M. Akter · Md. M. A. Soscho · S. Md. M. Hasan · M. Akter (✉)
Department of Apparel Engineering, Bangladesh University of Textiles, Dhaka, Bangladesh
e-mail: mahmuda.akter.mitu88@gmail.com

M. Shahid et al. (eds.), *Science and Technology of Silk*, Textile Science and Clothing Technology, https://doi.org/10.1007/978-3-032-26880-8_12

electrical performance even under bending or mechanical stress, which confirms its promise as a lightweight and durable platform for wearable and low-power flexible electronic and sensing devices.

Silk is an excellent material for capturing biological signals in devices such as EEGs and evoked potentials. It is soft, hydrophilic and electroconductive, which helps reduce stress on live tissue. Silk's unique features include biocompatibility, biodegradability and minimal cellular toxicity, making it an excellent choice for medicinal applications. Tsukada et al. first demonstrated silk-based biosensors by developing PEDOT silk fibres with excellent conductivity and durability and enable reliable ECG and EEG signal monitoring [10]. Zhang further showed that electrospun silk mats embedded with graphene could retain silk's mechanical and biofriendly properties while providing electrical conductivity [11]. Reizabal et al. incorporated carbon nanotubes into silk films, achieving electrically responsive and piezoresistive composites suitable for stress and physiological sensing applications [12]. Together, these studies highlight silk's strong potential as a flexible, conductive and biocompatible platform for wearable biomedical sensors.

Due to its strong triboelectric behaviour and ease of electron loss in the triboelectric series, silk is also a material that can be used to build a triboelectric nanogenerator. It ranks only below wool in this regard [13]. Kim et al. examined the impact of surface area shape on the voltage generated between the two contacting surfaces, taking into account silk's inherent triboelectric properties [14]. Huang et al. also demonstrated that silk could be incorporated in hybrid piezoelectric-triboelectric nanogenerators, which would result in high voltage and power generation that would be applicable in real-life energy harvesting systems [15].

The growing desire for green energy to tackle climate change has prompted scientists to look to nature for solutions, resulting in a reduction in coal and oil use for energy generation in the Western world, despite the substantial environmental cost of this method [16–19]. As energy consumption is expected to triple by 2050 and there is a push to transition away from fossil fuels, renewable energy sources will require innovative solutions to meet demand [20, 21]. Proteins and nanostructures are gaining popularity for their potential in material design, particularly for harnessing ion transport for bioelectric generation [22–24]. Because of their huge surface areas and high aspect ratios, silk nanofibrils derived from *B. mori* silkworm cocoons are appropriate for catalyzing oxygen reduction [25]. It is not surprising that silk is a crucial component in the creation of green energy, as silk nanofibrils are easily produced, naturally occurring and biocompatible and biodegradable [26]. Spider silk from Pholcus opilionoides was carbonized through simple heating and acid treatment to form nanofibre scaffolds with high oxygen reduction catalytic activity. The formation of small mesopores produces a large surface area as well as a very high number of activation sites. The method can also be applied to further improve the carbonized silk fibres, and the silk nanofibrils would be quite promising for green energy applications given their low cost, easy processing, biocompatibility and biodegradability [27].

12.2 Processing Techniques for Silk-Based Materials

12.2.1 Purification and Extraction of Silk Fibroin (SF)

Silk proteins are now produced in various expression systems. For the practical application of silk proteins as biological materials, it is crucial to produce them in large quantities using cost-effective techniques. Among the various available methods, two stand out for their industrial scalability which includes SF extraction and bioengineered silk fermentation [28]. Silk fibres from the *Bombyx mori* silkworm consist of a fibroin core coated with sericin. During Silk Fibroin (SF) extraction, it is essential to remove the sericin layer to obtain pure fibroin, which ensures that the resulting silk-based materials are non-allergenic and biocompatible [29]. To accomplish this purification, the regeneration of *B. mori* silk involves three key steps: degumming, rehydration and dialysis.

Firstly, Degumming is the process of removing sericin coating from silk fibres, which is typically achieved by alkaline treatment. Degumming process has been shown to improve the fibre's elasticity [30]. Several methods exist for degumming silk fibres (Fig. 12.1), including traditional techniques and solution-based approaches. These encompass extraction with water, boiling in soap solutions, alkaline or acidic treatments, the use of organic amines and enzymatic degumming. These methods are employed to obtain high-quality silk fibres [31–33]. Comparison of different degumming methods are shown in Table 12.1.

A common degumming method involves boiling silk for 30 min in a 0.02 M Na_2CO_3 (sodium carbonate) bath. More recently, recombinant cocoon enzymes derived from *B. mori* have been explored as an alternative degumming approach [38]. Secondly, during the dehydration process, degummed fibres are solubilized in a 9.3 molar lithium bromide (LiBr) solution at 60°celsius for a period of 4 h to

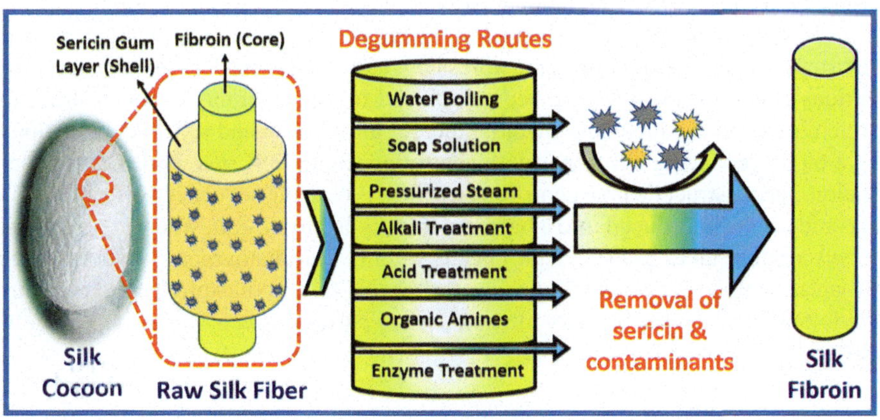

Fig. 12.1 Degumming methods of silk fibres. Reprint with permission from [37]. Copyright @2019 Elsevier

Table 12.1 Comparison of the methods utilized for degumming raw silk fibres [31–36]

Degumming methods	Processing conditions	Analysis
Water-based extraction (boiling)	Temp. 115 °C for 3 h	Removal of sericin is up to 96%
Boiling in soap solution	Temp: 90–100 °C for 90–120 min using soap solution (20–30%) (M:L ratio = 1:4)	Minimum degumming loss (depends on soap solution type), improved lustre to silk fabric
Alkali treatment (Na_2CO_3, $NaHCO_3$)	Temp: 96 °C for 20 min, pH:9.5–10.5 (alkali, non-ionic surfactant)	Improved degumming, but harsh yellowish silk
Acid treatment such as tartaric, lactic, citric acids	Temp: 110 °C for 60 min (Acid, non-ionic surfactant), pH:1.5–2	Least analysed
Treatment with organic amines such as ethyl, methyl, dimethyl or trimethyl amines	Temp: 80 °C, time:45 min, pH:5 (M:L ratio = 1:25)	Removal of sericin is up to 90%
Treatment with enzymes such as trypsin, papain, bacterial enzymes	Temp: 37–90 °C, time:1–4 h, pH:5	Minimum damage to silk

facilitate protein rehydration. This treatment disrupts the hydrogen bonds within the crystalline regions of the silk proteins, yielding a uniform starting material for further application. Finally, lithium bromide is removed through dialysis, producing a purified aqueous solution of regenerated SF, suitable for further application in advanced material development [29].

Subsequently, bioengineered silk fermentation involves advanced strategies of the precise regulation of the molecular structure of the end resultant silk proteins. Unlike extraction techniques, the genetic template allows for strict control over the sequence and molecular weight of bioengineered proteins, resulting in monodisperse, well-defined biopolymers. Production of bioengineered silks typically involves three key steps: genetic design, protein expression, and finally, product purification [29]. Various cloning strategies have been employed for the assembly of the desired silk protein gene. These include sequential directional ligation and recursive directional ligation for producing silk genes with controlled length [39], and while concatemerization can construct diverse gene libraries [40]. Following sequence verification, the plasmid can be introduced into a bacterial host, typically *E. coli*, wherein production of silk protein takes place. Metal chelate affinity chromatography, leveraging the strong attraction between histidine tags and Ni–NTA resin (nickel-charged nitrilotriacetic acid resin), is the primary method used to purify the recombinant silk proteins [39].

12.2.2 Manufacturing and Fabrication Techniques

12.2.2.1 Electrospinning

Electrospinning is a technique used to generate fine fibres with high surface area-to-volume ratios. While the equipment required for this process is relatively straightforward, it yields fibres with significantly smaller diameters than those produced by traditional spinning methods [41]. It is an efficient method for creating SF nanofibres. This technique utilizes electrical forces to draw charged threads of the SF solution into nanofibres with diameters in the nanometre range [42]. The typical electrospinning process for preparing SF nanofibres involves four key steps (Fig. 12.2). First, to enhance the solution's viscosity and its ability to be electrospun, a high-viscosity polymer, such as Polyethylene Oxide (PEO), is incorporated into the SF solution and thoroughly mixed until a homogeneous blend is achieved. This PEO addition is often necessary to produce uniform and continuous fibres during electrospinning. Second, the resulting SF/PEO solution is loaded into a syringe and electrospun onto a collector, forming a nonwoven fabric of SF/PEO fibres. Third, this fibre mat undergoes post-treatment, commonly involving either water vapour annealing or immersion in alcohol (e.g. methanol or ethanol). These annealing methods promote the formation of β-sheet crystals within the SF, which is important for the final fibre properties. Finally, the post-treated SF/PEO fibre mat is immersed in deionized water and agitated, typically overnight, to completely remove the PEO component, leaving behind the desired SF nanofibres [43]. Electrospun SF exhibits a high degree of polymer chain alignment along the fibre axis and is readily susceptible to electrical poling [44]. A key advantage of electrospinning is its ability to produce nonwoven fabrics where characteristics such as surface area-to-volume ratio, pore size and fibre morphology can be controlled [45].

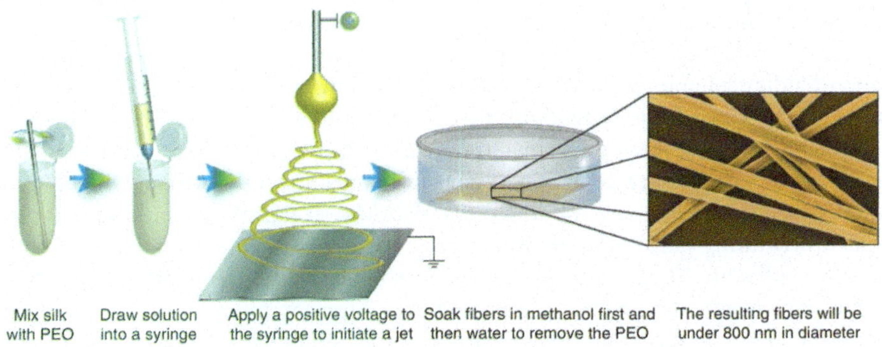

Fig. 12.2 Electrospinning process to prepare silk fibres. Reprint with permission from [29]. Copyright @2011 Nature Protocols

12.2.2.2 Crosslinking

To broaden the applications of SF, it's frequently necessary to add new capabilities by integrating functional materials. This can be achieved through two primary methods: intrinsic and post-functionalization. Intrinsic functionalization focuses on creating silk with inherent functionalities. This is accomplished by altering the silkworms themselves, either through genetic modifications or dietary adjustments, to produce silk with the desired properties directly [46]. Post-functionalization refers to modifying already-produced fibroin by adding functional components either to existing fibroin fibres or to regenerated fibroin solutions [47]. Introducing functional components into SF can be achieved by either physically attaching them to the surface or chemically bonding them within the material (crosslinking). These processes, however, can alter the silk's mechanical characteristics [48].

Crosslinking, a process that attaches several silk protein molecules together through chemical bonds, is a crucial step for altering silk properties post-production. It gives researchers the capability to select the silk properties for applications in nanofabrication and in making accurate micro- and nano-scale structures. Crosslinking techniques are generally classified into two categories. Physical crosslinking is accomplished by generating hydrogen-bonded beta-sheets in silk proteins through external forces such as heat, mechanical stress or electric field, while chemical crosslinking forms ionic or covalent bonds through chemical reactions—thus, a stable crosslinked network is obtained [29].

12.2.2.3 Nanofabrication

The burgeoning field of bio-nanotechnology represents the intersection of biological sciences, materials science and nanotechnology. The unification is bound to bring along considerable promises for the manufacture of new bioengineering applications. Progress in nanotechnology, including advancements in cryo-electron microscopy with high resolution, both top-down and bottom-up nanofabrication techniques and microfluidic technologies, enables the manipulation of biological structures, functions and processes. This capability facilitates the creation of innovative, functionally nanostructured biological materials and systems [28]. There are two major approaches used in the fabrication of nanostructures: the top-down and the bottom-up approaches. In general, top-down nanofabrication techniques involve reducing and restructuring bulk materials to create nanopatterned architectures. Some techniques applied to silk materials in this category include soft lithography, photolithography, ion beam lithography, nanoimprinting, liquid exfoliation and electrospinning. Perhaps the most important benefit of the top-down methods is being able to exert local control over creating features with the desired material properties. Bottom-up processes focus on the molecular self-assembly in which materials are built up from forces and interactions operating on the nanoscale, leading to the arrangement of nanoparticles in ordered nanodomains. The overriding advantage of a bottom-up approach is that it would allow one chemical control down to the atomic level. However, sometimes it suffers from a distinct disadvantage: scale-up problems regarding uniformity over large areas [28].

12.2.3 Modifications to Structure

Researchers have made remarkable strides in the last few decades in incorporating technology into silk-based materials, utilizing various techniques, including physical and chemical modifications to the material [37]. SF is frequently mixed with various polymers to modify its properties. The natural architecture of SF forms α-helices and sheets, which hold strong hydrogen bonds. The strong hydrogen bonds are classified as intramolecular, which are found within silk molecules, and intermolecular, which are found between different molecules. For the system to be amenable to additional modifications, such as attaching new functional groups or polymers, these inherent hydrogen bonds must be broken. Subsequent bonds will be formed with the new materials [5].

SF characteristics are heavily dependent on the structural arrangement and size of the molecules. Park et al. researched the relationship between molecular weight and electrospinning of reconstituted SF solutions. This study confirms that the molecular weight distribution exerts a significant influence on the physical characteristics of the fibres produced. Changes in pore size, shape and crystalline structure of the fibres were specifically observed. It was further noted that electrospun fibres with increased mechanical strength reflect a SF solution of higher average molecular weight viscosity. In essence, heavier fibroin molecules make thicker solutions and stronger fibres [49–51].

By manipulating the temperature of the water vapour, scientists can 'reshape' the internal makeup of silk molecules. Researchers such as Hu and co-workers have developed a new way of changing silk by carefully controlling exposure to water vapour at different temperatures. Temperature-monitored water vapour annealing enables the controlled alteration of silk's molecular architecture. Through specific means, this method can alter the crystalline structure of silk. At lower temperatures, such as 3 °C, the silk predominantly exhibits an α-helical structure. On the contrary, when conducted at higher temperatures, around 95–100 °C, the silk exhibits a bias towards a more β-sheet type of architecture with a crystallinity level of almost 60% [52].

While silk is prized for its beauty and comfort, it loses considerable strength when it gets wet. Researchers have now developed a technique to add water resistance and superhydrophobicity to silk fabrics. It is essentially an ion beam-based technique that involves oxygen etching followed by deposition of Fluorinated Diamond-Like Carbon (F-DLC). This approach is highly useful since it may produce an asymmetrical wettability pattern in which one side of the fabric is treated while the other side retains exposed hydrophilic water-absorbing characteristics. As a result, the fabric uses both hydrophobicity and hydrophilicity to regulate water within it [28].

12.3 Energy Harvesting Technologies Utilizing Silk

Energy harvesting from natural sources using biocompatible and biodegradable materials is both environmentally friendly and sustainable. This approach offers a promising alternative energy source for powering the next generation of portable biomedical devices [14]. Soft biomaterials like cellulose (derived from wood, bacteria), chitosan, chitin, gelatin, collagen and silk proteins from silkworms and spiders represent an excellent promise in bio-integrated soft electronics to be used for flexible di-, tri- or multifunctional biophysical/bioanalytical sensors and devices for energy harvesters [53]. Silk, a naturally abundant and eco-friendly biomaterial [14], has been extensively studied in various research fields due to its unique properties. There are four common energy harvesting mechanisms: photovoltaic, piezoelectric, triboelectric and thermoelectric [54].

12.3.1 Piezoelectric Materials Made from Silk

SF is a protein found in silk, and it can have two kinds of crystalline structures—silk I and silk II. Neither of these two crystalline types has a centre of symmetry, which is a basic requirement for piezoelectricity; it is most likely that SF would be a piezoelectric material, even though each crystal form may differ [55, 56].

Piezoelectric materials are known for their ability to generate energy from various mechanical and biomechanical movements [57, 58]. Novel development towards enhancing piezoelectric properties in (polyvinylidene fluoride) (PVDF) is the fabrication of flexible core–shell nanofibres from SF and PVDF through coaxial electrospinning—a well-known process in which two different materials are deposited in a controlled manner into a single nanofibre structure. The core of the nanofibre is made of SF, while its shell is made of PVDF.

The main outcome is that the interconnection between the molecules of PVDF and SF causes an increase in the piezoelectric β phase of PVDF. The phase β is significant for the piezoelectricity since it possesses a greater degree of alignment and molecular polarization than the other phases. Therefore, with the incorporation of SF at the core of the PVDF nanofibres, the devices made from these nanofibres will possess improved piezoelectric performance [59]. For preparing the PVDF solution, the powders are allowed to dissolve in a DMF-acetone solvent mixture (6:4 w/w) by stirring for three hours at 60 °C, which resulted in a clear solution. The coaxial electrospinning technique is utilized in the production of the SF/PVDF piezoelectric nanofibres. This method involves feeding different solutions through two separate needles, one inside the other, to form a composite nanofibre. In this case, the shell solution is composed of PVDF, while SF serves as the inner core solution. Subsequently, the electrospun materials are subjected to a drying process at 50 °C and followed by annealing at 160 °C for improvement of piezoelectric properties [59].

Table 12.2 Thermoelectric material coefficients (at T = 310 K) [62]

Thermoelectric materials	Seebeck coefficient (mVK^{-1})
Bi_2Te_3(n-type)	–0.0368
Sb_2Te_3(p-type)	0.1108

12.3.2 Thermoelectric Devices Using Silk

Silk is an ideal material for wearable devices due to its comfort and compatibility with human skin. In many studies, these properties have been utilized to develop wearable Thermoelectric (TE) power generators that can harness energy from body heat. Two high-performance TE materials can be used in the generator: bismuth telluride (Bi2Te3) for n-type conductivity and antimony telluride (Sb2Te3) for p-type conductivity [60, 61]. The Seebeck coefficient, which indicates how much voltage a material generates per unit temperature difference, is an important parameter for evaluating thermoelectric performance. The Seebeck coefficients of Bi_2Te_3 and Sb_2Te_3 are shown in Table 12.2. By the deposition of nanostructured Bi_2Te_3 and Sb_2Te_3 on the silk fabric renders it an ultra-flexible Thermoelectric Generator (TEG) [62].

Lu et al. developed a Thermoelectric (TE) generator by utilizing silk fabric as the substrate material (Fig. 12.3). To start with, silk fabric was coated with polyvinyl alcohol, and using a needle, some tiny holes were made at predetermined sites. By blending 80 mg of bismuth telluride (Bi_2Te_3) or 120 mg of antimony telluride (Sb_2Te_3) with liquid glue and deionized water, two types of thermoelectric paste were made. These pastes were applied to the silk fabric on the opposite sides to fill the holes and thus create thermoelectric columns. To completely remove moisture, the fabric was then placed in a vacuum oven. Ultimately, via silver paste and silver foil, electrical connections between the p-type (Sb_2Te_3) and n-type (Bi_2Te_3) columns were set up at the tops of the columns, leading to the creation of thermocouples. A functional prototype was built and its performance was evaluated through testing [62].

12.3.3 Triboelectric Nanogenerators (TENGs)

Triboelectric Nanogenerators (TENGs) are designed to convert mechanical energy from human activities in real time [63–67]. Triboelectric energy harvesting generates electricity by friction and electron transfer, and it has some advantages over other mechanical energy harvesting methods. These include high power, efficient energy conversion and device flexibility, making it ideal for integrating into textiles [68].

SF materials are being investigated in the context of Bio-Triboelectric Generators (Bio-TEGs). Silk nanofibre membranes have a larger surface area compared to cast silk films; hence, a higher open-circuit voltage was obtained by nanofibre-based Bio-TEG as compared to cast-film-based Bio-TEG [14]. In addition to 2D film-like structural Bio-TEGs, SF can be designed and fabricated along with other functional

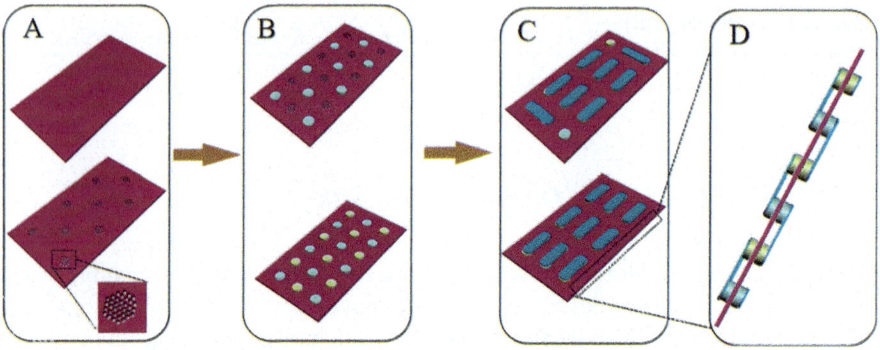

Fig. 12.3 Diagram of the fabrication process of a TE power generator based on silk. **a** Silk fabric pre-treatment; **b** Bi_2Te_3 and $Sb2Te_3$ nanomaterials deposition on the silk fabric; **c** Link of p-type and n-type columns with silver foils; **d** Side outlook of the silk-based TE power generator. Reprint with permission from [62]. Copyright @2016 Elsevier

components into triboelectric generator E-textiles. A 3D printing technique has been applied for the development of energy harvesting E-textile, which is based on core-sheath fibre. The coaxial printing of SF and carbon nanotube inks resulted in the fabrication of CNT@silk fibres [11].

A silk-triboelectric harvesting energy system provides a balance between the flexible triboelectric and mechanical performance desired for textile applications [69]. Both the electropositive and the electronegative sides of the system have core-spun yarns made up of a stainless-steel fibre core and a silk or PTFE fibre wrapped sheath, respectively. This is embodying woven and embroidered textile structures that could scale easily for production. With strength, the devices produce good generation and durability [68].

SF films, made by electrospinning technique, take the nanofiber-network form with an exceedingly high surface-to-volume ratio as well as a far rougher surface compared to the bare film surface, conducive to triboelectric power generation. The electrospun silk Bio-TEGs can generate high voltages, especially with a large surface area, and their power-generating performance can be fine-tuned for various working conditions. Structurally, this silk Bio-TEG consists of an electrospun SF nanofibre network film, a polyimide film, and aluminium foil, which serves as a current collector [14].

12.3.4 Photovoltaic Systems Based on Silk

Within the modern era, an enhancement of interest in photovoltaic devices has been found. The devices are either fibre-shaped or integrated into fabrics. Research has recently been done on the following examples of fibre-shaped cells: fibre-shaped organic photovoltaic cells [70, 71], fibre-shaped nanocrystalline solar cells (such as dye-sensitized solar cells) [72, 73], fibre-shaped perovskite solar cells. Such devices, based on energy fibres, have the main advantage of being lightweight and flexible because they can bend, twist and stretch [74].

Silk fibroin is the potential transient biomaterial. Its outstanding physicochemical properties that make it applicable in biophotoelectric. An Ag/silk protein/Si device can be produced by incorporating the volumes of silver nanoparticles (AgNPs) into a silk matrix. This construct exhibited a strong Lateral Photovoltaic Effect (LPE) with the highest sensitivity that is comparable to the performance of most existing inorganic materials [75].

Other investigations were performed to cast silk films for measuring surface and volume resistivity using a test fixture. The calculated values of resistivity were later used to calculate the resistance taking into consideration known values of the volume resistivity, film thickness and electrode dimensions. To further explain the electrical properties of silk, spin-coating was done on the Indium Tin Oxide (ITO) glass substrates and silver epoxy to form the electrical contacts. The finished devices were attached to a source meter that allowed current–voltage (I V) curves to be obtained automatically. Photoactive biological dopants, namely Bacteriorhodopsin (BR), as well as, the Photosystem I (PSI), were subsequently added to the silk solutions. Photovoltaic devices with heterojunction structures were fabricated from the doped solutions and then analyzed for their photocurrent response [76].

12.4 Sensors and Sensing Technologies Using Silk

Novel silk sensors and sensing devices are introduced as a new and eco-friendly approach to many applications such as biomedical, environmental and electronic devices. As SF has some exceptional properties, these make it a good polymer for implantable and wearable sensors. To increase flexibility, nanomaterials such as CNTs and graphene-hybridized silk have been shown to provide improved mechanical and sensitivity performances in applications including biosensing and other sensor-related functions [77]. Silk was also shown to be an ideal substrate for both capacitive and piezoresistive sensors, with enhanced performance in sensing motion and monitoring physiology. These sensors are used in wearable applications, bioelectronics, soft robotics and environmental monitoring to help create smarter and more sustainable technology [78].

12.4.1 Wearable Sensor Technologies

Wearable sensing technology, integral to the evolution of IoT and AI, requires materials that are flexible, breathable, biocompatible and multifunctional to meet the growing demands of advanced wearable electronics [79]. Due to the remarkable properties of Silk Fibroin (SF), it is considered as one of the best option for wearable technology as well as drug sensing and delivery [28, 80]. The advantages of silk such as its flexibility, adjustable mechanics and ability to work with conductive nanomaterials make it a perfect candidate for health monitoring, prosthetics, smart textiles and human–machine interface sensors, thus allowing the creation of flexible and eco-friendly wearable devices [28, 81]. However, its different forms of fibres, films, sponges and hydrogels make it suitable for efficient, comfortable and versatile devices [82–84].

Han et al. developed a silk fibre–based humidity by sensor using Graphene Oxide (GO) and electrostatic interactions between GO's negative groups and silk's positive sites enable direct sensing without adhesives, reducing costs and improving reliability. By tuning silk's charge properties, the sensor can be optimized for wearable and environmental monitoring applications [43, 85]. SF films find extensive use in sensor applications because of their stable surface, yet their conventional forms lack the elasticity required in wearable applications. To solve this, [86] developed a plasticized SF film by incorporating calcium chloride and employing ambient hydration. The process enhanced the stretchability of the film to over 400%, surpassing that of traditional SF films. Molecular dynamics simulations revealed that water molecules disrupted hydrogen bonds, thereby reducing stiffness, while $Ca2^{+}$ ions preferred α-helices and random coils, thereby enhancing extensibility. For the ease of demonstrating further capability, the film was coated with a 40 nm gold (Au) layer to create a low-electrical-conductivity, flexible electrode suitable for application on skin or outside the body [43, 86]. Electronic skin, as a wearable sensing platform, demands substrates that are biocompatible, heat resistant and flexible. Huang et al. also fabricated a polyurethane-doped SF Composite Membrane (SFCM) with attributes such as 200% stretchability, thermal resistance to 160 °C and greater than 90% clarity. The addition of silver (Ag) and platinum (Pt) nanofibre networks enabled SFCM to provide heating and temperature-measuring capabilities. Upon exposure to human skin for 20 days, the material showed no reaction or immune response, thereby demonstrating its biocompatibility and suitability for application. These qualities position the SFCM as the best candidate to transform electronic skins into wearable technologies and medicine (J. Huang et al., 2020; [43]). However, wearable electronics have the main challenge of skin adhesion, which has been made easier by Seo et al., mixing silk hydrogel with calcium ions ($Ca2^{+}$) and obtaining a material that is biocompatible, adhesive and soft at the same time. Adhesion is caused by the disruption of crystalline structures, metal-chelation cohesion and water-mediated bonding. The hydrogel was able to sustain four different kinds of wearable devices placed on human skin, thus emphasizing its medical potential [43, 87].

12.4.2 Biosensor

Biosensors are analytical instruments that utilize immobilized biomolecules, such as enzymes, antibodies, nucleic acids, hormones, organelles or even cells, to detect target molecules. The target analyte produces detectable signals through electrical, chemical or physical reactions upon contact with biological components, enabling accurate detection and analysis [88–90]. An analyte is any substance, for instance, glucose, urea, a drug or a pesticide, whose concentration needs to be measured. This is achieved by detecting and interpreting signals generated when in contact with a biosensor [91–93]. Biosensors are valued for their accuracy, sensitivity, affordability and ease of use, thereby assuming a central position in food safety, medicine and environmental monitoring. Electrochemical and medical biosensors require biocompatible and biodegradable materials. Silk, with its flexibility and strength, is proving to be a promising choice for a more efficient and safer material for electrochemical biosensors. Silk is a natural, biocompatible material that possesses immense strength and elasticity, with numerous applications. Due to silk being a versatile and satisfactory material in terms of mechanical properties, it is at the forefront of the evolution of biosensors, particularly electrochemical biosensors. Over the past few decades, biosensors based on silk have been a major area of interest and are now among the most significant areas of study in the scientific world [94, 95]. However, Silk Fibroin (SF), being naturally biocompatible and biodegradable, blends seamlessly with biological systems, making it ideal for medical applications where sensors need to operate safely inside the body [94]. Molinnus et al. fabricated a biosensor by employing an SF matrix containing the Glucose Oxidase (GOX) enzyme, which catalyses the oxidation of glucose to gluconic acid and hydrogen peroxide (H_2O_2). Thin- and thick-film technology is employed in sensor fabrication for the deposition of platinum (Pt) electrodes for glucose sensing and a silver/silver chloride (Ag/AgCl) electrode as a reference (Fig. 12.4). This fabrication renders the sensor biocompatible, flexible and very efficient in glucose electrochemical detection. At a functional level, glucose in the test solution is oxidized by GOX to form H_2O_2. H_2O_2 is further oxidized at a platinum electrode, producing a corresponding current that is proportional to the glucose concentration. The electric current can be measured with high accuracy to determine glucose concentration. It responds in approximately one minute and has a detection range of 0.5 mM to 3 mM glucose, with a sensitivity of approximately 0.24 ± 0.006 μA/mM [96].

The key performance and stability parameters of this silk-based biosensor are summarized in Table 12.3, highlighting its substrate material, enzyme, electrode configuration, fabrication method, detection mechanism, sensitivity, response time, mechanical stability and overall performance [94, 96].

In addition to its excellent biochemical performance, the silk biosensor is also highly resilient. It can last as long as 100 days, but sensitivity slowly diminishes over time due to the leaching of the enzyme, loss of enzyme activity or membrane peeling. The sensor also exhibits elevated mechanical stability with the ability to function even with up to 160 bend cycles. Despite the minor difference in resistance resulting from minor cracking of the platinum film, overall stability is maintained. This type of property makes silk-based biosensor a viable choice for flexible and wearable glucose sensing with an effective solution for real-time health monitoring [94, 96].

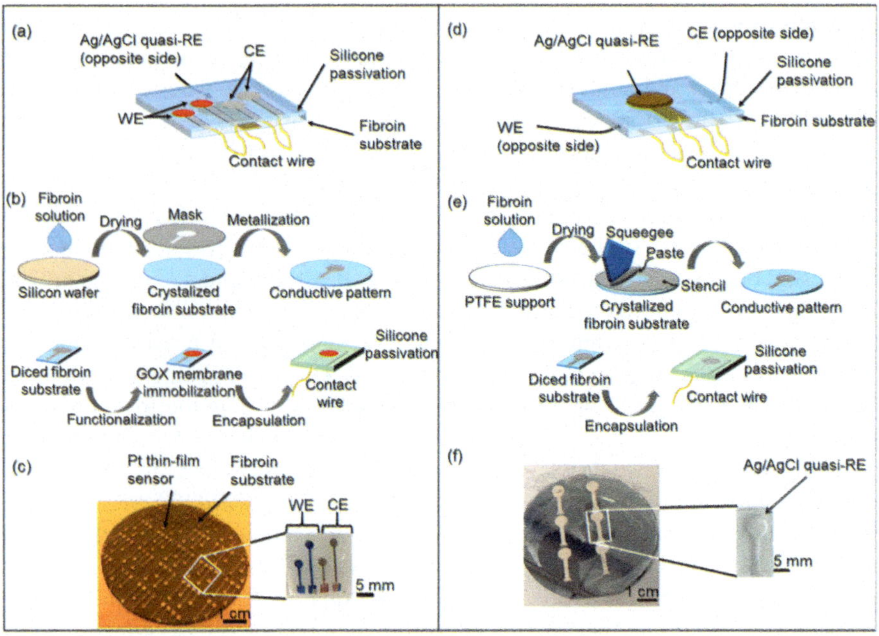

Fig. 12.4 **a** Schematic of the biocompatible glucose biosensor chip with Pt working and counter electrodes on the upper side and an Ag/AgCl quasi-reference electrode on the bottom. **b** Fabrication steps of the Pt electrodes using thin-film technology. **c** Light microscope image of a 4-inch fibroin substrate with Pt electrodes (left) and the final diced chip (right). **d** Flipped schematic of (**a**) with the Ag/AgCl electrode on top and Pt electrodes on the bottom. **e** Fabrication steps of the Ag/AgCl electrode using thick-film technology. **f** Microscopic image of a 4-inch fibroin substrate with Ag/AgCl electrodes (left) and the final Ag/AgCl electrode (right). Abbreviations: GOX—Glucose Oxidase, RE—Reference Electrode, WE—Working Electrode, CE—Counter Electrode. Reprint with permission from [96] Copyright @2021 Elsevier

12.4.3 Monitoring the Environment

Silk sensors can sense changes in temperature, humidity and even air contaminants and harmful pathogens. Even more exciting is that they can be designed to interface with future technologies, such as the Internet of Things (IoT), allowing real-time monitoring of environmental conditions. Already, researchers have demonstrated how such silk biosensors could be utilized in drones to rapidly and effectively detect environmental pathogens, thereby enhancing environmental monitoring even more efficiently than currently possible [97]. Besides the ability to identify toxic germs, silk sensors are also used to detect air and water quality. Scientists have also developed sunlight-powered, high-accuracy SF sensors for identifying pollutants in the environment [98]. Silk nanofibre sensors have also been investigated for toxic chemical sensing in industrial environments and in environmental monitoring devices to enhance the safety of the work environment and preserve the natural environment

Table 12.3 Key performance and stability parameters of the silk-based biosensor [94, 96]

Parameter	Details
Substrate material	SF, a natural, biocompatible polymer
Enzyme used	Glucose oxidase (GOX)
Electrodes	Platinum (Pt) working and counter electrodes; Silver/Silver Chloride (Ag/AgCl) quasi-reference electrode
Fabrication method	Thin- and thick-film technologies
Detection mechanism	Oxidation of glucose by GOX produces hydrogen peroxide (H_2O_2), which is oxidized at the Pt electrode, generating an electric current proportional to glucose concentration
Detection range	0.5 mM to 3 mM glucose
Sensitivity	0.24 ± 0.006 μA/mM
Response time	~1 min
Calibration curve	Linear within the measured concentration range
Long-term stability	Proven stable for over 100 days in enzyme-free moist conditions
Factors affecting sensitivity loss	Possible enzyme leaching, enzymatic activity loss or detachment of the enzymatic membrane
Mechanical stability	Maintains functionality after 160 bending cycles (180°)
Effect of bending on resistance	Resistance increased from ~80 Ω to ~115 Ω due to small crack formation in Pt layer
Overall performance	Biocompatible, flexible and reliable for glucose monitoring
Potential improvements	Replacement of synthetic reagents with other natural polymers for enhanced compatibility and stability

[99]. The results indicate that long-lasting, eco-friendly sensors can be fabricated from silk, contributing to a greener and safer world. They typically employ various sensing principles, including piezoresistive, capacitive and opto-chemical reactions.

12.4.3.1 Piezoresistive Silk-Based Sensors

Piezoresistive silk sensors have the ability to sense pressure and humidity in the environment. Upon the application of stress or strain on the silk, the resistance of the silk varies with the change in the flow of electrons. The variation is transformed into an electric signal that can be measured. The sensors are used directly in measuring humidity levels in farms and factories, where the levels of moisture need to be suitable in a bid to enhance productivity and safety. Even in recent times, innovation is witnessed where these sensors are integrated into wearable and flexible devices for environmental monitoring in real time. In addition, silk piezoelectric sensors have been innovated to develop autonomous environmental monitoring systems in a bid to eliminate the dependency on power sources and increase sustainability [100, 101].

12.4.3.2 Capacitive Silk-Based Sensors

Capacitive silk sensors operate through the detection of the change in the amount of charge that they can store as they are exposed to different environmental conditions like change in humidity and temperature change. They are made up of a sheet of SF sandwiched between two conductive materials in the form of a capacitor. The moment the moisture content changes, the properties of SF are altered, causing capacitance change. These too are quick and sensitive sensors, and they're great to utilize in the context of air quality monitoring, pollution detection and even wearable health sensors. And because they're flexible and biodegradable, they're ideal to use for disposable and green sensor technology [102].

12.4.3.3 Opto-Chemical Silk-Based Sensors

Opto-chemical silk sensors utilize the native light-sensitiveness of SF to capture environmental contamination. They operate via silk nanofibres and a discriminative material that darkens or alters illumination when exposed to harmful chemicals. Thus, they can effectively detect harmful gases, heavy metals and organic pollutants in water and air. Research has shown that SF is compatible with opto-chemical sensors, making it a superior option for industrial risk detection and environmental protection monitoring [99]. In addition, flexible electronics based on SF have also been integrated to detect environmental contaminants, pressure, humidity and industrial hazards, as well as human health, as indicated in Table 12.4.

Table 12.4 Major mechanisms and applications of silk-based sensors for environmental monitoring

Sensor type	Mechanism	Application	Reference
Optical biosensors	Light absorption and fluorescence changes	Detection of environmental contaminants	[98]
Piezoresistive sensors	Resistance changes due to mechanical stress	Humidity and pressure monitoring	[100]
Capacitive sensors	Dielectric property changes in response to moisture	Atmospheric monitoring	[102]
Piezoelectric sensors	Voltage generation from mechanical deformation	Self-powered environmental monitoring	[101]
Opto-chemical sensors	Silk nanofibres responding to chemical exposure	Industrial hazard detection	[99]

12.4.4 *Silk-Based Smart Textile Sensing System*

Silk-derived fabrics are embedded with sensors to monitor physiological and environmental signals and are used in healthcare, fitness and even environmental monitoring. New advancements include the functionalization of silk with functional nanomaterials to provide electrical conductivity and sensor capabilities. Evidence suggests that silk sensors have applications in pressure and strain sensing, gas and humidity sensing and bioelectronic interfaces for next-generation smart clothing [103].

12.4.4.1 Silk-Based All-Fibre Multifunctional Sensor for Smart Clothing

The SF-based All-Fibre Multifunctional Sensor (AFMS) has high potential for smart sensing applications, such as smart clothing, due to its multiple sensing functions and merit features. A two-layer structural design featuring a silk fibre film and an interdigital silver nanowire (AgNWs) electrode was envisioned to create an All-Fibre Multifunctional Sensor (AFMS). Here, the sensing element as well as the structural support is SF. Electrospinning and spray-coating techniques are used to form this fibre film. This fibre assembly is completely flexible and breathable because it's an all-fibre assembly, whereas SF enhances its biocompatibility. Among all the benefits of silk-based sensors, one of the major benefits is the ability to detect more than a single environmental condition due to its all-fibre platform and the unique dielectric properties of SF. It has the ability to detect pressure, temperature and humidity efficiently.

To illustrate its practical use, it has been successfully applied in a Respiratory Monitoring smart mask and a smart glove capable of detecting finger joint motion (Fig. 12.5). With its multifaceted sensing ability and improved flexibility, breathability and biocompatibility, AFMS has great potential in wearable intelligent textiles, such as intelligent clothing and future health monitoring equipment [79]. The pressure sensor leverages the dielectric property of SF, whereby capacitance changes in response to external pressure, enabling detection [104]. The sensor exhibited a good sensitivity of 2.27 pF/kPa and an excellent resolution of ~ 26 Pa at low pressure [79]. SF is also responsive to humidity and temperature naturally in the environment [105, 106]. SF is hydrophilic and will absorb moisture from the environment, causing a change in its dielectric constant and, consequently, a change in its capacitance. This renders the AFMS a potential humidity sensor. Similarly, the integration of silk nanofibres and AgNWs offers thermal sensing functionality by altering the material's properties upon exposure to heat and generating electrical signals to facilitate real-time temperature monitoring. All these sensing features collectively make AFMS useful in adaptive and climate-responsive intelligent textiles [79]. A highly innovative application of AFMS includes sensing bending angle and breathing patterns. By utilizing smart gloves, the sensing of finger joint motion becomes possible, making it suitable for human–machine interaction and rehabilitation monitoring.

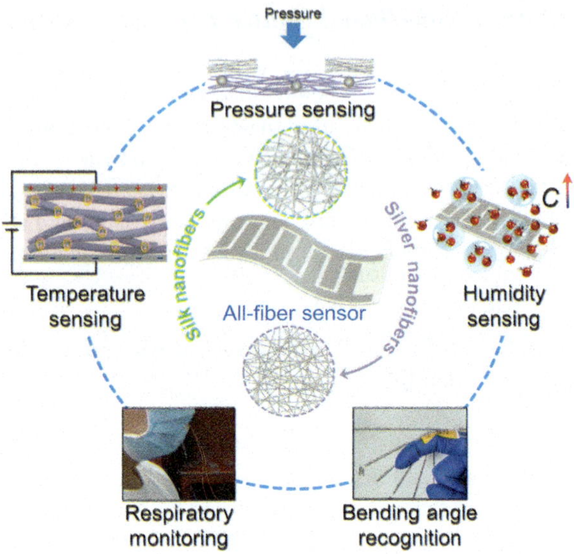

Fig. 12.5 SF-based wearable all-fibre multifunctional sensor for smart clothing. Reprint with permission from [79] Copyright @2022 Elsevier

AFMS has also been incorporated into smart masks to detect respiratory patterns. The sensor provides real-time respiratory monitoring, which has been extremely useful for medical diagnosis as well as tracking sports performances, based on the measurement of minute pressure changes caused by the inhalation and exhalation process [79, 80].

12.4.5 Collecting Real-Time Data

Silk has gradually become one of the most popular materials for creating devices that can monitor and record real-time physiological data. The silk-based devices outperform the conventional electronic sensors in terms of comfort, accuracy and environmental friendliness, as these sensors usually have many problems like poor skin adhesion after a while, limited mechanical strength and the need for non-sustainable environments. A recent study has reported Silk Fibroin (SF), the primary protein in silk, as a versatile foundation for the fabrication of biosensors with extremely high sensitivity to monitor human body changes like heart rate, muscle contractions and fluid balance. Besides the health care area, the silk-based sensors are also actively taking over the industrial and environmental monitoring, being very sensitive and responsive in detecting air quality, material strain and water content in real time [80, 101, 102].

12.4.5.1 Real-Time Health Monitoring

One of the brightest applications for silk-based sensors is in medical diagnosis and health monitoring, where comfort, precision and wearability over a protracted duration are paramount. Traditional wearable sensors, often made of rigid materials, are poorly tolerated and can be either unsanitary or inhibit movement, causing skin irritation or limited mobility. Silk sensors offer a soft, porous and skin-compliant alternative that naturally fits the shape of the body and is therefore best suited for continuous monitoring. The new sensors are wearable and may be inserted into soft patches or even woven into fabric so that they can be implantable in the skin. It allows real-time monitoring of such key vital signs as blood pressure, glucose levels, respiratory rate and even the activity in the brain, providing valuable health feedback without causing any pain [80, 107]. Also, the silk combined with ionic hydrogels has amplified the stretchability and durability of these sensors even more, enabling them to deliver accurate data under dynamic situations [108]. Indeed, the most powerful aspect of silk-based sensors is their sensitivity to a mere physiological change. Silk-based piezoelectric sensors, for example, take advantage of the inherent charge storage property of SF to sense very small body movement, i.e. muscle spasm or tension change in the skin, with incredible sensitivity. This makes them extremely useful for application in stroke rehabilitation, where sensing small muscle responses can provide critical feedback for rehabilitation [78]. Also, because silk is naturally biocompatible and biodegradable, the sensors can be fabricated for temporary or implantable use, meaning no surgical intervention to remove them after they have served their purpose. With advancing technology, silk-based sensors will increasingly be applied in personalized medicine to offer non-invasive, real-time health monitoring that is both comfortable and effective [101, 109].

12.4.5.2 Real-Time Environmental and Industrial Tracking

Apart from biomedical uses, silk-based sensors also have a significant role to play in environmental monitoring and industrial applications. SF's intrinsic ability to interact with volatile molecules has been used to design sensors that can sense air pollutants, allergens and toxicants. The sensors use a bio-responsive system, whose conductivity or fluorescence changes upon exposure to known pollutants, for online environmental monitoring. Silk-based humidity sensors are produced on an industrial scale for monitoring moisture levels in industries and warehouses to ensure product stability and prevent spoilage due to excess moisture. Because they are biodegradable and non-polluting, they can be applied in significant amounts without causing any environmental pollution [97, 99].

Silk sensors are also advantaged by the natural property of SF to be responsive to environmental temperature and humidity changes naturally [77, 110]. They thus have the capacity to detect environmental alteration in real time and with high accuracy. Silk sensors are best suited for applications that require continuous monitoring, such as weather prediction, environmental surveillance and plant monitoring.

Their lightweight and biodegradable qualities also make them an eco-friendly choice, without compromising usability and minimizing electronic waste. At the manufacturing stage, silk sensors are being utilized in warehouses, food processing and pharmaceutical manufacturing for maintaining optimal temperature and humidity. By maintaining invariable conditions, these sensors prevent wastage, ensure product quality and eliminate spoilage. Smart silk fabrics are being designed to monitor body perspiration rate and temperature, providing immediate feedback for medical treatment, defense personnel who encounter unfavourable climatic conditions and sportsmen and women. The biocompatibility and flexibility of 'silk-enabled sensors' make this technology a future star that can be effortlessly integrated into natural and industrial processes, delivering actual and innovative data gathering [102].

12.5 Hybrid Systems Combining Silk for Energy Harvesting and Sensing

12.5.1 Integrating Silk with Functional Nanomaterials

Recent studies have focused on the opportunity to hybridize silk with functional nanomaterials for enhanced performance, which would open new frontiers in bioelectronics, energy storage and wearable sensors. Such synergies between silk and nanomaterials yield superior, highly durable and multifunctional composites for use in next-generation smart textiles and biomedical applications. Among all such materials, the six well-known materials are carbon CNTs, MXenes, elastomers, metallic nanoparticles, graphene and conductive polymers, which are known to possess immense possibilities for enhancing device performance with high biocompatibility [111].

12.5.1.1 Carbon Nanotubes (CNTs) and Silk

Carbon Nanotubes (CNTs) are widely used in silk-based bioelectronic devices due to their high conductivity, mechanical strength and elasticity. The incorporation of CNTs into silk matrices has enabled scientists to fabricate conductive, flexible and biocompatible electronic skins and flexible sensors. The easy flow of charge is also enabled by the close interaction between CNTs and silk proteins; thus, the CNT-silk composites are best suited for use as strain sensors and flexible transistors. Additionally, silk acts as a stabilizer, preventing CNT agglomeration and ensuring uniform electrical behaviour in the composite material [111].

12.5.1.2 MXenes and Silk

Among emerging nanomaterials, Mxenes have attracted attention for their exceptional electrical conductivity and surface chemistry, making them ideal candidates for integration with silk-based bioelectronics. MXenes, when incorporated into silk, enhance the electrochemical properties of bioelectronic devices, enabling their utilization in biosensors, energy storage devices and implantable electronics. MXene combined with SF produces flexible electrodes with superior charge transfer properties. Additionally, the biocompatibility of silk enables it to be interfaced with human tissue without injury, facilitating real-time monitoring of health and wearable diagnosis [111–113].

12.5.1.3 Elastomers and Silk

Elastomers play a crucial role in the development of flexible and stretchable electronics, and their hybridization with silk yields ultra-soft bioelectronic interfaces. Silk-elastomer composites are designed to withstand large deformations without compromising electrical performance. The composites are extensively used in epidermal electronics and soft robotics. Silk is also an elastomeric material, serving as a biological scaffold, and when applied to transient electronic devices, it enhances the biodegradability and mechanical properties of the elastomeric material. The combination is particularly convenient for use in electronic skin (e-skin) and artificial nerves, where elastomeric properties with great conformability and flexibility are desired [111, 114, 115].

12.5.1.4 Metallic Nanoparticles and Silk

Metallic nanoparticles, such as gold (Au) and silver (Ag), are typically incorporated into silk matrices to enhance the electrical and antibacterial functionality of wearable bioelectronics. Silk provides a natural and stable environment for the dispersion of metallic nanoparticles, preventing agglomeration and maintaining homogeneous conductivity. These composites are utilized in a wide range of applications, including flexible electrodes, biosensors and antimicrobial coatings. For example, Ag nanoparticle silk composites have been prepared for wearable health monitors that can deliver enhanced signal transmission and real-time physiological monitoring [111, 116].

12.5.1.5 Graphene and Silk

Graphene is yet another highly promising nanomaterial for silk-based electronics, owing to its exceptional electrical conductivity, mechanical properties and large surface area. Silk-graphene composites are increasingly finding applications in flexible energy storage devices, such as batteries and supercapacitors. The blending of

SF with graphene sheets yields a light, conductive, stretchy and weightless material for next-generation bioelectronics. In addition, the incorporation of graphene into silk scaffolds renders them stable in aqueous environments, which is desirable for on-skin and implantable biosensors [111, 117, 118].

12.5.1.6 Conductive Polymers and Silk

Conductive polymers, such as Polyaniline (PANI) and poly(3,4-ethylenedioxythiophene), yield a conformable and stable substrate for silk electronics. The polymers offer electronic and ionic conduction and are therefore used in flexible displays, energy harvesting and bioelectronic interfaces. Blending conductive polymers with silk yields conformable, breathable and lightweight materials that are utilized in wearable sensors and electronic wear. The inherent biocompatibility of silk also ensures that conductive polymer devices are safe for prolonged skin exposure [111, 119, 120].

12.5.2 The Synergistic Effects of Silk and Conductive Polymers

Synergism between SF and conducting fillers is facilitated by molecular interactions, including hydrogen bonds and electrostatic forces, which enable efficient charge transport in addition to mechanical reinforcement [121]. Through the use of advanced functionalization techniques, SF-based materials become electrically conductive and triboelectric, making them suitable for bioelectronics and energy harvesting devices [122]. The mechanical strength of SF-polymer composites, based on hydrogen bonding and hybridization with graphene or MXene, renders them suitable for use in stretchable electronics, artificial skin and flexible supercapacitors [123, 124]. These features position SF-based conductive polymers as next-generation materials for biomedical and wearable devices (Table 12.5).

Table 12.5 Properties of silk-based conductive polymer composites

Material composition	Electrical conductivity	Mechanical strength	Flexibility	References
SF (SF)	Low	High	Moderate	[121]
Silk-PVA composite	Moderate	Very High	High	[121, 123]
Silk-graphene hybrid	High	High	Moderate	[123]
Silk-PANI composite	Very High	Moderate	Low	[108]

12.5.2.1 Conductive Properties of Silk-Polymer Composites

The incorporation of conductive materials, such as Graphene Oxide (GO), polypyrrole or poly(vinyl alcohol), as composites enhances the electric conductivity and mechanical stability of silk-polymer composites [85]. The studies have established that the hierarchical β-sheet architecture of SF facilitates charge transfer within polymer matrices, thus a valuable constituent in bioelectronic devices [122]. Silk-reduced GO composites, for instance, enhance the triboelectric property extensively, allowing the harvesting of energy in wearable devices [77, 85]. Additionally, the blending of SF with conductive hydrogels enhances their stretchability and adhesion and is therefore ideal for epidermal electronics and stretchable sensors [122]. Silk's degradability and biocompatibility also possess advantages over synthetic polymers, particularly for medical uses such as neural interfaces and bioelectronic scaffolds. A recent study of flexible neural interfaces confirmed that SF, with integrated conductive patterns, exhibits exceptional conformality alongside exceptional bioelectric conductivity. The blending of SF with conductive polymers such as polypyrrole and MXene has been shown to be more mechanically flexible and electrically conductive and has also been shown to be useful in next-generation bioelectronic interfaces [123]. Apart from that, the application of SF-based conductive hydrogels has improved strain sensors to be more stable and sensitive to make them suitable for real-time health monitoring and flexible electronic devices [124]. All these reveal a lot of things about how crucial silk-based conductive polymers are to the invention of new materials applied in industrial and biomedical applications.

12.5.2.2 Mechanical Strength and Flexibility of Silk-Conductive Polymer Composites

SF's β-sheet conformation exhibits high tensile strength and elasticity, greatly enhancing the mechanical toughness of polymer systems. SF and poly (vinyl alcohol) form a hydrogen bond network that greatly enhances elasticity with ultra-high mechanical toughness. The silk-graphene hybrid is also extremely fracture-tough and elastic and is used in wearable devices and flexible electronics [123]. The mechanical properties of silk-based composites are primarily governed by crosslink density and molecular structure, which enables their application in flexible sensors and bioelectronic devices [122]. This is also evident through the introduction of conductive polymers, such as PANI, which enhances both mechanical strength and electrical conductivity, thereby yielding prospects for future flexible energy storage devices [108].

Besides elasticity, silk-polymer composites exhibit strong self-healing properties, enhancing the stability of electronic and biomedical applications. Performing nanoparticle-doped hydrogels and SF were found to be soft upon application against mechanical stress and are hence best suited for artificial skin and soft robots [108, 124]. In addition, current studies reveal that silk-polymer nanocomposites are also mechanically stable in harsh environmental conditions, thereby increasing their shelf life in flexible electronics. Such synergetic integration between SF and conducting polymers thus yields a fascinating elasticity-mechanical stability relationship, which makes them applicable to a vast number of applications, ranging from implantable electronics and energy harvesting devices to smart textiles. Long-term innovation in this field will drive future advancements towards sustainable, high-performance materials for applications in emerging technologies [119].

12.6 Applications of Silk-Based Systems for Energy Harvesting and Sensing

12.6.1 Applications in Biomedicine

The field of biomedicine, dedicated to improving healthcare through the development of new therapeutics, diagnostics and regenerative medicine, is rapidly leveraging the unique properties of silk for energy utilization. Silk does not store meaningful energy, but it possesses the essential properties of being biocompatible, flexible and having tunable optical qualities, which make it an ideal platform for creating safe and energy-efficient biomedical devices [125]. Notably, silk is also a significant substrate for biosensors, enabling the accurate detection of physiological parameters with low energy consumption. The fact that silk-based sensors can be regenerated using external energy sources, such as UV light, indicates its contribution towards reusable and sustainable diagnostic devices. The ability to manipulate molecules in the silk film to restore sensor function, as demonstrated in this process, illustrates how energy can be harnessed to eliminate waste [95]. Additionally, the versatility of silk is also evident in drug delivery systems, where the controlled release of the drug can be triggered by external energy stimuli, such as light or heat [126]. This type of controlled drug release, facilitated by the biocompatible silk matrix, delivers specific therapy and minimizes side effects. In tissue engineering, silk scaffolds with electrical stimulation increase tissue regeneration and growth, demonstrating their potential to boost energy-based regenerative processes [127]. Controlled electrical fields and the structural integrity of silk augment cellular proliferation and differentiation, accelerating tissue repair. Silk's biocompatibility with wearable electronics also enables the design of health monitoring systems powered by energy harvesting devices, such as triboelectric nanogenerators, which utilize power from body motions or heat [128]. These silk-based devices are biocompatible and elastic, enabling doctors to monitor health readings non-invasively and in real time. In essence, silk plays a crucial role in developing energy-efficient medical technologies that enhance testing, therapy and healing, as it is compatible with the body and flexible, enabling the creation of improved and patient-centred healthcare technologies [129].

Fu et al. [130] reported silk-based conductive biomaterials for bio-interface technology, with potential applications in wearable electronics, implantable devices and the power needs of biomedical devices. Their article discusses the use of biomaterials in the incorporation of self-powered biosensors with TENGs or piezoelectric devices for biomechanical energy harvesting for real-time health monitoring. Additionally, conductive silk materials were determined to be suitable for implantable bioelectronics like neural stimulators and glucose monitors that need stable power sources. The research also proposed the employment of electrical stimulation for tissue engineering purposes to promote nerve, muscle and bone regeneration. Furthermore, silk scaffolds have been explored for use in biofuel cells, facilitating the harvesting of renewable energy in implantable devices. Consequently, silk-based conductive materials are confirmed to be viable candidates in energy-efficient biomaterials.

12.6.2 Environmental Monitoring Applications

Silk-based technology is one of the emerging green and highly viable systems of environmental monitoring. Silk Fibroin (SF) allows its design into various sensor formats, which minimize the environmental impact, yet enable real-time pollutant detection. As an example, Matthew et al. used eco-sensors made of silk origami, by folding silk films, coated with dye, into tiny boats that can be used as pH/colourimetric sensors of water pollutants. These devices do not require any external source of power, they generate a visible colour change in less than a minute and can be reused again [131]. However, recent reviews have reported SF and SF-based composites that can also sense humidity and toxic gases including ammonia and nitrogen oxides. To control conductivity and sensitivity, these composite sensors often use metal oxides, MXenes or graphene to make silk a versatile substrate and active gas and humidity detector [102]. Another major property of the silk is its ability to serve as a bioreceptor-immobilizing matrix. Due to the presence of functional groups and the ability to control their folding, SF has the capability to encapsulate enzymes or other sensing molecules without the need to use harsh chemical reactions. The property enables the creation of biosensors to detect pollutants either by enzymatic reactions or electrochemical reactions [132]. In addition, the materials made of silk would assist in the more sustainable environmental systems, since they can be degraded harmlessly. As an illustration, polymeric air filters (Fig. 12.6) made of silk fibroin have been demonstrated to break down in a biological environment, thus minimizing the waste production in the long run [133]. Future monitoring platforms will be able to be self-powered by incorporating energy harvesting devices like silk-based Triboelectric Nanogenerators (TENGs) or piezoelectric silk devices. Such self-powered sensors would be very beneficial especially in remote or inaccessible areas where power is inaccessible. Therefore, the potential of the synergistic nature of sensing and energy generation of silk is exciting and could result in a really green autonomous system aimed at environmental surveying.

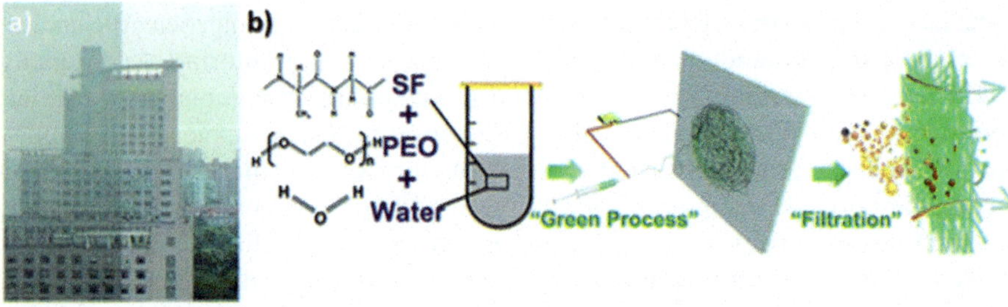

Fig. 12.6 **a** Clear and hazy day photograph of the same location in Shanghai and **b** a diagram showing the process of making silk fibroin nanofibres membranes for the purpose of air filtration [133]. Copyright© 2018 Gao et al., published by The Royal Society of Chemistry (under the Creative Commons Attribution 3.0 Unported Licence)

12.6.3 Smart Textile Applications

E-textiles, or smart textiles, are a materials science innovation in which the traditional softness and flexibility of textiles and the advanced electronic function are integrated. The future-generation textiles are engineered to sense and respond and interact with their wearer and environment in an active manner, possessing the capability to sense, respond and adapt to an extensive range of stimuli. This adaptability renders them very valuable in a broad spectrum of applications, ranging from but not limited to medical monitoring, sports performance, fashion in the lead and the new wearables technology [134]. The integration of conductive materials, high-performance sensors and efficient energy harvesting components within these fabrics enables real-time data acquisition, adaptive clothing designs responding to dynamic conditions and the formation of self-sustaining systems, thereby providing new frontiers for functional textile development [135]. Among the materials studied for intelligent textile applications, silk-based materials have been particularly promising candidates, known for their high mechanical strength, intrinsic biocompatibility, environmental biodegradability and unique structural adaptability. SF, being a silk protein, is designed into multifunctional structures, including TENGs and piezoelectric textiles, which enable the transformation of mechanical energy originating from body motion, friction forces and environmental vibrations into electrical energy usable for various applications [43]. This energy harvesting feature significantly reduces the need for batteries and external power sources, enhancing the sustainability and performance of smart textiles. Silk-based TENGs utilize the triboelectric effect, in which materials with sufficiently different electron affinities acquire an electric charge upon repeated contact and separation from each other [136]. Being perfectly embedded into fabric architectures, silk-based TENGs can be used to tap energy from almost any natural human motion, ranging from walking and running to minor body movements. This capability to harness power from everyday motion positions silk at the focal point of innovation, leading to the realization of next-generation, self-reliant intelligent apparel. This technology has the potential to revolutionize wearable technology with increased autonomy and sustainability [137].

Lu et al. [62] developed a stretchable TE power generator from commercially available silk fabric, with nanostructured Bi_2Te_3 and Sb_2Te_3 being deposited to form TE columns interconnected by silver foils. The generator effectively converted thermal energy into electricity under a temperature gradient of 5–35 K, with a maximum power and voltage of ~15 nW and 10 mV, respectively. It maintained stable performance even after undergoing 100 twisting and bending cycles. When strapped to the arm, the generator produced different voltage outputs before and after 30 min of walking, demonstrating its wearable energy harvesting feature. The work presented a novel solution to TENGs, towards self-powered wearable electronics. Chen et al. [138] presented a state-of-the-art survey of smart textiles, including silk-based textiles, that harvest energy from the body and environment to mitigate the energy crisis, environmental pollution and related health issues worldwide. The article categorizes such textiles based on their ability to harvest biomechanical, body temperature, biochemical, solar and hybrid energy sources. The article further discusses the greatest challenges and future direction, observing that material and chemical developments will play a crucial role in driving smart textiles into the mainstream and transforming day-to-day life in the Internet of Things era.

12.6.4 IoT Device Applications

Recent developments show how much potential Silk Fibroin (SF) can have in self-powered Internet of Things (IoT) devices. As an example, SF-based Triboelectric Nanogenerators (TENGs) using Polyethylenimine (PEI), MXene composites and porous silk aerogels were made to realize high-output, flexible and biodegradable energy harvesting. These types of generators inherently drive their own senses like humidity sensors, eliminating the use of external batteries [139–142]. At the same time, there has been a development of silk-based sensors. Pressure sensors made of SF/graphene nanofibre membranes have been applied to develop a better sensitivity and long-lasting sensor; degradable piezoresistive sensors made of pure silk fibroin have been used to make wearable health devices [100, 143, 144]. Moreover, conductive nanocellulose-reinforced silk fibres have been developed to sense strain and humidity and therefore have been used in smart textile development. It is worth noting that a silk fibroin ionic touch interface was recently presented in a biodegradable, flexible display with an integrated IoT capability [145]. These advances highlight the new nature of silk in terms of sustainable, self-powered and connected electronics, hence offering a viable route to IoT devices that are eco-friendly.

12.6.5 Industrial Applications

Energy is crucial in the various industries and emerging materials are helping to make industrial processes smoother and greener [146]. Several technologies have been developed that include Triboelectric Nanogenerators (TENGs), batteries, supercapacitors, solar cells and fuel cells relying on high-performance materials, which are flexible, durable and environmentally friendly [147]. Silk Fibroin (SF) has become one of the prominent candidates of such applications as it has the ability to satisfy those requirements [148]. In particular, TENGs made of silk could be used to transform mechanical movement into electrical energy, which makes it beneficial to wearable devices, robotics or self-driven sensors that are part of the industrial surveillance system [149]. Silk's flexibility facilitates the easy integration of silk into smart textiles to provide self-powered industrial uniforms capable of harnessing electricity from human motion, ultimately improving workplace safety and productivity. Silk Fibroin (SF) has also been considered a biodegradable binder in the field of energy storage, offering an alternative to synthetic polymers by maintaining the same level of energy density and long charge cycles and increasing the mechanical stability and electrical conductivity of lithium-ion batteries and supercapacitors [140, 150].

Liu et al. [149] introduces a silk-fibroin-based Triboelectric Nanogenerator (TENG) designed for industrial energy harvesting applications. By utilizing regenerated SF as a flexible and transparent triboelectric layer, the research overcomes the scalability challenges of traditional TENGs through a simple spray-coating fabrication process. The device demonstrates high energy output, with a maximum voltage of 213.9 V and a power density of 68.0 mW/m^2, making it suitable for self-powered industrial sensors, wearable electronics and environmental monitoring systems. Its ability to charge a 10 μF capacitor to 4.5 V in 6 min and power 100 LED bulbs instantaneously further showcases its potential for low-power industrial applications, such as wireless sensor networks and smart grids. Despite its low-cost and sustainable design, further research is required to enhance durability and energy storage integration for broader industrial adoption.

12.7 Challenges and Future Directions for Silk-Based Materials in Energy Harvesting and Sensing Technologies

Silk-hybrid materials encounter obstacles in scale production, notably spider silk harvesting, which is less efficient than synthetic polymers or polysaccharides such as cellulose. This has resulted in increased production of recombinant silk proteins. Long-term stability is also a worry, because silk, while resistant to thermal and environmental degradation, degrades in UV radiation, undergoing photo-oxidation and finally losing crystallinity. Despite these obstacles, silk-hybrid materials show promise for usage in hybrid energy materials [151]. Unlike other synthetic polymers, silk is non-conductive and needs to be pre- or post-treated, which limits its long-term use in light-irradiated devices [152]. When conductive fillers are applied to devices that don't have the mechanical qualities of silk, the conductive layers may break or deteriorate. This problem is critical to hybrid devices since it affects both function and efficiency. Silk will be remembered for its versatility and adaptability in spite of these disadvantages. For instance, silk-based sutures have several applications, such as antibacterial activity, medication encapsulation and release and wound stress monitoring. Silk is a useful material for many applications because of its adaptability and multipurpose nature [153]. A genetic mutation can insert a binding motif that specifically binds nanoparticles, altering its function in solar converging devices. This enables the construction of a more adaptable and efficient material. The real value of silk as a component of a hybrid material lies in its ability to be molecularly programmable, despite the fact that natural silk has several advantages such as mechanical strength, biocompatibility and renewableness [154].

12.8 Conclusion

Silk-derived materials are remarkable in sensing and energy harvesting technologies due to their compatibility with the human body, their flexibility and their being easy to produce. This chapter presents silk fibroin extraction, modification and its applications in different energy harvesting systems, like piezoelectric, thermoelectric, triboelectric and photovoltaic devices. The remarkable properties of silk, particularly when combined with functional nanomaterials and conductive polymers, permit the creation of hybrid systems that are more efficient and have better performance. In sensing applications, silk-based materials are employed in biosensors, wearable devices and environmental monitoring to provide data collection and multifunctionality in real time. Although there are still some problems to solve, such as long-term stability, scalability and commercialization, it is anticipated that the innovations in processing and material design will continuously expand the usage of silk in the Internet of Things, biomedicine and sustainable energy—thus, offering the industry and everyday life-promising advances.

Conflicts of Interest The author(s) declared no potential conflicts of interest concerning the research, authorship and/or publication of this article.

AI Usage Declaration While writing our article, the authors employed Quill Bot and Grammarly to correct grammatical errors and enhance the quality of the language. These technologies were created solely to increase the readability and clarity of language. All of the major concepts, conclusions and writings were initially created by the authors. We have carefully reviewed the final article and assume full responsibility for its integrity and accuracy.

References

1. Strassburg S, Zainuddin S, Scheibel T (2021) The power of silk technology for energy applications. Adv Energy Mater 11(43), Article 2100519. https://doi.org/10.1002/aenm.202100519
2. Badawy IM, Ali BA, Abbas WA, Allam NK (2021) Natural silk for energy and sensing applications: A review. Environ Chem Lett 19(3):2141–2155. https://doi.org/10.1007/s10311-020-01161-z
3. Drobysh M, Liustrovaite V, Baradoke A, Viter R, Chen C-F, Ramanavicius A, Ramanaviciene A (2022) Determination of rspike protein by specific antibodies with screen-printed carbon electrode modified by electrodeposited gold nanostructures. Biosensors 12(8):593. https://doi.org/10.3390/bios12080593
4. Eisoldt L, Smith A, Scheibel T (2011) Decoding the secrets of spider silk. Mater Today 14(3):80–86. https://doi.org/10.1016/S1369-7021(11)70057-8
5. Hardy JG, Scheibel TR (2010) Composite materials based on silk proteins. Prog Polym Sci 35(9):1093–1115. https://doi.org/10.1016/j.progpolymsci.2010.04.005
6. Kang M, Jin H-J (2007) Electrically conducting electrospun silk membranes fabricated by adsorption of carbon nanotubes. Colloid Polym Sci 285:1163–1167. https://doi.org/10.1007/s00396-007-1668-y
7. Bera S, Mondal SP, Naskar D, Kundu SC, Ray SK (2014) Flexible and transparent nanocrystal floating gate memory devices using silk protein. Org Electron 15(8):1767–1772. https://doi.org/10.1016/j.orgel.2014.04.038
8. Gogurla N, Kundu SC, Ray SK (2017) Gold nanoparticle-embedded silk protein-zno nanorod hybrids for flexible bio-photonic devices. Nanotechnology 28(14):145202. https://doi.org/10.1088/1361-6528/aa6144
9. Qi N, Zhao B, Wang S-D, Al-Deyab SS, Zhang K-Q (2015) Highly flexible and conductive composite films of silk fibroin and silver nanowires for optoelectronic devices. RSC Adv 5(63):50878–50882. https://doi.org/10.1039/C5RA03501E
10. Tsukada S, Nakashima H, Torimitsu K (2012) Conductive polymer combined silk fiber bundle for bioelectrical signal recording. PLoS ONE 7(4):e33689. https://doi.org/10.1371/journal.pone.0033689
11. Zhang H, Zhao J, Xing T, Lu S, Chen G (2019) Fabrication of silk fibroin/graphene film with high electrical conductivity and humidity sensitivity. Polymers 11(11):1774. https://doi.org/10.3390/polym11111774
12. Reizabal A, Gonçalves S, Brito-Pereira R, Costa P, Costa CM, Pérez-Álvarez L, Vilas-Vilela JL, Lanceros-Méndez S (2019) Optimized silk fibroin piezoresistive nanocomposites for pressure sensing applications based on natural polymers. Nanoscale Adv 1(6):2284–2292. https://doi.org/10.1039/C8NA00417J
13. Wang ZL (2013) Triboelectric nanogenerators as new energy technology for self-powered systems and as active mechanical and chemical sensors. ACS Nano 7(11):9533–9557. https://doi.org/10.1021/nn404614z
14. Kim HJ, Kim JH, Jun KW, Kim JH, Seung WC, Kwon OH, Park JY, Kim SW, Oh IK (2016) Silk nanofiber-networked bio-triboelectric generator: Silk bio-teg. *Adv Energy Mater*,*6*(8). https://doi.org/10.1002/aenm.201502329
15. Huang T, Zhang Y, He P, Wang G, Xia X, Ding G, Tao TH (2020) "Self-matched" tribo/piezoelectric nanogenerators using vapor-induced phase-separated poly (vinylidene fluoride) and recombinant spider silk. Adv Mater 32(10):1907336. https://doi.org/10.1002/adma.201907336
16. Jean-Baptiste P, Ducroux R (2003) Energy policy and climate change. Energy Policy 31(2):155–166. https://doi.org/10.1016/S0301-4215(02)00020-4
17. Karl TR, Trenberth KE (2003) Modern global climate change. Science 302(5651):1719–1723. https://doi.org/10.1126/science.1090228

18. Sathaye JA, Ravindranath NH (1998) Climate change mitigation in the energy and forestry sectors of developing countries. Annu Rev Energy Environ 23:387–437. https://doi.org/10.1146/annurev.energy.23.1.387
19. Sims REH (2004) Renewable energy: A response to climate change. Sol Energy 76(1–3):9–17. https://doi.org/10.1016/S0038-092X(03)00101-4
20. Lauri P, Havlík P, Kindermann G, Forsell N, Böttcher H, Obersteiner M (2014) Woody biomass energy potential in 2050. Energy Policy 66:19–31. https://doi.org/10.1016/j.enpol.2013.11.033
21. Sribna Y, Trokhymets O, Nosatov I, Kriukova I (2019) The globalization of the world coal market—contradictions and trends. In: E3S Web Conf, vol 123, Article 01044. https://doi.org/10.1051/e3sconf/201912301044
22. Borghei M, Lehtonen J, Liu L, Rojas OJ (2018) Advanced biomass-derived electrocatalysts for the oxygen reduction reaction. Adv Mater 30(24):1703691. https://doi.org/10.1002/adma.201703691
23. Guo C, Hu R, Liao W, Li Z, Sun L, Shi D, Li Y, Chen C (2017) Protein-enriched fish "biowaste" converted to three-dimensional porous carbon nano-network for advanced oxygen reduction electrocatalysis. Electrochim Acta 236:228–238. https://doi.org/10.1016/j.electacta.2017.03.169
24. Zhu X, Aoyama E, Birk AV, Onasanya O, Carr WH, Mourokh L, Minteer SD, Vittadello M (2020) Cytochrome c oxidase oxygen reduction reaction induced by cytochrome c on nickel-coordination surfaces based on graphene oxide in suspension. Biochim Biophys Acta Bioenerg, 1861(11):148262. https://doi.org/10.1016/j.bbabio.2020.148262
25. Yang W, Lv L, Li X, Han X, Li M, Li C (2020) Quaternized silk nanofibrils for electricity generation from moisture and ion rectification. ACS Nano 14(8):10600–10607. https://doi.org/10.1021/acsnano.0c04686
26. Wang C, Xia K, Zhang Y, Kaplan DL (2019) Silk-based advanced materials for soft electronics. Acc Chem Res 52(10):2916–2927. https://doi.org/10.1021/acs.accounts.9b00333
27. Zhou L, Fu P, Cai X, Zhou S, Yuan Y (2016) Naturally derived carbon nanofibers as sustainable electrocatalysts for microbial energy harvesting: A new application of spider silk. Appl Catal B 188:31–38. https://doi.org/10.1016/j.apcatb.2016.01.063
28. Huang W, Ling S, Li C, Omenetto FG, Kaplan DL (2018) Silkworm silk-based materials and devices generated using bio-nanotechnology. Chem Soc Rev 47(17):6486–6504. https://doi.org/10.1039/C8CS00187A
29. Rockwood DN, Preda RC, Yücel T, Wang X, Lovett ML, Kaplan DL (2011) Materials fabrication from bombyx mori silk fibroin. Nat Protoc 6(10):1612–1631. https://doi.org/10.1038/nprot.2011.379
30. Ranakoti L, Gupta MK, Rakesh PK (2019) Silk and silk-based composites: Opportunities and challenges. In: Rakesh PK, Singh I (eds) Processing of Green Composites. Materials Horizons: From Nature to Nanomaterials. Springer, Singapore, pp 91–106. https://doi.org/10.1007/978-981-13-6019-0_7
31. Chopra S, Gulrajani M (1994) Comparative evaluation of the various methods of degumming silk. Indian J Fibre Text Res 19:76–83
32. Gulrajani M (1992) Degumming of silk. Rev Prog Color Relat Top 22(1):79–89. https://doi.org/10.1111/j.1478-4408.1992.tb00091.x
33. Kim HJ, Kim MK, Lee KH, Nho SK, Han MS, Um IC (2017) Effect of degumming methods on structural characteristics and properties of regenerated silk. Int J Biol Macromol 104:294–302. https://doi.org/10.1016/j.ijbiomac.2017.06.019
34. Kusurkar TS, Tandon I, Sethy NK, Bhargava K, Sarkar S, Singh SK, Das M (2013) Fluorescent silk cocoon creating fluorescent diatom using a "water glass-fluorophore ferry". Sci Rep 3(1):3290, Article 3290. https://doi.org/10.1038/srep03290
35. Luong T-H, Dang T-NN, Ngoc OPT, Dinh-Thuy T-H, Nguyen T-H, Van Toi V, Duong HT, Le Son H (2015) Investigation of the silk fiber extraction process from the vietnam natural bombyx mori silkworm cocoon. In: 5th international conference on biomedical engineering in Vietnam, FMBE Proceedings, vol 46. Springer, Cham, pp 325–328. https://doi.org/10.1007/978-3-319-11776-8_79

36. Vyas SK, Shukla SR (2016) Comparative study of degumming of silk varieties by different techniques. J Text Inst 107(2):191–199. https://doi.org/10.1080/00405000.2015.1020670
37. Gore PM, Naebe M, Wang X, Kandasubramanian B (2019) Progress in silk materials for integrated water treatments: Fabrication, modification and applications. Chem Eng J 374:437–470. https://doi.org/10.1016/j.cej.2019.05.163
38. Unajak S, Aroonluke S, Promboon A (2015) An active recombinant cocoonase from the silkworm bombyx mori: Bleaching, degumming and sericin degrading activities. J Sci Food Agric 95(6):1179–1189. https://doi.org/10.1002/jsfa.6806
39. Huang W, Ebrahimi D, Dinjaski N, Tarakanova A, Buehler MJ, Wong JY, Kaplan DL (2017) Synergistic integration of experimental and simulation approaches for the de novo design of silk-based materials. Acc Chem Res 50(4):866–876. https://doi.org/10.1021/acs.accounts.6b00616
40. Wang Q, Xia X, Huang W, Lin Y, Xu Q, Kaplan DL (2014) High throughput screening of dynamic silk-elastin-like protein biomaterials. Adv Funct Mater 24(27):4303–4310. https://doi.org/10.1002/adfm.201304106
41. Jin H-J, Fridrikh SV, Rutledge GC, Kaplan DL (2002) Electrospinning bombyx mori silk with poly (ethylene oxide). Biomacromol 3(6):1233–1239. https://doi.org/10.1021/bm025581u
42. Li D, Xia Y (2004) Electrospinning of nanofibers: Reinventing the wheel? Adv Mater 16(14):1151–1170. https://doi.org/10.1002/adma.200400719
43. Wen D-L, Sun D-H, Huang P, Huang W, Su M, Wang Y, Han M-D, Kim B, Brugger J, Zhang H-X (2021) Recent progress in silk fibroin-based flexible electronics. Microsyst Nanoeng 7(1):35. https://doi.org/10.1038/s41378-021-00261-2
44. Sencadas V, Garvey C, Mudie S, Kirkensgaard JJ, Gouadec G, Hauser S (2019) Electroactive properties of electrospun silk fibroin for energy harvesting applications. Nano Energy 66:104106. https://doi.org/10.1016/j.nanoen.2019.104106
45. Bhardwaj N, Kundu SC (2010) Electrospinning: A fascinating fiber fabrication technique. Biotechnol Adv 28(3):325–347. https://doi.org/10.1016/j.biotechadv.2010.01.004
46. Koh L-D, Cheng Y, Teng C-P, Khin Y-W, Loh X-J, Tee S-Y, Low M, Ye E, Yu H-D, Zhang Y-W (2015) Structures, mechanical properties and applications of silk fibroin materials. Prog Polym Sci 46:86–110. https://doi.org/10.1016/j.progpolymsci.2015.02.001
47. Lawrence BD, Marchant JK, Pindrus MA, Omenetto FG, Kaplan DL (2009) Silk film biomaterials for cornea tissue engineering. Biomaterials 30(7):1299–1308. https://doi.org/10.1016/j.biomaterials.2008.11.018
48. Li G, Liu H, Zhao H, Gao Y, Wang J, Jiang H, Boughton R (2011) Chemical assembly of tio2 and tio2@ ag nanoparticles on silk fiber to produce multifunctional fabrics. J Colloid Interface Sci 358(1):307–315. https://doi.org/10.1016/j.jcis.2011.02.053
49. Cho HJ, Ki CS, Oh H, Lee KH, Um IC (2012) Molecular weight distribution and solution properties of silk fibroins with different dissolution conditions. Int J Biol Macromol 51(3):336–341. https://doi.org/10.1016/j.ijbiomac.2012.06.007
50. Gaviria A, Sanchez-Diaz S, Ríos A, Peresin M, Restrepo-Osorio A (2017) Silk fibroin from silk fibrous waste: Characterization and electrospinning. In: IOP conference series: Materials science and engineering, 254(10):102005. https://doi.org/10.1088/1757-899X/254/10/102005
51. Park BK, Um IC (2018) Effect of molecular weight on electro-spinning performance of regenerated silk. Int J Biol Macromol 106:1166–1172. https://doi.org/10.1016/j.ijbiomac.2017.08.115
52. Hu X, Shmelev K, Sun L, Gil E-S, Park S-H, Cebe P, Kaplan DL (2011) Regulation of silk material structure by temperature-controlled water vapor annealing. Biomacromol 12(5):1686–1696. https://doi.org/10.1021/bm200062a
53. Wang L, Chen D, Jiang K, Shen G (2017) New insights and perspectives into biological materials for flexible electronics. Chem Soc Rev 46(22):6764–6815. https://doi.org/10.1039/C7CS00278E
54. Huang L, Lin S, Xu Z, Zhou H, Duan J, Hu B, Zhou J (2020) Fiber-based energy conversion devices for human-body energy harvesting. Adv Mater 32(5):1902034. https://doi.org/10.1002/adma.201902034

55. Fukada E (1956) On the piezoelectric effect of silk fibers. J Phys Soc Jpn 11(12):1301A. https://doi.org/10.1143/JPSJ.11.1301A
56. Lotz B, Cesari FC (1979) The chemical structure and the crystalline structures of bombyx mori silk fibroin. Biochimie 61(2):205–214. https://doi.org/10.1016/S0300-9084(79)80067-X
57. Park K-I, Xu S, Liu Y, Hwang G-T, Kang S-JL, Wang ZL, Lee KJ (2010) Piezoelectric batio3 thin film nanogenerator on plastic substrates. Nano lett 10(12):4939–4943. https://doi.org/10.1021/nl102959k
58. Wang ZL, Song J (2006) Piezoelectric nanogenerators based on zinc oxide nanowire arrays. Science 312(5771):242–246. https://doi.org/10.1126/science.1124005
59. Reizabal A, Costa CM, Pérez-Álvarez L, Vilas-Vilela JL, Lanceros-Méndez S (2023) Silk fibroin as sustainable advanced material: Material properties and characteristics, processing, and applications. Adv Funct Mater 33(3):2210764. https://doi.org/10.1002/adfm.202210764
60. Huang B, Lawrence C, Gross A, Hwang G-S, Ghafouri N, Lee S-W, Kim H, Li C-P, Uher C, Najafi K (2008) Low-temperature characterization and micropatterning of coevaporated bi2te3 and sb2te3 films. J Appl Phys 104(11). https://doi.org/10.1063/1.3033381
61. Kim M-Y, Oh T-S (2014) Preparation and characterization of bi 2 te 3/sb 2 te 3 thermoelectric thin-film devices for power generation. J Electron Mater 43:1933–1939. https://doi.org/10.1007/s11664-013-2915-2
62. Lu Z, Zhang H, Mao C, Li CM (2016) Silk fabric-based wearable thermoelectric generator for energy harvesting from the human body. Appl Energy 164:57–63. https://doi.org/10.1016/j.apenergy.2015.11.038
63. Dong K, Peng X, Wang ZL (2020) Fiber/fabric-based piezoelectric and triboelectric nanogenerators for flexible/stretchable and wearable electronics and artificial intelligence. Adv Mater 32(5):1902549. https://doi.org/10.1002/adma.201902549
64. Dong K, Wang Y-C, Deng J, Dai Y, Zhang SL, Zou H, Gu B, Sun B, Wang ZL (2017) A highly stretchable and washable all-yarn-based self-charging knitting power textile composed of fiber triboelectric nanogenerators and supercapacitors. ACS Nano 11(9):9490–9499
65. Kwak SS, Yoon HJ, Kim SW (2019) Textile-based triboelectric nanogenerators for self-powered wearable electronics. Adv Funct Mater 29(2):1804533. https://doi.org/10.1002/adfm.201804533
66. Wang J, Li X, Zi Y, Wang S, Li Z, Zheng L, Yi F, Li S, Wang ZL (2015) A flexible fiber-based supercapacitor–triboelectric-nanogenerator power system for wearable electronics. Adv Mater 27(33):4830–4836. https://doi.org/10.1002/adma.201501934
67. Wu C, Wang AC, Ding W, Guo H, Wang ZL (2019) Triboelectric nanogenerator: A foundation of the energy for the new era. Adv Energy Mater 9(1):1802906. https://doi.org/10.1002/aenm.201802906
68. Dolez PI (2021) Energy harvesting materials and structures for smart textile applications: Recent progress and path forward. Sensors 21(18):6297. https://doi.org/10.3390/s21186297
69. Ye C, Dong S, Ren J, Ling S (2020) Ultrastable and high-performance silk energy harvesting textiles. Nano-Micro Lett 12:12. https://doi.org/10.1007/s40820-019-0348-z
70. Bedeloglu A, Demir A, Bozkurt Y, Sariciftci NS (2010) A photovoltaic fiber design for smart textiles. Text Res J 80(11):1065–1074. https://doi.org/10.1177/0040517509352520
71. Liu D, Li Y, Zhao S, Cao A, Zhang C, Liu Z, Bian Z, Liu Z, Huang C (2013) Single-layer graphene sheets as counter electrodes for fiber-shaped polymer solar cells. RSC Adv 3(33):13720–13727. https://doi.org/10.1039/C3RA40710A
72. Fan X, Chu Z, Wang F, Zhang C, Chen L, Tang Y, Zou D (2008) Wire-shaped flexible dye-sensitized solar cells. Adv Mater 20(3):592–595. https://doi.org/10.1002/adma.200701249
73. Hou S, Cai X, Fu Y, Lv Z, Wang D, Wu H, Zhang C, Chu Z, Zou D (2011) Transparent conductive oxide-less, flexible, and highly efficient dye-sensitized solar cells with commercialized carbon fiber as the counter electrode. J Mater Chem 21(36):13776–13779. https://doi.org/10.1039/C1JM12056E
74. Xiang S, Zhang N, Fan X (2021) From fiber to fabric: Progress towards photovoltaic energy textile. Adv fiber mater 3:76–106. https://doi.org/10.1007/s42765-020-00062-8

75. Cao Y, Zhao Z, Bao P, Gan Z, Wang H (2021) Lateral photovoltaic effect in silk-protein-based nanocomposite structure for physically transient position-sensitive detectors. Phys Rev Appl 15(5):054011. https://doi.org/10.1103/PhysRevApplied.15.054011
76. Eosakul S (2009) Electrical characterization of silk films for use in silk-based polymer photovoltaics. Tufts University
77. Lee C, Cho H, Choi J, Lee CB, Lee S, Kim DK, Kim S (2025) Multifunctional and flexible graphene oxide/silk sericin/carbon nanotube thin films on polyimide for real-time humidity sensing. Eur Polym J 225:113731. https://doi.org/10.1016/j.eurpolymj.2025.113731
78. Wu D, Weng L, Zhang X, Guan L, Wu Z (2024) Flexible and wearable multilayer structure fiber sensor for motion and human physiological signal monitoring. Polym Adv Technol 35(9):e6552. https://doi.org/10.1002/pat.6552
79. Wen DL, Pang YX, Huang P, Wang YL, Zhang XR, Deng HT, Zhang XS (2022) Silk fibroin-based wearable all-fiber multifunctional sensor for smart clothing. Adv Fiber Mater 4:873–884. https://doi.org/10.1007/s42765-022-00150-x
80. Song Y, Hu C, Wang Z, Wang L (2024) Silk-based wearable devices for health monitoring and medical treatment. iScience 27(5):109604. https://doi.org/10.1016/j.isci.2024.109604
81. Fan SN, Zhang Y, Huang XY, Geng LH, Shao HL, Hu XC, Zhang YP (2019) Silk materials for medical, electronic and optical applications. Sci China Technol Sci 62:903–918. https://doi.org/10.1007/s11431-018-9403-8
82. Chen F, Lu S, Zhu L, Tang Z, Wang Q, Qin G (2019) Conductive regenerated silk-fibroin-based hydrogels with integrated high mechanical performances. J Mater Chem B 7(10):1708–1715. https://doi.org/10.1039/C8TB02445F
83. Huang J, Xu Z, Qiu W, Chen F, Meng Z, Hou C, Guo W, Liu XY (2020) Stretchable and heat-resistant protein-based electronic skin for human thermoregulation. Adv Funct Mater 30(13):1910547. https://doi.org/10.1002/adfm.201910547
84. Liu Y, Xie Y, Liu Y, Song T, Zhang KQ, Liao L, Sun B (2015) Flexible organic light emitting diodes fabricated on biocompatible silk fibroin substrate. Semicond Sci Technol, 30, Article 104004. https://doi.org/10.1088/0268-1242/30/10/104004
85. Han KI, Kim S, Lee IG, Kim JP, Kim JH, Hong SW, Cho BJ, Hwang WS (2017) Compliment graphene oxide coating on silk fiber surface via electrostatic force for capacitive humidity sensor applications. Sensors 17(2):407. https://doi.org/10.3390/s17020407
86. Chen G, Matsuhisa N, Liu Z, Qi D, Cai P, Jiang Y, Wan C, Cui Y, Leow WR, Liu Z, Gong S, Zhang KQ, Cheng Y, Chen X (2018) Plasticizing silk protein for on-skin stretchable electrodes. Adv Mater 30(21):1800129. https://doi.org/10.1002/adma.201800129
87. Seo JW, Kim H, Kim KH, Choi SQ, Lee HJ (2018) Calcium-modified silk as a biocompatible and strong adhesive for epidermal electronics. Adv Funct Mater 28(36):1800802. https://doi.org/10.1002/adfm.201800802
88. Alaejos MS, Montelongo FJG (2004) Application of amperometric biosensors to the determination of vitamins and α-amino acids. Chem Rev 104(7):3239–3265. https://doi.org/10.1021/cr0304471
89. Alkire RC, Kolb DM, Lipkowski J (2012) Bioelectrochemistry: Fundamentals, applications and recent developments. 13:1–397. https://doi.org/10.1002/9783527644117
90. Fan Q, Li J, Zhu Y, Yang Z, Shen T, Guo Y, Wang L, Mei T, Wang J, Wang X (2020) Functional carbon quantum dots for highly sensitive graphene transistors for cu2+ ion detection. ACS Appl Mater Interfaces 12(4):4797–4803. https://doi.org/10.1021/acsami.9b20785
91. Hui N, Gao RF, Xq L, Sun W, Jiao K (2009) Direct electrochemistry of hemoglobin immobilized in polyvinylalcohol and clay composite film modified carbon ionic liquid electrode. J Braz Chem Soc 20(2):252–258. https://doi.org/10.1590/S0103-50532009000200009
92. Matsumoto T, Ohashi A, Ito N (2002) Development of a micro-planar ag/agcl quasi-reference electrode with long-term stability for an amperometric glucose sensor. Anal Chim Acta 462(2):253–259. https://doi.org/10.1016/S0003-2670(02)00334-3
93. Wang X, Kim HJ, Xu P, Matsumoto A, Kaplan DL (2005) Biomaterial coatings by stepwise deposition of silk fibroin. Langmuir 21(24):11335–11341. https://doi.org/10.1021/la051862m

94. Gonzalez OG, Gong C, Hossen MI, Huang X, Cui Z, Xiao X (2023) Role of silk in recent advances of the biosensors: A mini-review. Mater Plus 2(2):101–117. https://doi.org/10.37256/mp.2220232897
95. Ru M, Hai AM, Wang L, Yan S, Zhang Q (2023) Recent progress in silk-based biosensors. Int J Biol Macromol 224:422–436. https://doi.org/10.1016/j.ijbiomac.2022.10.134
96. Molinnus D, Drinic A, Iken H, Kröger N, Zinser M, Smeets R, Köpf M, Kopp A, Schöning MJ (2021) Towards a flexible electrochemical biosensor fabricated from biocompatible bombyx mori silk. Biosens Bioelectron 183:113204. https://doi.org/10.1016/j.bios.2021.113204
97. Seetharaman R, Mughuntha V, Harish B, Nithish KB, Prithivee S, Ambhika C, Shiney SA, Subramanian LH, Megala V (2024) Monitoring and detection of environmental pathogen using bio-polymer sensors with drone. In: 7th International Conference on Inventive Computation Technologies (ICICT 2024). IEEE, pp 1747–1751. https://doi.org/10.1109/ICICT60155.2024.10544446
98. Dong K, Zhou Q, Gao B (2024) New light-illuminated silk road: Emerging silk fibroin-based optical biomedical sensors. Analyst 149(17):4322–4342. https://doi.org/10.1039/D4AN00665H
99. Genovese ME, Caputo G, Nanni G, Setti C, Bustreo M, Perotto G, Athanassiou A, Fragouli D (2017) Light responsive silk nanofibers: An optochemical platform for environmental applications. ACS Appl Mater Interfaces 9(46):40707–40715. https://doi.org/10.1021/acsami.7b13372
100. Pang C, Li F, Hu X, Meng K, Pan H, Xiang Y (2024) Degradable silk fibroin based piezoresistive sensor for wearable biomonitoring. Discover Nano 19(1):55. https://doi.org/10.1186/s11671-024-04001-z
101. Rathinasamy SK, Maheswar R, Lorincz J (2023) Silk fibroin-based piezoelectric sensor with carbon nanofibers for wearable health monitoring applications. Sensors 23(3):1373. https://doi.org/10.3390/s23031373
102. Jain S, Vedavyas V, Prajwal RV, Shaji M, Nath VG, Angappane S, Achutharao G (2023a) Silk and its composites for humidity and gas sensing applications. Front Chem, 11:1141259. https://doi.org/10.3389/fchem.2023.1141259
103. Yang XC, Wang XX, Wang CY, Zheng HL, Yin M, Chen KZ, Qiao SL (2024) Silk-based intelligent fibers and textiles: Structures, properties, and applications. Chem Commun 60(61):7801–7823. https://doi.org/10.1039/D4CC02276A
104. Chen M, Liu J, Hu Y, Wu Y, Tang CY, Ke K, Yang W (2024) Silk fibroin-based flexible pressure sensors: Processing and application. Mater Futures, 3.https://doi.org/10.1088/2752-5724/ad5f48
105. Zhang M, Liu Z, Zhang Y, Zhang Y, Yang X, Zhang J, Yang J, Yuan L (2022) Spider dragline silk-based flexible temperature sensor. In: Optical Sensors and Sensing Congress 2022 (AIS, LACSEA, Sensors, ES), Technical Digest Series. Optica Publishing Group, paper SM3E.5. https://doi.org/10.1364/SENSORS.2022.SM3E.5
106. Zheng Y, Wang L, Zhao L, Wang D, Xu H, Wang K, Han W (2021) A flexible humidity sensor based on natural biocompatible silk fibroin films. Adv Mater Technol 6(1):2001053. https://doi.org/10.1002/admt.202001053
107. Shi C, Hu F, Wu R, Xu Z, Shao G, Yu R, Liu XY (2021) New silk road: From mesoscopic reconstruction/functionalization to flexible meso-electronics/photonics based on cocoon silk materials. Adv Mater 33(50):2005910. https://doi.org/10.1002/adma.202005910
108. Zhao Y, Zhao H, Wei Z, Yuan J, Jian J, Kong F, Xie H, Xiong X (2022) Silk fibroin-based hydrogel for multifunctional wearable sensors. J Renew Mater 10(11):2729. https://doi.org/10.32604/jrm.2022.019721
109. Meng G, Long F, Zeng Z, Kong L, Zhao B, Yan J, Yang L, Yang Y, Liu XY, Yan Z, Lin N (2023) Silk fibroin based wearable electrochemical sensors with biomimetic enzyme-like activity constructed for durable and on-site health monitoring. Biosens Bioelectron 228:115198. https://doi.org/10.1016/j.bios.2023.115198
110. Yang Z, Huang T, Cao P, Cui Y, Nie J, Chen T, Yang H, Wang F, Sun L (2022) Carbonized silk nanofibers in biodegradable, flexible temperature sensors for extracellular environments. ACS Appl Mater Interfaces 14(16):18110–18119. https://doi.org/10.1021/acsami.2c00384

111. Ahmed A, Bain S, Prottoy ZH, Morsada Z, Islam MT, Hossain MM, Shkir M (2022) Silk-templated nanomaterial interfaces for wearables and bioelectronics: Advances and prospects. ACS Mater Lett 4(1):68–86. https://doi.org/10.1021/acsmaterialslett.1c00618
112. Liang C, Fan Z, Zhu Y, Cao Y, Kang J, Tao J (2024) Synthesis and characterization of silk fibroin-mxene composite electrospun fibers for biomedical applications. Front Chem 12:1471148. https://doi.org/10.3389/fchem.2024.1471148
113. Pektas HK, Demidov Y, Ahvan A, Abie N, Georgieva VS, Chen S, Farè S, Brachvogel B, Mathur S, Maleki H (2023) Mxene-integrated silk fibroin-based self-assembly-driven 3d-printed theragenerative scaffolds for remotely photothermal anti-osteosarcoma ablation and bone regeneration. ACS Mater Au 3(6):711–726. https://doi.org/10.1021/acsmaterialsau.3c00040
114. Brooks AK, Pradhan S, Yadavalli VK (2023) Degradable elastomeric silk biomaterial for flexible bioelectronics. ACS Appl Bio Mater 6(10):4392–4402. https://doi.org/10.1021/acsabm.3c00593
115. Wang Y, Qu K, Li S, Zheng J, Qiu W, Ye F, Xiao Z, Xu Q, Xu J, Guo W (2023) Fully degradable, highly elastomeric and adhesive silk fibroin electronic skin for microdynamic pressure monitoring. Chem Eng J 469:143920. https://doi.org/10.1016/j.cej.2023.143920
116. de Lartigue C, Belda Marín C, Fitzpatrick V, Esposito A, Casale S, Landoulsi J, Guénin E, Egles C (2024) Silk foams with metallic nanoparticles as scaffolds for soft tissue regeneration. Int J Mol Sci 25(22):12377. https://doi.org/10.3390/ijms252212377
117. Qu J, Dai M, Ye W, Fang Y, Bian D, Su W, Li F, Sun H, Wei J, Li B (2021) Study on the effect of graphene oxide (go) feeding on silk properties based on segmented precise measurement. J Mech Behav Biomed Mater 113:104147. https://doi.org/10.1016/j.jmbbm.2020.104147
118. Zhang C, Shao H, Luo J, Hu X, Zhang Y (2018) Structure and interaction of silk fibroin and graphene oxide in concentrated solution under shear. Int J Biol Macromol 107:2590–2597. https://doi.org/10.1016/j.ijbiomac.2017.10.142
119. Pal RK, Farghaly AA, Wang C, Collinson MM, Kundu SC, Yadavalli VK (2016) Conducting polymer-silk biocomposites for flexible and biodegradable electrochemical sensors. Biosens Bioelectron 81:294–302. https://doi.org/10.1016/j.bios.2016.03.010
120. Wagholikar S, Mulay P, Wagholikar O, Mulay V (2024) Nanomaterials used in flexible electronics: Recent trends and future approaches. Recent Pat Eng 18(6):27–38. https://doi.org/10.2174/1872212118666230519161338
121. Ma H, Deng Y, Lin Y, Zhang Y, Zhou Z, Yu H, Jiang X (2024) Strong and tough conductive silk fibroin/poly (vinyl alcohol) composite hydrogel by a salting-in and salting-out synergistic effect. Iran Polym J 33(11):1527–1537. https://doi.org/10.1007/s13726-024-01334-5
122. Fu F, Liu C, Jiang Z, Zhao Q, Shen A, Wu Y, Gu W (2024) Polymeric silk fibroin hydrogel as a conductive and multifunctional adhesive for durable skin and epidermal electronics. Smart Med 3(3):e20240027. https://doi.org/10.1002/SMMD.20240027
123. Hu Z, Liang Y, Fan S, Niu Q, Geng J, Huang Q, Hsiao BS, Chen H, Yao X, Zhang Y (2024) Flexible neural interface from non-transient silk fibroin with outstanding conformality, biocompatibility, and bioelectric conductivity. Adv Mater 36(46):2410007. https://doi.org/10.1002/adma.202410007
124. Zhao K, Wang J, Wu Y, Yu X, Cao H, Yang X, Ni A, Ge Y, Liu Y, Zhuang S, Zhao Y, Gu X (2025) Silk fibroin-based hydrogels with low hysteresis, self-adhesion, and tunable ionic conductivity for wearable devices. Int J Biol Macromol 306:141597. https://doi.org/10.1016/j.ijbiomac.2025.141597
125. Janani G, Kumar M, Chouhan D, Moses JC, Gangrade A, Bhattacharjee S, Mandal BB (2019) Insight into silk-based biomaterials: From physicochemical attributes to recent biomedical applications. ACS Appl Bio Mater 2(12):5460–5491. https://doi.org/10.1021/acsabm.9b00576
126. Yucel T, Lovett ML, Kaplan DL (2014) Silk-based biomaterials for sustained drug delivery. J Control Release 190:381–397. https://doi.org/10.1016/j.jconrel.2014.05.059
127. Zheng N, Fitzpatrick V, Cheng R, Shi L, Kaplan DL, Yang C (2022) Photoacoustic carbon nanotubes embedded silk scaffolds for neural stimulation and regeneration. ACS Nano 16(2):2292–2305. https://doi.org/10.1021/acsnano.1c08491

128. Ramaraj SG, Durgadevi E, Tabata H, Zhang F, Liu X (2024) Biocompatible triboelectric energy generators (bt-tengs) for energy harvesting and healthcare applications. Nanoscale 16:18251–18273. https://doi.org/10.1039/D4NR01987C
129. Tat T, Chen G, Zhao X, Zhou Y, Xu J, Chen J (2022) Smart textiles for healthcare and sustainability. ACS Nano 16(9):13301–13313. https://doi.org/10.1021/acsnano.2c06287
130. Fu F, Liu D, Wu Y (2023) Silk-based conductive materials for smart biointerfaces. Smart Med 2(2):e20230004. https://doi.org/10.1002/SMMD.20230004
131. Matthew SAL, Egan G, Witte K, Kaewchuchuen J, Phuagkhaopong S, Totten JD, Seib FP (2022) Smart silk origami as eco-sensors for environmental pollution. ACS Appl Bio Mater 5(8):3658–3666. https://doi.org/10.1021/acsabm.2c00023
132. Prakash NJ, Mane PP, George SM, Kandasubramanian B (2021) Silk fibroin as an immobilization matrix for sensing applications. ACS Biomater Sci Eng 7(6):2015–2042. https://doi.org/10.1021/acsbiomaterials.1c00080
133. Gao X, Gou J, Zhang L, Duan S, Li C (2018) A silk fibroin based green nano-filter for air filtration. RSC Advances, 8(15):8181–8189. https://doi.org/10.1039/C7RA12879G
134. Younes B (2023) Smart e-textiles: A review of their aspects and applications. J Ind Text 53:15280837231215492. https://doi.org/10.1177/15280837231215493
135. Tseghai GB, Malengier B, Fante K, Nigusse AB, Van Langenhove L (2020) Integration of conductive materials with textile structures, an overview. Sensors 20(23):6910. https://doi.org/10.3390/s20236910
136. Gan N, Wei M, Zhou Q, Dong K, Gao B (2024) Silk fibroin-based triboelectric nanogenerators for energy harvesting and biomedical applications. ACS Appl Nano Mater 7(8):8407–8423. https://doi.org/10.1021/acsanm.4c00335
137. Niu Q, Wei H, Hsiao BS, Zhang Y (2022) Biodegradable silk fibroin-based bio-piezoelectric/triboelectric nanogenerators as self-powered electronic devices. Nano Energy 96:107101. https://doi.org/10.1016/j.nanoen.2022.107101
138. Chen G, Li Y, Bick M, Chen J (2020) Smart textiles for electricity generation. Chem Rev 120(8):3668–3720. https://doi.org/10.1021/acs.chemrev.9b00821
139. Gu Y, Niu Q, Cao Q, Zhou X, Zhao W, Hu Z, Fan S, Yao X, Zhang Y (2025) Biodegradable silk fibroin-based triboelectric nanogenerator with enhanced output regulated by interfacial and ionic polarization. Int J Biol Macromol 304(Pt 1):140832. https://doi.org/10.1016/j.ijbiomac.2025.140832
140. Huang S, Wang R, Lei M, Geng Q, Li Q, Zhang J, Zhang J (2025) Preparation of silk fibroin–carboxymethyl cellulose composite binder and its application in silicon-based anode for lithium-ion batteries. Nanomaterials 15(19):1509. https://www.mdpi.com/2079-4991/15/19/1509
141. Tan X, Huang Z, Chang L, Pei H, Jia Z, Zheng J (2024) High-performance triboelectric nanogenerator based on silk fibroin-mxene composite film for diagnosing insomnia symptoms. ACS Sens 9(11):5782–5791. https://doi.org/10.1021/acssensors.4c00852
142. Tan X, Huang Z, Pei H, Jia Z, Zheng J (2024) Highly porous, ultralight, biocompatible silk fibroin aerogel-based triboelectric nanogenerator. ACS Sens 9(8):3938–3946. https://doi.org/10.1021/acssensors.4c00401
143. Gong R, Dong Y, Ge D, Miao Z, Yu H-Y (2024) Wet spinning fabrication of robust and uniform intrinsically conductive cellulose nanofibril/silk conductive fibers as bifunctional strain/humidity sensor in potential smart dressing. Advanced Fiber Materials 6(4):993–1007. https://doi.org/10.1007/s42765-024-00404-w
144. Liu Z, Wang J, Zhang Q, Li Z, Li Z, Cheng L, Dai F (2022) Electrospinning silk fibroin/graphene nanofiber membrane used for 3d wearable pressure sensor. Polymers, 14(18):3875. https://www.mdpi.com/2073-4360/14/18/3875
145. Ye C, Zhang H, Yang Y, Shan Y, Fu J, Gao W, Ren J, Cao L, Ling S (2025) Sustainable silk fibroin ionic touch screens for flexible biodegradable electronics with integrated ai and iot functionality. Adv Mater 37(22):2412972. https://doi.org/10.1002/adma.202412972
146. Menghi R, Papetti A, Germani M, Marconi M (2019) Energy efficiency of manufacturing systems: A review of energy assessment methods and tools. J Clean Prod 240:118276. https://doi.org/10.1016/j.jclepro.2019.118276

147. Dehghan M, Ghasemizadeh M, Rashidi S, Doranehgard MH, Li LKB (2023) Introduction to solar energy harvesting and storage. In: Khalid M, Walvekar R, Panchal H, Vaka M (ed) Solar energy harvesting, conversion, and storage, pp. 1–23. Elsevier
148. Choi J, Lee S, Kim T, Min K, Lee S-N (2025) Capacitance and dielectric properties of spin-coated silk fibroin thin films for bioelectronic capacitors. Materials 18(7):1408. https://doi.org/10.3390/ma18071408
149. Liu C, Li J, Che L, Chen S, Wang Z, Zhou X (2017) Toward large-scale fabrication of triboelectric nanogenerator (teng) with silk-fibroin patches film via spray-coating process. Nano Energy 41:359–366. https://doi.org/10.1016/j.nanoen.2017.09.038
150. Liu X, Ostrovsky-Snider N, Lo Presti M, Kim T, Guidetti G, Omenetto FG (2024) Use of silk fibroin material composites for green, flexible supercapacitors. ACS Biomater Sci Eng 10(8):5390–5398. https://doi.org/10.1021/acsbiomaterials.4c00716
151. Lee S, Kim SH, Jo Y-Y, Ju W-T, Kim H-B, Kweon H (2021) Effects of ultraviolet light irradiation on silk fibroin films prepared under different conditions. Accessed 1
152. Hou J, Cao C, Idrees F, Ma X (2015) Hierarchical porous nitrogen-doped carbon nanosheets derived from silk for ultrahigh-capacity battery anodes and supercapacitors. ACS Nano 9(3):2556–2564. https://doi.org/10.1021/nn506394r
153. Liu M, Zhang Y, Liu K, Zhang G, Mao Y, Chen L, Peng Y, Tao TH (2021) Biomimicking antibacterial opto-electro sensing sutures made of regenerated silk proteins. Adv Mater 33(1):2004733. https://doi.org/10.1002/adma.202004733
154. Neubauer VJ, Kellner C, Gruen V, Schenk AS, Scheibel T (2021) Recombinant major ampullate spidroin-particles as biotemplates for manganese carbonate mineralization. Multifunct Mater 4(1):014002. https://doi.org/10.1088/2399-7532/abddc4
155. Zhang M, Zhao M, Jian M, Wang C, Yu A, Yin Z, Liang X, Wang H, Xia K, Liang X (2019) Printable smart pattern for multifunctional energy-management e-textile. Matter 1(1):168–179. https://doi.org/10.1016/j.matt.2019.02.003

Chapter 13
Cocoon to Climate Control: Silk-Based Smart Textiles for Thermal Management

Anik Das, Xu Lei, and Rony Mia

13.1 Introduction

The biogenesis of silk, a natural polymer with significant technological potential, commences within the specialized silk gland of the *Bombyx mori* larva, a process visually represented in Fig. 13.1, wherein the gland is functionally compartmentalized into posterior, middle, and anterior segments [1–3]. Specifically, the posterior region orchestrates the synthesis of fibroin, the proteinaceous backbone of the silk fiber [1]. Subsequently, the middle segment facilitates the deposition of sericin, a binding protein, onto the fibroin core, alongside the concentration of silk proteins and the exchange of critical metallic ions. The anterior region then executes the final preparatory steps, culminating in the extrusion of the silk solution through the spinneret, a specialized orifice located at the larval terminus [4]. This extrusion process yields a dual fibroin core, enveloped by a sericin coating, forming the nascent silk fiber. The epithelial lining of the silk gland, comprised of large hexagonal cells, serves a dual role in both protein synthesis and storage, ensuring the efficient production of this complex biomaterial. The inherent sophistication of this natural process has stimulated research into silk-based smart textiles, particularly for applications concerning thermal regulation [1, 3].

The escalating prevalence of intelligent wearable electronics and healthcare systems has precipitated a significant need for energy-efficient and robust personal thermoregulation strategies, aimed at mitigating thermal runaway and ensuring

A. Das · X. Lei
School of Textile Science and Engineering, Tiangong University, Tianjin, People's Republic of China
e-mail: xulei@tiangong.edu.cn

R. Mia (✉)
Department of Textile Engineering, Chemistry and Science, Wilson College of Textiles, North Carolina State University, Raleigh, NC, USA
e-mail: mroni_mia@yahoo.com

M. Shahid et al. (eds.), *Science and Technology of Silk*, Textile Science and Clothing Technology, https://doi.org/10.1007/978-3-032-26880-8_13

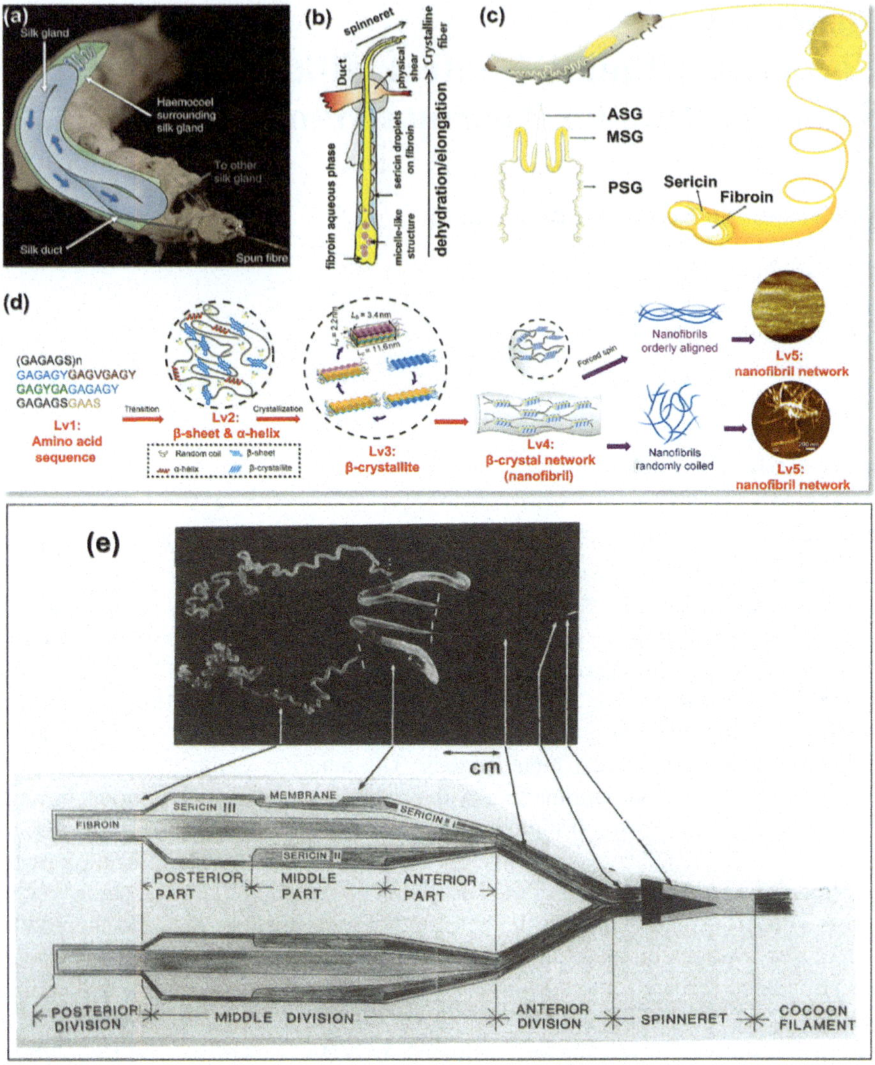

Fig. 13.1 Development and multi-tiered architecture of silk. **a** The spinning apparatus of mulberry silk. Reproduced with permission from Springer Nature. Copyright Springer Nature 2017. [5] **b** Spinning passageway. Reproduced with permission from Wiley. Copyright Wiley 2013. [6] **c** Diagrammatic representation of silk and its glandular system. Reproduced with permission from Springer Nature. Copyright Springer Nature 2022. [7] **d** The five-stage hierarchical organization of silk. Reproduced with permission from Wiley. Copyright Wiley 2019. [8] **e** Photomicrograph and schematic representation of the silk gland in the larvae of the silkworm, *Bombyx mori*. Reproduced with permission from ACS. Copyright ACS 1993 [3]

sustained human thermal comfort during prolonged device usage [9, 10]. Conventional thermoregulatory approaches, relying on alterations in garment thickness, prove insufficient to accommodate rapid environmental shifts and the stringent demands of advanced wearable healthcare applications [11–13]. In response, smart fibers and textiles have emerged as promising alternatives, characterized by their low mass, inherent mechanical flexibility, diverse responsive capabilities, and superior thermal management properties. These materials offer a pathway to resolve critical challenges, including the prevention of thermal runaway induced by abrupt temperature variations or heat accumulation within electronic components, and the alleviation of physiological and psychological discomfort arising from ambient temperature fluctuations [14–16]. Consequently, the development and integration of these advanced textile systems into wearable technologies represent a pivotal advancement toward achieving optimal thermal homeostasis and enhancing user experience in long-term applications.

The distinctive thermal regulatory capabilities of silk as a natural fiber stem from a confluence of inherent material properties, including its robust thermal stability, effective insulation, significant moisture absorption, and perceived comfort [17]. Specifically, the molecular arrangement of fibroin endows silk with thermal stability up to 100 °C. Demonstrating a specific heat of 1.38 J/gK, silk functions as a notable thermal insulator [18]. The material exhibits a high moisture regain, measured at 11% for raw mulberry silk, and possesses the capacity to absorb up to 30% of its weight in moisture without imparting a sensation of dampness [19]. While the low thermal conductivity and elevated moisture uptake of silk fibers may potentially compromise comfort in environments characterized by high temperatures and humidity, the fiber's heat of wetting, recorded at 69 J/g, coupled with a roughly 10% moisture regain, facilitates a buffer period for the wearer to acclimate to fluctuating environmental conditions. The presence of hydroxyl-containing amino acid residues within the silk structure contributes to its hygroscopic behavior, resulting in swelling upon exposure to moisture [20, 21].

Silk's thermoregulatory properties stem from its unique hierarchical structure, mechanical strength, and thermal characteristics, making it a promising material for smart textiles and biomedical applications [22]. The exceptional biocompatibility, biodegradability, and mechanical robustness of silk are attributed to its complex hierarchical organization, encompassing specific amino acid sequences, organized β-sheet structures, and interconnected nanofibrillar networks. This unique structural composition endows silk fibers with a notable combination of tensile strength, ranging from 300 to 740 MPa, a Young's modulus between 10 and 17 GPa, and an elongation at break of 4 to 26%, demonstrating a rare balance of strength, toughness, and flexibility [23]. Additionally, silk demonstrates high optical transparency (~90%) and reflective properties, enhancing its role in temperature regulation [24]. Its adaptability allows integration with nanomaterials, leading to advanced functionalities such as thermochromic silk fabrics for dynamic textile displays and temperature-sensitive clothing [22]. Beyond textiles, silk's hygroscopic nature enables its use in actuator devices, where volume expansion due to water absorption facilitates applications in artificial muscles and humidity-responsive materials [25]. Figure 13.2

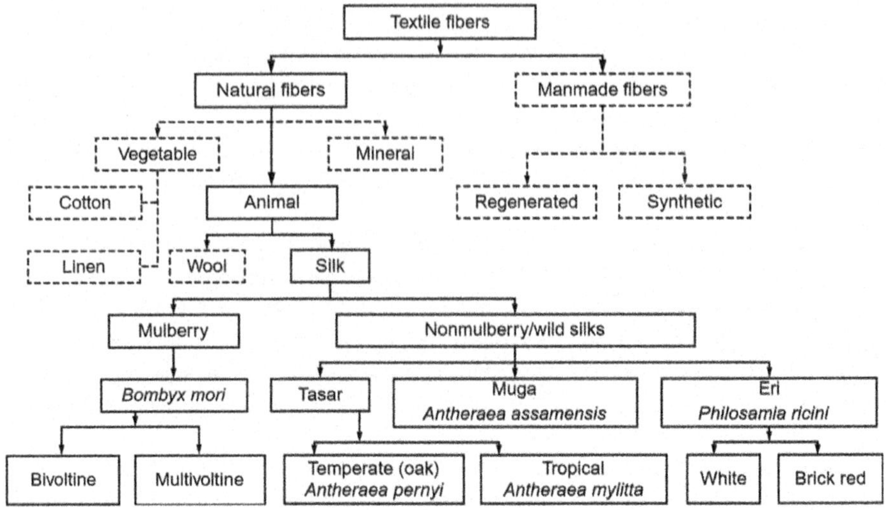

Fig. 13.2 Categorization of silk fibers. Republished with permission from Elsevier. Copyright Elsevier 2015 [19]

provides an overview of different types of textile fibers, with emphasis on the classification of silk fibers.

13.2 Structural and Chemical Characteristics of Silk Fiber

Figure 13.3 presents an illustrative diagram of the hierarchical and chemical structure of silk. The biopolymeric composition of silk is dominated by two proteinaceous components: fibroin, constituting the major structural element at roughly 75%, and sericin, a soluble adhesive protein comprising the remaining 25%. Fibroin's inherent mechanical robustness is attributed to its high content of β-sheet secondary structures, which contribute significantly to the material's overall strength [26, 27]. Specifically, fibroin's contribution to silk's tensile strength has been measured within the range of 0.5 to 1.3 GPa, alongside its notable toughness [28]. Conversely, sericin plays a critical role in maintaining the structural coherence of the silk fiber matrix. During textile processing, sericin is typically removed to enhance the comfort of silk and facilitate the weaving process [29–31]. The structural integrity of silk fiber remains reliably consistent, withstanding heat up to, but not exceeding, 100 °C [32]. Thermal analysis reveals that the material exhibits a specific heat capacity of 1.38 J/gK, a value closely aligned with those observed for established thermal insulators such as cotton (1.3 J/gK) and wool (1.36 J/gK). This empirically determined specific heat capacity indicates that the material possesses comparable thermal insulation properties to these commonly utilized fibers [33]. Silk starts to yellow when it is heated [32].

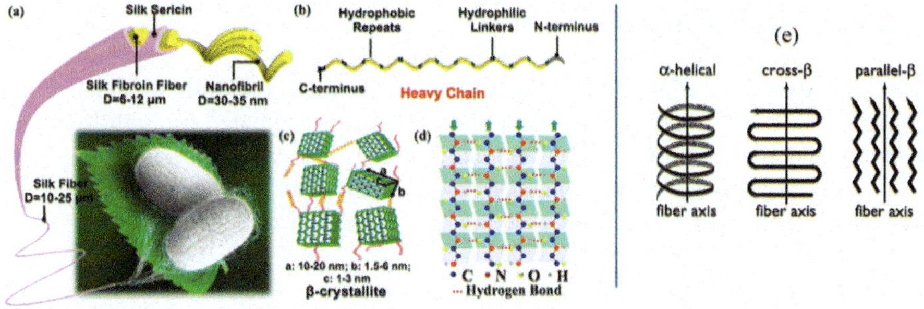

Fig. 13.3 Illustrative diagram of the hierarchical and chemical structure of silk. **a** The hierarchical organization of silk fibers. **b, c** Structural representation of the heavy chain **b** and β-crystallite **c** of silk fibroin. **d** Schematic depiction of the antiparallel β-sheet structure. Republished with permission from ACS. Copyright ACS 2019. [34] **e** Illustrative depiction of silk structures. Republished with permission from Annual Reviews. Copyright Annual Reviews 1997 [35]

13.2.1 Morphology and Microstructure

The performance characteristics of silk fibers, encompassing mechanical integrity, biological compatibility, and degradation profiles, are fundamentally governed by their intricate, multi-level architecture [1, 22]. This structural organization, essential for understanding silk's functional attributes, is characterized by a hierarchical arrangement comprising five distinct sections:

13.2.1.1 Primary Structure

The inherent elasticity and flexibility observed in protein chains are directly attributable to their constituent amino acid sequences [8, 36]. Specifically, the formation of polypeptide chains, which constitute these proteins, is facilitated by the covalent bonding of amino acids through amide linkages, thereby establishing the fundamental structural framework that dictates the material's mechanical properties [19].

13.2.1.2 Secondary Structure

The robust mechanical integrity of proteinaceous structures is attributed, in part, to the formation of stable hydrogen bonds within α-helical and β-sheet secondary structures [37, 38]. Specifically, the inter-chain hydrogen bonding between amine and carboxyl functional groups along the polypeptide backbone contributes significantly to the stabilization of β-sheet conformations [39].

13.2.1.3 Tertiary Structure

The establishment of a stable, three-dimensional matrix within the material is attributed to the formation of β-crystallites, a process mediated by both hydrogen bonding and various non-covalent interactions. This intricate network, arising from these intermolecular forces, contributes significantly to the material's structural integrity [36, 40, 41].

13.2.1.4 Quaternary Structure

Crystal networks are composed of interconnected β-crystallites, forming nanofibers and providing strength [42–44].

13.2.1.5 Quinary Structure

Silk's inherent microstructure, characterized by a network of interconnected nanofibrils, exhibits cross-sectional morphologies that are typically flat, elliptical, or triangular [24, 25, 45]. Furthermore, the presence of internal voids within silk fibers contributes to light scattering, effectively limiting light penetration into the cocoon [46]. Surface modification of degummed silk textiles can be achieved through etching with deep eutectic solvents, such as choline chloride/oxalic acid, which induces surface roughness and the formation of microfibrillar structures [47].

The fundamental composition of unprocessed silk is characterized by a dual-protein structure, wherein fibroin constitutes the fibrous core, responsible for the thread's structural integrity, and sericin, a glycoprotein matrix, acts as an adhesive agent, facilitating the cohesion of the fibroin filaments. Each strand of raw silk, known as a bave, is made up of a pair of fibroin filaments, called brins, which are coated and held together by sericin. This structure is depicted in Fig. 13.4a. Both fibroin and sericin are composed of amino acid chains, but they vary significantly in their composition and structure. Sericin contains randomly arranged amino acids that form amorphous regions, while fibroin features a more orderly arrangement, contributing to its crystalline nature. The fibroin filaments are made up of microfibrils—tiny, organized protein strands that cluster together to form the complete filament. Microscopic analysis, specifically utilizing Scanning Electron Microscopy (SEM) as illustrated in Fig. 13.4b, reveals that fibers derived from *Bombyx mori* silk exhibit a consistent longitudinal morphology characterized by a smooth surface and a unique triangular transverse profile. This structural uniformity, observed across the fiber's length, is a defining feature of mulberry silk as evidenced by the presented imaging data. In contrast, wild silk fibers, such as Tasar, tend to have a rougher surface and exhibit visible fibrillar patterns along their length, as shown in the SEM image in Fig. 13.4b. Their cross-sectional shape is usually flattened and triangular, often featuring internal voids, which are also observable in SEM images.

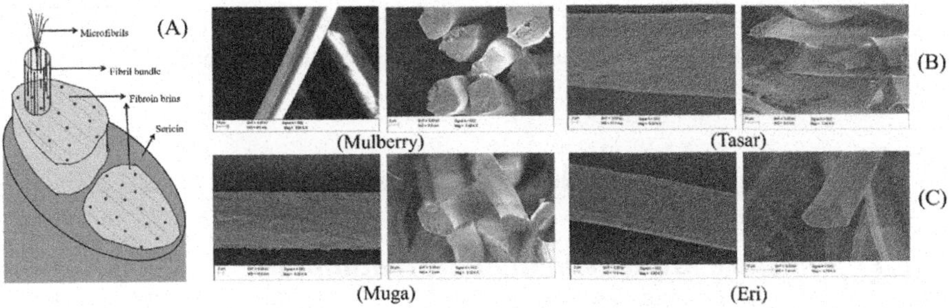

Fig. 13.4 **a** Structure of raw silk filament bave, **b** and **c** SEM images of silk fibers (longitudinal and cross-sectional). Republished with permission from Elsevier. Copyright Elsevier 2015 [19]

13.2.2 Comparative Thermoregulation of Silk and Other Fibers

13.2.2.1 Natural Fibers

Silk exhibits a sophisticated, multi-level structural organization that distinguishes it from cellulose and collagen, contributing to a unique combination of mechanical properties, including high tensile strength, elastic modulus, fracture toughness, and ductility. Furthermore, the inherent porosity of silk facilitates effective gaseous and aqueous vapor transport, indicating favorable permeability characteristics [19, 48, 49].

13.2.2.2 Synthetic Polymers

While conventional passive radiative cooling textiles frequently rely on synthetic polymeric materials, characterized by their non-renewable origins and limited biodegradability, silk fibroin presents a compelling alternative. This biopolymer demonstrates inherent advantages, including environmental sustainability through biodegradability and renewability, coupled with robust mechanical properties and significant hygroscopic capabilities, thereby offering a potentially more ecologically sound material platform for radiative cooling applications [50–52].

13.2.2.3 Thermal Conductivity

The intrinsic thermal transport properties of silk, notably lower than those observed in wool and cotton, play a significant role in its perceived comfort across varying environmental temperatures. Experimental characterization of mulberry silk fibers reveals a pronounced anisotropy in thermal conductivity, with values of 1.49 W/(mK) along the longitudinal axis and 0.119 W/(mK) in the transverse direction, indicating a directional dependence that influences heat dissipation and retention [19, 33, 53].

13.3 Silk Thermoregulation Mechanisms

13.3.1 Moisture Management and Breathability

13.3.1.1 Hygroscopic Nature

The inherent capacity of silk fibers to absorb moisture, a property known as hygroscopicity, stems from the molecular composition of their constituent chains, which feature polar amino acid residues, notably serine. These residues present hydrophilic functional groups, including amines, carboxyls, and hydroxyls, that facilitate the interaction and binding of water molecules. This resulting moisture absorption plays a crucial role in regulating the microclimate within silk textiles, thereby contributing to enhanced user comfort [19, 47, 54, 55].

13.3.1.2 Moisture Wicking

Silk can transport moisture through its structure. Janus silk fabrics, treated to have hydrophobic inner layers and hydrophilic outer layers, facilitate unidirectional moisture transport [56]. The described textile architecture facilitates efficient cutaneous perspiration transport toward the external fabric layer, thereby optimizing evaporative processes and maintaining a desiccated epidermal environment. This transport mechanism is characterized by a sequential progression involving initial moisture uptake, subsequent lateral and vertical dissemination within the fabric matrix, and accelerated desiccation at the fabric's exterior interface [56–58].

13.3.2 Heat Insulation and Retention

13.3.2.1 Air-Trapping Ability

The arrangement of silk fibers creates air pockets within the fabric, which reduces heat transfer through the material. The air trapped within the fabric structure acts as an insulator, which helps to maintain warmth in cooler environments [19, 59, 60].

13.3.2.2 Low Thermal Conductivity

Silk has a relatively low thermal conductivity, which means it does not efficiently transfer heat. This property helps in maintaining a stable temperature next to the skin, whether in warm or cool conditions [34, 60, 61]. Characterized by anisotropic thermal transport, mulberry silk fibers exhibit distinct conductivity values: longitudinal heat transfer is measured at 1.49 W/(mK), while transverse conduction registers at 0.119 W/(mK). This disparity underscores the material's directional thermal properties, a critical consideration for applications involving temperature regulation [33].

13.3.3 Infrared Reflectance and Emission

13.3.3.1 Passive Radiative Cooling

Human bodies, when exposed to ambient temperatures below 30 °C, release mid-infrared radiation within the 7–14 μm wavelength range, a phenomenon responsible for over half of total body heat dissipation [62, 63]. Given that natural silk fibers demonstrate elevated emissivity within the 8–13 μm mid-infrared spectral region, they offer potential for radiative cooling textile applications [46, 64, 65]. The inherent hierarchical microstructure of silk fibers facilitates both high solar reflectivity and thermal emissivity, resulting in a perceived cooling effect upon skin contact [1, 50, 66]. The efficacy of silk-based composites for thermoregulation has been experimentally validated, as evidenced by the results presented in Fig. 13.5(2) and Fig. 13.5(4).

13.3.3.2 Controlling Heat Loss

To achieve passive radiative heating for the human body by mitigating infrared heat loss, a promising avenue involves the fabrication of silk textiles exhibiting either enhanced internal mid-infrared reflectivity or diminished external mid-infrared emissivity. In a specific experimental approach, Yan et al introduced photothermal nanocomposites into the diet of *Bombyx mori* silkworms, resulting in silk fibroin with improved thermal insulation properties, as illustrated in Fig. 13.5(8). This study demonstrates a viable strategy for modifying silk's radiative properties through biomimetic material integration.

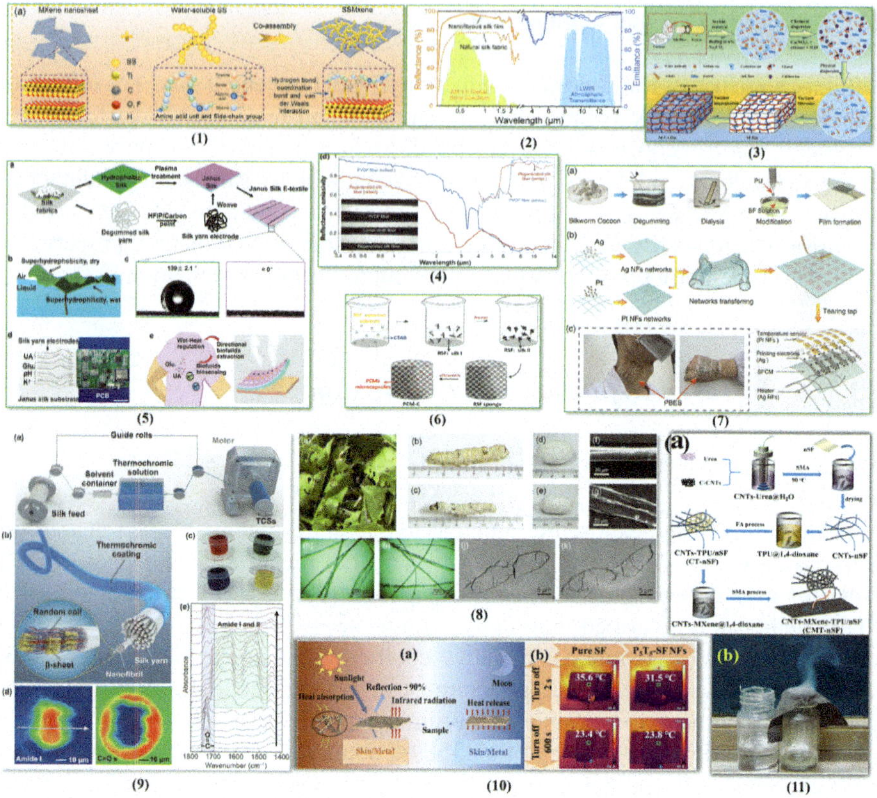
(1)
(2)
(3)
(4)
(5)
(6)
(7)
(8)
(9)
(10)
(11)
MXene nanosheet
Water-soluble SS
Co-assembly
SSMxene
Wavelength (μm)
Janus Silk E-textile
Silkworm Cocoon
Degumming
Dialysis
Modification
Film formation
Guide rolls
Motor
Thermochromic solution
Solvent container
Silk feed
Thermochromic coating
Random coil
β-sheet
Silk yarn
Nanofibril
Sunlight
Reflection ~ 90%
Infrared radiation
Heat absorption
Heat release
Pure SF
Turn off 2 s
Turn off 600 s

◂**Fig. 13.5** Modifications for efficient thermoregulation of silk: **(1)** Illustration of Surface Engineering: Schematic representation of MXene surface modification with sericin-infused silk. Copyright Elsevier 2023[47] **(2)** Optical Properties Comparison: Analysis of solar reflectance and mid-infrared emittance between nanofibrous silk film and natural silk fabric of equal thickness (200 μm). Copyright Elsevier 2023 [50] **(3)** Flexible Phase Change Material (PCM) Development: Schematic depiction of the synthesis process for a biocompatible silk fiber-based SF/CA-capric acid composite PCM. Copyright ACS 2022 [67] **(4)** Spectral Characteristics of Regenerated Silk: Integrated hemispherical reflectance and emissivity spectra of regenerated silk fibers versus a single PVDF fiber from the visible to mid-infrared range, with an inset showing fiber samples. Reprinted from [68]. Licensed under CC by 4.0. http://creativecommons.org/licenses/by/4.0/ **(5)** Janus Silk E-Textile for Smart Wearables: **a** Stepwise illustration of Janus silk e-textile fabrication, involving Janus silk biosupport creation, silk yarn electrode integration, and weaving onto the hydrophilic substrate. **b)** Conceptual depiction of Janus wettability, inspired by a lotus leaf's superhydrophobic and superhydrophilic surfaces. **c)** Water Contact Angle (WCA) comparison between hydrophobic and hydrophilic surfaces. **d)** Optical view of a PCB-connected Janus silk e-textile (Scale bar: 1 cm). **e)** Functional schematic demonstrating its comfort-enhancing properties, including biofluid extraction, wet-heat regulation, and multiplexed biosensing. Copyright ACS 2021. [56] **(6)** Formation of Regenerated Silk Fibroin (RSF) Sponge: Diagram of the synthesis pathway incorporating PCM into the RSF sponge. Copyright Wiley 2019. [69] **(7)** Fabrication of Protein-Based Electronic Skin (PBES): **a** Stepwise schematic of flexible and transparent Silk Fibroin Composite Membrane (SFCM) preparation. **b** Diagram depicting the PBES assembly for heating and temperature sensing. **c)** Photographic evidence of PBES closely adhered to the human neck and hand. Copyright Wiley 2020. [70] **(8)** Morphological Study of Silkworms and Silk Fibers: **a–e)** Photographs capturing different stages: feeding silkworms, silkworm-B, silkworm-P, cocoon-B, and cocoon-P. **f, g** SEM images of silk fibers from different sources. **h, i** Microscopy images of silk fibers. **j, k** Cross-sectional TEM images showcasing fiber structures. Copyright Springer Nature 2022. [71] **(9)** Thermoelectric Silk (TCS) Spinning and Properties: **a** Schematic of the continuous spinning setup for TCS fiber production. **b** Illustration of the hierarchical interfaces within TCS fibers. **c** Photographs displaying TCS fibers in red, green, blue, and yellow. **d** S-FTIR spectroscopic imaging of TCS fiber slices, highlighting silk's amide I peak (1656 cm^{-1}) and ink's $C=O$ stretching peak (1729 cm^{-1}). **e** Single-pixel spectra extracted along the specified direction. Reprinted from [72]. Licensed under CC by 4.0. http://creativecommons.org/licenses/by/4.0/ **(10)** Phase Change Nanofibers for Thermal Regulation: **a** Working principle schematic of TiO_2-Silk Fibroin@Phase Change Material nanofibers. **b** Infrared images showing thermal response of pure SF NFs and P5T5-SF NFs under light exposure. Copyright ACS 2024. [73] **(11)** Composite Silk Fabric for Breathability and Flexibility: **a** Schematic of surface adhesion and freeze-induced assembly of CMT nSF. **b** Air permeability and mechanical flexibility assessment of CMT-nSilk fabric composites. Copyright Elsevier 2024. [49]

13.3.4 Dynamic Thermal Adaptability

13.3.4.1 Response to Environmental Stimuli

Silk-based intelligent textile systems demonstrate adaptive capabilities, exhibiting property modulation in response to environmental fluctuations, specifically temperature and humidity. This dynamic adjustment allows for the attainment of sustained thermal comfort, wherein the material's characteristics are actively altered to counteract external stimuli, thus maintaining a stable microclimate [74, 75]. Thermoelectric material plays a vital role in this aspect as the material response to temperature stimuli; works of several researchers are depicted in Fig. 13.5(9).

13.3.4.2 Janus Fabrics

Janus textiles made from silk can adapt to different climatic conditions by modifying their hygroscopic properties. In warm environments, they facilitate sweat evaporation to keep the skin dry, while in cool conditions, they retain moisture to reduce heat loss and keep the skin warm [56, 76]. Janus fabrics are also used in wearable health monitoring applications where they serve both thermoregulation and sensing applications; schematic diagram and implementation idea are depicted in Fig. 13.5(5) and Fig. 13.5(7).

13.3.4.3 Shape Memory Polymers (SMPs)

Integrating SMPs with silk can create fabrics that change their structure in response to temperature, enhancing breathability and moisture permeability at higher temperatures and providing insulation at lower temperatures [60, 77, 78].

13.4 Silk's Thermoregulation Enhancement Through Modifications

Enhancing the thermoregulatory performance of silk fibers has become a key focus in advancing their functionality for high-performance applications. Table 13.1 provides a comprehensive overview of various modification methods employed to improve silk's thermal behavior. These methods include physical blending, chemical functionalization, nanoparticle incorporation, and structural engineering techniques [79, 80]. Each approach contributes uniquely to enhancing properties such as thermal conductivity, heat retention, and moisture responsiveness. By summarizing these modification strategies along with their respective mechanisms and outcomes, Table 13.1 highlights the innovative pathways through which silk's inherent thermoregulation can be significantly improved for advanced textile and material applications.

13.4.1 Chemical and Physical Treatments

13.4.1.1 Cross-Linking

Modifying silk fabrics using chemical cross-linking can enhance properties like shape retention and thermal stability [47].

13.4.1.2 Dyeing

Applying dyes can alter the thermal behavior of silk, potentially improving its ability to manage heat and moisture [47].

13.4.1.3 Surface Modifications

Surface treatments can significantly improve silk's functionality. For instance, etching silk fabrics with choline chloride/oxalic acid can create a rough surface, improving the bonding of coatings that enhance thermal management [82–86].

13.4.1.4 Solvent Treatment

Treating silk threads with solvents like 1,1,1,3,3,3-hexafluoro-2-propanol improves the adhesion of materials like carbon nanotubes, depicted in Fig. 13.5(11), enhancing properties like mechanical strength and thermal sensitivity [87–89].

13.4.2 Nano-Enhanced Silk Fibers

Integrating nanoparticles into silk fibers enhances their thermoregulatory properties [45, 88, 90–93].

13.4.2.1 Titanium Dioxide (TiO_2)

Coating silk with TiO_2 via Atomic Layer Deposition (ALD) improves UV resistance and cooling effects. TiO_2 is also known for its high solar reflectivity and cost-effectiveness, making it suitable for enhancing radiative cooling performance [94–98].

13.4.2.2 Aluminum Oxide (Al_2O_3)

The incorporation of aluminum oxide nanoparticles into silk matrices has been demonstrated to augment reflectivity within the ultraviolet spectrum, consequently elevating the material's total solar reflectance. This observed enhancement in reflectance directly contributes to an improved daytime radiative cooling effect, suggesting a potential application for thermal management in textiles [51, 59, 99].

13.4.3 Silk-Based Phase Change Materials (PCMs)

13.4.3.1 Integration with Microencapsulated PCMs

Combining silk with microencapsulated PCMs allows for controlled heat storage and release, enhancing thermoregulation, depicted in Fig. 13.5(6) [59, 60, 100].

13.4.3.2 SF/CA Composite PCM

Thermal analysis revealed that the synthesized Silk Fibroin/Cellulose Acetate (SF/CA) composite Phase Change Material (PCM) exhibited a melting enthalpy of 123.4 J/g and a melting temperature of 30.6 °C, as illustrated in Fig. 13.5(3). These thermal properties suggest the material's potential to buffer human body temperature variations when subjected to extreme thermal environments [67].

13.4.3.3 Applications

To achieve thermal regulation in textile applications, the incorporation of Phase Change Materials (PCMs) within silk matrices presents a viable strategy, leveraging the PCM's capacity for heat absorption during temperature elevation and subsequent release upon temperature decline. Notably, investigations have explored the fabrication of PCM-infused nanofibers utilizing silk fibroin, as evidenced in prior work, exemplified by the structural representation in Fig. 13.5(10) [67, 101, 102].

13.4.4 Hybrid and Composite Silk Fabrics

13.4.4.1 Blending with Functional Fibers

Combining silk with other fibers, such as carbon nanotubes, graphene, or polyethylene, can enhance its thermal regulation capabilities [72, 88, 103–106].

13.4.4.2 MXene Composites

Coating silk with MXene nanosheets enhances electrical conductivity and thermal management performance, creating textiles suitable for electrical heating and passive radiative heating, depicted in Fig. 13.5(1) [47, 60].

13.4.4.3 Aerogel Composites

Combining silk with aerogels, such as Kevlar nanofiber aerogels, results in materials with high thermal insulation properties. These composites can be used to create textiles with improved heat retention [107–109].

13.4.4.4 Janus Fabrics

Creating Janus fabrics with different surface properties (hydrophilic and hydrophobic) enhances moisture management and thermal comfort [51, 56, 81, 110].

13.5 Applications of Thermoregulating Silk Fibers

The diverse applications of thermoregulating silk fibers are largely driven by their exceptional ability to maintain thermal balance under varying environmental conditions. Table 13.2 presents a comparative analysis highlighting the functional performance of silk fibers across different application domains. This includes their use in textiles, biomedical materials, smart clothing, and advanced insulation systems. The table outlines how silk's thermoregulatory properties—such as moisture management, breathability, and thermal responsiveness—are harnessed in each application area. By comparing these aspects, Table 13.2 offers valuable insights into the versatility of silk fibers and their potential to enhance comfort, performance, and functionality in a wide range of end uses.

13.5.1 Textile and Apparel Industry

13.5.1.1 Adaptive Clothing

Thermoregulating silk fibers can be used in clothing designed to adapt to different environmental conditions. These adaptive textiles adjust heat and moisture dissipation based on external stimuli like light, heat, and moisture [1, 59, 81].

13.5.1.2 Sportswear

Silk's breathability, biodegradability, mechanical flexibility, and moisture absorption make it suitable for sportswear, enhancing comfort and performance [47, 56]. Investigations into silk-based thermal management textiles have yielded promising results. Yang and colleagues, as illustrated in Fig. 13.6(6), engineered a multifunctional coating on silk, termed M-silk, that exhibited a low mid-infrared emissivity of

Table 13.1 Methods for enhancing silk's thermoregulation through modifications

Modification type	Method	Materials/ techniques used	Thermoregulation enhancement	Refs
Chemical & physical treatments	Cross-linking	Chemical cross-linking agents	Enhances shape retention and thermal stability	[59]
	Dyeing	Dyes	Alters thermal behavior, potentially improving heat and moisture management	[81]
	Surface modifications	Etching with choline chloride/ oxalic acid, solvent treatments	Improves coating adhesion for thermal management, enhances mechanical strength and thermal sensitivity	[1, 47]
Nano-enhanced silk fibers	Incorporation of nanoparticles		Enhances thermoregulatory properties	[22]
		Titanium Dioxide (TiO_2)	Improves UV resistance and cooling effects via Atomic Layer Deposition (ALD). High solar reflectivity and cost-effectiveness for radiative cooling	[1, 22]
		Aluminum Oxide (Al_2O_3)	Improves reflectivity in the UV region, enhancing daytime radiative cooling	[1, 22]
Silk-based PCMs	Integration with microencapsulated PCMs	SF/CA composite PCMs	Regulates heat absorption and release, with a melting phase change enthalpy of 123.4 J/g at 30.6 °C, effectively minimizing temperature variations in both hot and cold conditions	[67]
Hybrid & composite silk fabrics	Blending with functional fibers	Carbon nanotubes, graphene, polyethylene	Enhances thermal regulation capabilities	[22, 60]
	MXene composites	MXene nanosheets	Improves electrical conductivity and thermal regulation, making it ideal for electrical applications and passive radiative heating	[47, 60]

(continued)

Table 13.1 (continued)

Modification type	Method	Materials/ techniques used	Thermoregulation enhancement	Refs
	Aerogel composites	Kevlar nanofiber aerogels	High thermal insulation properties for improved heat retention	[60]
	Janus fabrics	Hydrophilic and hydrophobic layers	Enhances moisture management and thermal comfort, facilitates sweat evaporation to keep skin dry in warm environments, retains moisture to reduce heat loss and keep skin warm in cool conditions	[56, 59]

0.24 within the 7–14 μm range, thereby indicating potential for reduced radiative heat loss; despite a primary focus on radiative heating, the material also displayed a water vapor transmission rate of 34.02 $mg{\cdot}cm^{-2}{\cdot}h^{-1}$, suggesting inherent breathability [47]. Conversely, He et al, presented in Fig. 13.6(9), developed a nanofibrous silk fibroin textile optimized for passive personal cooling, achieving a solar reflectance of 95% and a mid-infrared emissivity of 95%, which resulted in a daytime temperature reduction of approximately 3.8 °C below ambient conditions. Notably, the textile demonstrated a water absorption ratio of up to 170% and a 5.5 °C temperature decrease during outdoor sweat evaporation tests when compared to a nonhygroscopic control, emphasizing its efficacy in sweat-mediated cooling applications [50].

13.5.1.3 Winterwear and Summerwear

To expand the applicability of silk-based materials, strategic functionalization techniques are being explored to extend their operational range across diverse thermal environments, supported by empirical observations demonstrating this widened suitability. Specifically, the integration of passive radiative cooling properties into silk-derived intelligent textiles enables autonomous temperature regulation in warm climates, as evidenced by controlled experiments showing significant cooling effects, while analogous modifications have been shown to impart warming capabilities, expanding the utility of silk for cold-weather applications, thereby broadening its potential within advanced textile technologies [47, 60, 81].

Table 13.2 Comparative analysis based on applications of thermoregulating silk fibers

Application	Method/mechanism	Silk fibre modification/treatment	Refs
Wearable thermal management materials	Active electrical heating, passive radiative heating	Coating with Surface-Modified MXene (SSMXene) after DES micro-etching pretreatment	[111]
Active electrical heating textiles	Electrical conductivity of SSMXene coating	Coating with sericin-MXene (SSMXene)	[111]
Passive radiative heating textiles	Low mid-infrared emissivity of SSMXene coating	Coating with sericin-MXene (SSMXene)	[111]
Personal thermal management (cooling)	Radiative cooling, sweat management	Nanofibrous textile from silk fibroin with controlled nanofiber diameter for Mie scattering	[47]
	Increased UV and solar reflectivity, high IR emissivity	Nanofibrous structure of silk fibers	[47]
	Sweat absorption and evaporation	Hygroscopic property of silk fibroin in nanofibrous textile	[47]
Smart wearable textiles (thermal management)	Integration with flexible electronics	Silk as a substrate for conductive materials	[1]
Conductive fabrics for heating	Introduction of conductive materials (e.g., rGO)	Coating silk fibers with reduced graphene oxide (rGO)	[1]
Intelligent radiation refrigerated textiles	Radiative cooling	Electrospun nanosilk, Al_2O_3 nanoparticles on silk, ALD-TiO_2 coating on silk	[112]
Passive personal thermal management	Phase change material integration	Silk Fiber (SF)-based composite PCM with Capric Acid (CA)	[113]
Janus textile for wet-thermal comfort	Unidirectional biofluid transport	Hydrophobization of one side of silk textile with octadecyl trichlorosilane (OTS)	[114]
Thermochromic textiles	Temperature-dependent color change	Thermochromic inks coated on silk fibres with pre-solvent etching treatment	[72]
Flexible TENGs with thermal management	Triboelectric energy generation and temperature regulation	Titanium dioxide-silk fibroin@phase change microcapsule nanofiber films (TiO_2-SF@PCM NFs)	[115]
Temperature regulating sponge	Phase change material incorporation	Regenerated Silk Fibroin (RSF) sponge with Phase Change Microcapsules (PCMs)	[116]
Photothermal silk-based textiles	Conversion of photons (NIR region) into heat	Doping silk fibers with RGO/Bi_2S_3 PNCs	[71]
Water-responsive textiles for thermal management	Water-actuated structural changes	Regenerated silk fibers with semi-crystalline structure, reversible hydrogen bonds	[117]

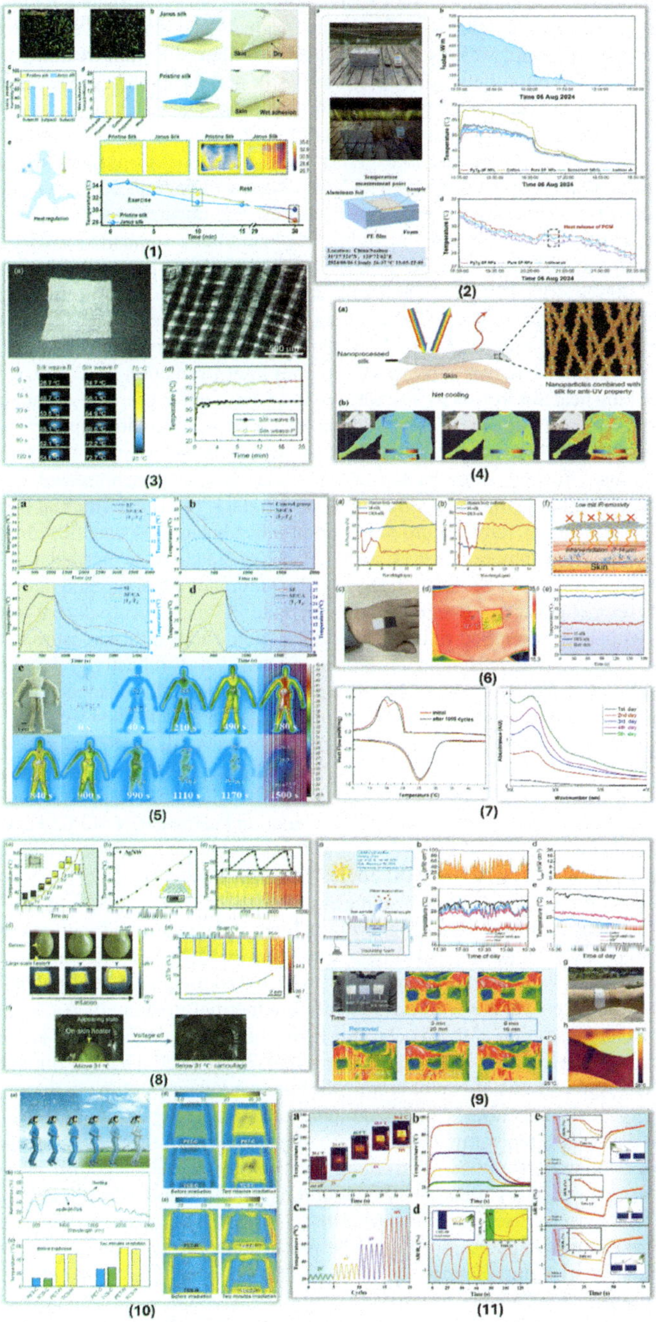
(1)
(2)
(3)
(4)
(5)
(6)
(7)
(8)
(9)
(10)
(11)

◀**Fig. 13.6** Performance of enhanced silk thermoregulation: **(1)** Performance Evaluation of Janus Silk Wearable Sensors: **a** Fluorescence microscopy image displaying the live/dead assay of NHDF cells incubated for 24 h with various silk materials (Scale bar: 200 μm). **b** Comparative analysis of on-body sweat absorption between conventional silk and Janus silk. **c** Measurement of local relative humidity on the skin when covered with both pristine and Janus silk. **d** Assessment of wet adhesion force for Janus silk compared to other commonly used fabrics. **e** Temperature variations of pristine and Janus silk in response to exercise and rest periods. Copyright ACS 2021. [56] **(2)** Outdoor Thermal Testing of TiO_2-SF@PCM Nanofibers: **a)** Diagram depicting the experimental setup for outdoor thermal measurements. **b** Solar irradiance during daylight hours. **c, d** Temperature monitoring of various materials, including P5 nanofibers, standard sunscreen fabric (90% nylon, 10% spandex; 70D × 70D; 180 g/m^2), T5-SF nanofibers, pure silk fibroin, cotton fabric (100% cotton; 40S × 40S; 150 g/m^2), and ambient air during both daytime and nighttime conditions. Copyright ACS 2024. [73] **(3)** Characterization of Silk Weaves Under Infrared Exposure: **a)** Photograph of silk weave-P. **b** Microscopy image illustrating the structural details of silk weave-P. **c** Infrared imaging of silk weaves subjected to 808 nm laser irradiation for 2 min. **d** Temperature–time graph tracking heat absorption by silk weaves under laser exposure. Copyright Springer Nature 2022. [71] **(4)** Cooling Efficiency of Nano-Engineered Silk: **a** Schematic representation of the net cooling effect achieved by nanoprocessing silk to enhance UV reflectance. **b** Infrared images and optical photographs (inset) of individuals wearing shirts made from nanotreated silk, regular silk, and cotton under sunlight in Nanjing. Copyright Nature 2021. [118] **(5)** Thermal Response of Silk Fibroin Composites: (a, b) Temperature variation of silk fibroin (SF) and silk fibroin/capric acid (SF/CA) composite when exposed to 35 °C and − 10 °C. **c)** Steady-state thermal performance of SF and SF/CA at 40 °C. **d** SF and SF/CA temperature fluctuation under simulated light conditions (T1 and T2 denote temperatures of SF and SF/CA composite PCMs, respectively). **e** Infrared imaging of a manikin with SF/CA applied to the abdominal region under light exposure. Copyright ACS 2022 [67] **(6)** Passive Radiative Heating Properties of M-Silk: **a** IR reflectivity comparison between M-silk and DES-silk. **b** IR emissivity of M-silk and DES-silk, with the shaded yellow region representing human body radiation at 33 °C. **c** Digital photograph of M-silk and DES-silk placed on bare skin. **d** Infrared image displaying thermal distribution over M-silk and DES-silk. **e** Temperature variations of M-silk, DES-silk, and bare skin over time once thermal equilibrium is established. **f** Schematic of the heat transfer mechanism governing M-silk's role in passive radiative heating. Copyright Elsevier 2023. [47] **(7)** Thermal and Degradation Performance of PCM and RSF Sponge: (Left) Thermal cycling analysis of PCM-3. (Right) Biodegradation behavior of RSF sponge in PBS solution. Copyright Wiley 2019. [69] **(8)** Joule Heating Characteristics of Ag Nanofiber (NFs) Heaters: **a** Temperature response under increasing voltage gradients (0.5 V to 4.5 V), with inset showing thermal images at various heating levels. **b** Temperature rise as a function of power density in Ag NFs heaters. **c** Cyclic heating performance at 3 V, with inset detailing temperature variations over time after voltage switching. **d** Photographs and infrared images of an Ag NFs heater adhered to an inflated balloon. **e** Average temperature shift and thermal imaging of Ag NFs heaters under tensile strain. **f** Demonstration of thermochromic behavior, with Ag NFs heaters applied to a human fist for skin camouflage by toggling voltage. Copyright Wiley 2020. [70] **(9)** Simulation of Sweat Evaporation and Applications of Nanofibrous Silk Films: **a** Diagram of the experimental setup for outdoor sweat evaporation studies, with topographic and meteorological data in the inset. **b–e** Solar irradiance and real-time temperature fluctuations under intense **b, c** and weak **d, e** sunlight exposure. **f** Comparative cooling performance analysis of nanofibrous silk films, cotton, and untreated silk fabric on a coat surface. **g, h** Utilization of nanofibrous silk films as a cooling component in adhesive bandages. Copyright Elsevier 2023 [50] **(10)** Temperature Regulation Using Thermochromic Silk (TCS) Fabrics: **a** Schematic illustrating thermochromic fabric response to varying temperatures. **b** Reflectance spectrum comparison of aq-Bl-28 TCS fibers before and after heat exposure. **c** Temperature shift in PET and TCS-based Chinese silk characters printed on polyester before and after two minutes of exposure in cold (C) and hot (H) conditions. **d, e** Infrared imaging of polyester fabric with PET and TCS Chinese silk characters before and after two minutes of irradiation in simulated cold **d** and hot **e** environments. Reprinted from [72]. Licensed under CC by 4.0. http://creativecommons.org/licenses/by/4.0/ **(11)** Electrothermal Properties of CMT-nSF Composites: **a** Joule heating performance of CMT-nSF under progressively increasing voltages. **b, c** Repetitive heating and cooling cycles at 2 V, 4 V, 6 V, and 10 V. **d** Thermal response cycles of CMT-nSF. **e** Sensor signal variations (sensor-a: orange, sensor-b: red) under different thermal source orientations. Copyright Elsevier 2024. [49]

13.5.2 Medical and Healthcare Textiles

13.5.2.1 Smart Bandages and Wound Dressings

The inherent biocompatibility and advantageous mechanical attributes of silk render it a promising material for biomedical textile applications, including bandages and wound dressings. Specifically, the fabrication of nanofibrous silk films allows for their incorporation into dermal-contacting dressings, facilitating localized thermal regulation and mitigating the risk of hyperthermia at wound sites, which is a critical factor in promoting optimal healing conditions [1, 50, 59]. Regenerated Silk Fibroin (RSF) sponges are promising for wound healing due to their biocompatibility and biodegradability, developed by Li et al., their thermal performance and thermal degradation of the sponges are depicted in Fig. 13.6(7). To enhance their functionality, RSF sponges incorporating Phase Change Microcapsules (PCMs) with an RSF wall (PCM-C) were developed for heat management. The PCM-C exhibited phase change enthalpies of 23.16 Jg^{-1} during heating and 23.32 Jg^{-1} during cooling, indicating its capacity for thermal regulation to create a stable microclimate. Thermal Gravimetric Analysis (TGA) of the PCMs microcapsules (PCM-3) revealed a two-stage degradation: an initial 5% weight loss between 200 °C and 270 °C, attributed partly to the evaporation of the n-octadecane core, followed by a second stage of degradation from 270 °C to 600 °C, corresponding to the degradation of the RSF wall. The RSF sponge component itself demonstrated continuous degradation in PBS solution over 5 days, as evidenced by an increasing absorption peak at 280 nm in UV–Vis spectroscopy. While the study confirms the degradation of the RSF sponge over time, the specific changes in the thermal regulation performance of the PCM-C composite as it degrades are not quantitatively measured beyond the initial characterization. However, the degradation of the RSF matrix and the potential leakage of the PCMs would likely lead to a decrease in the overall thermal buffering capacity and structural integrity of the wound dressing over the degradation period [69].

13.5.2.2 Temperature-Sensitive Fabrics

To harness silk's inherent biocompatibility for temperature-responsive therapeutic textiles, a methodology was developed to produce Thermochromic Silk (TCS) fibers and yarns by continuous coating with thermochromic inks [1, 59], as illustrated in Fig. 13.6(10). These TCS materials exhibit rapid, programmable color transitions within one second, maintaining stability across numerous cycles. Characterization using the CIELAB color space reveals a strong correlation between total color difference (ΔE^*) and temperature, accurately modeled by Gaussian functions for both heating and cooling phases. For example, a yellow TCS with a 28 °C transition temperature demonstrated the capacity to detect subtle thermal variations, such as finger proximity. Furthermore, when exposed to simulated heat, light-colored TCS fabrics displayed a 27.9% reduction in surface temperature compared to PET controls after two minutes of irradiation, indicating potential for thermal management. Integration with carbon fibers enabled dynamic textile displays, where Joule heating induced thermochromism at minimal current levels, such as 0.103 A for aq-Bl-31 ink, demonstrating its utility in advanced wearable applications [72].

13.5.3 Smart and Wearable Technologies

13.5.3.1 Integration in Biosensors and E-textiles

The inherent flexibility, biological compatibility, and ease of processing inherent to silk render it a promising matrix and structural support within wearable sensor platforms. Electronic skin applications leveraging silk matrices have demonstrated multifaceted functionalities, encompassing sensing, self-powered operation, and energy storage capabilities [1, 22, 34, 56, 59]. Notably, Huang et al. [70] engineered a Protein-Based Electronic Skin (PBES) employing Joule heating for precise thermoregulation, integrating silver nanofibers to achieve rapid and efficient surface temperature modulation via voltage control. This PBES, as detailed in Fig. 13.6(8), exhibited notable thermal stability and potential for applications in wearable hyperthermia and prolonged cutaneous thermal management, including arthritis alleviation. Similarly, Zhang et al [Fig. 13.6(11)] developed a conductive composite (CMT-nSF) utilizing a flexible porous Non-woven Silk Fabric (nSF), showcasing comparable efficacy in Joule heating performance [49].

13.5.3.2 Wearable Sensors

The fabrication of pliable sensing devices utilizing silk presents a promising avenue for addressing the demands of human motion and comfort, facilitating the acquisition of real-time physiological data and its subsequent wireless dissemination. This approach leverages the inherent flexibility and biocompatibility of silk to create sensors capable of seamlessly integrating with the human body, thereby enabling continuous monitoring and remote transmission of vital physiological parameters, which is crucial for applications in personalized healthcare and biomechanical analysis [1, 22, 34, 49, 66].

13.5.3.3 Thermal Management in Wearable Electronics

Exploiting silk's inherent properties, engineered materials derived from this biopolymer offer a promising avenue for thermal management in integrated wearable electronics, mitigating heat-induced degradation and enhancing device longevity [1]. Inspired by the lotus leaf's unique surface characteristics, Janus silk electronic textiles, possessing distinct hydrophobic and hydrophilic layers, enable both enhanced skin comfort and efficient biofluid manipulation for sensing applications. This layered design facilitates precise control of the skin's microclimate; the hydrophilic surface promotes rapid sweat evaporation during physical activity, yielding a discernible cooling effect relative to untreated silk, while the hydrophobic layer provides thermal insulation during periods of rest. Minimal skin adhesion further contributes to user comfort. For biofluid analysis, integrated silk yarn electrodes on the hydrophilic side of the Janus textile enable non-invasive sweat analysis, requiring minimal sample volumes (~5 μL) and achieving rapid response times (2 – 3 min), surpassing conventional fabrics. This system supports multiplexed detection of key sweat analytes, including uric acid, glucose, pH, and potassium, demonstrating accelerated sweat sampling and analysis in on-body trials compared to untreated silk regions, as illustrated in Fig. 13.6(1) [56].

13.5.4 Sustainable and Passive Cooling Applications

13.5.4.1 Radiative Cooling

Explorations into nanofibrous silk textiles reveal their potential for sub-ambient temperature attainment, yielding notable cooling properties. To enhance durability and radiative cooling, surface treatments, notably atomic layer deposition of titanium dioxide, have been employed, as evidenced in prior studies [1, 22, 50]. Expanding on this, Wang et al. [67] engineered a Silk Fiber/Capric Acid composite Phase Change Material (SF/CA PCM), demonstrating its thermal buffering capacity, depicted in Fig. 13.6(5). Through manikin-based experiments in controlled temperature environments, they observed a retardation of temperature fluctuations around 30 °C, with a peak temperature differential of 5.9 °C compared to unmodified silk at lower temperatures. Isothermal testing at 40 °C corroborated heat absorption by the SF/CA PCM, indicated by a temperature plateau, and a delayed cooling phase. Infrared thermography further confirmed the material's ability to moderate temperature variations under both heat and light exposure, showcasing its potential for thermal regulation. Additionally, Lu et al. [Fig. 13.6(2)] introduced titanium dioxide-silk fibroin films incorporating phase change microcapsules (TiO_2-SF@PCM NFs), contributing to the expanding repertoire of functionalized silk-based thermal management materials. In indoor thermal simulation tests under infrared light, Lu et al demonstrated a temperature reduction of approximately 10 °C beneath the TiO_2-SF@PCM NFs. After turning off the light, the temperature under the film was about 2.2 °C higher than the ambient temperature, showing thermal insulation. When applied to a model car under infrared light, Lu et al achieved a significant cooling effect of up to 22 °C compared to an uncovered car. The covered car also cooled down more slowly after the light was removed. Outdoor thermal simulation tests by Lu et al showed a daytime cooling effect of 1 to 4.3 °C with the TiO_2-SF@PCM NFs. At night, the material exhibited more stable temperature fluctuations and maintained a temperature higher than the ambient, indicating insulation. Experimental investigations utilizing a triboelectric sensor constructed from TiO_2-SF@PCM nanofibers, when affixed to the human arm, demonstrated a discernible capacity for thermal regulation. Specifically, Lu et al reported a temperature modulation range of approximately 0.2–4 °C, concurrently noting a significant enhancement in temperature curve stability relative to unshielded cutaneous tissue. These findings suggest the potential of this sensor configuration to provide controlled thermal management in contact with human skin [73]. Zhu et al reported that nanoprocessed silk exhibits effective thermoregulation, as demonstrated by various test results, Fig. 13.6(4). Empirical investigations, including field studies performed at Stanford, demonstrated that silk modified via nanoscale processing exhibited a significant thermal regulatory effect under solar irradiation, maintaining a temperature roughly 3.5 °C below ambient conditions, in contrast to unmodified silk which registered temperatures exceeding the surrounding environment. When applied as a covering over a skin-mimicking substrate, the nanoprocessed silk induced an approximate 8 °C temperature reduction compared to natural silk, with even more pronounced decreases relative to cotton and uncovered substrates. Consistent with these observations, human subject trials in Nanjing confirmed that apparel constructed from nanoprocessed silk displayed reduced outer surface temperatures when compared to garments fabricated from conventional silk and cotton, collectively indicating the material's viability for passive personal thermal management applications [118].

13.5.4.2 UV Resistance

Coatings like ALD-coated TiO_2 can significantly improve the UV resistance of silk, enhancing its radiation-cooling effect [1, 66].

13.6 Challenges and Future Perspectives

Several challenges and potential research directions exist for silk-based materials, particularly in the areas of thermoregulation, bioengineering, and innovative material combinations.

13.6.1 Limitations in Silk Thermoregulation

13.6.1.1 Durability

The longevity of silk-based radiative cooling textiles is compromised by the susceptibility of functional coatings to degradation via laundering and abrasive forces. Additionally, the inherent Ultraviolet (UV) sensitivity of silk manifests as yellowing and embrittlement upon solar irradiation, leading to a decline in mechanical integrity. This photodegradation is attributed to the cleavage of amide linkages within the protein structure following prolonged exposure to sunlight, ultimately diminishing the material's durability and wear resistance [119].

13.6.1.2 Processing

A significant challenge in the production of functional silk fibers lies in achieving durable integration of desired properties. While surface modification via functional material deposition is a common approach, the resulting coatings often exhibit susceptibility to environmental degradation and delamination. Furthermore, alternative strategies, such as carbonization, carbon nanotube application, or the inclusion of conductive additives, present potential drawbacks, as these processes can induce alterations to the intrinsic structural and mechanical characteristics of the silk, thereby limiting their practical applicability in scenarios demanding both enhanced functionality and preserved material integrity [47, 111, 114, 119].

13.6.1.3 Cost

Achieving widespread adoption of smart-responsive textiles is currently hindered by the intricate architectures and fabrication methods employed, which contribute to elevated production expenses. This cost factor presents a significant obstacle, particularly in the context of silk-based radiation-cooling apparel, where the imperative for scalable and economically viable manufacturing processes is paramount to facilitate broader accessibility and commercial viability [119].

13.6.2 Advancements in Bioengineered Silk

13.6.2.1 Recombinant Silk Production

Advancements in bioengineering and nanotechnology offer promising avenues for augmenting the mechanical resilience and biological activity of silk proteins, thereby broadening their utility within tissue engineering and regenerative medicine frameworks. Notably, genetic modification strategies have been successfully implemented to tailor silk protein functionalities, as exemplified by the production of fluorescent silk variants achieved via transgenic silkworm methodologies, demonstrating the potential for precise manipulation of silk properties [47, 72, 119, 120].

13.6.3 Potential Research Directions

13.6.3.1 Novel Material Combinations

Combining silk with other materials can create new functionalities. For example, incorporating nanomaterials can impart properties such as antimicrobial, UV resistance, flame retardancy, and self-heating capabilities [59, 112, 121, 122].

13.6.3.2 AI-Driven Textile Innovations

Integrating sensing capabilities, energy harvesting systems, and data communication components presents a significant avenue for advancing the functionality of intelligent textiles. Specifically, the incorporation of these modules enables the creation of fabrics capable of not only monitoring environmental or physiological parameters, but also autonomously powering themselves and relaying collected information, thus expanding the potential applications of smart textile technologies in diverse fields [1, 123, 124].

13.6.3.3 Silk-Based Radiative Cooling

A significant area of investigation within materials science concerns the optimization of silk's solar reflectance capabilities, aiming to augment its radiative cooling properties and ultraviolet radiation resistance. To facilitate the development of advanced silk-based radiative cooling apparel suitable for indoor thermal management, future research must focus on enhancing the performance characteristics of these textiles. This necessitates a comprehensive exploration of methodologies that effectively improve silk's ability to reflect solar radiation, thereby maximizing its cooling efficacy and protective attributes [125, 126].

13.6.3.4 Thermal and Moisture Management

The creation of textiles capable of dynamically adapting their thermal and moisture regulation across multiple environmental stimuli presents a promising avenue for research. This objective can be pursued through the strategic incorporation of diverse responsive elements, including but not limited to specialized coatings that modulate thermal and moisture transfer, substances that undergo phase transitions to absorb or release heat, water-retentive polymer networks, and materials engineered at the nanoscale. The synergistic integration of these components aims to yield advanced fabrics with enhanced environmental adaptability [127–131].

13.6.3.5 Wearability and Comfort

For the successful translation of adaptive thermal-moisture management fabrics from laboratory prototypes to commercially viable products, a critical consideration is the simultaneous optimization of several key performance metrics. Specifically, manufacturing processes must achieve a delicate equilibrium between effective thermal and humidity regulation, robust abrasion resistance, durable washability, application-specific suitability, and user-centric comfort. This multifaceted optimization is essential to ensure the practical utility and widespread adoption of these advanced textile materials in real-world applications [132–135].

13.6.3.6 Multifunctional Integration

Prospective advancements in silk-derived electronic skin technologies are directed toward incorporating diverse operational capabilities, creating integrated systems, and maximizing performance efficacy. This trajectory emphasizes the evolution of silk-based e-skins from singular-function devices to complex, high-performance platforms, signifying a shift toward comprehensive and efficient applications [1].

13.6.3.7 Improved Conductivity

A critical direction for future scientific advancement lies in the development of methodologies that enable the integration of substantial electrical conductivity within silk materials, without causing detrimental alterations to their inherent structural and functional properties. This necessitates the creation of processes that effectively balance the introduction of electrical conductance with the preservation of silk's native characteristics, representing a key challenge and opportunity for materials science [1].

13.6.3.8 Surface Modification

The successful surface modification and functionalization of natural silk demands meticulous process optimization to guarantee the robust binding of supplementary components without compromising the native biocompatibility and mechanical characteristics of the silk. These requirements restrict the expanded utilization of silk in clinical and industrial contexts [136].

13.7 Conclusion

Silk-based materials exhibit exceptional thermoregulatory properties, making them highly versatile for applications in smart textiles and wearable technologies. Recent advancements have enabled the engineering of silk to integrate passive radiative cooling, active heating, moisture management, and enhanced durability, thereby expanding its functional scope. For instance, sericin-MXene-coated silk fabrics demonstrate a combination of electrical heating, radiative cooling, and flame retardancy, while Thermally Adaptive Smart Textiles (TAST) incorporate Joule heating and capacitive sensing. Silk's hierarchical nanostructure enables superior solar reflectivity and mid-infrared emissivity, contributing to effective cooling performance. Textiles fabricated from silk, especially those of the Janus e-textile type, exhibit a capacity to adjust ambient moisture, improve thermal comfort in humid conditions, and manage biological fluids, which allows for health monitoring without invasive procedures. Structural modifications, including sericin surface treatment and Deep Eutectic Solvent (DES) pretreatment, have improved the stability and functionalization of silk fibers. Additionally, silk-based Phase Change Materials (PCMs) demonstrate excellent thermal stability, enabling precise temperature regulation.

Looking ahead, future research will focus on integrating stimulus-responsive materials, such as phase change composites, hydrogels, and nanomaterials, to develop real-time dynamic thermoregulating textiles. The introduction of two-way shape memory silk materials may further enhance adaptability by enabling reversible structural changes in response to environmental stimuli. Additionally, advancements in conductive silk textiles could bridge the gap between thermal management and wearable electronic applications, facilitating seamless integration with sensing systems and energy harvesting devices. However, addressing challenges related to durability, large-scale processing, and cost remains critical for the commercialization of silk-based smart textiles.

The impact of silk-based materials extends beyond textile innovation, offering sustainable solutions that align with environmental and energy efficiency goals. Silk's biodegradability and renewability provide an environmentally sustainable alternative to synthetic fibers in textiles, and its application in flexible electronics offers a means to reduce e-waste while maintaining functionality in wearable technologies. Passive radiative heating textiles derived from silk can reduce infrared heat dissipation from the human body, leading to lower energy consumption in climate control systems. Advancements in silk-based e-textiles also hold promises for remote health monitoring and smart healthcare applications, contributing to improved quality of life. With continued research and interdisciplinary collaboration, silk-based materials will play a transformative role in sustainable textile engineering, flexible electronics, and next-generation thermoregulatory wearables.

Declaration

The authors declare that they have no known competing financial interests or personal relationships that could have appeared to influence the work reported in this paper.

Generative AI and AI-assisted technologies have been used in the writing process to improve the readability and language of the work only. After using this tool, the author(s) reviewed and edited the content as needed and take(s) full responsibility for the content of the publication.

References

1. Xing T et al (2023) Silk-based flexible electronics and smart wearable textiles: progress and beyond. Chem Eng J 474:145534
2. Andersson M, Johansson J, Rising A (2016) Silk spinning in silkworms and spiders. Int J Mol Sci 17(8):1290
3. Magoshi J, Magoshi Y, Nakamura S (1993) Mechanism of fiber formation of silkworm. In: Silk polymers, vol 544. (ACS Symposium Series, no 544): American Chemical Society, ch 25, pp 292–310
4. Das A, Apu MMH, Akter A, Al Reza MM, Mia R (2024) An overview of phase change materials, their production, and applications in textiles. Results Eng 25:103603
5. Sparkes J, Holland C (2017) Analysis of the pressure requirements for silk spinning reveals a pultrusion dominated process. Nat Commun 8:594
6. Chae SK, Kang E, Khademhosseini A, Lee SH (2013) Micro/nanometer-scale fiber with highly ordered structures by mimicking the spinning process of silkworm. Adv Mater 25(22):3071–3078
7. Ma Y et al (2022) A single-cell transcriptomic atlas characterizes the silk-producing organ in the silkworm. Nat Commun 13:3316
8. Qiu W, Patil A, Hu F, Liu XY (2019) Hierarchical structure of silk materials versus mechanical performance and mesoscopic engineering principles. Small 15(51):1903948
9. Fang Y, Chen G, Bick M, Chen J (2021) Smart textiles for personalized thermoregulation. Chem Soc Rev 50(17):9357–9374
10. Li B, Wang M, Ao S, Lyu K, Su X, Sun F (2025) Knitting-stitching bifacial metafabrics with switchable thermal and moisture transmissibility for multimodal dynamic personal thermoregulation. Mater Horiz 12(2):642–653
11. Chithramol MK, Shine SR (2023) Review on modelling approaches of thermoregulation mechanisms. J Therm Anal Calorim 148:9343–9360
12. Katić K, Li R, Zeiler W (2016) Thermophysiological models and their applications: a review. Build Environ 106:286–300
13. Lim CL (2020) Fundamental concepts of human thermoregulation and adaptation to heat: a review in the context of global warming. Int J Environ Res Public Health 17(21):7795
14. Zuo X, Zhang X, Qu L, Miao J (2023) Smart fibers and textiles for personal thermal management in emerging wearable applications. Adv Mater Technol 8(6):2201137
15. Wang H, Zhang Y, Liang X, Zhang Y (2021) Smart fibers and textiles for personal health management. ACS Nano 15(8):12497–12508
16. Peng Y, Cui Y (2020) Advanced textiles for personal thermal management and energy. Joule 4(4):724–742
17. Zhao M, Qi Z, Tao X, Newkirk C, Hu X, Lu S (2021) Chemical, thermal, time, and enzymatic stability of silk materials with silk I structure. Int J Mol Sci 22(8):4136
18. Morton WE, Hearle JWS (2008) Physical Properties of Textile Fibres, Elsevier
19. Padaki N, Das B, Basu A (2015) Advances in understanding the properties of silk, In: Advances in Silk Science and Technology, pp. 3–16
20. Munro TR (2020) Reviewing thermophysical properties of silk fibers: a case study for the need for complementary measurement techniques. Int J Thermophys 41(9):133
21. Morikawa J et al (2016) Silk fibroin as a water-soluble bio-resist and its thermal properties. RSC Adv 6(14):11863–11869

22. Yang X-C et al (2024) Silk-based intelligent fibers and textiles: structures, properties, and applications. Chem Comm 60(61):7801–7823
23. Ren J et al (2019) Biological material interfaces as inspiration for mechanical and optical material designs. Chem Rev 119(24):12279–12336
24. Cheng L et al (2019) Quantum dots-reinforced luminescent silkworm silk with superior mechanical properties and highly stable fluorescence. J Mater Sci 54:9945–9957
25. Jia T et al (2019) Moisture sensitive smart yarns and textiles from self-balanced silk fiber muscles. Adv Funct Mater 29:1808241
26. Sahoo JK, Hasturk O, Falcucci T, Kaplan DL (2023) Silk chemistry and biomedical material designs. Nat Rev Chem 7(5):302–318
27. Bulmer, GS, De Andrade P, Field RA, van Munster JM (2021) Recent advances in enzymatic synthesis of β-glucan and cellulose. Carbohydr Res 508:108411
28. Cebe P et al (2013) Beating the heat-fast scanning melts silk beta sheet crystals. Sci Rep 3:1130
29. Nguyen AT, Huang QL, Yang Z, Lin N, Xu G, Liu XY (2015) Crystal networks in silk fibrous materials: from hierarchical structure to ultra performance. Small 11(9–10):1039–1054
30. Vepari C, Kaplan DL (2007) Silk as a biomaterial. Prog Polym Sci 32(8–9):991–1007
31. Volkov V, Sárria MP, Gomes AC, Cavaco-Paulo A (2015) Phosphorylated silk fibroin matrix for methotrexate release. Mol Pharm 12(1):75–86
32. Nakamura S, Magoshi J, Magoshi Y (1994) Thermal properties of silk proteins in silkworms. In: Silk Polymers: Materials Science and Biotechnology, pp. 211–221
33. Hearle JW, Morton WE (2008) Physical properties of textile fibres. Elsevier
34. Wang C, Xia K, Zhang Y, Kaplan DL (2019) Silk-based advanced materials for soft electronics. Acc Chem Res 52(10):2916–2927
35. Craig CL (1997) Evolution of arthropod silks. Annu Rev Entomol 42(1):231–267
36. Zhou CZ et al (2001) Silk fibroin: structural implications of a remarkable amino acid sequence. Proteins 44(2):119–122
37. Heim M, Römer L, Scheibel TJ (2010) Hierarchical structures made of proteins. The complex architecture of spider webs and their constituent silk proteins. Chem Soc Rev 39(1):156–164
38. Hofmaier M et al (2023) β-sheet structure formation within binary blends of two spider silk related peptides. Biomacromolecules 24(2):825–840
39. Hagler J, Peterson B, Murphy A, Leger JJ (2019) Performance of silk-polypyrrole bilayer actuators under biologically relevant conditions. J Appl Polym Sci 136(2):46922
40. Takahashi Y, Gehoh M, Yuzuriha KJ (1999) Structure refinement and diffuse streak scattering of silk (Bombyx mori). Int J Biol Macromol 24(2–3):127–138
41. Fossey SA, Némethy G, Gibson KD, Scheraga HA (1991) Conformational energy studies of β-sheets of model silk fibroin peptides. I. Sheets of poly (Ala-Gly) chains. Biopolym Orig Res Biomol 31(13):1529–1541
42. Xu X et al (2021) Seeded 2D epitaxy of large-area single-crystal films of the van der Waals semiconductor 2H MoTe2. Science 372(6538):195–200
43. Xu G, Gong L, Yang Z, and Liu X (2014) What makes spider silk fibers so strong? From molecular-crystallite network to hierarchical network structures. Soft Matter 10(13):2116–2123
44. Du F, Fischer JE, Winey KI (2005) Effect of nanotube alignment on percolation conductivity in carbon nanotube/polymer composites. Phys Rev B—Condensed Matter Mater Phys 72(12):121404
45. Ryan JD, Mengistie DA, Gabrielsson R, Lund A, Müller C (2017) Machine-washable PEDOT: PSS dyed silk yarns for electronic textiles. ACS Appl Mater Interfaces 9(10):9045–9050
46. Yang Z, Zhang J (2021) Bioinspired radiative cooling structure with randomly stacked fibers for efficient all-day passive cooling. ACS Appl Mater Interfaces 13(36):43387–43395
47. Yang H et al (2023) Multifunctional and durable thermal management coating from sericin-MXene biohybrid on silk fabric micro-etched by deep eutectic solvent. Appl Surface Sci 623:156962

48. Guo P et al (2024) Structural characterization and physiological role of Bombyx mori Fibroinase in the silk gland development, J Agric Food Chem 72(16):9457–9467
49. Zhang L et al (2024) Flexible porous non-woven silk fabric based conductive composite for efficient multimodal sensing. Chem Eng J 497:154445
50. He J, Zhang Q, Wu Y, Ju Y, Wang Y, Tang S (2023) Scalable nanofibrous silk fibroin textile with excellent Mie scattering and high sweat evaporation ability for highly efficient passive personal thermal management. Chem Eng J 466:143127
51. Zhu B et al (2021) Subambient daytime radiative cooling textile based on nanoprocessed silk. Nat Nanotechnol 16:1342–1348
52. Tao H, Kaplan DL, Omenetto FG (2012) Silk materials–a road to sustainable high technology. Adv Mater 24(21):2824–2837
53. Tong JK, Huang X, Boriskina SV, Loomis J, Xu Y, Chen G (2015) Infrared-transparent visible-opaque fabrics for wearable personal thermal management. Acs Photonics 2(6):769–778
54. Zhang Y-Q, Shen W-D, Xiang R-L, Zhuge L-J, Gao W-J, Wang W-BJ (2007) Formation of silk fibroin nanoparticles in water-miscible organic solvent and their characterization. J Nanoparticle Res 9:885–900
55. Venkatesan H, Chen J, Liu H, Liu W, Hu J (2020) A spider-capture-silk-like fiber with extremely high-volume directional water collection. Adv Funct Mater 30(30):2002437
56. He X, Fan C, Xu T, Zhang X (2021) Bioinsired Janus silk E-textiles with wet–thermal comfort for highly efficient biofluid monitoring. Nano Lett 21(20):8880–8887
57. Ma W et al (2021) Smart fibers for energy conversion and storage. Chem Soc Rev 50(12):7009–7061
58. Zheng S et al (2022) Moisture-wicking, breathable, and intrinsically antibacterial electronic skin based on dual-gradient poly (ionic liquid) nanofiber membranes. Adv Mater 34(4):2106570
59. Ma R et al (2024) Fabricated advanced textile for personal thermal management, intelligent health monitoring and energy harvesting. Adv Colloid Interface Sci 332:103252
60. Zuo X, Zhang X, Qu L, Miao J (2023) Smart fibers and textiles for personal thermal management in emerging wearable applications. Adv Mater Technol 8(6):2201137
61. Zhang Y, Lu H, Liang X, Zhang M, Liang H, Zhang Y (2022) Silk materials for intelligent fibers and textiles: potential, progress and future perspective. Acta Phys Chim Sin 38(2103034):10.3866
62. Peng Y et al (2018) Nanoporous polyethylene microfibres for large-scale radiative cooling fabric. Nat Sustain 1(2):105–112
63. Hsu P-C et al (2016) Radiative human body cooling by nanoporous polyethylene textile. 353(6303):1019–1023
64. Shi NN et al (2018) Nanostructured fibers as a versatile photonic platform: radiative cooling and waveguiding through transverse Anderson localization. Light Sci Appl 7:37
65. Choi SH et al (2018) Anderson light localization in biological nanostructures of native silk. Nat Commun 9:452
66. Ling X, Osotsi MI, Zhang W, Wu Y, Jin Q, Zhang D (2023) Bioinspired materials: from distinct dimensional architecture to thermal regulation properties. J Bionic Eng 20(3):873–899
67. Wang C et al (2022) Flexible and biocompatible silk fiber-based composite phase change material for personal thermal management. ACS Sustainable Chem Eng 10(49):16368–16376
68. Shi NN et al (2018) Nanostructured fibers as a versatile photonic platform: radiative cooling and waveguiding through transverse Anderson localization. Light Sci Appl 7(1):37
69. Li Y, Zhao L, Wang H, Li B (2019) Preparation of regenerated silk fibroin-based heat-management sponge for wound healing. J Appl Polym Sci 136(44):48173
70. Huang J et al (2020) Stretchable and heat-resistant protein-based electronic skin for human thermoregulation, Adv Funct Mater 30(13):1910547
71. Yan M et al (2022) Photothermal silk-based textiles. Fibers Polym 23(3):644–650
72. Wang Y, Ren J, Ye C, Pei Y, Ling S (2021) Thermochromic silks for temperature management and dynamic textile displays. Nano-Micro Lett 13:1–17

73. Lu S et al (2024) Flexible silk fibroin-based triboelectric nanogenerators with thermal management capabilities for temperature regulation and self-powered monitoring. Nano Lett 24(49):15933–15942
74. Wang H, Zhang Y, Liang X, Zhang Y (2021) Smart fibers and textiles for personal health management. ACS Nano 15(8):12497–12508
75. Shi J et al (2020) Smart textile-integrated microelectronic systems for wearable applications. Adv Mater 32(5):1901958
76. Wang S et al (2020) Environmentally-friendly and multifunctional graphene-silk fabric strain sensor for human-motion detection. Adv Mater Interfaces 7:1901507
77. Yu F, Xu X, Lin N, Liu XY (2015) Structural engineering of waterborne polyurethane for high performance waterproof coatings. RSC Adv 5(89):72544–72552
78. Mu J et al. (2018) Molecular-channel driven actuator with considerations for multiple configurations and color switching. 9(1):590
79. Mia R, Sultana S (2020) Fabrication and properties of silver nanowires (AgNWs) functionalized fabric. SN Appl Sci 2(12):2052
80. Zannat A, Uddin MN, Mahmud ST, Prithu PSS, Mia R (2023) Textile-based soft robotics for physically challenged individuals. J Mater Sci 58:12491–12536
81. Liu R et al (2024) Adaptive dynamic smart textiles for personal thermal-moisture management. Eur Polym J 206:112777
82. Shi M et al (2021) Ti3C2Tx MXene-decorated nanoporous polyethylene textile for passive and active personal precision heating. ACS Nano 15(7):11396–11405
83. Du X et al (2023) Visible transparent, infrared stealthy polymeric films with nanocoating of ITO@ MXene enable efficient passive radiative heating and solar/electric thermal conversion. Nano Res 16(2):3326–3332
84. Li L et al (2021) Ultrathin titanium carbide (MXene) films for high-temperature thermal camouflage. Adv Funct Mater 31:2101381
85. Zhang Y, Li L, Cao Y, Yang Y, Wang W, Wang J (2023) High-strength, low infrared-emission nonmetallic films for highly efficient Joule/solar heating, electromagnetic interference shielding and thermal camouflage. Mater Horiz 10(1):235–247
86. Liu X et al (2020) Air-permeable, multifunctional, dual-energy-driven MXene-decorated polymeric textile-based wearable heaters with exceptional electrothermal and photothermal conversion performance. J Mater Chem A 8(25):12526–12537
87. Larsen ST, Nielsen GD (2012) Acute airway irritation of methyl formate in mice. Arch Toxicol 86:285–292
88. Ye C et al (2019) Design and fabrication of silk templated electronic yarns and applications in multifunctional textiles. Matter 1(5):1411–1425
89. Weiss F et al (2023) Detecting hexafluoroisopropanol using soft chemical ionization mass spectrometry and analytical applications to exhaled breath. J Am Soc Mass Spectrom 34(5):958–968
90. Lu Z, Mao C, Zhang H (2015) Highly conductive graphene-coated silk fabricated via a repeated coating-reduction approach. J Mater Chem C 3(17):4265–4268
91. Chiu W-T et al (2018) Platinum coating on silk by a supercritical CO2 promoted metallization technique for applications of wearable devices. Surf Coat Technol 350:1028–1035
92. Gao Y-J et al (2024) Polyaniline/chitosan coating as the novel and sustainable flame retardant and UV protection route for silk fabric. Prog Org Coat 186:108036
93. Guo J, Lu S, Zhou Y, Yang Y, Yao X, Wu G (2024) Heat-insulated regenerated fibers with UV resistance: silk fibroin/Al2O3 nanoparticles. Molecules 29(9):2023
94. Deng B, Yan X, Wei Q, Gao W (2007) AFM characterization of nonwoven material functionalized by ZnO sputter coating. Mater Charact 58(10):854–858
95. Liang Z et al (2021) Multi-functional silk fibers/fabrics with a negligible impact on comfortable and wearability properties for fiber bulk. Chem Eng J 415:128980
96. Fan W, Gao Q, Xiang J, Yan J, Chen Y, Fan H (2023) Synergistic effect of silica aerogel and titanium dioxide in porous polyurethane composite coating with enhanced passive radiative cooling performance. Prog Org Coat 183:107763

97. Dong S, Quek JY, Van Herk Am, Jana S (2020) Polymer-encapsulated TiO2 for the improvement of NIR reflectance and total solar reflectance of cool coatings. Ind Eng Chem Res 59(40):17901–17910
98. Huang Z, Ruan X (2017) Nanoparticle embedded double-layer coating for daytime radiative cooling. Int J Heat Mass Transf 104:890–896
99. Jing W et al (2021) Scalable and flexible electrospun film for daytime subambient radiative cooling. ACS Appl Mater Interfaces 13(25):29558–29566
100. Li Y, Zhao L, Wang H, Li B (2019) Preparation of regenerated silk fibroin-based heat-management sponge for wound healing. J Appl Polym Sci 136(44):48173
101. Pielichowska K, Pielichowski K (2014) Phase change materials for thermal energy storage. Prog Mater Sci 65:67–123
102. Wu S, Yan T, Kuai Z, Pan W (2020) Thermal conductivity enhancement on phase change materials for thermal energy storage: a review. Energy Storage Mater 25:251–295
103. Wang H, Dong Q, Yao J, Shao Z, Ma J, Chen X (2020) Colorless silk/copper sulfide hybrid fiber and fabric with spontaneous heating property under sunlight. Biomacromolecules 21(4):1596–1603
104. Li Q et al (2021) Highly thermal-wet comfortable and conformal silk-based electrodes for on-skin sensors with sweat tolerance. ACS Nano 15(6):9955–9966
105. Zhou J et al (2020) Multi-walled carbon nanotubes functionalized silk fabrics for mechanical sensors and heating materials. Mater Design 191:108636
106. Bhattacharjee S et al (2020) Reduced graphene oxide and nanoparticles incorporated durable electroconductive silk fabrics. Adv Mater Interfaces 7(20):2000814
107. Si Y, Yu J, Tang X, Ge J, Ding B (2014) Ultralight nanofibre-assembled cellular aerogels with superelasticity and multifunctionality. Nat Commun 5(1):5802
108. Si Y, Wang X, Dou L, Yu J, Ding B (2018) Ultralight and fire-resistant ceramic nanofibrous aerogels with temperature-invariant superelasticity. Sci Adv 4(4):8925
109. Liu Z, Lyu J, Fang D, Zhang X (2019) Nanofibrous kevlar aerogel threads for thermal insulation in harsh environments. ACS Nano 13(5):5703–5711
110. Shao Z, Wang Y, Bai H (2020) A superhydrophobic textile inspired by polar bear hair for both in air and underwater thermal insulation. Chem Eng J 397:125441
111. Yang H et al. (2023) Multifunctional and durable thermal management coating from sericin-MXene biohybrid on silk fabric micro-etched by deep eutectic solvent. Appl Surf Sci 623:156962
112. Liang Z et al (2021) Multi-functional silk fibers/fabrics with a negligible impact on comfortable and wearability properties for fiber bulk. Chem Eng J 415:128980
113. Wang C, Dong H, Cheng C, Sun K, Jin T, Shi Q (2022) Flexible and biocompatible silk fiber-based composite phase change material for personal thermal management. ACS Sustain Chem Eng 10(49):16368–16376
114. He X, Fan C, Xu T, Zhang X (2021) Bioinspired janus silk E-textiles with wet-thermal comfort for highly efficient biofluid monitoring. Nano Lett 21(20):8880–8887
115. Lu S et al (2024) Flexible silk fibroin-based triboelectric nanogenerators with thermal management capabilities for temperature regulation and self-powered monitoring. Nano Lett 24(49):15933–15942
116. Li Y, Zhao L, Wang H, Li B (2019) Preparation of regenerated silk fibroin-based heat-management sponge for wound healing. J Appl Polym Sci 136(44):48173
117. Xu X et al (2024) Biomimetic design of hydration-responsive silk fibers and their role in actuators and self-modulated textiles. Adv Funct Mater 34:2401732
118. Zhu B et al (2021) Subambient daytime radiative cooling textile based on nanoprocessed silk. Nat Nanotechnol 16(12):1342–1348
119. Xing T et al (2023) Silk-based flexible electronics and smart wearable textiles: progress and beyond. Chem Eng J 474:145534
120. Yang XC et al (2024) Silk-based intelligent fibers and textiles: structures, properties, and applications. Chem Comm 60(61):7801–7823

121. Pan J et al (2024) Construction of Janus structures on thin silk fabrics via misting for wet–thermal comfort and antimicrobial activity. J Colloid Interface Sci 656:587–596
122. Gao Y-J et al (2024) Polyaniline/chitosan coating as the novel and sustainable flame retardant and UV protection route for silk fabric. Prog Org Coat 186:108036
123. Yu S, Zhang Q, Liu L, Ma R (2023) Thermochromic conductive fibers with modifiable solar absorption for personal thermal management and temperature visualization. ACS Nano 17(20):20299–20307
124. Wang Y, Yu Y, Zhao F, Feng Y, Feng W (2023) Multi-functional and multi-responsive layered double hydroxide-reinforced polyacrylic acid composite hydrogels as ionic skin sensors. Adv Compos Hybrid Mater 6(2):65
125. Xing T et al (2023) Silk-based flexible electronics and smart wearable textiles: progress and beyond. Chem Eng J 474:145534
126. Yang H et al (2019) Facile and effective fabrication of highly UV-resistant silk fabrics with excellent laundering durability and thermal and chemical stabilities. ACS Appl Mater Interfaces 11(30):27426–27434
127. Guo N, Yang R, Chen M, Yan H, Chen W (2023) Self-adaptive colored radiative cooling by tuning visible spectra. Solar RRL 7(21):2300512
128. Yang M, Zhong H, Li T, Wu B, Wang Z, Sun D (2023) Phase change material enhanced radiative cooler for temperature-adaptive thermal regulation. ACS Nano 17(2):1693–1700
129. Qu J et al (2023) All-in-one strain-triboelectric sensors based on environment-friendly ionic hydrogel for wearable sensing and underwater soft robotic grasping. Nano Energy 111:108387
130. Persson NK, Martinez JG, Zhong Y, Maziz A, Jager EW (2018) Actuating textiles: next generation of smart textiles. Adv Mater Technol 3(10):1700397
131. Kamal A, Ashmawy M, Shanmugam S, Algazzar AM, Elsheikh AH (2022) Fabrication techniques of polymeric nanocomposites: a comprehensive review. Proc Inst Mech Eng, Part C J Mech Eng Sci 236(9):4843–4861
132. Peng Y et al (2021) Hierarchically structured and scalable artificial muscles for smart textiles. ACS Appl Mater Interfaces 13(45):54386–54395
133. Jia T et al (2019) Moisture sensitive smart yarns and textiles from self-balanced silk fiber muscles. Adv Funct Mater 29(18):1808241
134. Li X et al (2021) Metalized polyamide heterostructure as a moisture-responsive actuator for multimodal adaptive personal heat management. Sci Adv 7(51):eabj7906
135. Wang Y, Yan W, Li R, Jia X, Cheng Y (2019) Impact of deamidation on gliadin-based nanoparticle formation and curcumin encapsulation. J Food Eng 260:30–39
136. Wen D et al (2022) Silk fibroin-based wearable all-fiber multifunctional sensor for smart clothing. Adv Fiber Mater 4:873–884

The manufacturer's authorised representative in the EU is Springer Nature Customer Service Centre GmbH, Europaplatz 3, 69115 Heidelberg, Germany. If you have any concerns regarding our products, please contact ProductSafety@springernature.com

Printed and bound by CPI Group (UK) Ltd, Croydon, CR0 4YY
17/09/2026
02218437-0009